Polymer Electrolyte Fuel Cell Durability

Felix N. Büchi • Minoru Inaba
Thomas J. Schmidt
Editors

Polymer Electrolyte Fuel Cell Durability

Editors
Felix N. Büchi
Electrochemistry Laboratory
Paul Scherrer Institut
5232 Villigen PSI
Switzerland
felix.buechi@psi.ch

Minoru Inaba
Department Molecular Science
and Technology
Faculty of Engineering
Doshisha University
Kyotanabe, Kyoto 610-0321, Japan
minaba@mail.doshisha.ac.jp

Thomas J. Schmidt
BASF Fuel Cell GmbH
Industrial Park Hoechst, G865
D-65926 Frankfurt am Main
Germany
Thomas.justus.schmidt@basf.com

ISBN: 978-0-387-85534-9 e-ISBN: 978-0-387-85536-3
DOI: 10.1007/978-0-387-85536-3

Library of Congress Control Number: 2008944002

Printed on acid-free paper

springer.com

Foreword

This book covers a significant number of R&D projects, performed mostly after 2000, devoted to the understanding and prevention of performance degradation processes in polymer electrolyte fuel cells (PEFCs). The extent and severity of performance degradation processes in PEFCs were recognized rather gradually. Indeed, the recognition overlapped with a significant number of industrial demonstrations of fuel cell powered vehicles, which would suggest a degree of technology maturity beyond the resaolution of fundamental failure mechanisms. An intriguing question, therefore, is why has there been this apparent delay in addressing fundamental performance stability requirements. The apparent answer is that testing of the power system under fully realistic operation conditions was one prerequisite for revealing the nature and extent of some key modes of PEFC stack failure. Such modes of failure were not exposed to a similar degree, or not at all, in earlier tests of PEFC stacks which were not performed under fully relevant conditions, particularly such tests which did not include multiple on–off and/or high power–low power cycles typical for transportation and mobile power applications of PEFCs.

Long-term testing of PEFCs reported in the early 1990s by both Los Alamos National Laboratory and Ballard Power was performed under conditions of constant cell voltage, typically near the maximum power point of the PEFC. Under such conditions, once the effects of residual CO in the hydrogen feed stream had been addressed, for example, by air bleed, the loss in performance over thousands of hours was found to be very small. In such early testing at Los Alamos National Laboratory, loss of platinum surface area on the cathode, clearly assigned by transmission electron microscopy to platinum particle agglomeration, was found to have a minimal effect on the performance of the cell and this was explained by the expected increase of activity per square centimeter of platinum as the platinum particle size increases from about 2 to 4–5 nm. In other words, no significant issues with platinum corrosion, carbon loss, or poly(perfluorosulfonic acid) membrane longevity were observed in 5,000-h-long tests (a 11,000-h-long test under similar conditions was reported later by Ballard Power). Such reports must have created at the time an impression of very good, if not excellent, long-term stability of PEFCs based on poly(perfluorosulfonic acid) membranes and Pt/C catalysts of precious metal loading under 1.0 mg cm^{-2} of cell area, operating at a temperature as high

80 °C. Such findings may have consequently led to the conclusion that the combination of performance and performance stability demonstrated was sufficient to start building complete, PEFC-based power systems, incorporate them in vehicles, and test-drive such vehicles to pursue reduction in practice.

As we know today from what is extensively covered in this book, some serious issues of PEFC and PEFC stack stability emerged only as prototype system testing began, where a realistic duty cycle, including stop and restart, was included. In conjunction with such testing, a more realistic form of testing, the deleterious effects of frequent changes in the design operation point, and, particularly, the effects of power on–off cycles, could be fully realized. It became clear that, much like the case of phosphoric acid fuel cells, the condition of open circuit carries much more risk in terms of cathode carbon and platinum corrosion versus the maximum power point of the cell, which was chosen for earlier longevity tests. In fact, the assumption for some time was that carbon corrosion rates in PEFCs should be negligible compared with the case in the PAFC, because of the 100 °C difference in the temperature of operation. And then, just a few years ago, came the finding that the routine of refilling the anode compartment with hydrogen, as the first step in cell restart from standby, could have a destructive effect on the air-side cathode, resulting from massive carbon corrosion in the cathode catalyst layer under open circuit conditions. Diagnosis of the latter phenomenon was highly challenging, but it was eventually deciphered (at UTC Fuel Cells) and proper prevention tactics followed next. Side by side with realization of the vulnerability of the carbon material in PEFC cathodes under open circuit conditions came also the realization that the rate of cathode platinum corrosion could be high at open circuit and, particularly so, when the transition from stack "power on" to "power off," or even from high power to low power operation point, is abrupt. Under the latter conditions, platinum is temporarily exposed to a high open circuit potential, while not having been passivated as yet by the thin surface oxide that will form on it at open circuit given sufficient time. Again, it is the *change* in operation conditions, rather than the design operation point by itself, that exacerbates the material loss process. Similarly, poly(perfluorosulfonic acid) membranes that were considered highly stable earlier on were shown to be highly vulnerable to variable operation conditions, particularly humidity cycling. The stress associated with dry-out of the membrane was shown to cause tearing of a membrane clamped between adjacent ribs of a flow field after several dry-out and rehumidification cycles. Moreover, the attack of such membranes by OH radicals became a major concern and, again, open circuit conditions turned out to be the worst in this regard, because the highest rates of gas crossover under such conditions increase the probability of OH radical formation.

What is then the present state of the art regarding PEFC stack performance loss, and what is the possible conclusion from the relatively late discovery of a list of material property barriers on the way to a successful product? As to the first question, it is fair to say that, to a large degree thanks to a strong commitment of resources worldwide to PEFC technology, very able technical teams encountering these problems provided in a relatively short time a list of prevention and remediation steps to address to a large degree the various modes of PEFC performance loss.

Significantly, such methods of performance loss prevention were developed on the basis of preceding steps of careful diagnosis and, in many cases, quantitative connection between known material properties and measured material loss as function of PEFC operation conditions. This does not mean that all performance stability issues have been fully resolved. One indication, for example, is the continued search for a replacement for carbon as a catalyst support material in PEFC cathodes and the other is the continuous attempt to improve the membrane electrode assembly and seal combination to minimize membrane failure along the border of the active area. All in all, however, we seem to have passed a phase in the technology and product development during which a complete failure of a PEFC stack could still take place after a few days of operation with a mysterious cause. The understanding of the failure and performance degradation modes has been so much improved during the last 5 years, as can be seen from the highly informative chapters included in this book. The result is much better confidence in the state of the art and in the type of commitments that can and cannot be made at this point. Most importantly, early market entry of PEFCs did, in fact, start to take off around 2006. The 50-W direct methanol fuel cell technology developed and commercialized by Smart Fuel Cell (Brunnthal, Germany) for auxiliary power applications in recreational vehicles and yachts and the several kilowatt hydrogen fueled PEFCs for backup power applications deployed by ReliOn (Spokane, WA, USA) are two encouraging examples where units have been sold to customers, i.e., where "testing" is now in the form of actual use of the power source by the consumer. The sales of Smart Fuel Cell increased from €7 million in 2006 to €14 million in 2007, providing some testimony to a positive experience of early users. Indeed, the probability of performance deterioration is a function of the specific application and one cannot draw a perfectly parallel conclusion on the readiness of PEFC technology for the demanding application in transport. However, there is something to be said for the apparent readiness for market entry achieved in these two fields as a general indicator for the height reached on the PEFC performance–stability learning curve.

And what about the apparent delay in the discovery and full description of a number of degradation and failure modes at the technology core level of the single cell? PEFC technology is interdisciplinary in the full sense of the word. Tension between resolution of issues at the level of cell electrochemistry and materials and faster pursuit of building and testing complete power systems for some specific application is unavoidable. Moreover, considering the actual phase of development of this technology, one has to recognize that, although the number of man-years of development work invested may now be close to 10^5, it is still a very new technology compared, for example, with mainstream automotive power technology. And if the latter is to serve as an example, great improvements taking place much later in the history of mainstream automotive technology have been witnessed by those of us who drove a passenger vehicle back in the 1970s and experienced the limited road reliability of the power system and, as another example, the extensive corrosion of the chassis. These failure and degradation modes, which were insufficiently addressed 70 years into the development of mainstream automotive technology, were indeed solved during the next 30 years. With such a perspective, uncovering of serious material and cell operation challenges during the phase of prototype system testing is clearly not unexpected and should not lead to overreactions regarding the remaining road to implementation of PEFC technology, particularly when considering all potential applications in the wide power ranges of 10^0–10^5 W.

In summary, this book provides a collection of high-quality chapters written by top experts in the respective specific fields. The take-home message is that many major technical problems on the way to PEFC technology implementation have been identified and carefully studied and characterized to define the root cause(s) for each degradation and failure mechanism. Most importantly, with this knowledge, prevention and remediation tools were quick to follow. The final outcome of this phase of PEFC technology development is, therefore, better readiness for a future in which efficient, clean, and green power sources are to play an ever-increasing role, as evidenced by the recent strong increase of public awareness and by a corresponding significant new flow of venture capital investment into the wide field named by the investment sector "cleantech."

Niskayuna, USA/Caesarea, Israel Shimshon Gottesfeld

Contents

Contributors

M. Adler
Freudenberg Forschungsdienste KG, Höhnerweg 2-4, 69469 Weinheim, Germany

R. Adzic
Chemistry Department, Brookhaven National Laboratory, P.O. Box 5000,
Upton, NY 11973-5000, USA

R. Bieringer
Freudenberg Forschungsdienste KG, Höhnerweg 2-4, 69469 Weinheim, Germany

F. Blank
GR/VFS, Daimler AG

R.L. Borup
Materials Physics and Applications, MPA-11, MS J579, P.O. Box 1663,
Los Alamos National Laboratory, Los Alamos, NM 87545, USA

F.N. Büchi
Electrochemistry Laboratory, Paul Scherrer Institut, 5232 Villigen PSI,
Switzerland

F.D. Coms
Fuel Cell Activities, General Motors, 10 Carriage St., Honeoye Falls,
NY 14472-0603, USA

R.M. Darling
UTC Power, 195 Governors Highway, South Windsor, CT 06074, USA

B. Du
Plug Power Inc., 968 Albany Shaker Road, Latham, NY 12110, USA

J.F. Elter
Empire Innovation Professor of Nanoengineering & Executive Director Center
for Sustainable Ecosystem Nanotechnologies, College of Nanoscale Science and
Engineering, State University of New York at Albany, Albany, New York 12203.

E. Endoh
Research Center, Asahi Glass Co., Ltd., 1150 Hazawa-cho, Kanagawa-ku,
Yokohama 221-8755, Japan

H.A. Gasteiger
Fuel Cell Activities, General Motors, 10 Carriage St., Honeoye Falls, NY 14472-0603, USA

S. Geiss
Freudenberg Forschungsdienste KG, Höhnerweg 2-4, 69469 Weinheim, Germany

S. Gottesfeld
Cellera Technologies, 2 Hatochen St., Caesarea Industrial Park North, Israel 38900

W. Gu
Fuel Cell Activities, General Motors, 10 Carriage St., Honeoye Falls, NY 14472-0603, USA

L. Gubler
Electrochemistry Laboratory, Paul Scherrer Institut, 5232 Villigen PSI, Switzerland

M. Hicks
Ida Tech, LLC, 63065 N.E. 18th Street, Bench, OR 97701, USA

S. Hommura
Research Center, Asahi Glass Co., Ltd., 1150 Hazawa-cho, Kanagawa-ku, Yokohama 221-8755, Japan

M. Inaba
Department Molecular Science and Technology, Doshisha University, Kyotanabe, Kyoto 610-0321, Japan

S. Kandoi
UTC Power, 195 Governors Highway, South Windsor, CT 06074, USA

Y.S. Kim
Materials Physics and Applications Division, Los Alamos National Laboratory, Los Alamos, NM 87545, USA

A.B. LaConti
Giner, Inc., Giner Electrochemical Systems, LLC, 89 Rumford Avenue, Newton, MA 02466, USA

H. Liu
Giner, Inc., Giner Electrochemical Systems, LLC, 89 Rumford Avenue, Newton, MA 02466, USA

R. Makharia
Fuel Cell Activities, General Motors, 10 Carriage St., Honeoye Falls, NY 14472-0603, USA

J.P. Meyers
Department of Mechanical Engineering, College of Engineering, The University of Texas at Austin, 1 University Station C2200, Austin, TX 78712, USA

T. Mitani
Advanced Technology R&D Center, Mitsubishi Electric Corporation, 8-1-1 Tsukaguchi-Honmachi, Amagasaki, Hyogo 661-8661, Japan

K. Mitsuda
Advanced Technology R&D Center, Mitsubishi Electric Corporation, 8-1-1 Tsukaguchi-Honmachi, Amagasaki, Hyogo 661-8661, Japan

S. Miyata
The New Energy and Industrial Technology Development Organization (NEDO), MUZA Kawasaki Central Tower, 1310 Omiya-cho, Saiwai-ku, Kawasaki City, Kanagawa 212-8554, Japan

D. Münter
European Fuel Cell Support Center, DANA Holding Corporation Sealing Products, Reinz-Dichtungs-GmbH, Reinzstraße 3-7, 89233 Neu-Ulm, Germany

T. Okada
National Institute of Advanced Industrial Science and Technology (AIST), Higashi 1-1-1, Central 5, Tsukuba, Ibaraki 305-8565, Japan

Y. Oomori
Residential Cogeneration Development Department, Osaka Gas Co., Ltd., Science Center Bldg. 3rd, Chudoji Awatacho 93, Shimogyo-ku, Kyoto 600-8815, Japan

T.W. Patterson
UTC Power, 195 Governors Highway, South Windsor, CT 06074, USA

M.L. Perry
United Technologies Research Centre, 411 Silver Lane, East Hartford, CT 06108, USA

D. Pierpont
Fuel Cell Components Program, 3M Company, 3M Center, Building 0201-BN-33, St. Paul, MN 55144-1000, USA

R. Pollard
Shell Exploration & Production Co., Bellaire Technology Centre, 3737 Bellaire Blvd, Houston, TX 77025, USA

M. Ramani
Plug Power Inc., 968 Albany Shaker Road, Latham, NY 12110, USA

C. Reiser
UTC Power, 195 Governors Highway, South Windsor, CT 06074, USA

K. Sasaki
Chemistry Department, Brookhaven National Laboratory, P.O. Box 5000, Upton, NY 11973-5000, USA

G.G. Scherer
Electrochemistry Laboratory, Paul Scherrer Institut, 5232 Villigen PSI, Switzerland

J. Scherer
European Fuel Cell Support Center, DANA Holding Corporation Sealing Products, Reinz-Dichtungs-GmbH, Reinzstraße 3-7, 89233 Neu-Ulm, Germany

T.J. Schmidt
BASF Fuel Cell GmbH, Industrial Park Hoechst, G865, 65926 Frankfurt am Main, Germany

M. Shao
Chemistry Department, Brookhaven National Laboratory, P.O. Box 5000, Upton, NY 11973-5000, USA

K. Shinohara
Fuel Cell Laboratory, Nissan Research Center, Nissan Motor Co., Ltd., 1 Natsushima-cho, Yokosuka-shi, Kanagawa 237-8523, Japan

H. Shintaku
Residential Cogeneration Development Department, Osaka Gas Co., Ltd., Science Center Bldg. 3rd, Chudoji Awatacho 93, Shimogyo-ku, Kyoto 600-8815, Japan

J. St-Pierre
Department of Chemical Engineering, Future Fuels Initiative, University of South Carolina, Swearingen Engineering Center, 301 Main Street, Columbia, SC 29208, USA

R. Ströbel
European Fuel Cell Support Center, DANA Holding Corporation Sealing Products, Reinz-Dichtungs-GmbH, Reinzstraße 3-7, 89233 Neu-Ulm, Germany

S. Sugawara
Fuel Cell Laboratory, Nissan Research Center, Nissan Motor Co., Ltd., 1 Natsushima-cho, Yokosuka-shi, Kanagawa 237-8523, Japan

T. Tabata
Fuel Cell Development Department Osaka Gas Co., Ltd., 3-4 Hokko Shiratsu 1-Chome, Konohana-ku, Osaka 554-0041, Japan

M. Viol
Freudenberg Dichtungs- und Schwingungstechnik GmbH & Co. KG, Höhnerweg 2-4, 69465 Weinheim/Bergstraße, Germany

F.T. Wagner
Fuel Cell Activities, General Motors, 10 Carriage St., Honeoye Falls, NY 14472-0603, USA

D.L. Wood III
Materials Physics and Applications, MPA-11, MS D429, P.O. Box 1663,
Los Alamos National Laboratory, Los Alamos, NM 87545, USA

S. Yamamoto
Fuel Cell Laboratory, Nissan Research Center, Nissan Motor Co., Ltd., 1
Natsushima-cho, Yokosuka-shi, Kanagawa 237-8523, Japan

O. Yamazaki
Residential Cogeneration Development Department, Osaka Gas Co., Ltd.,
Science Center Bldg. 3rd, Chudoji Awatacho 93, Shimogyo-ku,
Kyoto 600-8815, Japan

K. Yasuda
Advanced Fuel Cell Research Group, Research Institute for Ubiquitous Energy
Devices, National Institute of Advanced Industrial Science and Technology
(AIST), 1-8-31 Midorigaoka, Ikeda, Osaka 563-8577, Japan

P.T. Yu
Fuel Cell Activities, General Motors, 10 Carriage St., Honeoye Falls,
NY 14472-0603, USA

P. Zelenay
Materials Physics and Applications Division, Los Alamos National Laboratory,
Los Alamos, NM 87545, USA

J. Zhang
Fuel Cell Activities, General Motors, 10 Carriage St., Honeoye Falls,
NY 14472-0603, USA

Part I
Stack Components

1. Introduction

Editors

Durability is one of the most important issues for commercialization of polymer electrolyte fuel cell (PEFC) based automotive and stationary applications, as well as cost and hydrogen storage. The current lifetimes of fuel cell vehicles and stationary cogeneration systems are approximately 1,000 h and approximately 10,000 h (2008), respectively, and should be improved before they are commercialized in the near future. The 2010/1015 US DOE lifetime target for automotive applications is 5,000 h, which is equivalent to 150,000 driven miles, and the Japanese NEDO's lifetime targets for stationary applications are 40,000 and 90,000 h in 2010 and 2015, respectively.

PEFCs consist of a number of different components, such as catalysts, membranes, gas-diffusion layers, bipolar plates, and sealings. To achieve the durability targets for PEFC systems, it is essential that each of the components has the required durability. To close the gap between today's status and the required targets it is important to understand the degradation phenomena of each of the components used in PEFCs and to improve their durability. The comprehensive analysis and understanding of the degradation phenomena of each component in a cell or stack is not easy. This difficulty arises from the fact that the stability of each component greatly depends on a wide variety of operation conditions, such as temperature, cell voltage, and the degree of humidification of the cell or stack. In addition, degradation of one component in a cell often causes parasitic degradation of other components. This therefore requires a detailed separation, analysis, and understanding of the individual degradation processes of the various components. Another issue is the long durability goals, which require the development of accelerated test methods to facilitate the development of highly durable materials. Here again we encounter a great difficulty because elevating temperature, a strategy which is widely employed as an accelerating parameter in many fields, usually changes not only the properties of a material itself, but also correlations with other materials.

In Part I, degradation phenomena of stack components, catalysts, membranes, gas-diffusion layers, membrane–electrode assemblies, bipolar plates, and sealings are discussed on the basis of their materials chemistry. Accelerating methods and recent progress in durability improvement are also reviewed by prominent authors in the field.

F.N. Büchi et al. (eds.), *Polymer Electrolyte Fuel Cell Durability*,

2
Catalysts

Dissolution and Stabilization of Platinum in Oxygen Cathodes

Kotaro Sasaki, Minhua Shao, and Radoslav Adzic

Abstract In this brief review of the dissolution and solubility of platinum under equilibrium conditions and the degradation of platinum nanoparticles at the cathode under various operating conditions, we discuss some mechanisms of degradation, and then offer recent possibilities for overcoming the problem. The data indicate that platinum nanoparticle electrocatalysts at the cathode are unstable under harsh operating conditions, and, as yet, often would be unsatisfactory for usage as the cathode material for fuel cells. Carbon corrosion, particularly under start/stop circumstances in automobiles, also entails electrical isolation and aggregation of platinum nanoparticles. We also discuss new approaches to alleviate the problem of stability of cathode electrocatalysts. One involves a class of platinum monolayer electrocatalysts that, with adequate support and surface segregation, demonstrated enhanced catalytic activity and good stability in a long-term durability test. The other approach rests on the stabilization effects of gold clusters. This effect is likely to be applicable to various platinum- and platinum-alloy-based electrocatalysts, causing their improved stability against platinum dissolution under potential cycling regimes.

1 Introduction

One critical issue facing the commercialization of low-temperature fuel cells is the gradual decline in performance during operation, mainly caused by the loss of the electrochemical surface area (ECA) of carbon-supported platinum nanoparticles at the cathode. The major reasons for the degradation of the cathodic catalyst layer are the dissolution of platinum and the corrosion of carbon under certain operating conditions, especially those of potential cycling. Cycling places various loads on

K. Sasaki, M. Shao, and R. Adzic (✉)
Department of Chemistry, Building 555, Brookhaven National Laboratory,
P.O. Box 5000, Upton, NY 11973-5000, USA
e-mail: ksasaki@bnl.gov, adzic@bnl.gov

F.N. Büchi et al. (eds.), *Polymer Electrolyte Fuel Cell Durability*,

fuel cells; in particular, stop-and-go driving, and fuel starvation in vehicular applications can generate high voltage loads. The coalescence of platinum nanoparticles through migration also results in the loss of surface area. Hence, a detailed understanding of degradation mechanisms of platinum nanoparticles will help in designing durable materials for the oxygen reduction reaction (ORR).

In this review, we briefly discuss the dissolution and solubility of platinum (Sect. 2), the degradation of platinum nanoparticles in fuel cells (Sect. 3), and carbon corrosion (Sect. 4). We then describe new cathode electrocatalysts wherein the platinum content can be dramatically reduced, while offering possibilities for enhancing catalytic activity and stability (Sect. 5).

2 Platinum Dissolution

2.1 Bulk Material

The thermodynamic behavior of a platinum bulk material as a function of electrolyte pH and electrode potential is guided by potential–pH diagrams (also known as Pourbaix diagrams) (Pourbaix 1974). The main pathways for platinum dissolution at 25°C involve either direct dissolution of metal,

$$\mathrm{Pt} \rightarrow \mathrm{Pt}^{2+} + 2e^{-} \quad E_0 = 1.19 + 0.029\log[\mathrm{Pt}^{+}], \tag{1}$$

or production of an oxide film and a subsequent chemical reaction,

$$\mathrm{Pt} + \mathrm{H_2O} \rightarrow \mathrm{PtO} + 2\mathrm{H}^{+} + 2e^{-} \quad E_0 = 0.98 - 0.59\mathrm{pH}, \tag{2}$$

$$\mathrm{PtO} + 2\mathrm{H}^{+} \rightarrow \mathrm{Pt}^{2+} + \mathrm{H_2O} \quad \log[\mathrm{Pt}^{2+}] = -7.06 - 2\mathrm{pH}. \tag{3}$$

The potential–pH diagram suggests that platinum is fairly stable thermodynamically; there is only a small corrosion region around 1 V at pH −2 to 0 (we note that the potential–pH diagram is valid only in the absence of ligands with which platinum can form soluble complexes or insoluble compounds). However, as described below, platinum actually dissolves under fuel cell operating conditions, which is unsatisfactory especially for its usage in the cathode of fuel cells for automotives.

Conway et al. (Angerstein-Kozlowska et al. 1973; Conway 1995) summarized earlier work on platinum oxide formation; in the potential region of 0.85–1.10 V in an H_2SO_4 solution, adsorbed species (OH_{ads}) are formed by the oxidation of H_2O molecules (4). The OH_{ads} and platinum surface atoms then undergo place-exchange, forming a quasi-3D lattice (5). At higher potentials (1.10–1.40 V), the OH species in this lattice are oxidized, generating a Pt–O quasi-3D lattice (6).

$$\mathrm{Pt} + \mathrm{H_2O} \rightarrow \mathrm{Pt{-}OH_{ads}} + \mathrm{H}^{+} + e^{-} \quad (0.85\,\mathrm{V} < E < 1.10\,\mathrm{V}) \tag{4}$$

$$\mathrm{Pt-OH_{ads}} \xrightarrow{\text{place exchange}} \mathrm{(OH-Pt)_{quasi\text{-}3D\ lattice}} \quad (5)$$

$$\mathrm{(OH-Pt)_{quasi\text{-}3D\ lattice}} \rightarrow \mathrm{(Pt-O)_{quasi\text{-}3D\ lattice}} + \mathrm{H^+} + e^- \quad (1.10\,\mathrm{V} < E < 1.40\,\mathrm{V}) \quad (6)$$

However, recent studies using an electrochemical quartz crystal nanobalance revealed that the platinum oxides are not hydrated (Birss et al. 1993; Harrington 1997; Jerkiewicz et al. 2004); thus, at 0.85–1.15 V a half monolayer (0.5 ML) forms from chemisorbed oxygen (O_{chem}), rather than OH_{ads}. Figure 1 depicts the proposed mechanism. First, H_2O molecules adsorb on the platinum electrode at low potentials (0.27–0.85 V) since the platinum surface partial positive charge can be compensated for by the negatively charged oxygen end of the H_2O molecule (Fig. 1a). At potentials between 0.85 and 1.15 V, the discharge of 0.5 ML of H_2O molecules occurs, thereby leading to the formation of 0.5 ML of O_{chem} (Fig. 1b). Above a potential of 1.15 V, further discharge of H_2O results in the formation of the second 0.5 ML of O_{chem}, with the first 0.5 ML O_{chem} (Fig. 1c). During this process, strong dipole–dipole lateral repulsions cause the initial 0.5 ML O_{chem} adatoms to undergo place-exchange with platinum atoms, so forming a quasi-3D surface Pt–O lattice (Fig. 1d). You et al.

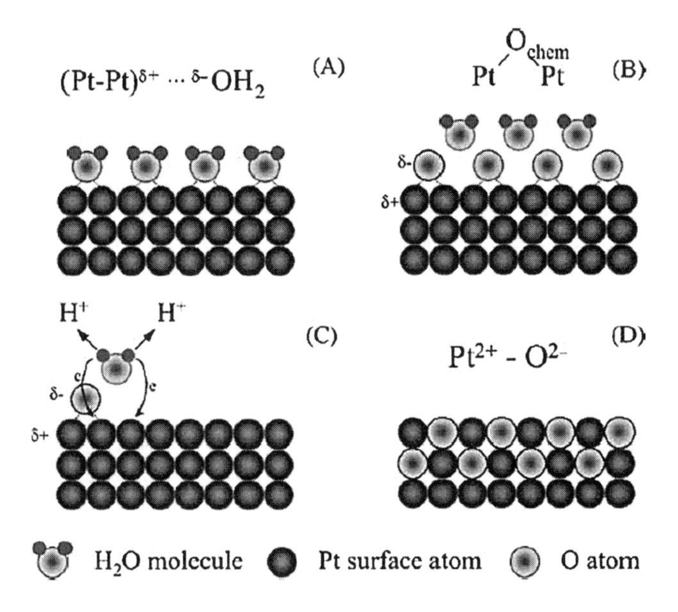

Fig. 1 The PtO growth mechanism (Jerkiewicz et al. 2004). (**a**) Interactions of H_2O molecules with the platinum surface that occur in the $0.27 \le E \le 0.85$ V range. (**b**) Discharge of a half monolayer of H_2O molecules and subsequent formation of a half monolayer of chemisorbed oxygen (O_{chem}) in the $0.85 \le E \le 1.15$ V range. (**c**) Discharge of the second half monolayer of H_2O molecules at $E > 1.15$ V; the process is accompanied by the development of repulsive interactions between $(\mathrm{Pt-Pt})^{\delta+} - O^{\delta-}_{chem}$ surface species that stimulate an interfacial place-exchange of O_{chem} and platinum surface atoms. (**d**) Formation of a quasi-3D surface PtO lattice comprising Pt^{2+} and O^{2-} moieties through the place-exchange process

found that only 0.3 ML of platinum atoms exchange places with the oxygen-containing species (You et al. 2000; Nagy and You 2002). This mechanism exposes platinum to the electrolyte, thereby promoting its oxidation and dissolution (Wang et al. 2006b). Nagy and You (Nagy et al. 2002) demonstrated that PtO is mobile and can diffuse to energetically favorable sites on the platinum surface, so increasing the exposure of the underlying platinum atoms to the electrolyte. At higher potentials, more of the PtO film is oxidized to PtO_2, which also is mobile (You et al. 2000).

2.2 *Equilibrium Solubility of Platinum*

The solubility of platinum changes with various factors, including potentials, electrolyte components, pH, and temperature. Azaroual et al. (2001) examined the solubility of platinum wires and particles in several buffer solutions of pH 4–10 at 25°C. They found that the solubility of platinum increases with increasing pH, suggesting that the hydroxylated complex $PtOH^+$ is a major determinant of platinum solubility in this pH range. They also reported that the solubility of platinum particles (diameter 0.27–0.47 μm) is two orders of magnitude higher than that of platinum wires. By contrast, platinum solubility increases with decreasing pH in sulfuric acid (pH < 1.5); apparently, dissolution of platinum in the acidic medium follows an acidic dissolution mechanism (Mitsushima et al. 2007b). Platinum solubility strongly depends on the temperature (Dam and de Bruijn 2007), rising with increasing temperature, following the Arrhenius relationship (Mitsushima et al. 2007a, b).

There are extensive studies of the effect of potential on platinum solubility in acidic solutions (Bindra et al. 1979; Ferreira et al. 2005; Wang et al. 2006a, b Dam et al. 2007 ; Mitsushima et al. 2007a, b). Figure 2 shows a plot of the published data on dissolved platinum concentrations as a function of applied potential, summarized by Mitsushima et al. (2007a, b), Borup et al. (2007), and Shao-Horn et al. (2007). We also include those data calculated from (1), i.e., the two-electron dissolution process, as we indicate by the Pourbaix's diagram at two different temperatures (dashed-dotted line at 25°C, and solid line at 196°C). Overall, the equilibrium concentration of dissolved platinum increases with increasing potentials up to 1.1 V. The solubility data in a concentrated H_3PO_4 solution (18.6 M) at 196°C obtained by Bindra et al. (1979) (open circles in Fig. 2) agree well with those calculated from (1) at the same temperature, suggesting that platinum dissolution underwent a two-electron reaction pathway in the experiment. Except for this case, however, the slopes of all other experimental data are much less than those calculated from the Pourbaix diagram. Although the origin of the discrepancy between the solubility data and the Pourbaix model is unclear, platinum dissolution might involve chemical processes, such as the dissolution of PtO in other electrochemical reactions (Mitsushima et al. 2007a, b). Another notable feature in Fig. 2 is that the solubility of platinum wire in 0.57 M $HClO_4$ at 23°C starts to decline at potentials over 1.1 V. As we described in Sect. 2.1, above 1.15 V the place-exchange process starts and a 3D PtO film is formed; accordingly, this retardation in platinum solubility can be attributed to the formation of a protective overlying oxide film (Mitsushima et al. 2007a, b).

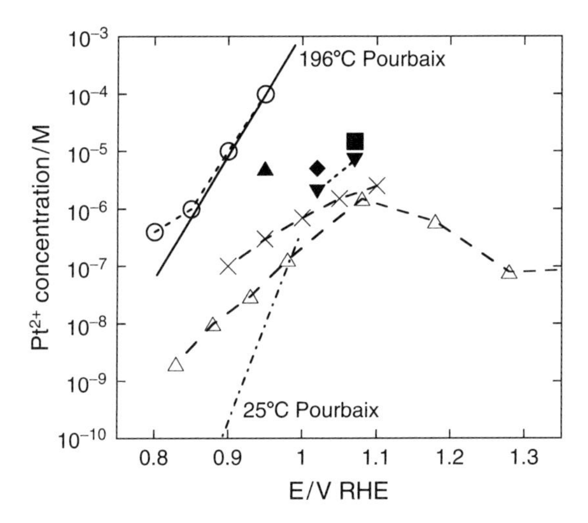

Fig. 2 Dissolved platinum concentrations as a function of potential. *Crosses* 76°C in 1 M H_2SO_4 (Ferreira et al. 2005), *open triangles* 23°C in 0.57 M $HClO_4$ (Wang et al. 2006), *circles* 196°C in concentrated H_3PO_4 (Bindra et al. 1979), *inverted filled triangles* 23°C in 1 M H_2SO_4, *upright filled triangle* 35°C in 1 M H_2SO_4, *square* 51°C in 1 M H_2SO_4, *diamond* 76°C in 1 M H_2SO_4 (Mitsushima et al. 2007). The *dashed-dotted lines* and the *solid lines* were calculated from the Pourbaix diagram at 25°C and 196°C, respectively. *RHE* reversible hydrogen electrode

Platinum dissolution also depends on crystallographic orientation. Komanicky et al. (2006) studied the dissolution on different low-index facets in 0.6 M $HClO_4$ solution at three different potentials (0.65, 0.95, 1.15 V). Even at the low 0.65-V potential, they observed dissolution at edges and pits on the Pt(111) surface, while the terrace was stable. Surprisingly, dissolution was inhibited at 0.95 V owing to the formation of oxide films at the edges, but it started again at 1.15 V, when the terrace plane was irreversibly roughened with multiple pits on its surface. For Pt(100) and (110) facets, dissolution declined with increasing potential owing to the formation of a passive layer on their surfaces. These authors also explored the changes on a nanofaceted platinum surface consisting of alternating (111) rows and (100) facets of several nanometer width, to simulate the behavior of platinum nanoparticles. They reported that the extent of the nanofaceted surface that dissolved rose with increasing potentials and almost entire nanofacets had been dispersed at 1.15 V, suggesting that the edges and corners on platinum nanoparticles, with their low coordinate sites, might have a higher tendency to dissolve away compared with the terraced facets.

2.3 Dissolution Under Potential Cycling

Potential cycling conditions accelerate platinum dissolution compared with potentiostatic conditions (Johnson et al. 1970; Rand and Woods 1972; Kinoshita et al. 1973; Ota et al. 1988). For example, Wang et al. (2006a, b) reported that the dissolution rates under potential cycling are 3–4 orders of magnitude higher than those determined for potentiostatic conditions. The dissolution rate during triangular potential cycling reportedly was around 2–5.5 ng cm^{-2} per cycle, with the upper potential limit between 1.2 and 1.5 V and various potential scanning rates (Johnson et al. 1970; Rand et al. 1972; Kinoshita et al. 1973; Wang et al. 2006a, b). The dissolution rate increased with a rise in the upper potential limit (Rand et al. 1972). Meyers and Darling (Darling and Meyers 2003, 2005) developed a mathematical model based on the reactions in (1)–(3) to study the dissolution and movement of platinum in a proton exchange membrane fuel cell (PEMFC) during potential cycling from 0.87 to 1.2 V. The oxide film that developed was found to retard dissolution markedly. Dissolution was severe when the potential switched to an upper limit of 1.2 V, but then it stopped once a monolayer of PtO had accumulated. However, we lack detailed knowledge of mechanisms of dissolution, and of the particular species dissolved during potential cycling.

Rand and Woods (1972) detected both Pt(II) and Pt(IV) ions after 200 triangular potential cycles between 0.41 and 1.46 V at 40 mV s^{-1} in 1 M H_2SO_4. They found that the charge difference between anodic and cathodic cycles in oxygen adsorption and desorption regions ($Q_O^a - Q_O^c$) was positive and consistent with the amount of platinum dissolved. Therefore, they considered that anodic dissolution during anodic scans, partly either via the reaction in 1 or via that in (2) and (3), is the main cause.

On the other hand, a "cathodic" dissolution mechanism has been suggested; Johnson et al. (1970) detected Pt(II) in a rotating ring-disk electrode study during the negative-going potential scan in a 0.1 M $HClO_4$ solution. The Pt(II) species formed due to the reduction of PtO_2 (Johnson et al. 1970; Mitsushima et al. 2007a):

$$PtO_2 + 4H^+ + 2e^- \rightarrow Pt^{2+} + 2H_2O \quad E_0 = 0.84 + 0.12\text{pH} + \log[Pt^{2+}]. \qquad (7)$$

In this case, the charge difference between the anodic and cathodic scans $Q_O^a - Q_O^c$ is also positive because the charge is less than that needed to reduce adsorbed oxygen to water since only one electron is used for each oxygen atom (Rand and Woods 1972). Mitsushima et al. (2007a, b) compared the dissolution rates of platinum in sulfuric acid during potential cycling with four different potential profiles. Among them, the slow cathodic triangular sweep (20 mV s^{-1} anodic and 0.5 mV s^{-1} cathodic) showed a significantly enhanced dissolution rate, attaining over 20 ng cm^{-2} per cycle and an electron transfer number of 2 [indicating that the dissolved species is Pt(II)]; at the other potential-wave modes the dissolution rate remained around a few nanograms per square centimeter per cycle with an electron transfer number of 4, indicating that the dissolved species is Pt(IV). The enhanced platinum dissolution during the slow cathodic scans is considered to follow the cathodic-dissolution mechanism represented in (7).

3 Degradation of Platinum Nanoparticles in Fuel Cells

Extensive studies on catalyst degradation in PEMFCs and phosphoric acid fuel cells (PAFCs) demonstrated that its cause can be attributed mainly to a loss of ECA in the cathode. PEMFCs and PAFCs use similar catalysts, although the degradation in PAFCs generally is severer because of the relatively higher cell temperature (200°C) at which they operate, and the use of a more corrosive electrolyte. Shao-Horn et al. (2007) proposed classification of four mechanisms for the decrease in ECA (1) crystallite migration on carbon supports forming larger particles, (2) platinum dissolution and its redeposition on larger particles (electrochemical Ostwald ripening), (3) platinum dissolution and precipitation in ion conductors, and (4) the detachment and agglomeration of platinum particles caused by carbon corrosion. The fourth mechanism is discussed in Sect. 4.

3.1 Crystallite Migration and Coalescence

This mechanism involves the migration of platinum nanoparticles on the carbon support and their coalescence during fuel cell operations (Fig. 3a); no platinum dissolution is involved. The underlying driving force minimizes the total surface energy as the surface energy of the nanosized particles declines with the particles' growth. This mechanism of particle growth generates a specific particle size distribution, which peaks at small sizes, tailing toward larger sizes. In fact, it was observed in both PAFC (Bett et al. 1976; Blurton et al. 1978; Aragane et al. 1988) and PEMFC (Wilson et al. 1993) studies. Wilson et al. recorded such a size distribution of aged platinum nanoparticles in the cathode by X-ray diffraction analysis. They suggested that particle growth in PEMFCs is caused by the crystalline migration mechanism, not by a dissolution–redeposition process that results in a different size distribution, as we discuss below (Wilson et al. 1993). Other evidence supporting the particle-migration mechanism arises from the demonstrated insensitivity of ECA loss during the operation of PAFCs to the potential (Blurton et al. 1978; Gruver et al. 1980),

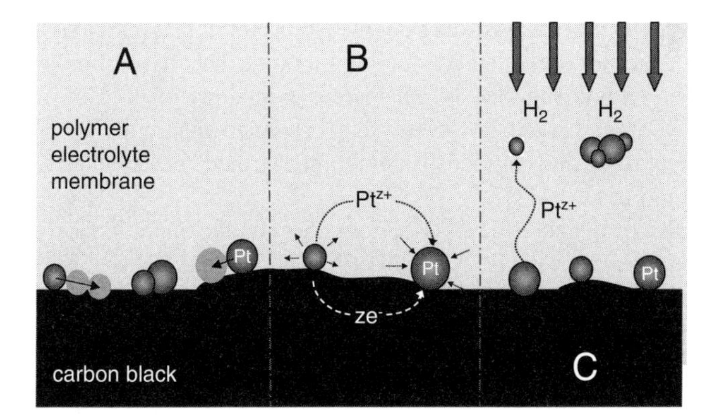

Fig. 3 Three mechanisms for the degradation of carbon-supported platinum nanoparticles in low-temperature fuel cells. (**a**) Particle migration and coalescence. (**b**) Dissolution of platinum from smaller particles and its redeposition on larger particles (electrochemical Ostwald ripening). (**c**) Dissolution of platinum and its precipitation in a membrane by hydrogen molecules from the anode

suggesting that dissolution–redeposition is not the dominant mechanism. We note that most of the data supporting the migration mechanism were obtained below 0.8 V, at which point the solubility of platinum is low, and, therefore, platinum dissolution is not considered dominant at these relatively lower potentials.

3.2 Dissolution and Redeposition: Electrochemical Ostwald Ripening

The second mechanism involves the dissolution and redeposition of platinum on large particles. If platinum is partially soluble in electrolytes/ionomers, smaller particles will dissolve preferentially as their chemical potential is higher than that of larger platinum particles (Voorhees 1985; Virkar and Zhou 2007). Dissolved platinum species move to the surfaces of larger particles through the electrolyte/ionomer, while electrons are transported through the carbon supports to larger particles. Thus, platinum is precipitated at the surfaces of larger particles. As particle size falls, the chemical potential increases, and so dissolution accelerates, resulting in the growth of large particles at the expense of small ones with the necessary concomitant decrease in the system's total energy. Honji et al. (1988) recorded potential-dependent particle growth at 205°C in PAFCs. They showed that the platinum particle's size starts to increase and the amount of platinum in the electrode decreases above 0.8 V, suggesting that platinum dissolution and redeposition is the main mechanism for particle growth at high potentials (Tseung and Dhara 1975; Honji et al. 1988). Their notion is supported by Bindra et al.'s (1979) experiment that demonstrated an exponential increase in the solubility of a platinum foil in phosphoric acid at 176–196°C above 0.8 V. Virkar and Zhou (2007) found that electronically conductive supports, such as carbon black, are critical for this process. Platinum particles did not grow when they were supported on nonconductive Al_2O_3 because this growth process involves a coupled transport of platinum ions and electron transport via water/ionomer and conductive supports. The dissolution–redeposition of platinum on carbon supports is considered an analogy of the Ostwald ripening mechanism (Voorhees 1985); however, this mechanism apparently accompanies the electrochemical reactions, and thus is sometimes termed "electrochemical Ostwald ripening" (Honji et al. 1988).

It is difficult to rationalize the contributions of crystallite migration and coalescence and of Oswald ripening to the ECA loss in PEMFCs. The size distributions of platinum particles are sometimes employed to differentiate them, as the electrochemical Ostwald ripening process is characterized by an asymmetric particle distribution with a tail toward the small particle end owing to the consumption of smaller particles (Wilson et al. 1993), while crystallite migration and coalescence has a tail toward large particles, as we described in Sect. 3.2. However, some researchers observed a bimodal distribution of particle size during potential cycling (Xie et al. 2005a; Garzon et al. 2006), suggesting that a combination of these two processes takes place (Bindra et al. 1979).

3.3 *Platinum Dissolution and Precipitation in Membranes*

The third mechanism also involves the dissolution of platinum; however, thereafter, the dissolved platinum species diffuse into a membrane and are reduced chemically by hydrogen permeating from the anode (Fig. 3c). The direct evidence supporting this mechanism is the observation of platinum particles in the matrix (Aragane et al. 1988) and membrane (Patterson 2002) after fuel cell operation. The driving force underlying the crossover of dissolved platinum species into the membrane could be electro-osmotic drag and/or the concentration gradient diffusion (Guilminot et al. 2007a, b). The particular counteranions of Pt^{z+} have not yet been established. Membrane degradation products, such as fluoride (Healy et al. 2005; Xie et al. 2005b) and sulfate (Xie et al. 2005a, b; Teranishi et al. 2006) anions were detected during the operation of PEMFCs, and could possibly be the complexing ligands for Pt^{z+}. In fact, strong evidence from Guilminot et al. (2007b) demonstrated that the concentration of fluoride around the platinum nanoparticles in an aged membrane was higher than that in a new membrane. Another possible source is other halide ions, such as chloride and bromide, left on carbon and platinum surfaces during the synthesis of the catalyst (Guilminot et al. 2007a, b). Both mobile Pt(II) and Pt(IV) species were detected in the fresh/aged membrane electrode assemblies. The concentrations of these species increase upon ageing, and these ions are highly mobile in the membrane (Guilminot et al. 2007a, b). The platinum band or large particles form in the membrane near the interface of the membrane/cathode during cycling with H_2/N_2 (Ferreira et al. 2005; Yasuda et al. 2006a, b; Ferreira and Shao-Horn 2007) and somewhere away from the cathode with H_2/O_2(air) (Patterson 2002; Yasuda et al. 2006a, b; Bi et al. 2007; Zhang et al. 2007a). Platinum can move into the anode with absence of H_2 (Yasuda et al. 2006a, b). These results strongly point to the chemical reduction by H_2 of Pt^{z+} in the membrane. Some studies combining experimental data and mathematical models revealed that the location of the platinum band in the membrane under open circuit voltage (OCV) and cycling conditions depends on the partial pressure of H_2 and O_2 and also their permeability and that of Pt^{z+} through the membrane (Bi et al. 2007; Zhang et al. 2007a, b).

4 Carbon Corrosion

4.1 *General Aspects of Carbon Corrosion*

Corrosion of carbon supports may cause the electrical isolation and aggregation of platinum nanoparticles, causing a decrease in the ECA in the catalyst layers too. The electrochemical behaviors of carbon in different forms have been studied under a variety of conditions; there is a comprehensive review of works conducted two decades ago in Kinoshita's classic book (Kinoshita 1988). In aqueous solutions, the general carbon corrosion reaction can be written as (Kinoshita 1988)

$$C + 2H_2O \rightarrow CO_2 + 4H^+ + 4e^-, \tag{8}$$

with a standard potential of 0.207 V versus the standard hydrogen electrode, indicating that carbon can be thermodynamically oxidized at potentials above 0.207 V. To a lesser extent, the heterogeneous water–gas reaction (Stevens et al. 2005) also can occur with a standard potential of 0.518 V:

$$C + H_2O \rightarrow CO + 2H^+ + 2e^-. \tag{9}$$

However, the corrosion rate of carbon at potentials lower than 0.9 V is reasonably slow at the typical operating temperatures (60–90°C) of PEMFCs. Despite the sluggishness of the reaction, long-term operations can cause a decrease in carbon content in the catalyst layers. Furthermore, under some circumstances when electrode potentials are raised extremely high, there is a rapid degradation of carbon supports as well as of platinum.

4.2 Carbon Corrosion in PEMFCs

Several recent studies of carbon corrosion in a membrane electrode assembly of a PEMFC system found the process to be much more complicated than in aqueous solutions. The corrosion rate depends, among other factors, on the type of carbon, operating potential, temperature, humidity, and uniformity of fuel distribution. Generally, three types of carbon corrosion were identified (Fuller and Gray 2006).

The first one occurs under normal operating conditions. As (8) and (9) show, carbon is thermodynamically unstable owing to its low standard potential, which is much lower than the operating potential range of PEMFCs, and thus carbon corrosion can occur during the fuel operation at elevated temperatures. In practice, however, corrosion of conventional carbon supports, such as Vulcan, is insignificant at cell voltages lower than 0.8 V, but becomes a serious problem at voltages over 1.1 V (Roen et al. 2004). Furthermore, supported platinum nanoparticles catalyze carbon corrosion (Roen et al. 2004). Mathias et al. (2005) studied the kinetics of carbon corrosion as functions of temperature, potential, and time, estimating that 5 wt% of carbon (Ketjen black) would be lost over several thousand hours at open circuit voltage (0.9 V). Their study suggested that the stability of Ketjen black does not meet the requirements for automotive applications.

The second type of carbon corrosion in PEMFCs is elicited in the cathode by the partial coverage of hydrogen and oxygen in the anode. This phenomenon would be the most hazardous one in PEMFCs, especially for automotive applications wherein frequent start/stop cycling is expected. The mechanism of start/stop-induced carbon corrosion is as follows (Fig. 4) (Reiser et al. 2005; Stuve and Gastaiger 2006). During the start and shutdown of a cell, the anode would be partially covered by air from outside or from the cathode through the membrane. Thus, in an anode compartment, hydrogen oxidation reactions and ORRs can take place simultaneously.

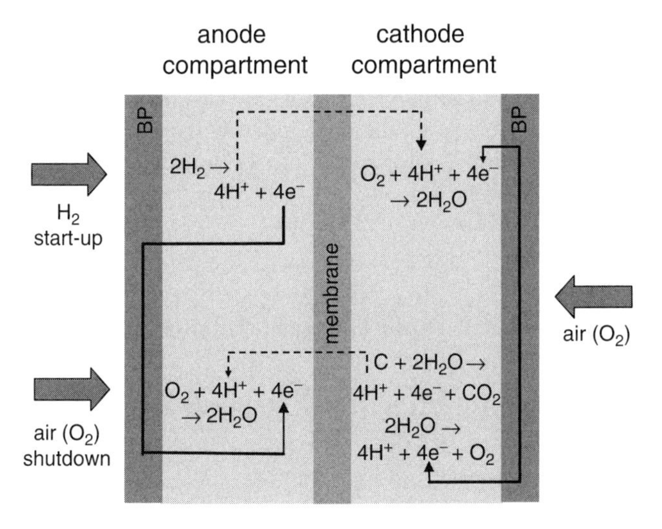

Fig. 4 Start/stop-induced carbon corrosion in the cathode ($C + 2H_2O \rightarrow 4H^+ + 4e^- + CO_2$) when air is partially introduced in the anode (Reiser et al. 2005; Stuve et al. 2006). *Solid line*: electron path, *dashed line*: proton path

In a cathode compartment, the ORR and concomitant two anodic reactions, i.e., carbon dissolution (8) and oxygen evolution ($2H_2O \rightarrow 4H^+ + 4e^- + O_2$), can occur at the same time. The flows of the electrons and the protons generated are designated by the solid line and the dashed line in Fig. 4, respectively. These reactions create a cathode interfacial potential difference of 1.44 V and higher in the hydrogen-starved region, triggering the evolution of oxygen and carbon corrosion on the cathode. We note that this "reverse-current" mechanism also exists during normal cell operation under localized fuel starvation even in a very small region, which likely happens simply by interrupting the fuel supply or blocking fuel in a single-flow channel (Satija et al. 2004; Fuller and Gray 2006; Patterson and Darling 2006). The very high potential generated by this mechanism, even for a short time, can severely damage the cathode by dissolving platinum and corroding carbon.

Fuel starvation causes the third mode of carbon corrosion that occurs on the anode (Reiser et al. 2005; Stuve et al. 2006). Supposing that no hydrogen is supplied to the anode but a cell potential of 0.7 V is maintained, then without hydrogen no anodic reaction at low potentials is feasible. Should the anode potential shift far above that of the cathode, then carbon dissolution and oxygen evolution reactions will occur on the anode.

High potentials may be loaded on the cathode and anode during the startup/shutdown process and hydrogen starvation. Consequently, carbon supports and platinum degrade rapidly, thereby leading to significant loss in performance. To ensure the long-term life of PEMFCs, alternative supports have been considered, including carbon nanotubes (Shao et al. 2006, 2007; Wang et al. 2006a), oxides (Dieckmann and Langer 1998; Wu et al. 2005), carbides (Meng and Shen 2005; Nie et al. 2006), polymers (Lefebvre et al. 1999), boron-doped diamond (Wang and Swain 2002,

2003; Hupert et al. 2003; Fischer and Swain 2005; Fischer et al. 2007), and nonconductive whiskers (Parsonage and Debe 1994; Bonakdarpour et al. 2005; Debe et al. 2006). These materials have some advantages compared with carbon black, but also have issues and limitations hindering their applications as cathode supports in PEMFCs. Further information can be found in the recent review of Borup et al. (2007).

5 Stability of New Cathode Electrocatalysts

5.1 Platinum Monolayer on Metal Nanoparticle Electrocatalyst

Platinum monolayer electrocatalysts offer a dramatically reduced platinum content while affording considerable possibilities for enhancing their catalytic activity and stability. These electrocatalysts comprise a monolayer of platinum on carbon-supported metal or metal alloy nanoparticles. The platinum monolayer approach has several unique features, such as high platinum utilization and enhanced activity, making it very attractive for practical applications with their potential for resolving the problems of high platinum content and low efficiency apparent in conventional electrocatalysts (Adzic et al. 2007). Further, long-term tests of these novel electrocatalysts in fuel cells demonstrated their reasonably good stability. The synthesis of platinum monolayer electrocatalysts was facilitated by a new synthetic method, i.e., a monolayer deposition of platinum on a metal nanoparticle by the redox replacement of a copper monolayer (Brankovic et al. 2001). Three types of platinum monolayer electrocatalysts for the ORR were synthesized (1) platinum on carbon-supported palladium nanoparticles (Zhang et al. 2004), (2) mixed-metal platinum monolayers on palladium nanoparticles (Zhang et al. 2005a), and (3) platinum monolayers on noble/nonnoble core–shell nanoparticles (Zhang et al. 2005b). We discuss the results from the first option.

The experimentally measured electrocatalytic activity of platinum monolayers for the ORR shows a volcano-type dependence on the center of their *d*-bands, as determined by density functional theory (DFT) calculations (Zhang et al. 2005c). A monolayer of platinum on a palladium substrate was shown to have higher activity than the bulk platinum surface (Zhang et al. 2004), which partly reflects the decreased Pt–OH coverage in comparison with bulk platinum (PtOH, derived from H_2O oxidation on platinum, blocks the ORR). In addition, the small compression of a platinum deposit on a palladium substrate, causing a downshift of the *d*-band center, lowers the reactivity of platinum and slightly decreases platinum interaction with the intermediates in the ORR. Both effects enhance the ORR rates (vide infra) (Zhang et al. 2004, 2005a–c).

Long-term fuel cell tests were conducted using electrodes of 50 cm^2 with the Pt/Pd/C cathode catalyst containing 77 $\mu g\ cm^{-2}$ of platinum (0.21 $g_{Pt}\ kW^{-1}$) and 373 $\mu g\ cm^{-2}$ of palladium. A commercial Pt/C electrocatalyst (180 $\mu g\ cm^{-2}$ of platinum) was used for the anode. Figure 5a illustrates the trace of the cell voltage at a constant

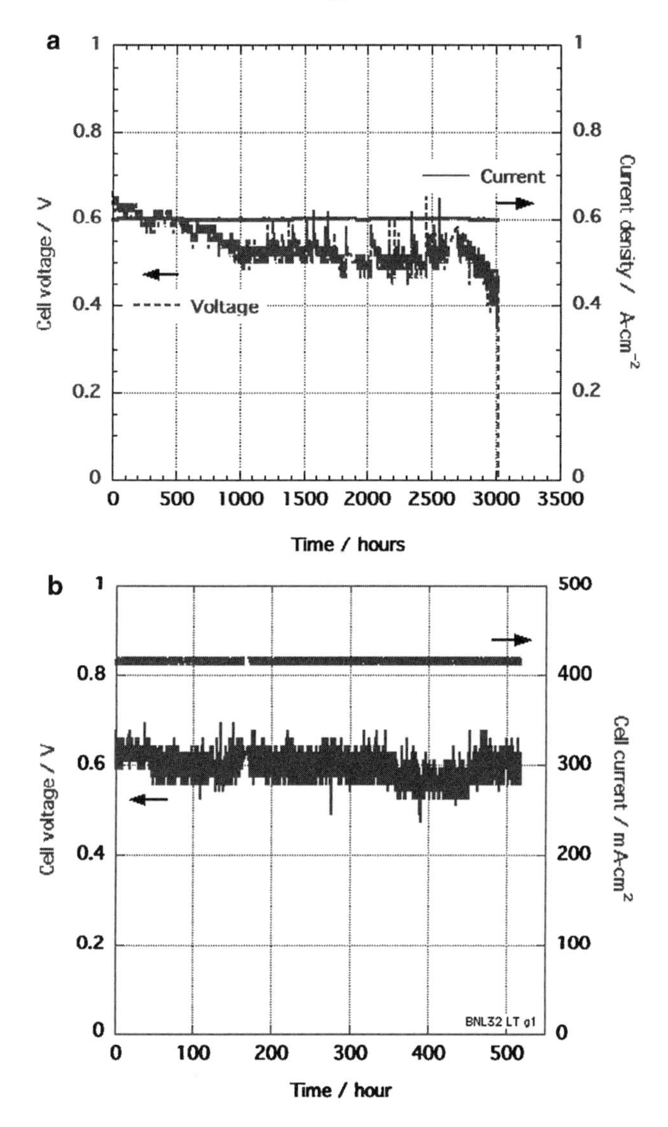

Fig. 5 The long-term stability tests of the Pt/Pd cathode electrocatalyst in an operating fuel cell at 80°C. (**a**) Commercial Pt/C anode catalyst. (**b**) Brookhaven National Laboratory's $PtRu_{20}/C$ anode catalyst

current of 0.6 A cm^{-2} at 80°C against time. Up to 1,000 h, the cell voltage fell by about 120 mV, and then showed a much lower decrease with time. The total loss was approximately 140 mV by the end of the testing at 2,900 h. Compared with initial surface-area values, the losses of the platinum and palladium established by voltammetry were approximately 26% at 1,200 h and 29% at 2,000 h. The losses in platinum can be caused by dissolution of platinum at the operating potential or the embedment of platinum atoms into the palladium substrate; the latter mode is predicted by a weak antisegregation of platinum on a palladium host according to DFT calculations (Greeley et al. 2002). We note that part of the observed drop in voltage also may be due to losses at the anode (a commercial Pt/C electrocatalyst) since the losses at the anode and cathode in these fuel cell measurements were not separated. The results are promising and improving further the electrocatalyst's long-time durability seems necessary.

Another long-term stability test was carried out using a fuel cell comprising both anode and cathode with platinum monolayer electrocatalysts of 5-cm^2 area (Fig. 5b). The cathode catalyst was Pt/Pd/C with platinum loading of 99 $\mu g\ cm^{-2}$, while the anode catalyst was Pt/Ru_{20}/C with platinum loading of 50 $\mu g\ cm^{-2}$. The latter, synthesized at Brookhaven National Laboratory by the electroless deposition of submonolayer platinum on ruthenium nanoparticles, exhibited enhanced activity for hydrogen oxidation, and had a low platinum loading (one-tenth of the standard loading) (Sasaki et al. 2004). When the cell was kept at a constant current density of 0.417 A cm^{-2}, it ran for 450 h with an average voltage of 0.602 V, with no significant loss of voltage. Its catalytic performance was 0.47 $g_{Pt}\ kW^{-1}$. Membrane failure caused the termination of the test after 450 h of operation. Again, the results are promising, but further improvement is needed in the long-time durability of the electrocatalyst.

The stability of platinum monolayer electrocatalysts under conditions of potential cycling is higher than that of intrinsic platinum surfaces. We ascribed this advantage to a shift in the surface oxidation (PtOH formation) of monolayer catalysts to more positive potential than those for Pt/C. The shift is occasioned by the electronic effect of the underlying substrates, as discussed next.

5.2 Stabilization of Platinum Electrocatalysts Using Gold Clusters

The loss of platinum surface area in the cathode caused by the dissolution of platinum under the electrode potential cycling remains a serious obstacle for a widespread application of PEMFCs. In a recent report, Zhang et al. (2007b) demonstrated that platinum oxygen reduction electrocatalysts can be stabilized (exhibit negligible dissolution under potential cycling regimes) when platinum nanoparticles are modified with small gold clusters. Such increased stability was observed under the oxidizing conditions of the ORR and potential cycling between 0.6 and 1.1 V in 30,000 cycles. There were insignificant changes in the activity and the surface area of gold-modified platinum, in contrast to sizeable losses observed with platinum only under the same conditions. Also, these data offered the first evidence that small gold clusters can affect the properties of metal supports. The gold clusters were deposited on a Pt/C catalyst by the galvanic displacement by gold of a copper monolayer on platinum. The stabilizing effect of gold clusters on platinum was determined in an accelerated stability test by continuously applying linear potential sweeps from 0.6 to 1.1 V that cause the surface oxidation/reduction cycles of platinum.

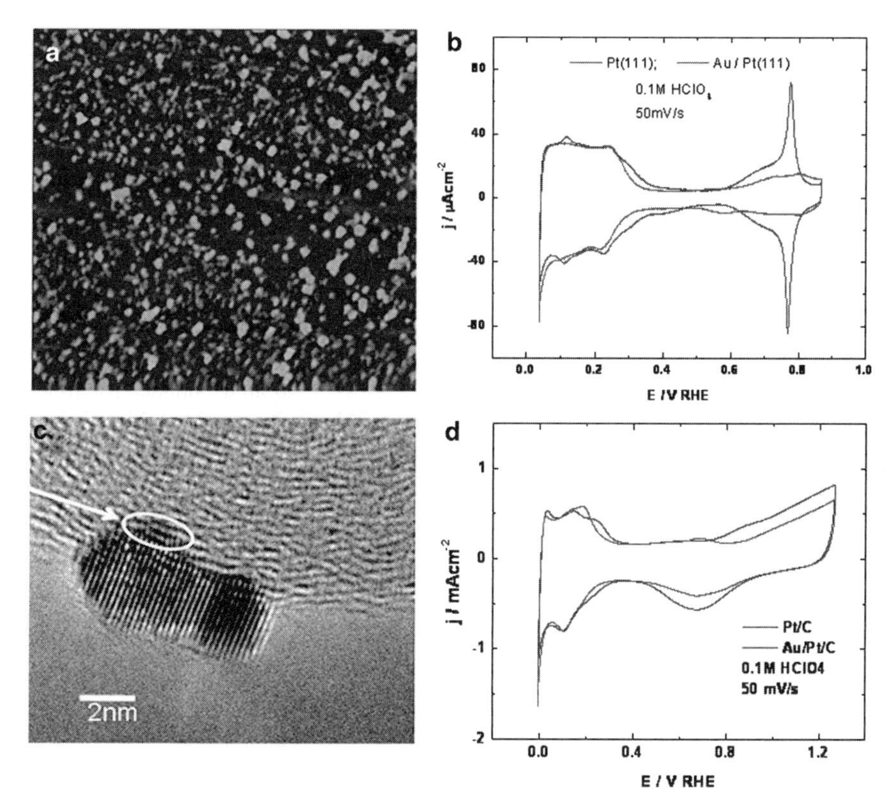

Fig. 6 (**a**) Scanning tunneling microscope image of gold clusters (two-thirds monolayer equivalent charge) and (**b**) the corresponding voltammetry curves for Pt(111) and Au/Pt(111). (**c**) Transmission electron microscope image of a gold-modified platinum nanoparticle and (**d**) the corresponding voltammetry curves

Figure 6a shows a scanning tunneling microscope image of gold clusters (two-thirds monolayer equivalent charge) and the corresponding voltammetry curves are shown in Fig. 6b, revealing that the gold clusters inhibit PtOH formation. Figure 6c displays a transmission electron microscope image of gold-modified platinum nanoparticles and the corresponding voltammetry curves are shown in Fig. 6d, revealing the same effect of gold as in Fig. 6b.

Figure 7 compares the catalytic activities of Au/Pt/C and Pt/C measured as the currents of O_2 reduction obtained before and after 30,000 potential cycles at the rate of 50 mV s^{-1} of a thin-layer rotating disk electrode in an O_2-saturated 0.1 M $HClO_4$ solution at room temperature. There is only 5-mV degradation in the half-wave potential for Au/Pt/C after this period of potential cycling (Fig. 7a). The platinum surface area of the Au/Pt/C and Pt/C electrodes was determined by measuring hydrogen adsorption before and after potential cycling (Fig. 7b). Integrating the charge between 0 and 0.36 V associated with hydrogen adsorption for Au/Pt/C revealed no difference, indicating that there was no recordable loss of the platinum

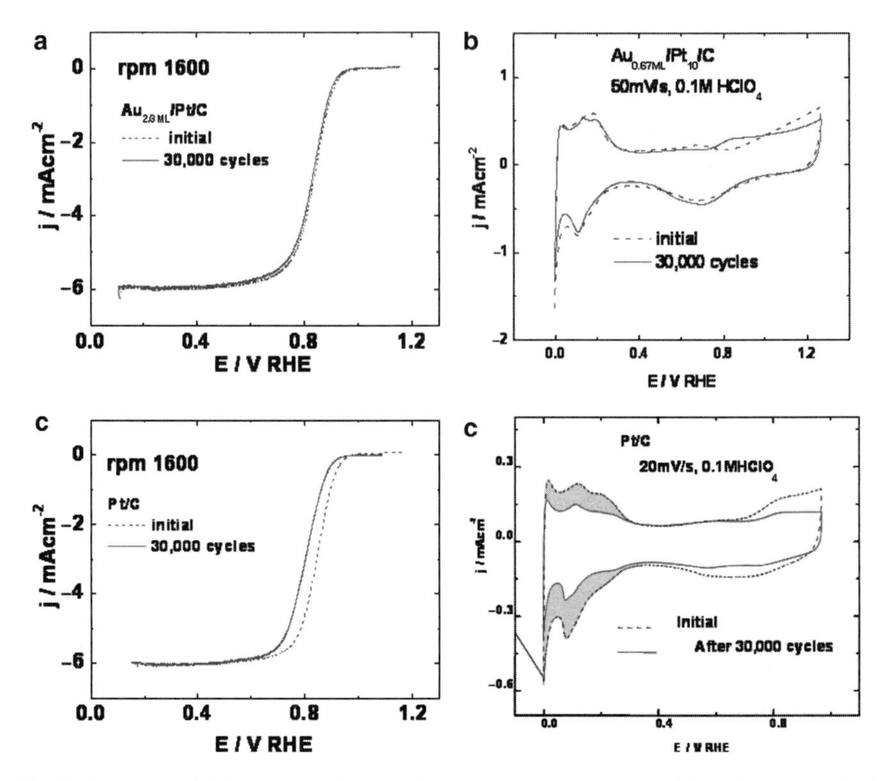

Fig. 7 Catalytic activities of Au/Pt/C and Pt/C measured as the currents of O_2 reduction obtained before (**a**) and after (**c**) 30,000 potential cycles from 0.6 to 1.1 V at the rate of 50 mV s^{-1} of a thin-layer rotating disk electrode in an O_2-saturated 0.1 M $HClO_4$ solution at room temperature. Corresponding voltammetry curves for the Au/Pt/C and Pt/C electrodes before (**b**) and after (**d**) potential cycling

surface area. However, for Pt/C, only about 55% of platinum surface area remained after such potential cycling (Fig. 7d), while the corresponding change in activity for Pt/C amounts to a loss of 39 mV (Fig. 7c). The same experiment with Au/Pt/C at 60°C showed no loss in activity, giving us additional evidence for the stabilization effect of gold clusters on the platinum support.

On the basis of an in situ X-ray absorption near-edge structure measurement, which explored the oxidation state of platinum as a function of potential for the Au/Pt/C and Pt/C surfaces, the origin of the observed stabilization effect of gold clusters was ascribed to a shifting of the oxidation of gold-covered platinum nanoparticles to more positive potentials than those without gold clusters (not shown). Another possibility for explaining the effect of gold clusters involves gold atoms diffusing to the kink- and step-platinum sites and blocking their interaction with the solution phase, consequently decreasing the platinum dissolution rate. The requisite mobility of gold on various surfaces has often been verified (Roudgar and Groβ 2004).

We can infer the electronic effects from theoretical work. Roudgar and Groβ (2004) used DFT calculations to demonstrate the coupling of *d*-orbitals of small palladium clusters to the Au(111) substrate. An equivalent type of interaction between gold and platinum can account for the observed stabilization of platinum that, hence, can became "more noble" in its interactions with gold. Since clusters of a softer metal, gold, are placed on the surface of one that is considerably harder, platinum, there is practically no mixing between them; Del Popolo et al. (2005) earlier reached a similar conclusion about palladium on an gold system. The surface alloying of gold with platinum, although unlikely, also would modulate the electronic structure of platinum toward a lower surface energy, or lower-lying platinum *d*-band states. Thus, the interaction of gold clusters with metal surfaces differs from their interactions with the oxide supports. These findings hold promise for resolving the problem of platinum dissolution under potential cycling regimes, a feature that is critical for using fuel cells in electric vehicles.

6 Concluding Remarks

In this brief review of the dissolution and solubility of platinum under equilibrium conditions and the degradation of platinum nanoparticles at the cathode under various operating conditions, we discussed some mechanisms of degradation, and then offered recent possibilities for overcoming the problem. The data indicate that platinum nanoparticle electrocatalysts at the cathode are unstable under harsh operating conditions, and, as yet, often would be unsatisfactory for usage as the cathode material for fuel cells. Carbon corrosion, particularly under start/stop circumstances in automobiles, also entails electrical isolation and aggregation of platinum nanoparticles. We also discussed new approaches to alleviate the problem of stability of cathode electrocatalysts. One involves a class of platinum monolayer electrocatalysts that, with adequate support and surface segregation, demonstrated enhanced catalytic activity and good stability in a long-term durability test. The other approach rests on the stabilization effects of gold clusters. This effect is likely to be applicable to various platinum- and platinum-alloy-based electrocatalysts, causing their improved stability against platinum dissolution under potential cycling regimes. Such electrocatalysts, if supported on carbon with high corrosion stability (various graphitized carbons, carbon nanotubes, some oxides), look promising for resolving both stability problems so that fuel cells can be used in transportation in the near future.

Acknowledgments This work was supported by the US Department of Energy, Divisions of Chemical and Material Sciences, under contract no. DE-AC02-98CH10886.

References

Adzic, R. R., Zhang, J., Sasaki, K., Vukmirovic, M. B., Shao, M., Wang, J. X., Nilekar, A. U., Mavrikakis, M., Valerio, J. A. and Uribe, F. (2007) Platinum monolayer fuel cell electrocatalysts, Top. Catal. 46, 249–262.

Angerstein-Kozlowska, H., Conway, B. E. and Sharp, W. B. A. (1973) The real condition of electrochemially oxidized platinum surfaces, J. Electroanal. Chem. 43, 9–36.
Aragane, J., T. Murahashi and Odaka, T. (1988) Change of Pt distribution in the active components of phosphoric acid fuel cell, J. Electrochem. Soc. 135, 844–850.
Azaroual, M., Romand, B., Freyssinet, P. and Disnar, J.-R. (2001) Solubility of platinum in aqueous solutions at 25°C and pHs 4 to 10 under oxidizing conditions, Geochim. Cosmochim. Acta 65, 4453–4466.
Bett, J. A. S., Kinoshita, K. and Stonehart, P. (1976) Crystallite growth of platinum dispersed on graphitized carbon black: II. Effect of liquid environment, J. Catal. 41, 124–133.
Bi, W., Gray, G. E. and Fuller, T. F. (2007) PEM fuel cell Pt/C dissolution and deposition in nafion electrolyte, Electrochem. Solid-State Lett. 10, B101–B104.
Bindra, P., Clouser, S. J. and Yeager, E. (1979) Platinum dissolution in concentrated phosphoric-acid, J. Electrochem. Soc. 126, 1631–1632.
Birss, V. I., M. Chang and J. Segal (1993) Platinum oxide film formation-reduction: an in-situ mass measurement study, J. Electroanal. Chem. 355, 181–191.
Blurton, K. F., Kunz, H. R. and Rutt, D. R. (1978) Surface area loss of platinum supported on graphite, Electrochim. Acta 23, 183–190.
Bonakdarpour, A., Wenzel, J., Stevens, D. A., Sheng, S., Monchesky, T. L., Lobel, R., Atanasoski, R. T., Schmoeckel, A. K., Vernstrom, G. D., Debe, M. K. and Dahn, J. R. (2005) Studies of transition metal dissolution from combinatorially sputtered, nanostructured $Pt_{1-x}M_x$ (M = Fe, Ni; $0 < x < 1$) electrocatalysts for PEM fuel cells, J. Electrochem. Soc. 152, A61–A72.
Borup, R., Meyers, J., Pivovar, B., Kim, Y. S., Mukundan, R., Garland, N., Myers, D., Wilson, M., Garzon, F., Wood, D., Zelenay, P., More, K., Stroh, K., Zawodzinski, T., Boncella, J., McGrath, J. E., Inaba, M., Miyatake, K., Hori, M., Ota, K., Ogumi, Z., Miyata, S., Nishikata, A., Siroma, Z., Uchimoto, Y., Yasuda, K., Kimijima, K. I. and Iwashita, N. (2007) Scientific aspects of polymer electrolyte fuel cell durability and degradation, Chem. Rev. 107, 3904–3951.
Brankovic, S. R., Wang, J. X. and Adzic, R. R. (2001) Metal monolayer deposition by replacement of metal adlayers on electrode surfaces, Surf. Sci. 474, L173–L179.
Conway, B. E. (1995) Electrochemical oxide film formation at noble metals as a surface-chemical process, Prog. Surf. Sci. 49, 331–345.
Dam, V. A. T. and de Bruijn, F. A. (2007) The stability of PEMFC electrodes – platinum dissolution vs. potential and temperature investigated by quartz crystal microbalance, J. Electrochem. Soc. 154, B494–B499.
Darling, R. M. and Meyers, J. P. (2003) Kinetic model of platinum dissolution in PEMFCs, J. Electrochem. Soc. 150, A1523–A1527.
Darling, R. M. and Meyers, J. P. (2005) Mathematical model of platinum movement in PEM fuel cells, J. Electrochem. Soc. 152, A242–A247.
Debe, M. K., Schmoeckel, A. K., Vernstrorn, G. D. and Atanasoski, R. (2006) High voltage stability of nanostructured thin film catalysts for PEM fuel cells, J. Power Sources 161, 1002–1011.
Del Popolo, M. G., Leiva, E. P. M., Mariscal, M. and Schmickler, W. (2005) On the generation of metal clusters with the electrochemical scanning tunneling microscope, Surf. Sci. 597, 133–155.
Dieckmann, G. R. and Langer, S. H. (1998) Comparison of Ebonex and graphite supports for platinum and nickel electrocatalysts, Electrochim. Acta 44, 437–444.
Ferreira, P. J. and Shao-Horn, Y. (2007) Formation mechanism of Pt single-crystal nanoparticles in proton exchange membrane fuel cells, Electrochem. Solid-State Lett. 10, B60–B63.
Ferreira, P. J., la O', G. J., Shao-Horn, Y., Morgan, D., Makharia, R., Kocha, S. and Gasteiger, H. A. (2005) Instability of Pt/C electrocatalysts in proton exchange membrane fuel cells – a mechanistic investigation, J. Electrochem. Soc. 152, A2256–A2271.
Fischer, A. E. and Swain, G. M. (2005) Preparation and characterization of boron-doped diamond powder – a possible dimensionally stable electrocatalyst support material, J. Electrochem. Soc. 152, B369–B375.
Fischer, A. E., Lowe, M. A. and Swain, G. M. (2007) Preparation and electrochemical characterization of carbon paper modified with a layer of boron-doped nanocrystalline diamond, J. Electrochem. Soc. 154, K61–K67.

Fuller, T. F. and Gray, G. (2006) Carbon corrosion induced by partial hydrogen coverage, ECS Trans. 1, 345.

Garzon, F. H., Davey, J. and Borup, R. (2006) Fuel cell catalyst particle size growth characterized by X-ray scattering methods, ECS Trans. 1, 153.

Greeley, J., Norskov, J. K. and Mavrikakis, M. (2002) Electronic structure and catalysis on metal surfaces, Annu. Rev. Phys. Chem. 53, 319–348.

Gruver, G. A., Pascoe, R. F. and Kunz, H. R. (1980) Surface area loss of platinum supported on carbon in phosphoric acid electrolyte, J. Electrochem. Soc. 127, 1219–1224.

Guilminot, E., Corcella, A., Charlot, F., Maillard, F. and Chatenet, M. (2007a) Detection of Pt^{z+} ions and Pt nanoparticles inside the membrane of a used PEMFC, J. Electrochem. Soc. 154, B96–B105.

Guilminot, E., Corcella, A., Chatenet, M., Maillard, F., Charlot, F., Berthome, G., Iojoiu, C., Sanchez, J. Y., Rossinot, E. and Claude, E. (2007b) Membrane and active layer degradation upon PEMFC steady-state operation, J. Electrochem. Soc. 154, B1106–B1114.

Harrington, D. A. (1997) Simulation of anodic Pt oxide growth, J. Electroanal. Chem. 420, 101–109.

Healy, J., Hayden, C., Xie, T., Olson, K., Waldo, R., Brundage, A., Gasteiger, H. and Abbott, J. (2005) Aspects of the chemical degradation of PFSA ionomers used in PEM fuel cells, Fuel Cells 5, 302–308.

Honji, A., Mori, T., Tamura, K. and Hishinuma, Y. (1988) Agglomeration of platinum particles supported on carbon in phosphoric acid, J. Electrochem. Soc. 135, 355–359.

Hupert, M., Muck, A., Wang, R., Stotter, J., Cvackova, Z., Haymond, S., Show, Y. and Swain, G. M. (2003) Conductive diamond thin-films in electrochemistry, Diamond Relat. Mater. 12, 1940–1949.

Jerkiewicz, G., Vatankhah, G., Lessard, J., Soriaga, M. P. and Park, Y. S. (2004) Surface-oxide growth at platinum electrodes in aqueous H_2SO_4 reexamination of its mechanism through combined cyclic-voltammetry, electrochemical quartz-crystal nanobalance, and Auger electron spectroscopy measurements, Electrochim. Acta 49, 1451–1459.

Johnson, D. C., Napp, D. T. and Bruckenstein, S. (1970) A ring-disk electrode study of the current/potential behaviour of platinum in 1.0 M sulphuric and 0.1 M perchloric acids, Electrochim. Acta 15, 1493–1509.

Kinoshita, K. (1988) *Carbon: Electrochemical and Physicochemical Properties*, Wiley, New York, NY.

Kinoshita, K., Lundquist, J. T. and Stonehart, P. (1973) Potential cycling effects on platinum electrocatalyst surfaces, J. Electroanal. Chem. 48, 157–166.

Komanicky, V., Chang, K. C., Menzel, A., Markovic, N. M., You, H., Wang, X. and Myers, D. (2006) Stability and dissolution of platinum surfaces in perchloric acid, J. Electrochem. Soc. 153, B446–B451.

Lefebvre, M. C., Qi, Z. G. and Pickup, P. G. (1999) Electronically conducting proton exchange polymers as catalyst supports for proton exchange membrane fuel cells – electrocatalysis of oxygen reduction, hydrogen oxidation, and methanol oxidation, J. Electrochem. Soc. 146, 2054–2058.

Mathias, M. F., Makharia, R., Gasteiger, H. A., Conley, J. J., Fuller, T. J., Gittleman, C. J., Kocha, S. S., Miller, D. P., Mittelsteadt, C. K., Xie, T., Yan, S. G. and Yu, P. T. (2005) Two fuel cell cars in every garage?, Interface 14, 24–35.

Meng, H. and Shen, P. K. (2005) The beneficial effect of the addition of tungsten carbides to Pt catalysts on the oxygen electroreduction, Chem. Commun. 35, 4408–4410.

Mitsushima, S., Kawahara, S., Ota, K.-I. and Kamiya, N. (2007a) Consumption rate of Pt under potential cycling, J. Electrochem. Soc. 154, B153–B158.

Mitsushima, S., Koizumi, Y., Ota, K. and Kamiya, N. (2007b) Solubility of platinum in acidic media (I) – in sulfuric acid, Electrochemistry 75, 155–158.

Nagy, Z. and You, H. (2002) Applications of surface X-ray scattering to electrochemistry problems, Electrochim. Acta 47, 3037–3055.

Nie, M., Shen, P. K., Wu, M., Wei, Z. D. and Meng, H. (2006) A study of oxygen reduction on improved Pt-WC/C electrocatalysts, J. Power Sources 162, 173–176.

Ota, K.-I., Nishigori, S. and Kamiya, N. (1988) Dissolution of platinum anodes in sulfuric acid solution, J. Electroanal. Chem. 257, 205–215.
Parsonage, E. E. and Debe, M. K., (1994), U.S. Patent 5,338,430.
Patterson, T. W. (2002) *AIChE Spring National Meeting*, New Orleans, LA, pp. 313–318.
Patterson, T. W. and Darling, R. M. (2006) Damage to the cathode catalyst of a PEM fuel cell caused by localized fuel starvation, Electrochem. Solid-State Lett. 9, A183–A185.
Pourbaix, M. (1974) *Atlas of Electrochemical Equilibria*, 2nd ed., NACE, Houston, TX.
Rand, D. A. J. and Woods, R. (1972) A study of the dissolution of platinum, palladium, rhodium and gold electrodes in 1 M sulphuric acid by cyclic voltammetry, J. Electroanal. Chem. 35, 209–218.
Reiser, C. A., Bregoli, L., Patterson, T. W., Yi, J. S., Yang, J. D., Perry, M. L. and Jarvi, T. D. (2005) A reverse-current decay mechanism for fuel cells, Electrochem. Solid-State Lett. 8, A273–A276.
Roen, L. M., Paik, C. H. and Jarvi, T. D. (2004) Electrocatalytic corrosion of carbon support in PEMFC cathodes, Electrochem. Solid-State Lett. 7, A19–A24.
Roudgar, A. and Groβ, A. (2004) Local reactivity of supported metal clusters: Pd_n on Au(111), Surf. Sci. 559, L180–L186.
Sasaki, K., Wang, J. X., Balasubramanian, M., McBreen, J., Uribe, F. and Adzic, R. R. (2004) Ultra-low platinum content fuel cell anode electrocatalyst with a long-term performance stability, Electrochim. Acta 49, 3873–3877.
Satija, R., Jacobson, D. L., Arif, M. and Werner, S. A. (2004) In situ neutron imaging technique for evaluation of water management systems in operating PEM fuel cells, J. Power Sources, 129, 238–245.
Shao, Y., Yin, G., Zhang, J. and Gao, Y. (2006) Comparative investigation of the resistance to electrochemical oxidation of carbon black and carbon nanotubes in aqueous sulfuric acid solution, Electrochim. Acta 51, 5853.
Shao, Y., Yin, G. and Gao, Y. (2007) Understanding and approaches for the durability issues of Pt-based catalysts for PEM fuel cell, J. Power Sources 171, 558–566.
Shao-Horn, Y., Sheng, W. C., Chen, S., Ferreria, P. J., Hollby, E. F. and Morgan, D. (2007) Instability of supported platinum nanoparticles in low-temperature fuel cells, Top. Catal. 46, 285–305.
Stevens, D. A., Hicks, M. T., Haugen, G. M. and Dahn, J. R. (2005) Ex situ and in situ stability studies of PEMFC catalysts, J. Electrochem. Soc. 152, A2309–A2315.
Stuve, E. M. and Gastaiger, H. A., (2006), PEMFC short course, 210th Meeting of The Electrochemical Society, Cancun, Mexico.
Teranishi, K., Kawata, K., Tsushima, S. and Hirai, S. (2006) Degradation mechanism of PEMFC under open circuit operation, Electrochem. Solid-State Lett. 9, A475–A477.
Tseung, A. C. C. and Dhara, S. C. (1975) Loss of surface area by platinum and supported platinum black electrocatalyst, Electrochim. Acta 20, 681–683.
Virkar, A. V. and Zhou, Y. K. (2007) Mechanism of catalyst degradation in proton exchange membrane fuel cells, J. Electrochem. Soc. 154, B540–B546.
Voorhees, P. W. (1985) The theory of Ostwald ripening, J. Stat. Phys. 38, 231–252.
Wang, J. and Swain, G. M. (2002) Dimensionally stable Pt/diamond composite electrodes in concentrated H3PO4 at high temperature, Electrochem. Solid-State Lett. 5, E4–E7.
Wang, J. and Swain, G. M. (2003) Fabrication and evaluation of platinum/diamond composite electrodes for electrocatalysis – Preliminary studies of the oxygen-reduction reaction, J. Electrochem. Soc. 150, E24–E32.
Wang, X., Li, W. Z., Chen, Z. W., Waje, M. and Yan, Y. S. (2006a) Durability investigation of carbon nanotube as catalyst support for proton exchange membrane fuel cell, J. Power Sources 158, 154–159.
Wang, X. P., Kumar, R. and Myers, D. J. (2006b) Effect of voltage on platinum dissolution relevance to polymer electrolyte fuel cells, Electrochem. Solid-State Lett. 9, A225–A227.
Wilson, M. S., Garzon, F. H., Sickafus, K. E. and Gottesfeld, S. (1993) Surface area loss of supported platinum in polymer electrolyte fuel cells, J. Electrochem. Soc. 140, 2872–2877.

Wu, G., Li, L., Li, J. H. and Xu, B. Q. (2005) Polyaniline-carbon composite films as supports of Pt and PtRu particles for methanol electrooxidation, Carbon 43, 2579–2587.
Xie, J., Wood, D. L., More, K. L., Atanassov, P. and Borup, R. L. (2005a) Microstructural changes of membrane electrode assemblies during PEFC durability testing at high humidity conditions, J. Electrochem. Soc. 152, A1011–A1020.
Xie, J., Wood, D. L., Wayne, D. M., Zawodzinski, T. A., Atanassov, P. and Borup, R. L. (2005b) Durability of PEFCs at high humidity conditions, J. Electrochem. Soc. 152, A104–A113.
Yasuda, K., Taniguchi, A., Akita, T., Ioroi, T. and Siroma, Z. (2006a) Characteristics of a platinum black catalyst layer with regard to platinum dissolution phenomena in a membrane electrode assembly, J. Electrochem. Soc. 153, A1599–A1603.
Yasuda, K., Taniguchi, A., Akita, T., Ioroi, T. and Siroma, Z. (2006b) Platinum dissolution and deposition in the polymer electrolyte membrane of a PEM fuel cell as studied by potential cycling, Phys. Chem. Chem. Phys. 8, 746–752.
You, H., Chu, Y. S., Lister, T. E., Nagy, Z., Ankudiniv, A. L. and Rehr, J. J. (2000) Resonance X-ray scattering from Pt(1 1 1) surfaces under water, Physica B 283, 212–216.
Zhang, J., Mo, Y., Vukmirovic, M. B., Klie, R., Sasaki, K. and Adzic, R. R. (2004) Platinum monolayer electrocatalysts for O_2 reduction: Pt monolayer on Pd(111) and on carbon-supported Pd nanoparticles, J. Phys. Chem. B 108, 10955–10964.
Zhang, J. L., Vukmirovic, M. B., Sasaki, K., Nilekar, A. U., Mavrikakis, M. and Adzic, R. R. (2005a) Mixed-metal Pt monolayer electrocatalysts for enhanced oxygen reduction kinetics, J. Am. Chem. Soc. 127, 12480–12481.
Zhang, J., Lima, F. H. B., Shao, M. H., Sasaki, K., Wang, J. X., Hanson, J. and Adzic, R. R. (2005b) Platinum monolayer on nonnoble metal-noble metal core-shell nanoparticle electrocatalysts for O_2 reduction, J. Phys. Chem. B 109, 22701–22704.
Zhang, J. L., Vukmirovic, M. B., Xu, Y., Mavrikakis, M. and Adzic, R. R. (2005c) Controlling the catalytic activity of platinum-monolayer electrocatalysts for oxygen reduction with different substrates, Angew. Chem. Int. Ed. 44, 2132–2135.
Zhang, J., Litteer, B. A., Gu, W., Liu, H. and Gasteiger, H. A. (2007a) Effect of hydrogen and oxygen partial pressure on Pt precipitation within the membrane of PEMFCs, J. Electrochem. Soc. 154, B1006–B1011.
Zhang, J., Sasaki, K., Sutter, E. and Adzic, R. R. (2007b) Stabilization of platinum oxygen reduction electrocatalysts using gold clusters, Science 315, 220–222.

Carbon-Support Requirements for Highly Durable Fuel Cell Operation

Paul T. Yu, Wenbin Gu, Jingxin Zhang, Rohit Makharia, Frederick T. Wagner, and Hubert A. Gasteiger[a]

Abstract Owing to its unique electrical and structural properties, high surface area carbon has found widespread use as a catalyst support material in proton exchange membrane fuel cell (PEMFC) electrodes. The highly dynamic operating conditions in automotive applications require robust and durable catalyst support materials. In this chapter, carbon corrosion kinetics of commercial conventional-carbon-supported membrane electrode assemblies (MEAs) are presented. Carbon corrosion was investigated under various automotive fuel cell operating conditions. Fuel cell system start/stop and anode local hydrogen starvation are two major contributors to carbon corrosion. Projections from these studies indicate that conventional-carbon-supported MEAs fall short of meeting automotive the durability targets of PEMFCs. MEAs made of different carbon support materials were evaluated for their resistance to carbon corrosion under accelerated test conditions. The results show that graphitized-carbon-supported MEAs are more resistant to carbon corrosion than nongraphitized carbon materials. Fundamental model analyses incorporating the measured carbon corrosion kinetics were developed for start/stop and local hydrogen starvation conditions. The combination of experiment and modeling suggests that MEAs with corrosion-resistant carbon supports are promising material approaches to mitigate carbon corrosion during automotive fuel cell operation.

1 Introduction

Carbon in various forms is commonly used in phosphoric acid fuel cells and proton exchange membrane fuel cells (PEMFCs) as a catalyst support, gas-diffusion media (GDM), and bipolar plate material (Dicks 2006). Among these carbon materials, carbon black is used as a catalyst support for PEMFC application because of its unique properties

---

P.T. Yu (✉)
Fuel Cell Activities, General Motors, 10 Carriage Street Honeoye Falls, NY 14472-0603, USA
e-mail: taichiang.yu@gm.com

F.N. Büchi et al. (eds.), *Polymer Electrolyte Fuel Cell Durability*,

(1) high electronic conductivity of carbon-black-based electrodes (ca. 1–10 S cm^{-1}); (2) high Brunauer–Emmett–Teller (BET) surface areas (ca. 100–1,000 $m^2\ g_C^{-1}$) to enable high dispersion of platinum or platinum-alloy catalysts; (3) formation of high-porosity electrodes owing to the *high-structure* characteristics of primary carbon-black aggregates (Medalia and Heckman 1969; Medalia 1967; Gruber et al. 1993). Although it has been documented that carbon black corrodes in the phosphoric acid fuel cell operation condition (150–200°C) (Gruver 1978), carbon-support corrosion has only recently received attention in PEMFCs since the lower operating temperatures (60–80°C) of PEMFCs were thought to render carbon-support corrosion insignificant (Roen et al. 2004). However, in the automotive fuel cell application, the highly dynamic operation conditions (Makharia et al. 2006) in combination with approximately 38,500 start/stop cycles could result in excursions to high potentials (Mathias et al. 2005) that would reduce the lifetime of conventional-carbon-support membrane electrode assemblies (MEAs). Progress on the durability issue is essential for the hydrogen-powered fuel cell vehicle to be commercially viable. The issue has motivated more extensive research on the corrosion of carbon catalyst supports in PEMFC systems.

In this chapter, we attempt to evaluate state-of-the-art commercial conventional-carbon-support MEAs for their carbon corrosion kinetics, the relationship between cell voltage loss and carbon-support weight loss, and the life projection of the catalyst support under automotive operating conditions. These operational conditions include steady-state operation, transient, start/stop, and unintended deviations from nominal run parameters. On the basis of these analyses, we elucidate (1) which operational conditions result in severe carbon corrosion, (2) whether current conventional-carbon-support MEAs are robust enough to meet automotive durability targets, and (3) if a state-of-the-art corrosion-resistant carbon-support MEA is absolutely required for improving automotive fuel cell durability.

2 Carbon Corrosion Mechanism, Kinetics, and its Correlation to Cell Voltage Loss

Typically, electrochemical carbon-support corrosion (oxidation) proceeds as in (1). The mechanism and kinetics of electrochemical oxidation of carbon as a function of potential (usually on the basis of the reversible hydrogen electrode (RHE) potential) have mainly been investigated using potentiostatic experiments in phosphoric acid (Kinoshita 1988), and are described by

$$C + 2H_2O \rightarrow CO_2 + 4H^+ + 4e^- \quad E = 0.207\,V\ (\text{vs. RHE}). \tag{1}$$

The generation of CO_2 is believed to proceed through the intermediate formation of carbon surface oxides, as suggested in earlier studies in phosphoric acid fuel cells (Kinoshita and Bett 1973). It was reported that formation of surface oxides and evolution of CO_2 occur simultaneously during the initial corrosion of high surface area carbon blacks. Similar conclusions are also reported by recent studies of PEMFCs (Kangasniemi et al. 2004). These studies showed that the surface oxides on the commonly used Vulcan XC-72 carbon support were observed at potentials higher than 0.8 V at 65°C and higher than 1.0 V at room temperature (potentials vs. RHE). Owing to the slow buildup of surface oxides, the measured carbon corrosion current becomes equivalent to the CO_2 formation rate only after hundreds to thousands of minutes (Kinoshita and Bett 1973), so quantification of CO_2 formation rates cannot be done on the basis of simple corrosion current measurements.

Therefore, to determine the carbon corrosion kinetics in terms of the equivalent CO_2 current generated from (1) at time scales shorter than hundreds of minutes, direct measurement of the CO_2 evolution rate is required, as was done previously in alkaline (Ross and Sokol 1984) or acidic (Kinoshita and Bett 1973) aqueous electrolytes as well as in a PEM single cell (Roen et al. 2004). Even at longer time scales, the overall measured current densities during carbon corrosion experiments in a PEM single cell (using the so-called driven cell mode) are generally not suitable for kinetic studies, since the current of CO_2 evolution (usually <1 A g_C^{-1}, corresponding to <1 mA cm_{MEA}^{-2} at loadings of typically <1 mg_C cm_{MEA}^{-2}; see, e.g., Fig. 3) is on the same order of magnitude as the parasitic current densities from H_2 crossover (1–5 mA cm_{MEA}^{-2}; Kocha 2003) and ohmic shorting currents through the membrane (Kocha 2003). Thus, to deconvolute the carbon corrosion current (1) from the total current (i.e., the sum of carbon surface oxidation, H_2 crossover, and ohmic shorting currents), we measured the CO_2 concentration at the exit of a 50-cm^2 cell using a gas chromatograph equipped with a methanizer and a flame ionization detector (Yu et al. 2006a). The CO_2 evolution rate from carbon corrosion can be converted to an equivalent carbon corrosion current (assuming $4e^-$ per CO_2 molecule) and integrated over time to yield a curve of percentage of carbon weight loss versus time. Figure 1 shows the time dependence of the concentration of evolved CO_2 and the accumulated carbon weight loss of a conventional-carbon-supported MEA potentiostated at 1.2 V versus the RHE at 95°C. In this 50 cm^2 single-cell experiment, the N_2-purged working electrode (50 cm^3 min^{-1}) was first held at 0 V versus the H_2-purged counter/reference electrode (200 cm^3 min^{-1}), during which time no significant CO_2 was found. The CO_2 concentration due to carbon corrosion rose to near 1,000 ppm as soon as the potential of the working electrode was set to 1.2 V versus the RHE (i.e., vs. the H_2-purged counter/reference electrode). The CO_2 concentration, i.e., the CO_2 evolution rate, decreased with time. The accumulated carbon weight loss based on the integrated CO_2 evolution rate amounted to around 20% over 24 h.

Detailed kinetic studies of a commercial conventional-carbon-supported MEA were conducted to predict its lifetime. The carbon corrosion rates of conventional-carbon-support MEAs as a function of time at 80°C with potential hold at 1.1, 1.2, and 1.3 V versus the RHE, respectively, are shown in Fig. 2. As might be expected, the carbon corrosion current increases as the potential increases. The response of the CO_2 current versus corrosion time under these experimental conditions follows a linear log–log relation, which is consistent with the description of the corrosion current of carbon blacks in H_3PO_4 by Kinoshita (Kinoshita 1988; Kinoshita and Bett 1973). Detailed studies of corrosion currents of conventional-carbon-support

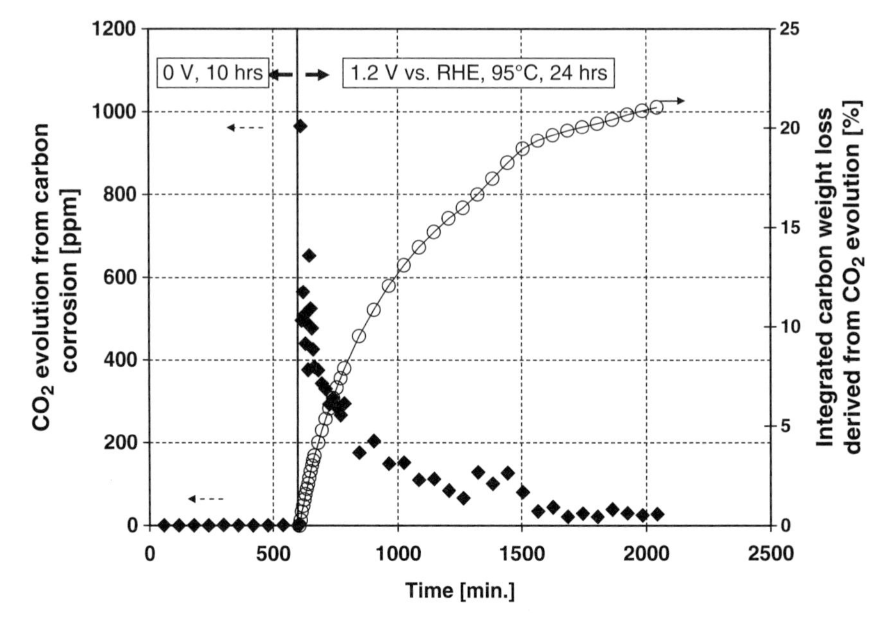

Fig. 1 CO_2 concentration (*solid symbols*) and accumulative carbon weight loss (*open symbols*, integrated from the CO_2 flux at the exit of the working electrode) as a function of time over a commercial conventional-carbon-supported membrane electrode assembly (MEA). The N_2-fed working electrode (50 cm^3 min^{-1}) was held at a potential of 1.2 V versus the H_2-fed counter/reference electrode (200 cm^3 min^{-1}) at 95°C and 80% inlet relative humidity (RH_{inlet}). *RHE* reversible hydrogen electrode

MEAs with respect to potential and temperature resulted in the following empirical equation (Mathias et al. 2005), determined for approximately 100% relative humidity (RH)

$$i_{CO_2} \propto 10^{E/TS} e^{-E_a/RT} t^{-m}, \tag{2}$$

where i_{CO_2} is the carbon corrosion current at time t, E is the applied potential, and R and T represent the universal gas constant and temperature, respectively. By fitting the measured carbon corrosion rates to (2), we obtained an activation energy (E_a) of 67 kJ mol^{-1}, a Tafel slope (TS) of 152 mV per decade, and a power-law parameter of 0.32 for the time decay. Figure 3 shows a log–log plot of experimental carbon corrosion current versus the carbon corrosion kinetic fit. The model agrees fairly well with data points measured over wide ranges of temperature, potential, and corrosion time.

The functional form of the time dependence of the carbon-support corrosion rate is consistent with previous observations of carbon-black corrosion in H_3PO_4 (Stonehart 1984; Kinoshita and Bett 1973), and is most likely related to the inhomogeneous graphitization of commonly used carbon supports, which generally consist of primary particles with diminishing graphitic order toward the particle center, as shown in Fig. 4 (Heckmann and Harling 1966; Wissler 2006). Since the corrosion resistance of carbon within a graphitic plane is approximately 1–2 orders of magnitude

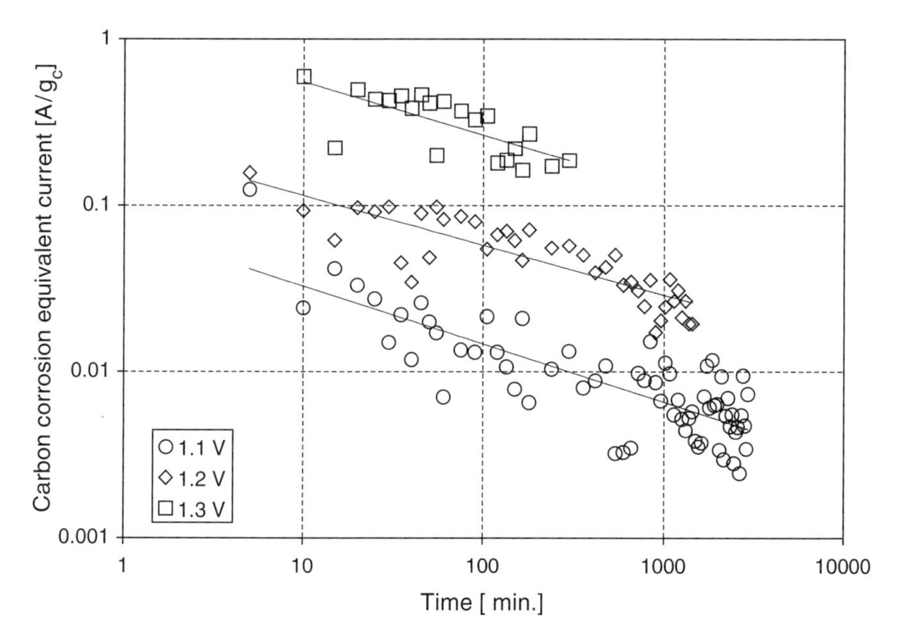

Fig. 2 The effect of potential on carbon corrosion rate versus time over commercial conventional-carbon-supported MEAs measured at 80°C and 80% RH_{inlet}. The working electrode was fed with N_2 (50 cm^3 min^{-1}), while the counter/reference electrode was purged with H_2 (200 cm^3 min^{-1})

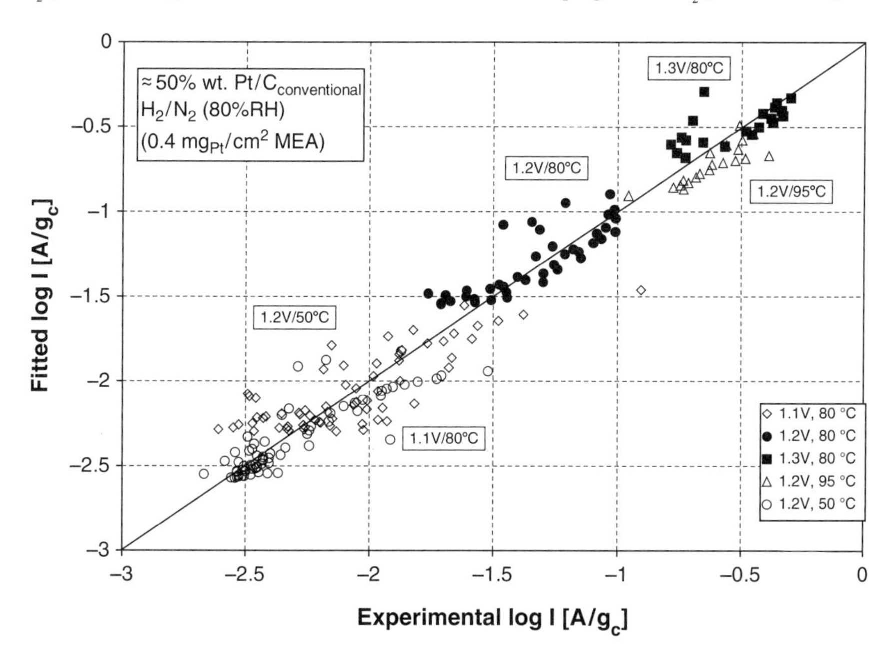

Fig. 3 The log–log plot of the fitted carbon corrosion kinetics (2) versus experimental carbon corrosion rates determined by CO_2 evolution from a N_2-purged working electrode at various temperatures, corrosion times, and potentials (vs. RHE). (Reproduced by permission of Makharia et al. (2006), The Electrochemical Society)

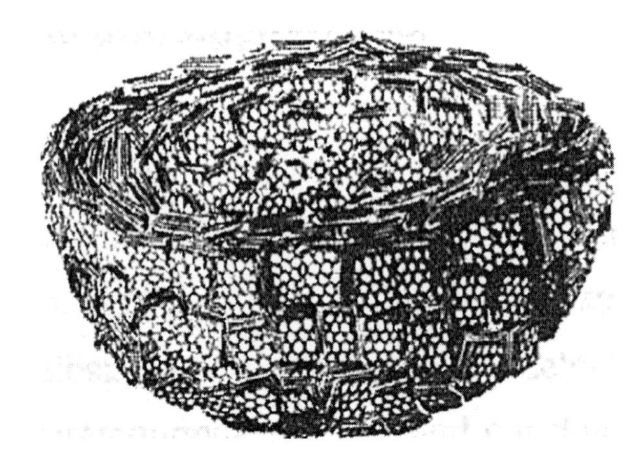

Fig. 4 Cutaway model of a carbon-black primary particle with concentric layers. Parallel orientation of ordered layer groupings and diminishing graphitic order near particle center. (Reproduced with permission of Wissler (2006))

higher than that of carbon along the edges of graphitic planes (Kinoshita 1988), Fig. 4 suggests that the corrosion of conventional carbon supports such as Vulcan XC-72 should progress from the center region of the primary carbon-black particles (high density of graphitic plane edge sites and/or amorphous carbon), progressing over time toward the more graphite-like (and thus more corrosion resistant) shell region that has a lower density of graphitic plane edge sites. This indeed was observed by transmission electron microscopy (TEM) of Vulcan XC-72 supported cathode catalysts corroded in H_3PO_4, revealing a shell-like structure of the corroded carbon-black particles (Fig. 5) in contrast to the homogeneous TEM contrast which is obtained for the initial, noncorroded carbon blacks (Gruver 1978).

Using the kinetic parameters above, one can estimate the carbon weight loss of a conventional-carbon-supported MEA (ignoring the transient effects, which will be discussed later) by integrating (2) with respect to time under various automotive fuel cell operation conditions. To understand the impact of the carbon weight loss on automotive fuel cell durability, one also has to correlate this carbon weight loss to cell voltage loss. Figure 6 shows the cell voltage loss versus the carbon weight loss at 0.8 A cm^{-2} with H_2/air at 2:2 stoichiometric flows, measured at 80°C, 175 kPa(abs), and 50%/50% inlet RH. The results show cell voltage losses of approximately 50 mV at 0.8 A cm^{-2} for a carbon weight loss of 11 ± 2% for a variety of graphitized and nongraphitized carbon supports.

The large performance losses at less than 15% carbon weight loss are due to a substantial loss of electrode porosity, visualized by the observed *electrode thinning* (Fig. 7) once 5–10% of carbon weight losses have occurred. While the porosity of a typical MEA for H_2/air PEMFCs is on the order of 65% (Gasteiger et al. 2003), partial corrosion of the carbon support leads to a collapse of the initially very porous electrodes produced by the *high-structure* carbon blacks used in PEMFC electrodes such as Vulcan XC-72C and Ketjenblack (Kinoshita 1988; Medalia and Heckman 1969). The decreasing electrode porosity (i.e., increasing electrode thinning)

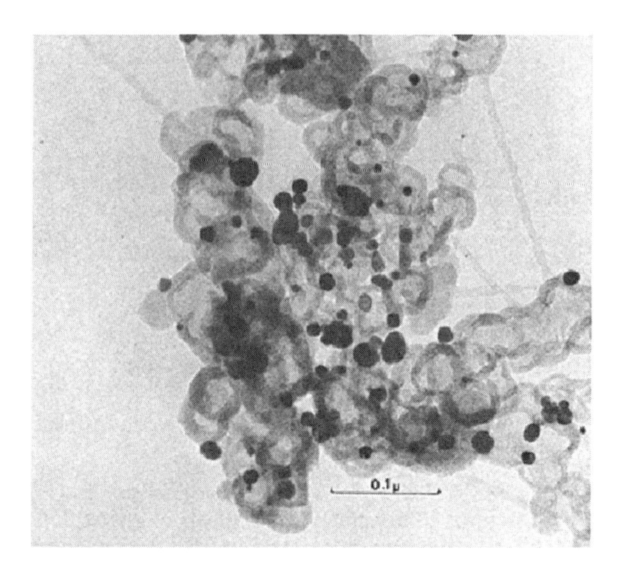

Fig. 5 Transmission electron microscope image of a Vulcan XC-72C supported platinum catalyst which was corroded for 1,000 h at 191°C in 95% H_3PO_4 at a potential of 835 mV (ohmic resistance corrected voltage vs. RHE). (Reproduced with permission of Gruver (1978))

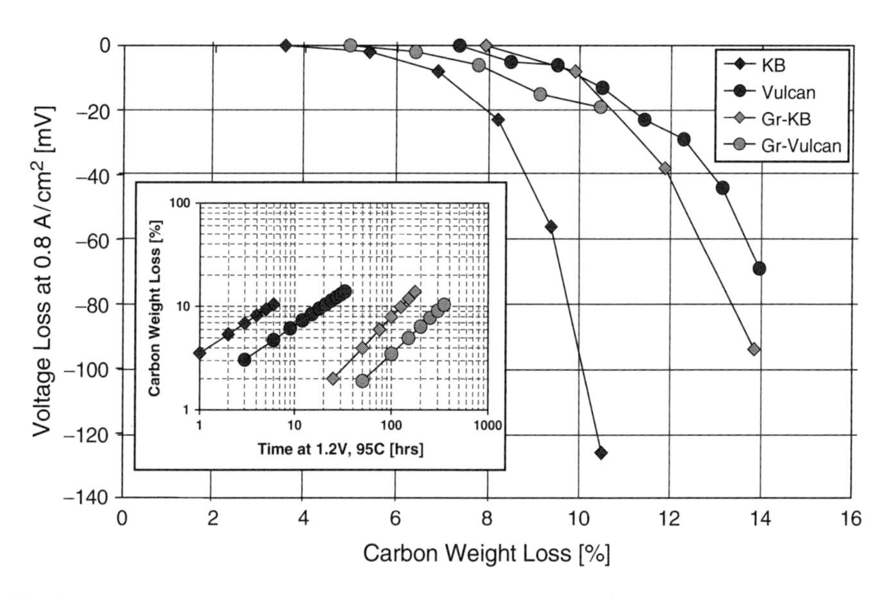

Fig. 6 Cell voltage loss versus carbon weight loss measured at 0.8 A cm^{-2} with H_2/air (2:2 stoichiometric flows) at 80°C, 175 kPa(abs), and 50%/50% RH_{inlet}. The *inset* shows the carbon weight loss versus time at 1.2 V (vs. RHE), 95°C, and 80% RH. *KB* Ketjenblack, *Gr-KB* Graphitized Ketjenblack, *Gr-Vulcan* Graphitized Vulcan XC-72C. (Reproduced with permission of Yu et al. (2006a), The Electrochemical Society)

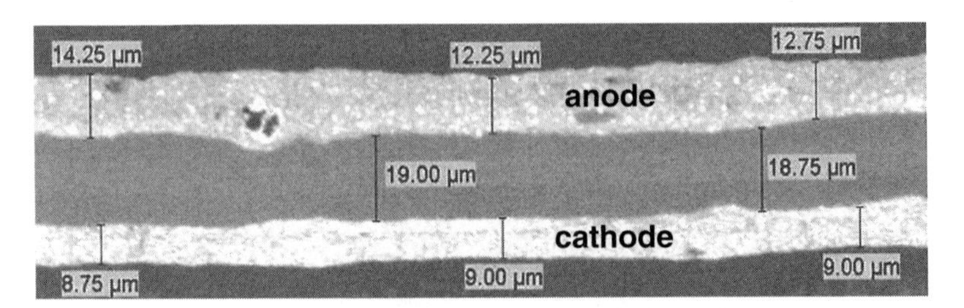

Fig. 7 Scanning electron microscope cross section of an MEA of which 8 wt% of the cathode carbon support had been corroded. The initial cathode electrode thickness was identical to the anode electrode thickness shown

with increasing carbon weight loss results in increasing gas transport losses through the electrode. Therefore, it is not surprising that the same cell voltage loss of approximately 50 mV occurs at carbon weight losses as low as $9 \pm 2\%$ at the higher current density of $1.2\,A\ cm^{-2}$ (Makharia et al. 2006).

3 Carbon-Support Corrosion in Automotive PEMFC Systems

Automotive PEMFC systems experience significant dynamic operation conditions. Some of the operation conditions could result in excursions to high (i.e., positive) potentials of the electrodes, both anode and cathode. The following analyses demonstrate that automotive applications involve anode and cathode potential ranges which exceed the stability region of the carbon-support catalyst.

3.1 Steady-State Operation Conditions

Within the 5,500 operational hours of automotive PEMFC systems, the stack could sit for a long time at *idle* condition, i.e., when no power is supplied to the propulsion system, and the only power drawn from the fuel cell stack is that required to support the fuel cell system ancillaries (e.g., cooling pumps). This results in cathode potentials of approximately 0.90 V (Mathias et al. 2005). This idle time could total over several thousand hours depending on the particular usage profile. The stack could also be at *open circuit voltage* (OCV) condition, where no current is drawn from the stack and the cathode potential increases to approximately 0.95 V. This can occur just before fuel cell system shutdown, immediately after startup, as well as in the case of battery/fuel cell hybrid systems in which the battery provides the electrical load under low power or idle conditions. The accumulated time at OCV condition for 38,500 start/stop cycles could be over 100 h (38,500 cycles × 10 s per cycle). Figure 8 shows the predicted carbon weight loss as a function of time at these two potentials, based on the carbon corrosion

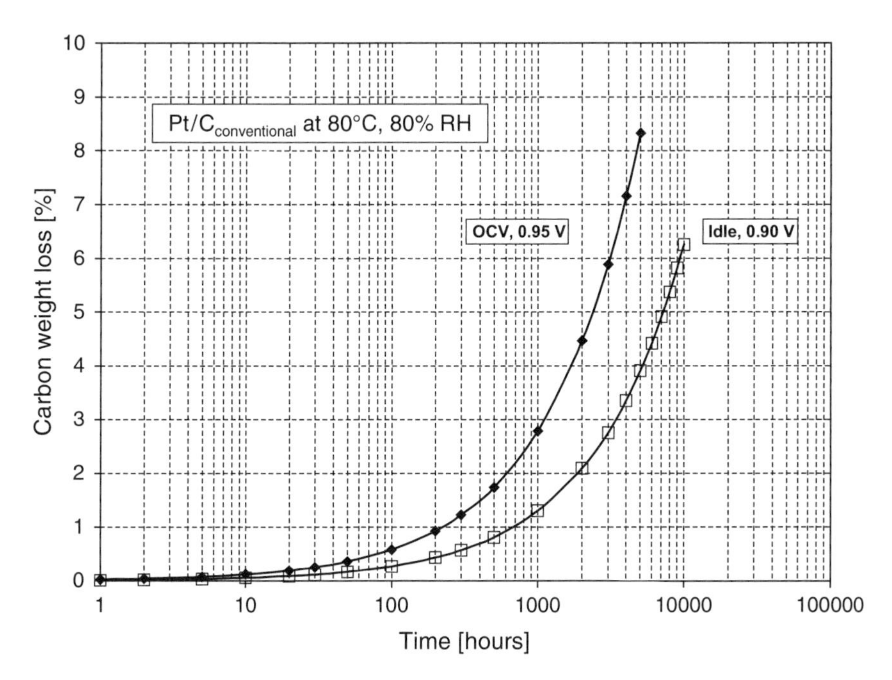

Fig. 8 Predicted carbon weight loss versus time at 0.90 V (idle conditions) and 0.95 V (open circuit voltage, *OCV*, conditions) for a commercial conventional-carbon-supported MEA at 80°C and 80% RH_{inlet}

kinetics listed in (2) and the kinetic parameters listed following (2). The total amount of carbon weight loss is about 5% after approximately 5,500 h at idle conditions (i.e., 0.90 V) and after only approximately 2,500 h at OCV conditions (i.e., 0.95 V). These results suggest that, even at these relatively low potentials, the carbon support could lose 5% weight, which is right on the threshold point beyond which significant electrode damage and strong voltage loss could take place as was shown in Fig. 6.

3.2 Transient Operation Conditions

An automotive PEMFC system operates with frequent dynamic power changes, which translate into cell voltage changes. Figure 9 shows a typical vehicle operation potential profile for a typical drive cycle. While some voltage cycles between 0.50 and 0.95 V do exist, the most frequent large voltage changes are from 0.55 to 0.90 V. Of these, it is estimated that there are over 300,000 cycles in a vehicle's 5,500 h operation time. Therefore, it is essential to evaluate the impact of dynamic voltage cycling on the corrosion of the carbon support.

CO_2 evolution was observed when cycling the potential between 0.1 and 0.6 V (triangular wave at 10 mV s^{-1}) (Ball et al. 2006), even though one would expect little carbon-support corrosion at such low potentials. On the basis of this reported acceleration of carbon-support corrosion under potential cycling conditions and the large number of expected voltage cycles in automotive applications (Fig. 9), experiments were conducted for voltage cycles between 0.6 and 0.9 V (square waves at 40 cycles per hour at 80°C and 80% RH). The measured carbon corrosion rate and the cumulative carbon weight loss, extrapolated to 5,500 h of operation are shown in Fig. 10 (Zhang et al. 2007). It is seen that the carbon corrosion rate decreases dramatically with time, following a similar power-law dependency as was seen under steady-state conditions (see Fig. 2). However, voltage cycling alone can indeed lead to close to approximately 8% carbon weight loss for a commercial conventional-carbon-supported cathode electrode, at which point substantial electrode thinning (see Fig. 7) and significant cell voltage losses are expected (see Fig. 6).

To estimate the accelerating effect of voltage cycling on carbon-support corrosion, the approximately 8% carbon weight loss estimated for 5,500 h of voltage cycling between 0.6 and 0.9 V may be compared with the carbon weight loss expected for 2,750 h at 0.9 V (see Fig. 8). The latter would amount to approximately 2.5%, which is approximately three times less than what one would expect under cycling conditions. The reason for the enhanced carbon corrosion during voltage cycling is not entirely clear and remains a topic for further research. It is generally attributed to the repeated reduction and oxidation of platinum and/or carbon surface groups during voltage cycles (Ball et al., 2006). It is clear, however, that the effect of voltage cycling needs to be considered when estimating the durability of carbon supports for automotive applications on the basis of steady-state data. Fortunately, our more recent studies showed that more-corrosion-resistant carbons have significantly reduced voltage-cycling-induced losses (Zhang et al. 2007).

3.3 *Start/Stop Operation Conditions*

An automotive PEMFC system undergoes an estimated 38,500 start/stop cycles (seven cycles per hour over 5,500 h) over the life of a vehicle. If not properly mitigated, these start/stop events can lead to short-term potential excursions of the cathode electrode to 1.5 V versus the adjacent electrolyte (Reiser et al. 2005, Tang et al. 2005) owing to the existence of H_2/air fronts in the anode compartment. Figure 11 shows a schematic of a PEMFC in the presence of a H_2/air front in the anode while the cathode is filled with air. Since in-plane proton transport can be neglected (ca. 100-µm transport length compared with hundreds of millimeters of active area), the H_2/air (anode/cathode) segment shown on the left-hand-side of Fig. 11 acts as a power source that drives a current through the air/air segment (right-hand-side of Fig. 11), which acts as a load. Therefore, based on the required charge conservation (see Fig. 11), we obtain

$$(i_{HOR} = i_{ORR})_{H_2/air} = (i_{ORR} = i_{COR} + i_{OER})_{air/air}, \tag{3}$$

where i_{COR} and i_{OER} represent the carbon oxidation reaction (COR) current and the oxygen evolution reaction (OER) current on the cathode side of the air/air segment, which is

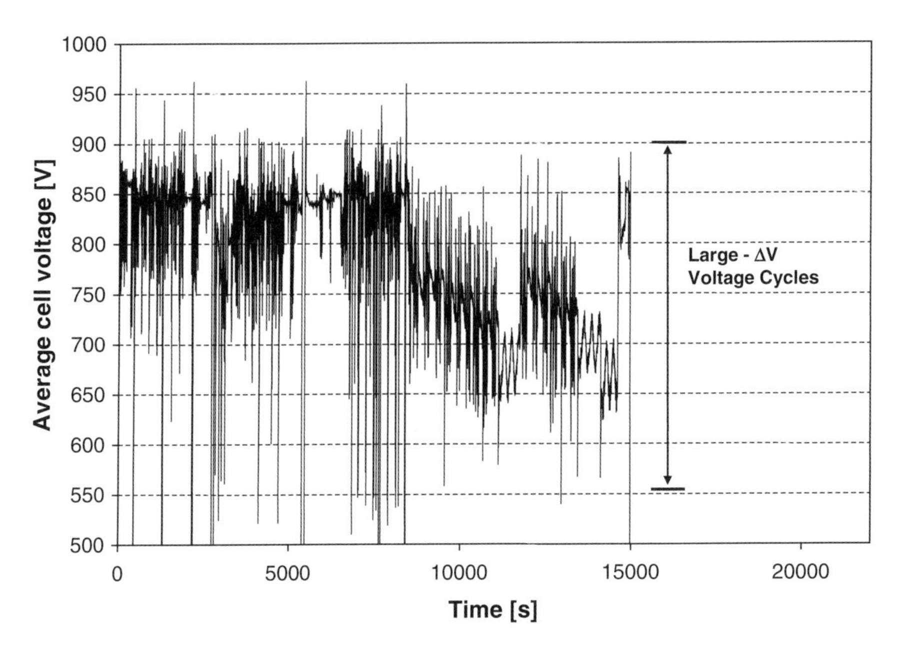

Fig. 9 Typical automotive fuel cell vehicle drive cycle, showing average cell voltage versus time

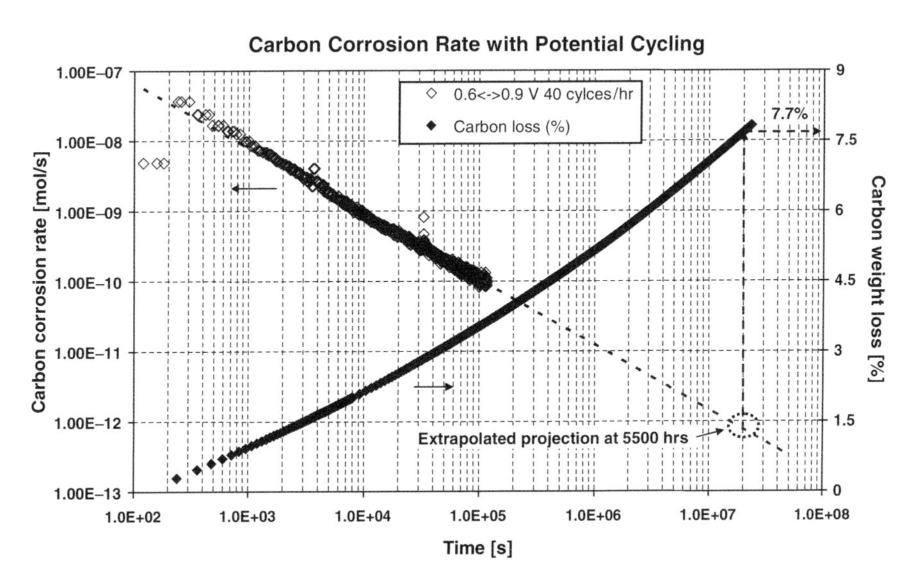

Fig. 10 Carbon corrosion rate versus cycle time (*open symbols*) of a commercial conventional-carbon-supported MEA under an accelerated test condition: cycles between 0.6 and 0.9 V (40 cycles per hour) at 80% RH_{inlet} and 80°C. The *solid symbols* are for carbon weight loss versus time, projected out to 5,500 h

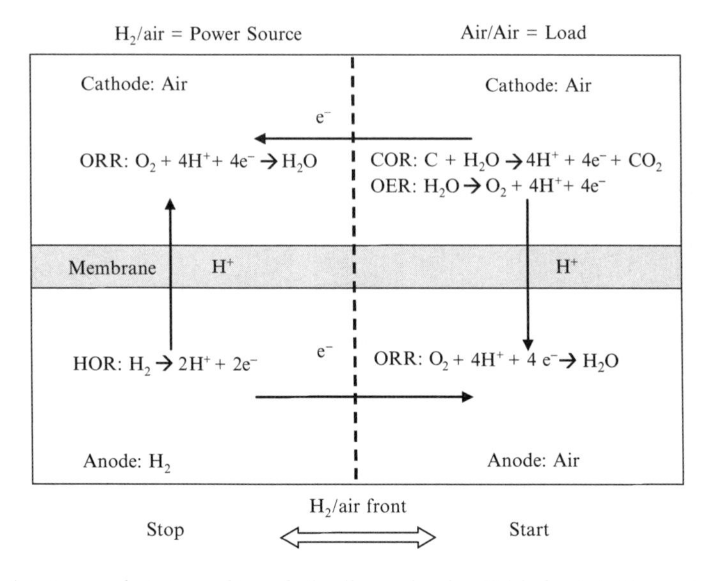

Fig. 11 A proton exchange membrane fuel cell experiencing H_2/air front start/stop, showing the major electrochemical reactions considered during start/stop operations. *ORR* Oxygen reduction reaction, *COR* Carbon oxidation reaction, *OER* Oxygen evolution reaction, and *HOR* Hydrogen oxidation reaction

balanced by the oxygen reduction reaction (ORR) current on the opposing anode side. This current through the air/air segment is then equal to the current generated by the H_2/air segment. From this, it should be clear that in the case of sufficiently corrosion resistant carbon supports, a larger fraction of the current through the air/air segment can be made up by the OER, resulting in less overall damage. The above-described kinetic model has been augmented by including proton and gas transport resistances (Meyers and Darling 2006) and, more recently, by also considering capacitive effects (Gu et al. 2007).

Figure 12 shows the cell voltage loss over 200 H_2/air front start/stop cycles at various current densities. The degradation rates (defined as 100-mV voltage loss divided by the number of start/stop cycles needed to produce 100-mV loss) are up to 1,500 μV per cycle at 1.5 A cm^{-2}, far exceeding the automotive start/stop durability target of 1 μV per cycle. Although operating and control strategies have been shown to effectively mitigate voltage degradation (Yu et al. 2004; Perry et al. 2006) during start/stop, significant improvement of corrosion-resistant carbonsupport materials (Yu et al. 2006a) is also required to meet automotive fuel cell start/stop durability targets.

3.4 Nonoptimal Operation Conditions

3.4.1 Global Anode Hydrogen Starvation

Under normal operation conditions, current and power are generated when both H_2 and air are flowing into anode and cathode compartments in the fuel cell system. However,

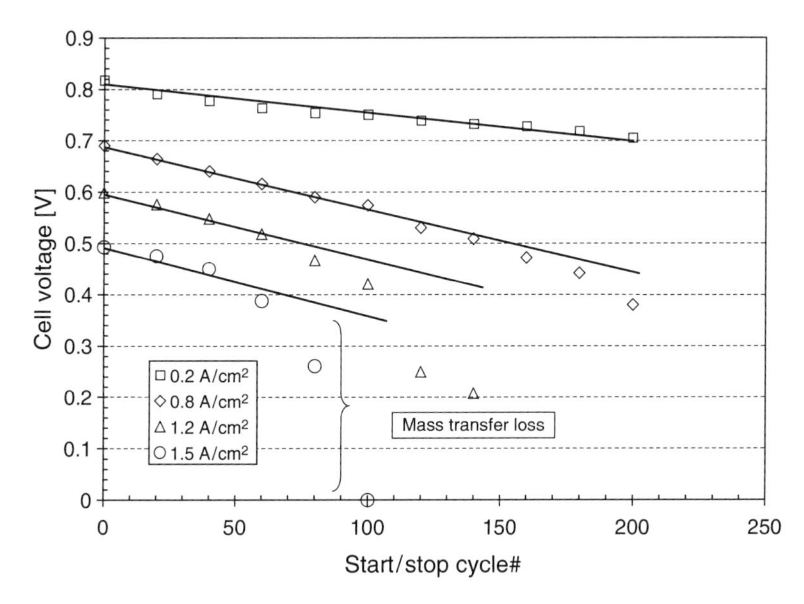

Fig. 12 Cell voltage decays for different current densities versus the number of start/stop cycles over a commercial conventional-carbon-supported MEA at 80°C and 66% RH_{inlet} at a H_2/air front residence time of 1.3 s. (Reproduced with permission of Yu et al. 2006a), The Electrochemical Society)

hydrogen flow maldistributions, which can be caused by pressure-drop variations between the hundreds of anode flow fields in a fuel cell stack, by system control failures, or by insufficiently fast response of the H_2 flow control during power-up transients, could result in a fuel cell system experiencing *global* H_2 *starvation* in some of the cells. In that case, insufficient supply of H_2 to some of the cells of a stack would lead to an increase in the anode potential until carbon oxidation and oxygen evolution occur in order to sustain the stack current (for a detailed explanation, see Ralph et al. 2006). Such carbon corrosion current could quickly consume all the carbon in the anode electrode. Although advanced control strategies seek to avoid the damage from global anode starvation, imperfect system control could lead to extensive carbon corrosion on the anode.

Figure 13 depicts a simple anode carbon lifetime projection in the case of global H_2 starvation. It is derived for a 50 wt% $Pt/C_{conventional}$ anode catalyst and a platinum and carbon loading of 0.2 mg cm^{-2}, which correspond to a total carbon inventory charge of 6.4 A s cm^{-2}. For these types of electrodes, and for anode operation with pure H_2, we have found that a 30% carbon weight loss results in approximately 80 mV cell voltage loss at 1.5 A cm^{-2} (Zhang 2007). Therefore, the maximum allowable accumulated charge from anode carbon corrosion is 1.9 A s cm^{-2}. On this basis, a fuel cell system operating at 0.2 A cm^{-2} would only last for approximately 100 s in a case of global H_2 starvation, if 10% of the current in a specific cell has to be supplied by carbon-support oxidation (Fig. 13). Although very reliable fuel cell system control strategies are desirable, corrosion-resistant carbon-support materials and/or the use of an additional oxygen evolution catalyst in the anode (Ralph et al. 2006) can potentially relax the system control requirements.

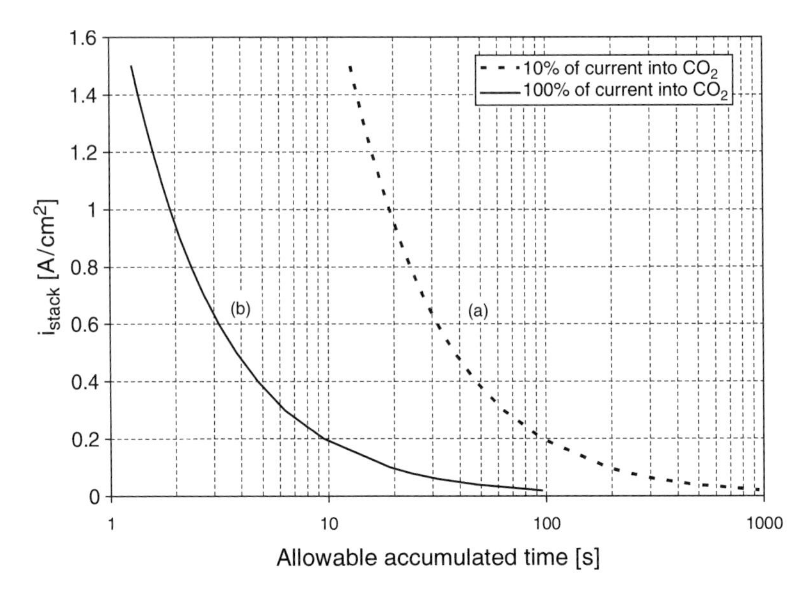

Fig. 13 Allowable accumulated time resulting in 30% carbon weight loss in the anode during *global H_2 starvation* for a 50 wt% Pt/$C_{conventional}$ catalyst at a platinum and carbon loading of 0.2 mg cm^{-2}. (**a**) Ten percent of the cell current is supplied by anode carbon corrosion and (**b**) Hundred percent of the cell current is supplied by anode carbon corrosion

3.4.2 Local Anode Hydrogen Starvation

Under normal cell operation the hydrogen oxidation reaction (HOR) occurs at the anode, and the ORR takes place at the cathode. However, carbon corrosion can also take place at the cathode when H_2 depletes locally at the anode (Patterson and Darling 2006, Meyers and Darling 2006, Gu et al. 2006). In this case, oxygen permeating through the membrane from the cathode side (*O_2 crossover*) will be reduced owing to the low solid phase potential at the anode side, resulting in a so-called *reverse current* of protons, i.e., protons flowing from the cathode to the anode side. This leads to an increase of the electrolyte phase potential of the adjacent cathode electrode until protons can be provided by either oxygen evolution from water or carbon-support oxidation (for details, see Meyers and Darling 2006).

One possible scenario which can lead to *local H_2 starvation* is that one or several adjacent anode flow channels are flooded with water or filled by water slugs, as shown schematically in Fig. 14. This results in a wide region with restricted H_2 flow, equivalent to a very wide land in the anode flow field. In this case, H_2 will be transported in-plane within the anode diffusion medium, by both diffusion- and pressure-driven flow, depending on the in-plane permeability of the diffusion medium and depending on the rate of N_2 pressure buildup from nitrogen permeation through the membrane from the cathode side (*N_2 crossover*). Oxygen reaching the anode via O_2 crossover

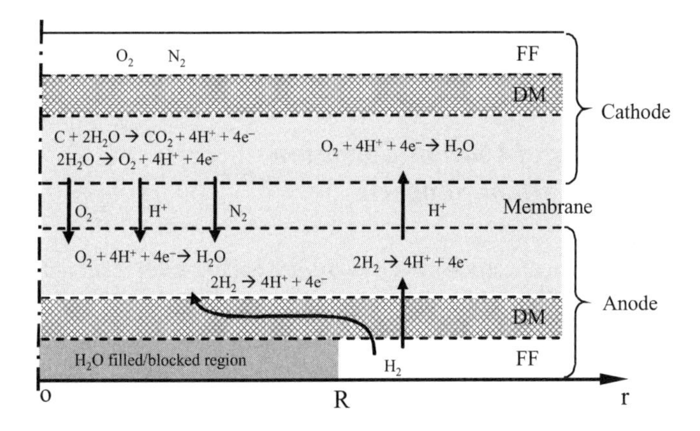

Fig. 14 The process of local H_2 starvation, leading to cathode carbon corrosion in the case where the H_2 supply rate to a portion of the anode electrode is less than the O_2 permeation rate through the membrane (from cathode to anode). *FF* Flow field and *DM* Diffusion medium

will be reduced on the anode side (ORR) and, in the absence of H_2, the protons required for the ORR will be provided by the OER or the COR on the adjacent cathode electrode (Fig. 14). Recent model experiments have nicely illustrated this mechanism by locally blocking H_2 access to the anode electrode, either via blocking sections of the anode flow field (Carter et al. 2007) or via rendering sections of the diffusion medium impermeable (Patterson and Darling 2006).

Although stack hardware design and sophisticated operating strategies help to reduce the frequencies of local anode H_2 starvation, corrosion-resistant carbon materials and an active OER catalyst may be required, especially under very dynamic automotive fuel cell operation conditions.

3.5 *Summary*

Several scenarios in automotive fuel cell operation leading to carbon-support corrosion have been reviewed in this section. It was found that the total weight loss of conventional carbon supports can exceed 10% in the lifetime of automotive fuel cell systems, even under regular operation conditions (i.e., steady state plus voltage cycling). This weight loss of carbon would result in a 20–100-mV voltage loss at 0.8 A cm^{-2} as shown in Fig. 6, which would not meet automotive fuel cell durability targets. Furthermore, fuel cell system start/stop, anode local H_2 starvation, and global anode H_2 starvation would lead to short-term potential excursions up to approximately 1.5 V, resulting in severe carbon corrosion. Hence, the carbon-support materials that have higher corrosion resistance are required in automotive fuel cell applications, even after system mitigation strategies have been applied.

4 Corrosion-Resistant Carbon Supports for PEMFC MEAs

4.1 Comparisons of Carbon Corrosion of Various Carbon Supports

The gas-phase oxidation of carbon blacks by oxygen and/or water is strongly catalyzed by the presence of catalytically active metals, such as platinum (Rewick et al. 1974, Stevens and Dahn 2005), whereby several weight percent of platinum on carbon can increase the gas-phase oxidation rate by orders of magnitude. This, however, is not the case for the electrochemical oxidation of carbon blacks, where at potentials of 0.8 V and higher (vs. RHE) the carbon corrosion rate is within a factor of 2 between that for noncatalyzed and platinum-catalyzed carbon blacks (Roen et al. 2004, Passalacqua et al. 1992, Kinoshita 1988). Therefore, gas-phase oxidation tests to screen potential carbon-black supports is not a reliable method for predicting their stability in the electrochemical environment, so it is essential to measure the carbon corrosion rates directly in an electrochemical cell.

Since the electrochemical corrosion of carbon supports is accelerated by high potentials and high temperatures (see Fig. 3), we conducted the screening of various carbon supports at 95°C and 1.2 V (vs. RHE). The results are shown in Fig. 15, comparing the corrosion rates of Black Pearls, Ketjenblack, and Vulcan XC-72C carbon blacks catalyzed with approximately 50 wt% platinum, tested in 50-cm^2 single cells operated with N_2 (working electrode) and H_2 (counter/reference electrode). Analogous to what was observed in Fig. 2 and consistent with the literature (Kinoshita and Bett 1973), the carbon corrosion rates decrease with time, following a power-law behavior with very similar time dependence. The carbon corrosion rates of these three types of MEAs were in the order Pt/Black Pearls > Pt/Ketjenblack > Pt/Vulcan XC-72C. If these carbon corrosion rates are normalized by the respective carbon-support BET surface areas (Table 1), very similar values are observed (Fig. 16), suggesting that the carbon corrosion rates are determined by the BET surface area for these commonly used nongraphitized fuel cell catalyst carbon supports.

To examine the differences in carbon corrosion rates between a nongraphitized and a graphitized carbon support, the carbon-support corrosion rates of an MEA using an approximately 50 wt% Pt/Ketjenblack catalyst was compared with that of an MEA using an approximately 50 wt% Pt/graphitized Ketjenblack catalyst. As shown in Fig. 17, the BET-normalized carbon corrosion rate of the graphitized carbon support is essentially invariant over time (up to a total carbon weight loss of approximately 7%), while that of the nongraphitized carbon support decreases with time (up to a total carbon weight loss of approximately 20%) as already shown in Figs. 2, 15, and 16. This observation may be understood if one considers that the inhomogeneous distribution of the graphitic planes in nongraphitized carbon supports from the outer shell to the core of the primary carbon particles (see Fig. 4) is more time-independent carbon corrosion rate. After extended corrosion times

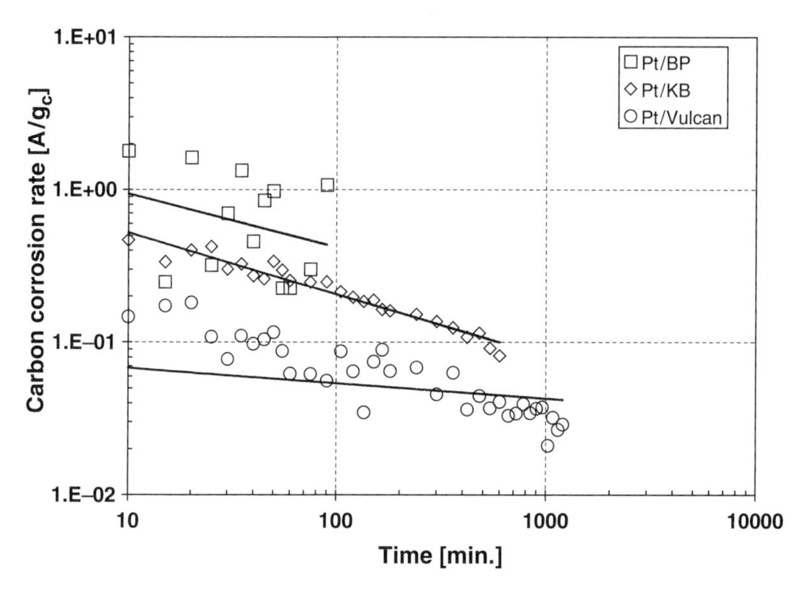

Fig. 15 Carbon corrosion currents of approximately 50% Pt/C catalysts normalized by carbon mass versus time at 95°C, 1.2 V (vs. RHE) and 80% RH_{inlet}. The working electrode was fed with N_2, while the counter/reference electrode was pure H_2 at the same temperature. *BP* Black Pearls, *KB* Ketjenblack, and *Vulcan* Vulcan XC-72C

Table 1 Brunauer–Emmett–Teller (BET) surface areas of carbon supports used in the approximately 50 wt% Pt/C catalysts examined

Catalyst-support	Abbreviation	BET ($m^2\ g_C^{-1}$)
Black Pearls	Pt/BP	1,600
Ketjenblack	Pt/BP	800
Vulcan XC-72C	Pt/Vulcan	240
graphitized Black Pearls	Pt/graph-BP	240
graphitized Ketjenblack	Pt/graph-KB	160
graphitized Vulcan XC-72C	Pt/graph-Vulcan	80

(more than 100 min), Fig. 17 shows that the difference in the BET-normalized carbon-support corrosion rates between graphitized and nongraphitized carbon blacks is approximately a factor of 2–3, consistent with previous reports in concentrated H_3PO_4 at 135–210°C (Kinoshita 1988, Stonehart 1984). Therefore, when modeling carbon-support corrosion, one must consider the differences between the time dependence of the corrosion rates of various carbon supports, particularly between graphitized and nongraphitized supports.

As we discussed in the previous section (see Fig. 6), the cell voltage typically loses approximately 50 mV at 0.8 A cm^{-2} at 11 ± 2% carbon weight loss for a variety of graphitized

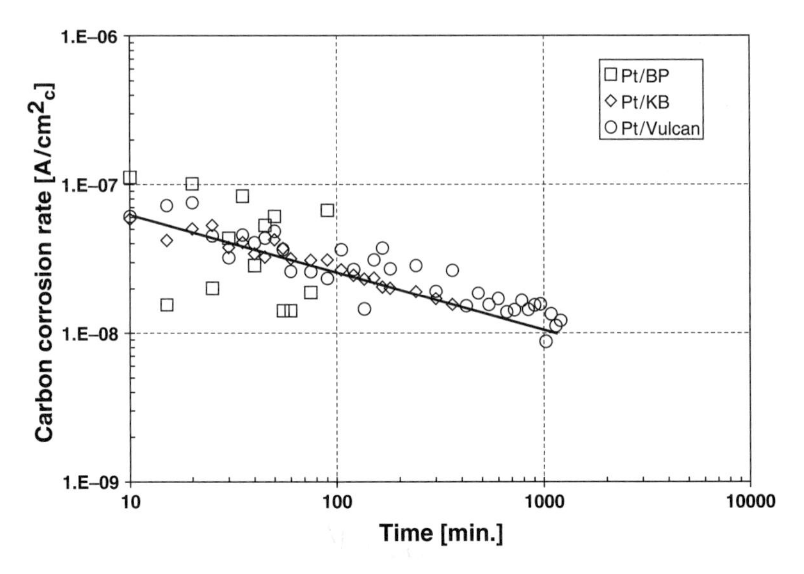

Fig. 16 Carbon corrosion rates of approximately 50% Pt/C catalysts normalized by carbon-support BET surface areas versus time at 95°C, 1.2 V (vs. RHE) with 80% RH_{inlet}. The working electrode was fed with N_2, while the counter/reference electrode was pure H_2 at the same temperature

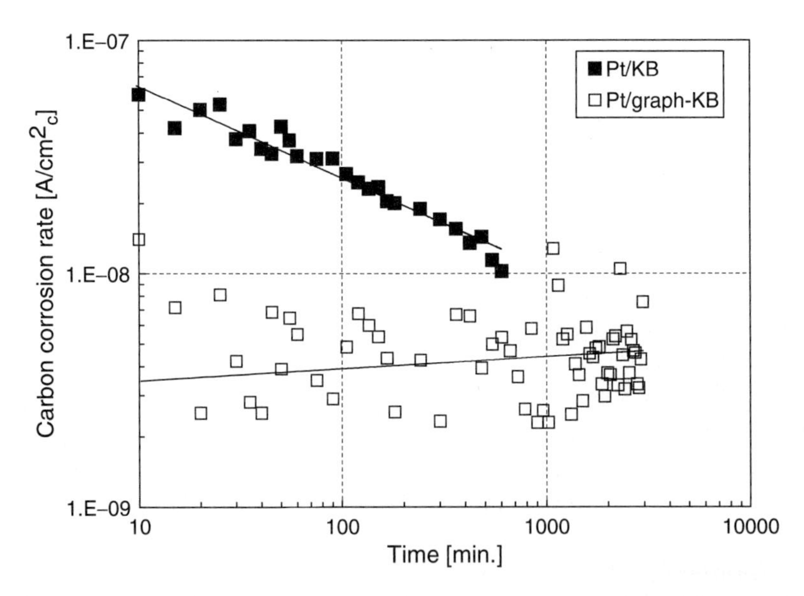

Fig. 17 Comparisons of BET surface normalized carbon corrosion rates of approximately 50 wt% Pt/KB with approximately 50 wt% Pt/graphitized KB (*graph-KB*) catalysts at 95°C, 1.2 V (vs. RHE) and 80%RH_{inlet}. The working electrode was fed with N_2; the counter/reference electrode was H_2 at the same temperature

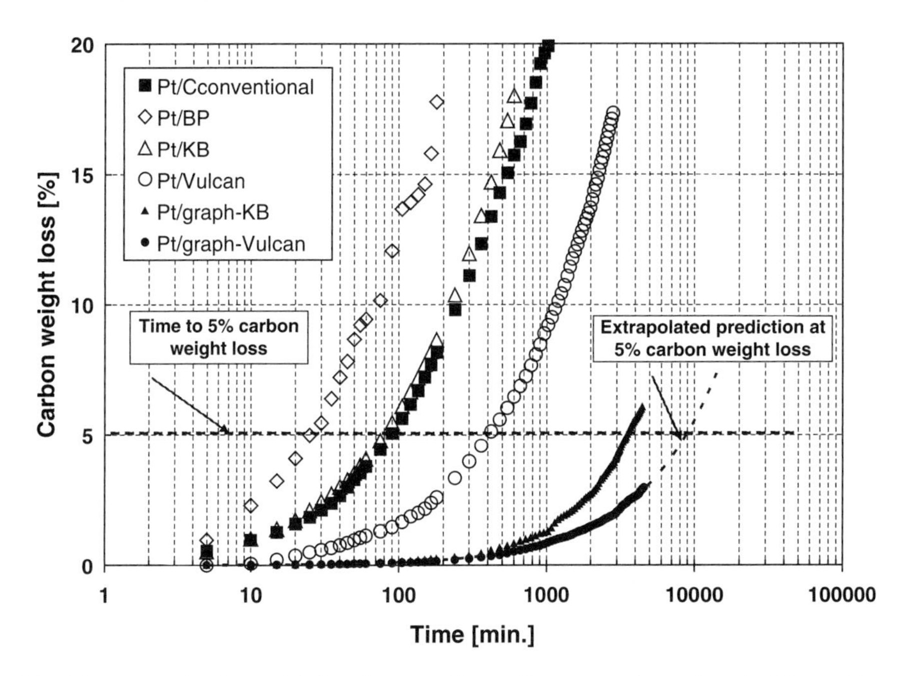

Fig. 18 Integrated carbon weight loss (integrated from CO_2 concentration at the exit of the working electrode) of approximately 50 wt% platinum catalysts supported on various carbon blacks as a function of time at 95°C, 1.2 V (vs. RHE), and 80% RH_{inlet}

and nongraphitized carbon supports, while the same voltage loss can be observed at only 9 ± 2% carbon weight loss at the higher current density of 1.2 A cm^{-2} (Makharia et al. 2006). Therefore, the time required for the first 5% carbon weight loss should be a good index to compare the corrosion resistance of different carbon supports. Such a comparison is made in Fig. 18, clearly illustrating the superior corrosion resistance of catalysts supported on graphitized carbon blacks (see Pt/graphitized Ketjenblack and Pt/graphitized Vulcan XC-72C in Fig. 18) compared with more conventional nongraphitized carbon supports. The results show that the extrapolated time for 5% weight loss of a graphitized Vulcan XC-72C supported MEA is longer than that of a commercial conventional-carbon-supported MEA ($Pt/C_{conventional}$) by two orders of magnitude. Similarly, a 35-fold improvement in corrosion resistance is observed between approximately 50 wt% platinum catalysts supported on graphitized Ketjenblack versus conventional (nongraphitized) Ketjenblack. Section 4.2 considers whether graphitized carbon supports would also give such large improvements for start/stop and local anode H_2 starvation.

4.2 *Impact of Carbon Supports on PEMFC Durability*

Previous analysis has shown that for start/stop and local H_2 starvation, the sum of the currents for oxygen evolution and carbon-support corrosion at the cathode electrode equals the oxygen reduction current at the anode electrode (3). Therefore, the relative rates of carbon corrosion to oxygen evolution have significant implications for mitigating carbon-support corrosion. Figure 19 shows the carbon corrosion kinetics for various carbon supports (averaged between 0 and 5 wt% carbon weight loss in order to account for the time dependence) and the oxygen evolution kinetics (OER) for an approximately 50 wt% Pt/C catalyst. Clearly, in the case of nongraphitized carbon supports, the OER currents at any given potential are always negligible compared with the carbon-corrosion rates (see lines A and B vs. E in Fig. 19), and any reverse proton current during start/stop or local H_2 starvation will be carried by carbon-support corrosion (1) rather than by O_2 evolution. On the other hand, in the case of graphitized carbon supports, a significant fraction of the reverse proton current will be provided by the OER (see lines C and D vs. E in Fig. 19).

The impact of carbon-support materials on the two most severe cathode carbon corrosion cases in automotive operation, i.e., start/stop and local anode H_2 starvation, can in general be determined by a simple kinetic model on the basis of (3), as was suggested previously in the literature (Reiser et al. 2005, Meyers and Darling 2006). However, it needs to be stressed that the predictions of such models very strongly depend on the fidelity of the kinetic parameters required to describe the COR and the OER. Unfortunately, these are not very well known, so the rates of the COR and the OER which have been used for previous models (see Meyers and

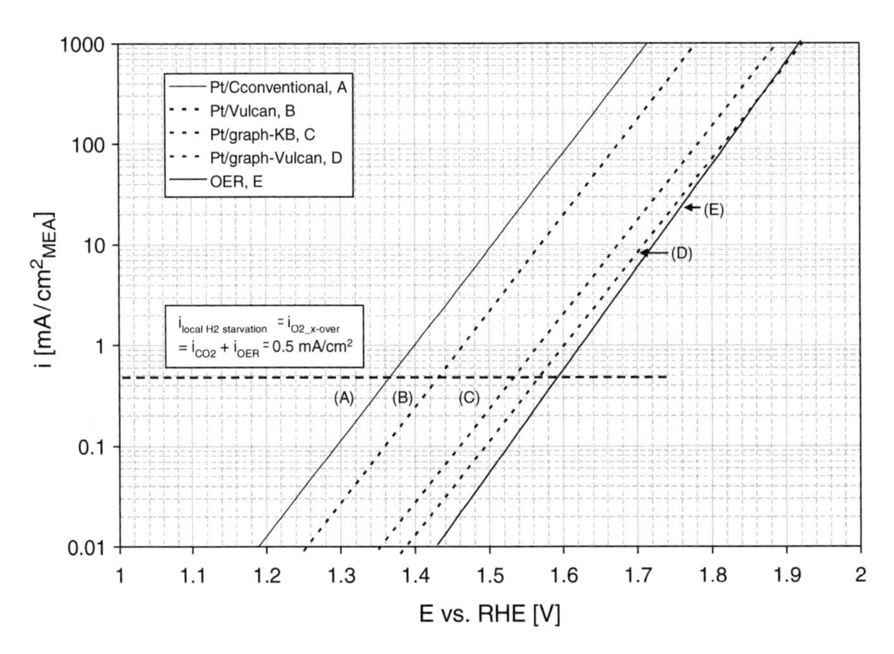

Fig. 19 Kinetics of carbon corrosion for approximately 50 wt% platinum catalysts on various carbon-supports (A–D) and of the OER (E). The *horizontal dashed line* indicates the maximum local H_2 starvation current, equivalent to the O_2 crossover current, for fully developed local H_2 starvation. MEA specifications and test conditions were as follows: 0.4 mg_{Pt} $cm^{-2}_{electrode}$ using an approximately 50 wt% Pt/C catalyst, 80°C, 80% RH (for detail see Yu et al. 2006a)

Darling 2006) are very different from our measurements as shown in Fig. 19. Therefore, in general, it is advisable to measure the relevant COR and OER kinetics which are used for the start/stop and local H_2 starvation models discussed next.

4.2.1 Start/Stop Operations

As seen in Fig. 12, uncontrolled PEMFC system start/stop results in significant voltage degradation owing to carbon corrosion. The use of carbon-supported MEAs with higher corrosion resistance can effectively force more current to be taken by the OER instead of the carbon corrosion reaction as discussed above and also reduce the overall start/stop current since higher potentials are required to drive a reverse proton current. Figure 20 compares experimentally determined start/stop degradation rates for MEAs with three different cathode catalyst carbon supports (normalized to that of $Pt/C_{conventional}$) with kinetic model estimates based on (3), using the carbon-support corrosion kinetics and the OER kinetics as shown in Fig. 19 (Yu et al. 2006b). The results show that using a cathode catalyst with graphitized Ketjenblack support provides a start/stop durability enhancement of a factor of 5–7

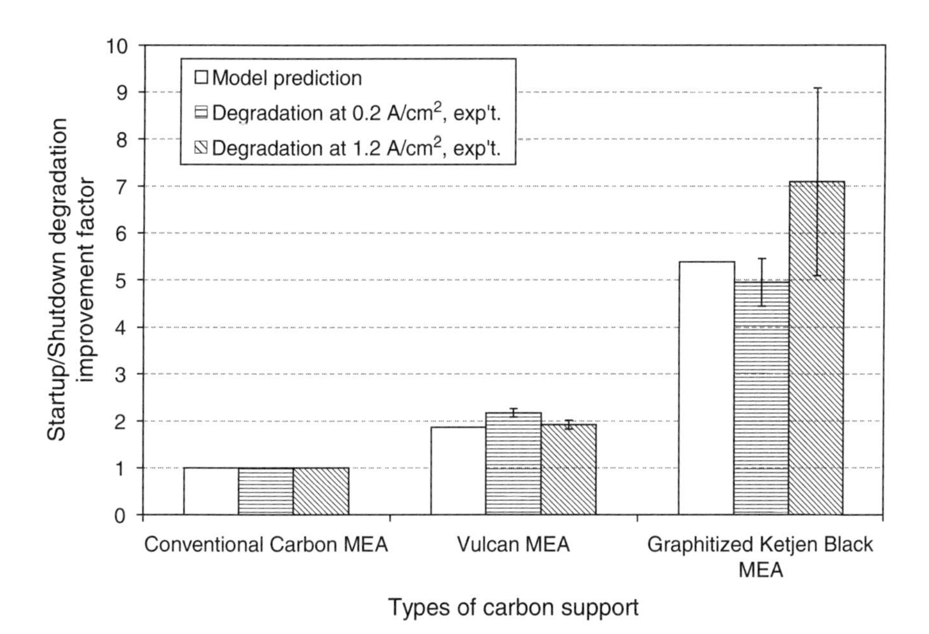

Fig. 20 Predicted relative carbon corrosion rate compared with the measured relative voltage degradation rates at two applied current densities and a H_2/air front residence time of 1.3 s. The startup/shutdown improvement factor is defined as the inverse of the degradation rate of a conventional carbon MEA over the degradation rate of a Vulcan MEA or a graphitized carbon MEA. The model predictions were based on HOR kinetics (Neyerlin et al. 2007), ORR kinetics (Neyerlin et al. 2006), COR kinetics (Yu et al. 2006a), and OER kinetics (Yu et al. 2006b). (Reproduced with permission of Yu et al. (2006a), The Electrochemical Society)

compared with that of a conventional-carbon-supported MEA. In general, the start/stop model predictions based on (3) agree well with experimental data, as long as the H_2/air front residence time is short enough not to deplete the flow field from reactants during passing of the front (typically less than 2–3 s) but long enough to not be dominated by capacitive effects (typically more than 0.5 s). In practical applications, the H_2/air front residence time is usually less than 0.5 s to reduce the overall voltage degradation, and a model including capacitive effects is required (Gu et al. 2007).

4.2.2 Local Anode Hydrogen Starvation

Local anode H_2 starvation can be considered as a special case of start/stop operation, the difference being that the anode ORR current is limited by the oxygen crossover rate, which depends on temperature, RH, and membrane thickness (Kocha 2003). *Fully developed* local H_2 starvation occurs if the length scale of the H_2-starved region (R in Fig. 14) is sufficiently large (Gu et al. 2006), in which case the entire O_2 crossover current will have to be balanced by the COR and the OER on the adjacent cathode electrode. For an 18-μm-thick perfluorosulfonic acid membrane in a cell operating at 80°C, H_2/air at 150 kPa(abs), and 100% RH, this O_2 crossover current amounts to approximately 0.5 mA cm^{-2}, which is also indicated by the horizontal dashed line in Fig. 19. Thus, it is clear from Fig. 19 that in the case of fully developed local H_2 starvation essentially all of the O_2 crossover current of a conventional-carbon-supported MEA is provided by the COR (line A in Fig. 19) since the OER kinetics (line E in Fig. 19) is much slower. Only in the case of graphitized carbon supports can some of the O_2 crossover current be provided by the OER. Table 2 shows that the cathode carbon-support corrosion is reduced by a factor of approximately 1.5–2 if graphitized carbon supports are used. Further improvements can be achieved by enhancing the OER (Gu et al. 2006), as should be evident from Fig. 19.

Table 2 Model predictions of local anode H_2 starvation durability improvement for approximately 50 wt% platinum catalysts on various carbon supports

Types of electrodes	i_{CO_2} (mA cm^{-2})	Improvement factor of carbon corrosion[a]
Conventional carbon MEA	0.48	1.00
Vulcan MEA	0.46	1.04
Graphitized KB MEA	0.34	1.41
Graphitized Vulcan MEA	0.26	1.85

Conditions were as follows: H_2/air at 150 kPa(abs), 80°C, and 100% RH. The membrane electrode assembly (*MEA*) was 18-μm-thick perfluorosulfonic acid, 0.4 $mg_{Pt} \cdot cm^{-2}_{electrode}$. *KB* Ketjenblack

[a]Improvement factor is defined as i_{CO_2} (MEA) / i_{CO_2} Conventional MEA

5 Conclusions

Carbon corrosion kinetics of commercial conventional-carbon-supported MEAs were studied at various potentials and temperatures. The lifetime projection of conventional-carbon-supported MEAs in the automotive fuel cell system was then analyzed using the kinetics shown in this chapter. It is found that these conventional-carbon-supported MEAs are not likely to meet automotive fuel cell durability targets under the severely dynamic automotive operational conditions. Automotive fuel cell system start/stop and local anode H_2 starvation are believed to be two of the major contributors to fuel cell voltage degradation.

Carbon corrosion rates of various carbon blacks were evaluated in this chapter. The carbon corrosion rates normalized by the BET surface area of conventional carbon blacks (Black Pearls, Ketjenblack, and Vulcan XC-72) are similar, while the carbon-mass-normalized corrosion rates are in the order Black Pearls > Ketjenblack > Vulcan XC-72. Carbon corrosion rates normalized by the BET surface area of graphitized-carbon-supported MEAs were approximately 2–3 times lower than those of non-graphitized-carbon-supported MEAs. It was also shown that significant voltage losses can be observed once approximately 5–10% of the original carbon-support weights have been corroded, independent of whether the carbon supports are graphitized or nongraphitized. Therefore, the durability of different carbon supports can be estimated by comparing the differences in corrosion time required to corrode 5% of the carbon support by weight. The estimated time to 5% carbon weight loss of graphitized Vulcan XC-72C supported MEAs is approximately 1–2 orders of magnitude longer than those of conventional-carbon-supported MEAs, including Pt/Black Pearls, Pt/Ketjenblack, and Pt/Vulcan XC-72C.

Fundamental model analyses incorporating the measured carbon corrosion kinetics were developed for conditions of start/stop or local H_2 starvation. The combination of experiment and modeling suggests that MEAs with corrosion-resistant carbon supports are one of the major material approaches to mitigate carbon corrosion during automotive fuel cell operation. With further development, corrosion-resistant non-carbon catalyst supports could provide alternative approaches to achieving adequate automotive durability (Debe et al. 2006; Van Zee 2007).

Acknowledgments The authors would like to thank Susan Yan and Chunxin Ji for providing MEA samples. Furthermore, we would like to acknowledge Mike Budinski, Brian Brady, and Brian Litteer for their analysis of carbon corrosion induced electrode thinning, as well as Robert Moses and Kannan Subramanian for their experiments on the effect of potential cycling on carbon corrosion.

References

Ball, S., Hudson, S., Theobald, B., and Thompsett, D. (2006) The effect of dynamic and steady state voltage excursions on the stability of carbon supported Pt and PtCo catalyst, ECS Trans., 3(1), 595.

Carter, R.N., Brady, B.K., Subramanian, K., Tighe, T., and Gasteiger, H.A. (2007) Spatially resolved electrode diagnostic technique for fuel cell applications, ECS Trans., 11(1), 423.

Debe, M.K., Schmoeckel, A.K., Vernstrom, G.D., and Atanasoski, R. (2006) High voltage stability of nanostructured thin film catalysts for PEM fuel cells. J. Power Sources, 161, 1002.
Dicks, A.L. (2006) The role of carbon in fuel cells. J. Power Sources 156, 128.
Gasteiger, H.A., Gu, W., Makharia, R., Mathias, M.F., and Sompalli, B. (2003) Beginning-of-life MEA performance – Efficiency loss contributions, in: *Handbook of Fuel Cells – Fundamentals, Technology, and Applications*, (Eds.: Vielstich, W., Lamm, A., Gasteigerr, H.A.), Wiley (Chichester, UK), volume 3, p. 593.
Gruber, T.C., Zerda, T.W., and Gerspacher, M. (1993) Three dimensional morphology of carbon black aggregates, Carbon, 31, 1209.
Gruver, G.A. (1978) The corrosion of carbon black in phosphoric acid, J. Electrochem. Soc., 125, 1719.
Gu, W., Makharia, R., Yu, P.T., and Gasteiger, H.A. (2006) Prediction of Local Hydrogen Starvation in a PEM Fuel Cell: Origin and Materials Impact, American Chemical Society 232nd National Meeting, Div. Fuel Chem. Fuel Chemistry Preprints Vol. 52/No. 2, San Francisco, CA.
Gu, W., Carter, R.N., Yu, P.T., and Gasteiger, H.A. (2007) Start/stop and local H_2 starvation mechanisms of carbon corrosion: Model vs. experiment, ECS Trans., 11(1), 963.
Heckman, F. and Harling, D. (1966) Progressive oxidation of selected particles of carbon black: Further evidence for a new microstructural model, Rubber Chem. Technol., 39, 1.
Kangasniemi, K.H., Condit, D.A., and Jarvi, T.D. (2004) Characterization of vulcan electrochemiclly oxidized under simulated PEM fuel cell conditions, J. Electrochem. Soc. 151(4), 125.
Kinoshita, K. (1988) *Carbon*, Wiley (New York, NY), pp. 316–333.
Kinoshita, K. and Bett, J. (1973) Electrochemical oxidation of carbon black in concentrated phosphoric acid at 135°C, Carbon, 11, 237.
Kocha, S.S. (2003) Preparation principles of MEA preparation, in: *Handbook of Fuel Cells – Fundamentals, Technology, and Applications*, (Eds.: Vielstich, W., Lamm, A., Gasteigerr, H.A.), Wiley (Chichester, UK), volume 3, pp. 538.
Makharia, R., Kocha, S.S., Yu, P.T., Sweikart, M., Gu, W., Wagner, F.T., and Gasteiger, H.A. (2006) Durability PEM fuel cell electrode materials: Requirements and benchmarking methodologies, ECS Trans., 1(8), 3.
Mathias, M.F., Makharia, R., Gasteiger, H.A., Conley, J.J., Fuller, T.J., Gittleman, C.J., Kocha, S.S., Miller, D.P., Mittlelsteadt, C.K., Xie, T., Yan, S.G., and Yu, P.T. (2005) Two fuel cell cars in every garage, ECS Interface, Fall, 24.
Medalia, A.I. (1967) Morphology of aggregates I. Calculation of shape and bulkiness factors; application to computer-simulated flocs, J. Colloid Interface Sci., 24, 393.
Medalia, A.I. and Heckman, F.A. (1969) Morphology of aggregates – II. Size and shape factors of carbon black aggregates from electron microscopy, Carbon, 7, 567.
Meyers, J.P. and Darling, R.M. (2006) Model of carbon corrosion in PEM fuel cells, J. Electrochem. Soc., 153, A1432.
Neyerlin, K.C., Gu, W., Jorne, J., and Gasteiger, H.A. (2006) Determination of catalyst unique parameters for the oxygen reduction reaction in a PEM fuel cell, J. Electrochem. Soc., 153, A1955.
Neyerlin, K.C., Gu, W., Jorne, J., and Gasteiger, H.A. (2007) Study of the exchange current density for the hydrogen oxidation and evolution reactions, J. Electrochem. Soc. 154(7), B631.
Passalacqua, E., Antonucci, P.L., Vivaldi, M., Patti, A., Antonucci, V., Giordano, N., and Kinoshita, K. (1992) The influence of Pt on the electrooxidation behavior of carbon in phosphoric acid, Electrochim. Acta, 37, 2725.
Patterson, T.W. and Darling, R.M. (2006) Damage to the cathode catalyst of a PEM fuel cell caused by localized fuel cell starvation. Electrochem. Solid-State Lett., 9(4), A183.
Perry, M.L., Patterson, T.W., and Reiser, C. (2006) System strategies to mitigate carbon corrosion in fuel cells, ECS Trans., 3(1), 783.
Ralph, T.R., Hudson, S., and Wilkinson, D.P. (2006) Electrocatalyst stability in PEMFCs and the role of fuel starvation and cell reversal tolerant anodes, ECS Trans., 1(8), 67.
Reiser, C.A., Bregoli, L., Patterson, T.W., Yi, J.S., Yang, J.D., Perry, M.L., and Jarvi, T.D. (2005) A reverse-current decay mechanism for fuel cells. Electrochem. Solid-State Lett., 8(6), A273.

Rewick, R.T., Wentrcek, P.R., and Wise, H. (1974) Carbon gasification in the presence of metal catalysts, Fuel, 53, 274.
Roen, L.M., Paik, C.H., and Jarvi, T.D. (2004) Electrocatalytic corrosion of carbon support in PEMFC cathodes, Electrochem. Solid-State Lett., 7(1), A19.
Ross, P.N. and Sokol, H. (1984) The corrosion of carbon black anodes in alkaline electrolyte, J. Electrochem. Soc., 131, 1742.
Stevens, D.A. and Dahn, J.R. (2005) Thermal degradation of the support in carbon-supported platinum electrocatalysts for PEM fuel cells, Carbon, 43, 179.
Stonehart, P. (1984) Carbon substrates for phosphoric acid fuel cell cathodes, Carbon, 22, 423.
Tang, H., Qi, Z., Ramani, M., and Elter, F.E. (2005) PEM fuel cell cathode carbon corrosion due to the formation of air/fuel boundary at the anode. J. Power Sources, 158, 1306.
Van Zee, J.W. (2007) Fuel Cell Research at University of South Carolina. DOE Program Review May 2007, Project ID: FCP#8.
Wissler, M. (2006) Graphite and carbon powders for electrochemical applications, J. Power Sources, 156, 142.
Yu, P.T., Kocha, S., Paine, L., Gu, W., and Wagner, F.T. (2004) The Effect of Air Purge on the Degradation of PEM Fuel Cell during Startup and Shutdown Procedures, AIChE 2004 Annual Meeting, New Orleans, LA, April 25–29.
Yu, P.T., Gu, W., Makharia, R., Wagner, F.T., and Gasteiger, H.A. (2006a) The impact of carbon stability on PEM fuel cell startup and shutdown voltage degradation, ECS Trans., 3(1), 797.
Yu, P.T., Gu, W., and Gasteiger, H.A. (2006b) GM Internal Experimental Data, to be published.
Zhang, J. (2007) GM-Internal Experimental Data, unpublished.
Zhang, J., Moses, R., and Subramanian, K. (2007) GM-Internal Experimental Data, unpublished.

3
Membranes

Chemical Degradation of Perfluorinated Sulfonic Acid Membranes

Minoru Inaba

Abstract Chemical degradation of perfluorinated sulfonic acid (PFSA) membranes during operation is a serious problem to be overcome before commercialization of electric vehicles and stationary cogeneration systems based on polymer electrolyte fuel cells. In this chapter, the mechanism for peroxide/radical formation, which works as an active species for membrane degradation, in the cell is comprehensively introduced, including recent findings. Accelerated testing methodology and the chemistry of PFSA membrane degradation are also outlined to improve the durability of PFSA membranes in polymer electrolyte fuel cells.

1 Introduction

Membranes are one of the key components of polymer electrolyte membrane fuel cells (PEMFCs). Perfluorinated sulfonic acid (PFSA) membranes are used as electrolyte membranes in most state-of-the-art PEMFCs, and they have long been believed to be mechanically, thermally, and chemically stable. However, stack manufacturers have disclosed that degradation of the membrane has been a serious failure mode in long-term operation of PEMFCs since the late 1990s. Degradation modes of the PFSA membranes are classified into two categories. One is physical degradation, and the other is chemical degradation. The physical degradation is caused by time-dependent deformation (creep) under compressive force, changes in temperature, relative humidity (RH), etc. The manufacturers have greatly improved physical durability using reinforced membranes, multilayered membrane-electrode assemblies (MEAs), etc. In contrast to this, chemical degradation of the electrolyte membranes is still a major problem in long-term operation of PEMFCs, and therefore membrane durability should be improved before commercialization of fuel cell vehicles and stationary cogeneration systems in the near future. To realize this,

M. Inaba
Faculty of Engineering, Doshisha University
e-mail: minba@mail.doshisha.ac.jp

F.N. Büchi et al. (eds.), *Polymer Electrolyte Fuel Cell Durability,*

understanding of the mechanism for membrane degradation as well as factors in membrane degradation is absolutely indispensable.

The purpose of this review is to summarize historical and the latest understanding of the mechanism for chemical degradation. Accelerated testing methodology and recent efforts to improve the chemical durability are also discussed. Further details of membrane durability, including that of hydrocarbon-based membranes, are described in the chapters "Chemical Degradation, Correlation Between Electrolyzer and Fuel Cell Findings; Improvement of Membrane and Membrane Electrode Assembly Durability; and Durability of Radiation-Grafted Fuel Cell Membranes."

2 Chemical Degradation Mechanism

Very early hydrocarbon-based membranes tested as electrolytes in PEMFCs for Gemini space missions, such as sulfonated phenol-formaldehyde resins, sulfonated poly(styrene-divinylbenzene) copolymers, and grafted polystyrene sulfonic acid membranes, were chemically weak, and therefore PEMFCs using these membranes showed poor performance and had only lifetimes of several hundred hours (LaConti et al. 2003). Nafion®, a PFSA membrane, was developed in the mid-1960s by DuPont (LaConti et al. 2003). It is based on an aliphatic perfluorocarbon sulfonic acid, and exhibited excellent physical properties and oxidative stability in both wet and dry states. A PEMFC stack using Nafion 120® (250-μm thickness, equivalent weight = 1,200) achieved continuous operation for 60,000 h at 43–82°C (LaConti et al. 2003, 2006). A Nafion®-based PEMFC was used for the NASA 30-day Biosatellite space mission (LaConti et al. 2003).

Several kinds of PFSA membranes are presently commercially available from several membrane manufacturers: DuPont, Gore, Asahi Glass, Asahi Kasei, Solvay, and 3M. The structures of some of these PFSA membranes are shown in Fig. 1. They are manufactured by copolymerization of tetrafluoroethylene and perfluorinated vinyl ether sulfonyl fluoride, followed by hydrolysis of the sulfonyl fluoride groups (Banergee and Curtin 2004). As mentioned above, PFSA membranes show excellent chemical and oxidative stability, and their carboxylate analogs have shown lifetimes of more than 5 years under harsh conditions in chlor-alkali electrolyzers. Although the operating conditions are less oxidative than those encountered in chlor-alkali electrolyzers, serious chemical degradation of PFSA membranes during operation of PEMFCs has recently been presented (Cleghorn et al. 2003; Collier et al. 2006; Endoh et al. 2004; Healy et al. 2005; Inaba et al. 2006a, 2007; Kinumoto et al. 2006; Luo et al. 2006; Ohma et al. 2006; Oomori et al. 2003; Pozio et al. 2003; Qiao et al. 2006; Teranishi et al. 2006; Xie et al. 2005; Yoshioka et al. 2005; Yu et al. 2005). The degree of membrane degradation depends on many factors, such as temperature, humidification, current density, cell voltage, MEA structure, and catalyst. It has often been reported that inadequate humidification (i.e., low-RH operation) leads to cell failure through membrane degradation (Endoh et al. 2004; Inaba et al. 2007; Luo et al. 2006; Yoshioka et al. 2005; Yu et al. 2005), whereas

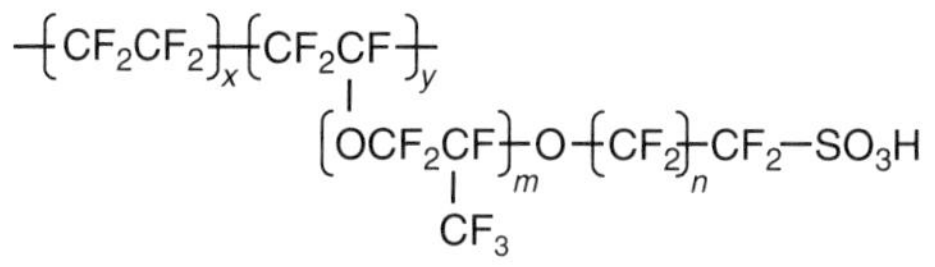

$$-\!\left[CF_2CF_2\right]_x\!\left[CF_2CF\right]_y-$$
$$\left[OCF_2CF(CF_3)\right]_m-O-\left[CF_2\right]_n-CF_2-SO_3H$$

Nafion® ($m \geq 1, n = 2, x = 5$-13.5)
Flemion® ($m = 0, 1, n = 1$-5)
Aciplex® ($m = 0, 3, n = 2$-$5, x = 1.5$–14)

Fig. 1 Chemical structures of commercially available perfluorinated sulfonic acid membranes

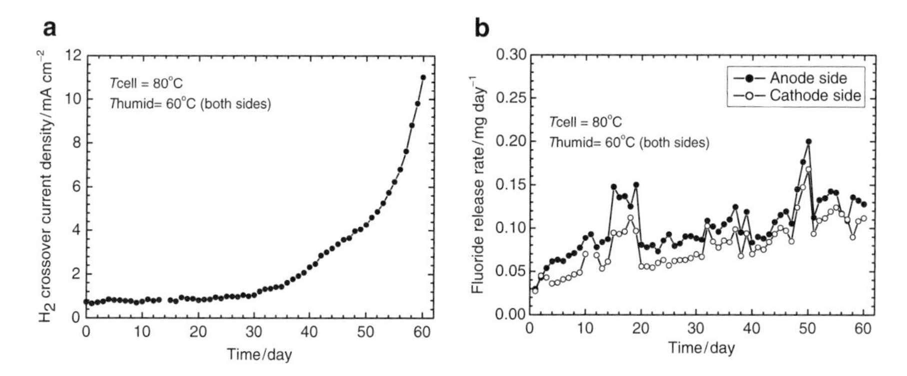

Fig. 2 Variations of H_2 permeability in current density (**a**) and fluoride ion release rates (FRR) (**b**) during the open circuit voltage durability test at 80°C. Electrode area, 25 cm². H_2 and air were humidified at 60°C. H_2/air = 150/150 mL min⁻¹ at atmospheric pressure (Reproduced from Inaba et al. (2006a). Copyright 2006, Elsevier Ltd)

sufficient prehumidification of the reactant gases greatly improves the lifetime of the membrane (Cleghorn et al. 2006; Oomori et al. 2003). It is also reported that membrane degradation is greatly accelerated under open circuit voltage (OCV) conditions (Inaba et al. 2006a; Ohma et al. 2006; Qiao et al. 2006; Teranishi et al. 2006). Figure 2 shows the variations of H_2 permeability (in current density) across the membrane and the fluoride ion release rate (FRR) in drain water when a single cell was left under open-circuit conditions. The membrane degradation is usually accompanied by emission of fluoride ions in drain water, and the resulting membrane thinning and pinhole formation lead to an abrupt increase in gas crossover and finally to cell failure. The mechanism for membrane degradation therefore must be fully understood to improve the durability of PEMFCs.

Chemical degradation of the membranes during operation was recognized in the early R&D era for space missions (LaConti et al. 2003, 2006). A diagnostic life testing was conducted for 60,000 h by General Electric for a stack using Nafion® 120 and platinum black catalysts, and 10% degradation of the membrane was determined by cumulative fluoride ion analysis in drain water (LaConti et al. 2003, 2006). Because PEMFCs employ very thin electrolyte membrane (250 μm for Nafion®

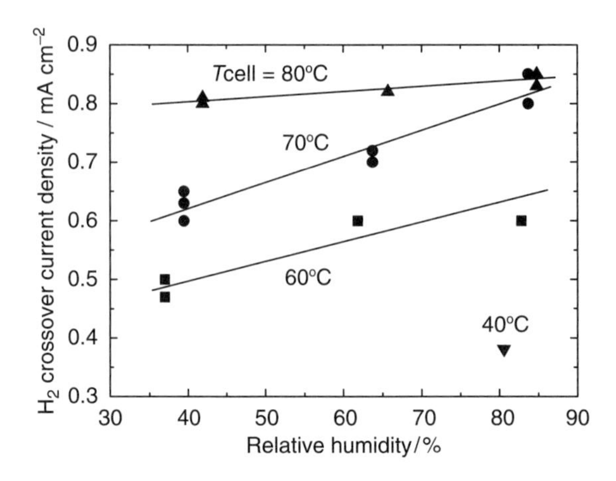

Fig. 3 Effect of cell temperature and humidification on H_2 crossover current density at atmospheric pressure. H_2 and Ar gases were humidified at the same temperature in each case. H_2/Ar = 300/300 mL min^{-1} (Reproduced from Inaba et al. (2006a). Copyright 2006, Elsevier Ltd)

120, typically 25–50 μm for membranes in state-of-art PEMFCs) to reduce the ionic resistance, permeation of the reactant gases, hydrogen and oxygen, across the electrolyte membranes, so-called gas crossover, is mostly inevitable in PEMFCs (Inaba et al. 2006a). Figure 3 shows hydrogen crossover rates estimated electrochemically in current density in a single cell at different temperatures and RHs. The hydrogen crossover rate increases with increasing temperature and humidification. The oxygen crossover rate has a similar tendency, though the rate is lower, about a half of the hydrogen crossover rate.

LaConti et al. proposed a mechanism where oxygen molecules permeate through the membrane from the cathode side, and are reduced at the anode platinum catalyst to form H_2O_2 (LaConti 1982; LaConti et al. 1977, 2003) in the early R&D era for space missions. This mechanism for H_2O_2 formation in a PEMFC is shown schematically in Fig. 4a.

$$H_2 + Pt \rightarrow Pt\text{–}H\,(\text{at the anode}) \tag{1}$$

$$Pt\text{–}H + O_2\,(\text{diffused through the polymer electrolyte membrane to the anode}) \rightarrow \bullet OOH \tag{2}$$

$$\bullet OOH + Pt\text{–}H \rightarrow H_2O_2 \tag{3}$$

It is widely known that H_2O_2 formation in the oxygen reduction reaction (ORR) on polycrystalline platinum (Damjanovic 1969; Tarasevich et al. 1983) and platinum single crystals (Markovic et al. 1995; Markovic and Ross 2002) as well as a Pt/C catalyst (Antoine and Durand 2000; Inaba et al. 2006b; Paulus et al. 2001) is greatly enhanced in the anode potential region (below 0.2 V), where atomic hydrogen is

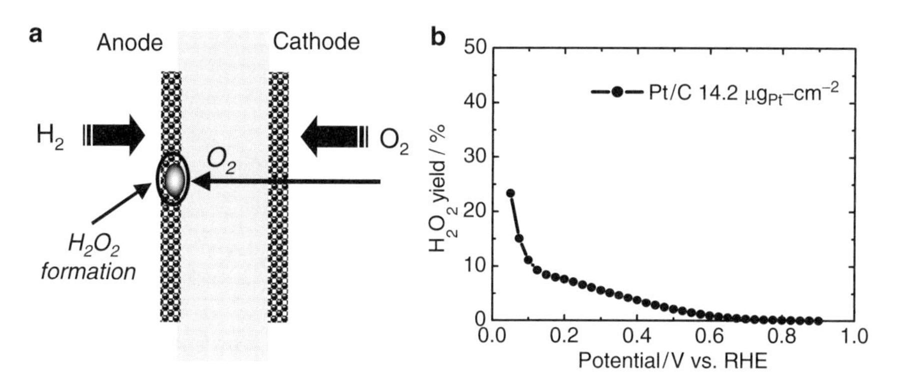

Fig. 4 (**a**) Mechanism for H_2O_2 formation in a polymer electrolyte fuel cell suggested by LaConti et al. (**b**) Variation of H_2O_2 yield in the oxygen reduction reaction on a Pt/C catalyst (14.2 μg cm^{-2} on glassy carbon) in 1.0 M $HClO_4$. *RHE* reversible hydrogen electrode

adsorbed on platinum (Markovic et al. 1995). A typical example for the variation of H_2O_2 yield on a commercially available Pt/C catalyst in the ORR is shown in Fig. 4b. These facts supported the mechanism proposed by LaConti et al. A recent rotating ring-disk electrode study suggested that H_2O_2 yield exceeds 80% at Pt/C catalysts in the anode potential range (approximately 0 V when highly dispersed on a glassy carbon disk; Inaba et al. 2004). The presence of H_2O_2 has been confirmed in exhaust gas (Teranishi et al. 2006), in drain water (Inaba et al. 2006a; Panchenko et al. 2004a), and directly in the membrane (Liu and Zuckerbrod 2005) during operation of PEMFCs.

PFSA membranes are stable against 30% H_2O_2 even at 80°C in the absence of impurity metal ions. Laconti et al. postulated that the H_2O_2 formed could react with minor impurities, forming hydroxyl (•OH) and hydroperoxy (•OOH) radicals that could attack the membrane (LaConti 1982; LaConti et al. 1977, 2003).

$$H_2O_2 + M^{2+}(\text{found in MEA}) \rightarrow M^{3+} + {\bullet}OH + OH^- \tag{4}$$

$$\bullet OH + H_2O_2 \rightarrow H_2O + {\bullet}OOH \tag{5}$$

It has been reported that the presence of Fe^{2+} ion and Cu^{2+} ion greatly accelerates the membrane degradation rate in the Fenton test of PFSI membranes (Kinumoto et al. 2006). Figure 5 shows effects of metal cations on membrane degradation 30% H_2O_2 at 80°C.

Pozio et al. (2003) reported that iron contamination from end plates accelerates the rate of membrane degradation. It was recently reported that the use of a Pt–Fe alloy catalyst accelerates membrane degradation under OCV conditions, whereas the use of a Pt–Ni alloy catalyst elongates the life of the membrane (Sulek et al. 2008). Electron spin resonance (ESR) signals were detected from deteriorated MEAs and the signals were attributed to carbon radicals (Endoh 2006; Endoh et al.

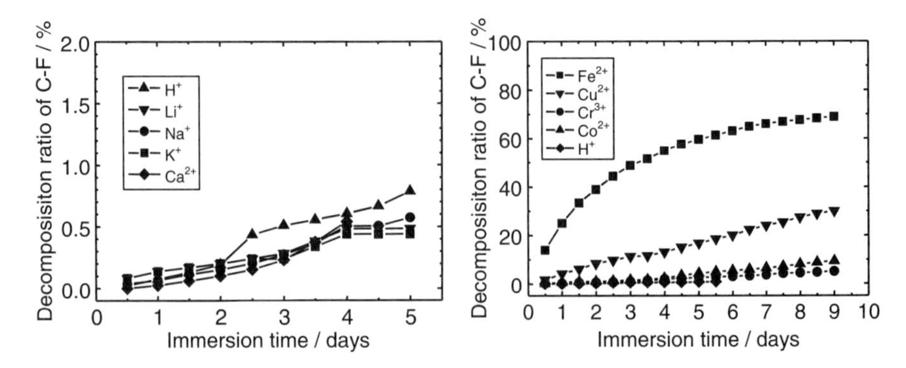

Fig. 5 Decomposition rates of the C–F bonds of Nafion with various metal cations as counterions in 30% H_2O_2 at 80°C (Fenton test) (Reproduced from Kinumoto et al. (2006). Copyright 2006, Elsevier Ltd)

2004). It was suggested that the carbon radicals are formed from •OH and •OOH radicals, though direct evidence of peroxide radicals was not found.

3 Accelerated Testing Methodology

The Fenton test has been employed as ex situ accelerated tests for membrane durability (Curtin et al. 2004; Kinumoto et al. 2006), in which a H_2O_2 solution containing a trace amount of Fe^{2+} ion is used as a test solution (Schumb et al. 1955). This may be reasonable as an accelerated test because a compound, HOOC–CF(CF_3)–O–$CF_2CH_2SO_3H$, is commonly detected as the primary short-chain degradation product of Nafion® in both PEMFC operation and the Fenton test, and was identified by ^{19}F-NMR and mass spectroscopy (Healy et al. 2005). Very harsh conditions are employed in the Fenton test. For example, PFSA membranes are immersed in 30% H_2O_2 solution containing 20 ppm Fe^{2+} ion at 85°C for 16–20 h, and their durability is evaluated as the total number of fluoride ions released into the solution (Curtin et al. 2004), while about 3% H_2O_2 solution containing several parts per million of Fe^{2+} ion at room temperature is employed for hydrocarbon membranes. Sulfate ions are also detected in the solution, and can be used to evaluate the decomposition of the sulfonic acid moieties (Kinumoto et al. 2006). The drawback of the Fenton test is difficulty in evaluating the accelerating factor, i.e., correlating the test results with the durability of membranes in PEMFC operation, because the concentrations of H_2O_2 and Fe^{2+} in the MEAs greatly depend on the catalysts, MEAs, gas-feeding systems, and operating conditions.

Alternative ex situ accelerated tests have been reported. Aoki et al. (2005, 2006) developed a novel ex situ method for membrane durability tests, in which mixed gases of H_2 and air were supplied at given ratios to a water suspension of a Pt/C catalyst coated with Nafion®. This method has an advantage that it can simulate the atmospheres on the anode side (H_2 rich) and the cathode side (O_2 rich) by changing the ratio of H_2 to air. They observed a larger number of fluoride ions (0.27% of the total fluorine in the membrane) in the H_2-rich atmosphere than in the O_2-rich atmosphere (0.16%), which supported the degradation mechanism proposed by LaConti et al. (2003) described earlier. Hommura et al. (2005a, b) developed another ex situ method, in which PFSA membranes are exposed directly to H_2O_2 vapor at 120°C.

PFSA membranes in PEMFCs are most drastically deteriorated at OCV with hydrogen and oxygen (or air) flow at the anode and cathode, respectively, and thus OCV tests have been used as in situ accelerated tests for membrane durability (Inaba et al. 2006a; Qiao et al. 2006), as shown in Fig. 2. The FRR in drain water or the gas crossover rate across the membrane is used as a measure of membrane degradation. The cell temperature and the humidity of the gases greatly affect membrane durability and typically a cell temperature of 80–90°C and a RH of approximately 30% are employed (Miyake et al. 2006; Ohma et al. 2006). The accelerating factor can be obtained by comparing the variation of the gas crossover rate or accumulative fluoride ion release with that achieved under standard conditions.

4 Recent Understanding of Degradation Mechanisms

As mentioned earlier, insufficiently low humidification greatly enhances membrane degradation during operation as well as under OCV. This behavior seems peculiar, because gas permeation through a PFSA membrane decreases with decreasing gas humidification, as seen in Fig. 3. The reason for the enhanced degradation rate under low humidification has not been fully understood so far, and is the subject of much debate. Inaba et al. (2007) reported that sulfate ions and ferrous ions are accumulated in low-RH operation and they are washed out in drain water in highly humidified conditions as shown in Fig. 6. The boiling temperature of H_2O_2 (150°C) is higher than that of water, and hence they attributed the high degradation rate to accumulation of impurities and H_2O_2 in the membrane under low humidification. A high activity of H_2O_2 in the vapor state may be raised as a reason for the enhanced membrane degradation rate (Hommura et al. 2005a, b).

Several groups have tested membrane durability of single-side-catalyzed MEAs under open-circuit conditions (Liu et al. 2006; Mittal et al. 2006a, b; Miyake et al. 2006). They commonly have reported that the membrane degradation rate is higher for a MEA catalyzed only at the cathode side than for one catalyzed only at the anode side as shown schematically in Fig. 7. This fact is inconsistent with the mechanism suggested by LaConti et al., in which H_2O_2 is formed at the anode catalyst. Two mechanisms have been suggested: one is H_2O_2 formation upon reduction of adsorbed oxygen on the cathode catalyst with permeating H_2 (Miyake et al. 2006), and the other is direct formation of •OH or •OOH radicals from oxygen-containing species on platinum (e.g., Pt–OH and Pt–OOH) (Liu et al. 2006; Mittal et al. 2006b). However, these mechanisms should be carefully verified because the potential of the cathode (approximately 1 V) under open-circuit conditions is more positive than the potential for H_2O_2 formation ($E_{H_2O_2} = 0.695$ V). Although the possibility

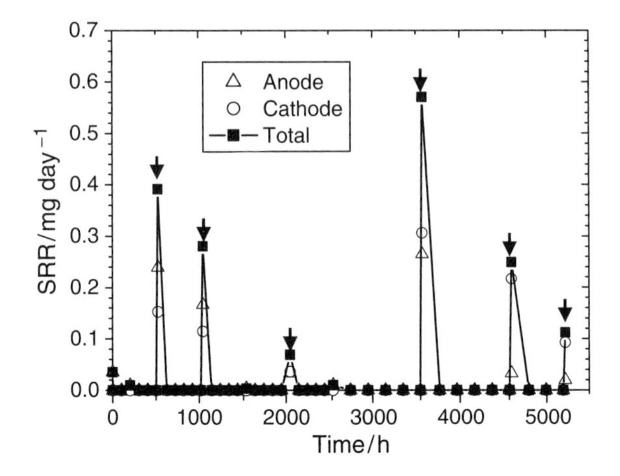

Fig. 6 Variation of sulfate ion release rate (*SRR*) in drain water from the anode and the cathode during the durability test at 80°C. Electrode area, 25 cm^2. H_2 and air were humidified at 60°C. H_2/air = 150/350 mL min^{-1} ($U_f = U_o = 40\%$) at atmospheric pressure. *Arrows* show temporary changes in operating conditions to fully humidified ones (cell, 70°C, H_2 and air humidification, 67°C) (Reproduced with permission from Inaba et al. (2007). Copyright 2007, The Electrochemical Society, Japan)

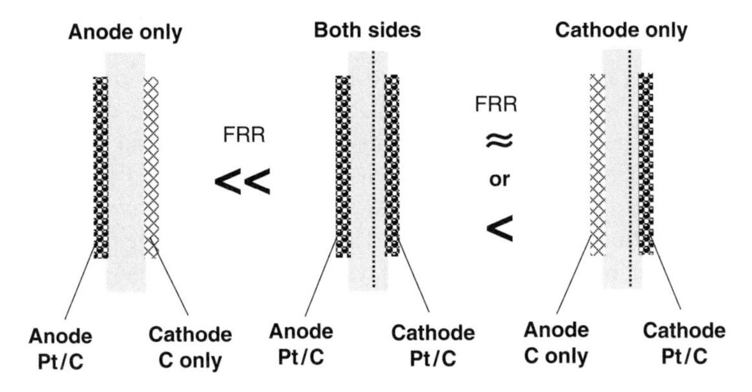

Fig. 7 Open-circuit voltage durability tests using single-side-catalyzed membrane-electrode assemblies. The presence of platinum at the cathode greatly enhances the FRR

for •OH or •OOH radical formation on platinum in an oxygen atmosphere has been confirmed by ab initio calculations (Atrazhev et al. 2005; Panchenko et al. 2004b), the lifetime of the radicals would be too short for them to attack the bulk of the electrolyte membrane (Endoh et al. 2004). Liu et al. (2006) observed that the interaction of permeating H_2 with an adsorbed oxide layer formed electrochemically on platinum, at 1.0–1.5 V in a nitrogen atmosphere, does not lead to a significant increase in the FRR, and showed that gaseous O_2 is needed for degradation to occur.

Platinum dissolution from the cathode and particle deposition inside the bulk membrane, called the platinum band, is another serious degradation phenomenon

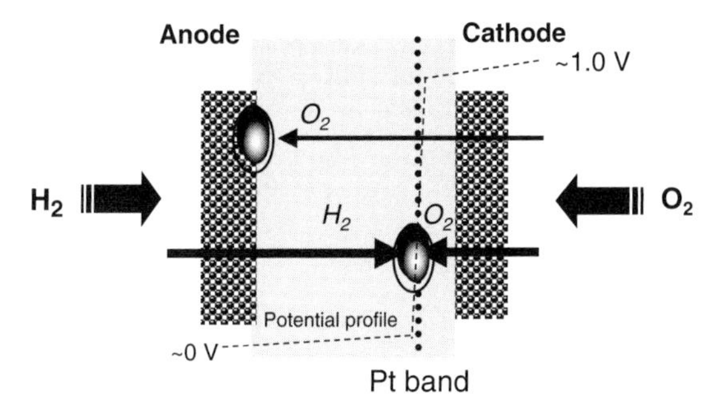

Fig. 8 Enhancement of H_2O_2 formation by platinum-band formation

under open-circuit conditions (Ferreira et al. 2005; Yasuda et al. 2006). Ohma et al. (2006, 2007) found that the platinum band is formed in a relatively short time (several tens to a few hundreds of hours) depending on the test conditions, and suggested that the platinum band formation greatly enhances H_2O_2 formation and membrane degradation. The position where the platinum band is formed inside the membrane greatly depends on the operating conditions, especially hydrogen and oxygen partial pressures at the anode and the cathode, respectively (Zhang et al. 2007). The theoretical potential profile in the membrane suddenly changes at the position of the platinum band from the cathode side (approximately 1 V) to the anode side (approximately 0 V) as shown in Fig. 8; hence the potential requirement for H_2O_2 formation (less than 0.2 V as seen in Fig. 4) is satisfied (Ohma et al. 2006). In addition, oxygen flux is greatly enhanced, especially under H_2/air conditions, because the platinum band is formed in the vicinity of the cathode catalyst layer under H_2/air conditions (Ohma et al. 2006, 2007; Zhang et al. 2007). It should be noted that the platinum-band mechanism of Ohma et al. explains why membrane degradation is accelerated greatly at OCV. It also explains the discrepancy in the results of one-side-catalyzed MEAs mentioned above, because the platinum band originates from the dissolution of the cathode catalyst. The FRR gradually increased with time in an OCV test at 80°C, while the gas crossover rate did not change appreciably for the initial 30 days (Inaba et al. 2007), and this discrepancy may be attributed to the growth of the platinum band.

5 Decomposition Mechanism of PFSA Membranes

During operation of PEMFCs, fluoride ions, sulfate ions, and low molecular weight perfluorosulfonic acid are found in drain water. Direct gas mass spectroscopy of the cathode outlet gas indicated the formation of HF, H_2O_2, CO_2, SO, SO_2, H_2SO_2, and H_2SO_3 in OCV durability tests (Teranishi et al. 2006). PFSA membranes contain no α-hydrogens, which are vulnerable to radical attacks, and hence the membranes would be stable against radical attacks (Mitov et al. 2005) if they have perfectly fluorinated structures as shown in Fig. 1. The susceptibility to peroxide radical attack has been attributed to a trace amount of polymer end groups with residual hydrogen-containing terminal bonds (Curtin et al. 2004). Hydroxy or hydroperoxy radicals attack the polymer at the end-group sites and initiate decomposition. An example of attack on an end group such as $-CF_2X$, where X is COOH, is as follows (Curtin et al. 2004; Hommura et al. 2005b):

$$R_f\text{–}CF_2COOH + \bullet OH \rightarrow R_f\text{–}CF_2\bullet + CO_2 + H_2O \tag{6}$$

$$R_f\text{–}CF_2\bullet + \bullet OH \rightarrow R_f\text{–}CF_2OH \rightarrow R_f\text{–}COF + HF \tag{7}$$

$$R_f\text{–}COF + H_2O \rightarrow R_f\text{–}COOH + HF \tag{8}$$

It should be noted that –COOH is regenerated in the reaction in (8). Hence, once decomposition begins at one end group, a complete PFSA unit is decomposed to HF, CO_2, and low molecular weight compounds by the radical depolymerization reactions (called the unzipping mechanism). Degradation studies using model compounds have proved that decomposition starts from $-CHF_2$ and $-CF_2COOH$ groups (Healy et al. 2005). The decomposition rate of model compounds without –COOH end groups is 2–3 orders of magnitude lower than those with –COOH end groups (Schiraldi et al. 2006). Curtin et al. (2004) reported that these reactive end groups of Nafion® could be minimized during extrusion processes by pretreating the polymer with fluorine gas. After more than 50h of exposure, they could reduce 61% of the remaining hydrogen-containing end groups and found a 56% decrease in the number of released fluoride ions in the Fenton test, as compared with an untreated polymer.

Even when the residual hydrogen-containing end groups are completely fluorinated, the degradation rate of PFSA membranes cannot be reduced to "zero." There should be another mechanism for membrane degradation. Hommura et al. (2005b) carried out durability tests of Flemion® membranes, and reported that the average molecular weight decreased, while the number of –COOH groups increased with time when the membranes were exposed to vapor-phase H_2O_2 at 120°C. They suggested that not only the unzipping reactions, but also scission of the main chain is involved in the mechanism for membrane decomposition. It has been suggested that the ether linkages are the weakest sites of the side chains for radical attack (Schiraldi et al. 2006).

6 Improvement of Membrane Durability

Membrane manufacturers have developed PFSI membranes with improved chemical and thermal stability as well as improved water management at high temperatures. DuPont has reported that the number of remaining hydrogen-containing end groups can be reduced by treating Nafion® polymer with fluorine gas, which improves the chemical stability against the radicals as mentioned already (Curtin et al. 2004). Asahi Glass has developed new polymer composites (NPCs), which have a highly durable PFSA-based membrane under high-temperature and low-humidity conditions, though details of the membrane structure have not been disclosed (Endoh 2006). The NPC membrane showed excellent stability over 1,000 h in an OCV test at 120°C and 18% RH. The FRR was less than 1% of the FRR in an OCV test using their standard MEA. They also demonstrated continuous operation using a new perfluorinated polymer composite based MEA for more than 4,000 h at 120°C, 200 kPa, 0.2 A cm^{-2}, and 50% RH (see the chapter "Improvement of Membrane and Membrane Electrode Assembly Durability").

Another approach to improve the durability of the membranes is to place an additional layer for H_2O_2 decomposition or trapping of radicals (Burlastsky et al. 2004; Tsurumaki 2006). It has been reported that the durability of electrolyte membrane is improved by placing a catalyst layer based on platinum, palladium, iridium, etc. inside the membrane between the anode and the cathode (Burlastsky et al. 2004). Radical-trap layers, which are based on rare earth metal oxides, on both sides of the membrane are also effective, and it has been reported that the durability of the MEA in OCV tests is improved by one order of magnitude compared with the standard MEA (Tsurumaki 2006).

Finally, oxygen crossover across the membrane is the essential factor for H_2O_2 formation in a cell as described earlier and hence the simplest mitigation to suppress membrane degradation is a reduction of gas permeation. In this respect, the use of hydrocarbon-based membranes, which have been extensively studied lately to reduce the cost of the membranes, might be effective because they have much less gas permeability (one order of magnitude lower than that for PFSA membranes), although their chemical stability against oxygen radicals is much weaker than that of PFSA membranes.

7 Concluding Remarks

Membrane durability of >20,000 h has recently been proved in continuous operation of state-of-the-art PEMFC stacks for stationary cogeneration systems. This has been achieved mainly by employing nearly full humidification conditions, because of severe membrane degradation under insufficient humidification. However, the high humidification conditions cause another problem, that is, difficulty in water management. Careful water management raises the cost of the total systems, and simultaneously reduces the energy conversion efficiency. Robustness with much less careful water management is needed for membranes to simplify the system and to reduce the cost for PEMFCs. To obtain robustness of cell stacks, development of highly durable membranes is inevitable. As described in this chapter, understanding of membrane degradation phenomena has greatly progressed lately, and several promising techniques to improve membrane durability have been suggested. The author believes that highly durable membranes will be developed and commercialized in a few years, and hopes that PEMFCs free from membrane degradation even under harsh operating conditions will be realized in the near future.

Acknowledgments The author greatly acknowledge financial support from "Degradation Mechanisms of Polymer Electrolyte Fuel Cells (FY2001–2004)" and "Fundamental Research of Degradation of PEFC Stacks (FY2005–2007)" from the New Energy and Industrial Technology Development Organization (NEDO), Japan, and from the Academic Frontier Research Project on "Next Generation Zero-Emission Energy Conversion System (FY2003–7)" from the Ministry of Education, Culture, Sports, Science and Technology (MEXT), Japan.

References

Antoine O. and Durand R. (2000) *J. Appl. Electrochem.* 30, 839–844.

Aoki A., Uchida H., and Watanabe M. (2005) *Electrochem. Commun.* 7, 1434–1438.

Aoki M., Uchida H., and Watanabe M. (2006) *Electrochem. Commun.* 8, 1509–1513.

Atrazhev V., Burlatsky S. F., Cipollini N. E., Condit D. A., and Erikhman N. (2005) *208th Meeting of the Electrochemical Society*, Los Angels, CA, Abstract No. 1189.

Banergee S. and Curtin D. E. (2004) *J. Fluorine Chem.* 125, 1211–1216.

Burlastsky S. F., Hertzberg J. B., Copollini N. E., Condit D. A., Jarvi T. D., Leistra J. A., Perry M. L., and Madden T. H. (2004) US Patent Application US 2004/0224216 A1.

Cleghorn S., Kolde J., and Liu W. (2003) In: W. Vielstich, H. A. Gasteiger, and A. Lamm (Eds.) *Handbook of Fuel Cells-Fundamentals, Technology and Application*, Wiley, New York, NY, pp. 566–675.

Cleghorn S. J. C., Mayfield D. K., Moore D. A., Moore J. C., Rusch G., Sherman T. W., Sisofo N. T., and Beuscher U. (2006) *J. Power Sources* 158, 446–454.

Collier A., Wang H., Yuan X. Z., Zhang J., and Wilkinson D. P. (2006) *Int. J. Hydrogen Energy* 31, 1838–1854.

Curtin D. E., Lousenberg R. D., Henry T. J., Tangeman P. C., and Tisack M. E. (2004) *J. Power Sources* 131, 41–48.

Damjanovic A. (1969) *Modern Aspects of Electrochemistry*, Plenum, New York, NY, pp. 369–483.

Endoh E. (2006) *ECS Trans.* 3(1), 9–18.

Endoh E., Terazono S., Widjaja H., and Takimoto Y. (2004) *Electrochem. Solid-State Lett.* 7, A209–A211.

Ferreira P. J., la O' G. J., Shao-Horn Y., Morgan D., Makharia R., Kocha S., and Gasteiger H. A. (2005) *J. Electrochem. Soc.* 152, A2256–A2271.

Healy J., Hayden C., Xie T., Olson K., Waldo R., Brundage M., and Gasteiger H. A. (2005) *Fuel Cells* 5, 302–308.

Hommura S., Kawahara K., and Shimohira T. (2005a) *Extended Abstracts of the 207th Electrochemical Society Meeting*, Quebec, Canada, Abstract No. 803.

Hommura S., Kawahara K., and Shimohira T. (2005b) *Polymer Preprints, Japan* 54, 4517–4518

Inaba M., Yamada H., Tokunaga J., and Tasaka A. (2004) *Electrochem. Solid-State Lett.* 7, A474.

Inaba M., Taro K., Kiriake M., Umebayashi R., Tasaka A., and Ogumi Z. (2006a) *Electrochim. Acta* 51, 5746–5753.

Inaba M., Yamada H., Tokunaga J., Matsuzawa K., Hatanaka A., and Tasaka A. (2006b) *ECS Trans.* 1(8), 315–322.

Inaba M., Yamada H., Umebayashi R., Sugishita M., and Tasaka A. (2007) *Electrochemistry* 75, 207–212.

Kinumoto T., Inaba M., Nakayama Y., Ogata K., Umebayashi R., Tasaka A., Iriyama Y., Abe T., and Ogumi Z. (2006) *J. Power Sources* 158, 1222–1228.

LaConti A. B. (1982) *ACS Polymer Division Topical Workshop on Perfluorinated Ionomer Membranes*, Lake Buena Vista, FL.
LaConti A. B., Fragala A. R., and Boyack J. R. (1977) In: S. Srinivasan, J. D. E. McIntyre, and F. G. Will (Eds.), *Proceeding of the Symposium on Electrode Materials and Process for Energy Conversion and Storage*, The Electrochemical Society, Inc., Princeton, NJ, pp. 354–374.
LaConti A. B., Hamdan M., and McDonald R. C. (2003) In: W. Vielstich, H. A. Gasteiger, and A. Lamm (Eds.), *Handbook of Fuel Cells-Fundamentals, Technology and Application*, Wiley, New York, NY, pp. 647–662.
LaConti A. B., Liu H., Mittelsteadt C., and McDonald R. C. (2006) *ECS Trans.* 1(8), 199–219.
Liu W. and Zuckerbrod D. (2005) *J. Electrochem. Soc.* 152, A1165–A1170.
Liu H., Gasteiger H. A., Laconti A., and Zhang J. (2006) *ECS Trans.* 1(8), 283–293.
Luo Z., Li D., Tang H., Pan M., and Ruan R. (2006) *Int. J. Hydrogen Energy* 31, 1831–1837.
Markovic N. M. and Ross P. N. (2002) *Surface Sci. Rep.* 45, 117–229.
Markovic N. M., Gasteiger H. A., and Ross P. N. (1995) *J. Phys. Chem.* 99, 3411–3415.
Mitov S., Panchenko A., and Roduner E. (2005) *Chem. Phys. Lett.* 402, 485–490.
Mittal V., Kunz R., and Fenton J. (2006a) *ECS Trans.* 1(8), 275–282.
Mittal V. O., Kunz H. R., and Fenton J. M. (2006b) *Electrochem. Solid-State Lett.* 9, A299–A302.
Miyake N., Wakizoe M., Honda E., and Ohta T. (2006) *ECS Trans.* 1(8), 249–261.
Ohma A., Suga S., Yamamoto S., and Shinohara K. (2006) *ECS Trans.* 3(1), 519–529.
Ohma A., Suga S., Yamamoto S., and Shinohara K. (2007) *J. Electrochem. Soc.* 154, B757–B760.
Oomori Y., Yamazaki O., and Tabata T. (2003) *The11th FCDIC Fuel Cell Symposium Proceedings*, FCDIC, Tokyo, pp. 99–102.
Panchenko A., Dilger H., Moeller H., Sixt T., and Roduner E. (2004a) *J. Power Sources* 127, 325–330.
Panchenko A., Koper M. T. M., Shubina T. E., Mitchell S. J., and Roduner E. (2004b) *J. Electrochem. Soc.* 151, A2016–A2007.
Paulus U. A., Schmidt T. J., Gasteiger H. A., and Behm R. J. (2001) *J. Electroanal. Chem.* 495, 134–145.
Pozio A., Silva R. F., Francisco M. D., and Giorgi L. (2003) *Electrochim. Acta* 48, 1543–1549.
Qiao J., Saito M., Hayamizu K., and Okada T. (2006) *J. Electrochem. Soc.* 153, A967–A974.
Schiraldi D. A., Zhou C., and Zawodzinski T. A. (2006) *Extended Abstracts of the 210th Electrochemical Society Meeting*, Cancun, Mexico, Abstract No. 443.
Schumb W. C., Satterfield C. N., and Wentworth R. L. (1955) *Hydrogen Peroxide, American Chemical Society Monograph Series*, Reinhold Publishing, New York, NY.
Sulek M. S., Mueller S. A., and Paik C. H. (2008) *Electrochem. Solid-State Lett.* 11, B79–B82.
Tarasevich M. R., Sadkowski A., and Yeager E. (1983) *Comprehensive Treatise of Electrochemistry*, Plenum, New York, NY, pp. 301–398.
Teranishi K., Kawata K., Tsushima S., and Hirai S. (2006) *Electrochem. Solid-State Lett.* 9, A475–A477.
Tsurumaki S. (2006) *Annual Review of NEDO R&D for Fuel Cells and Hydrogen Technology*, NEDO, Tokyo, pp. 64–67.
Xie J., Wood D. L., Wayne D. M., Zawodinski T. A., Atanassov P., and Borup R. L. (2005) *J. Electrochem. Soc.* 152, A104–A109.
Yasuda K., Taniguchi A., Akita T., Ioroi T., and Siroma Z. (2006) *Phys. Chem. Chem. Phys.* 8, 746–752.
Yoshioka S., Yoshimura A., Fukumoto H., Horii O., and Yoshiyasu H. (2005) *Fuel Cell Bull.* 3, 11–15.
Yu J., Matsuura T., Yoshikawa Y., Islam M. N., and Hori M. (2005) *Electrochem. Solid-State Lett.* 8, A156–A158.
Zhang J., Litteer B. A., Gu W., Liu H., and Gasteiger H. A. (2007) *J. Electrochem. Soc.* 154, B1006–B1011.

Chemical Degradation: Correlations Between Electrolyzer and Fuel Cell Findings

Han Liu, Frank D. Coms, Jingxin Zhang, Hubert A. Gasteiger, and Anthony B. LaConti

Abstract Membrane chemical degradation of polymer electrolyte membrane fuel cells (PEMFCs) is summarized in this paper. Effects of experimental parameters, such as external load, relative humidity, temperature, and reactant gas partial pressure, are reviewed. Other factors, including membrane thickness, catalyst type, and cation contamination, are summarized. Localized degradations, including anode versus cathode, ionomer inside the catalyst layer, degradation along the Pt precipitation line, gas inlets, and edges are discussed individually. Various characterization techniques employed for membrane chemical degradation, Fourier transform IR, Raman, energy-dispersive X-ray, NMR, and X-ray photoelectron spectroscopy are described and the characterization results are also briefly discussed. The detailed discussion on mechanisms of membrane degradation is divided into three categories: hydrocarbon, grafted polystyrene sulfonic acid, and perfluorinated sulfonic acid. Specific discussion on the radical generation pathway, and the relationship between Fenton's test and actual fuel cell testing is also presented. A comparison is made between PEMFCs and polymer electrolyte water electrolyzers, with the emphasis on fuel cells.

1 Introduction

Among various types of fuel cell technologies, such as alkaline, molten carbonate, solid oxide, phosphoric acid, and polymer electrolyte membrane (PEM), the PEM fuel cell (PEMFC) is the most prevalent technology owing to its potentially versatile applications from portable devices, transportation vehicles to stationary power generation (Hickner et al. 2004; Smitha et al. 2005). According to current guidelines from the US Department of Energy from 2007, fuel cell applications in automobiles require a lifetime of about 5,000 h, while stationary application requires operation

H. Liu (✉)
Giner Electrochemical Systems, LLC, 89 Rumford Avenue, Newton, MA 02466, USA
e-mail: hliu@ginerinc.com

F.N. Büchi et al. (eds.), *Polymer Electrolyte Fuel Cell Durability*,

exceeding 40,000 h. Durability concern of PEMFC systems is one of the major factors that impedes the commercialization of fuel cell technology (LaConti et al. 2003). Different stack components, from hardware, catalyst to membrane, limit the lifetime of the fuel cells (Healy et al. 2005). Chemical degradation of the membrane, combined with mechanical and thermal degradation, can lead to membrane thinning and pinhole formation, causing performance degradation and severe stack failures.

PEM water electrolyzers (PEMELCs), which are used as H_2/O_2 generators, have been developed parallel to PEMFCs. High-power electrolyzer stacks (approximately 30–50 kW) have demonstrated stack and system life of over 50,000 h without incident (LaConti et al. 2005). Based on similar operating fluids (i.e., H_2O, H_2, O_2), PEMELCs face similar durability challenges compared with PEMFCs. Thus, lessons and experiences obtained from PEMELC chemical degradation can be used to illustrate the underlying mechanisms for both of the electrochemical devices.

In this paper, a comparison between PEMFCs and PEMELCs will be made, with an emphasis on PEMFCs. Recent findings in degradation mechanisms, impact factors, characterization techniques, and mitigation strategies will be discussed.

2 Fundamentals Governing the Mechanism for Membrane Chemical Degradation

Before discussing membrane chemical degradation in detail, the factors governing the degradation mechanism must be identified. Among three major types of membrane materials, hydrocarbon, partially fluorinated, and perfluorinated ionomers, perfluorinated sulfonic acid (PFSA) is the most widely used membrane material owing to its high chemical stability (Schiraldi 2006).

Certain sulfonated hydrocarbon membranes have hydrolysis stability issues, which can lead to membrane disintegration and loss of acid functionality (Genies et al. 2001; Asano et al. 2006). Condensation reaction of the sulfonic groups can also cause cross-linking and increase of equivalent weight. PFSA and partially fluorinated ionomers are generally immune to such hydrolytic instability. Thus, membrane degradation due to hydrolysis will not be discussed in the following sections.

For perfluorinated and partially fluorinated ionomers, hydrofluoric acid (HF) is one of the major products of chemical degradation, which can be easily quantified by a fluoride ion selective electrode. The fluoride release rate (FRR) has long been used as the parameter measuring the chemical degradation rate (LaConti et al. 1968; Chludzinski 1982; Healy et al. 2005). The fluoride loss rate determined by the analysis of effluent water has been used to predict the lifetime of fluorocarbon-based PEMFCs and PEMELCs (Baldwin et al. 1990). FRR has been extensively used by numerous researchers as the general gauge for membrane chemical degradation and lifetime. Under certain conditions, however, local membrane area degradation may lead to premature membrane failures (Liu et al. 2001).

A comprehensive study of the key elements of PEMFC membrane chemical degradation was recently conducted for PFSA-type ionomers (Table 1; Liu et al. 2005).

Table 1 Key elements for membrane chemical degradation. All tests employed Nafion® 112 with 0.4 $mg_{Pt}cm^{-2}$ per electrode and were conducted at 95°C, 300 kPa_{abs}, 100% relative humidity, 525 sccm gas flows, open-circuit voltage

Configuration	(H)F$^-$-loss rate (g_f cm^{-2} h^{-1})	Expected lifetime
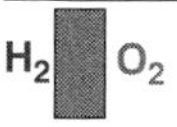 H_2 [membrane] O_2	<2 × 10^{-8}	>17 kh
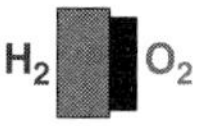 H_2 [membrane][catalyst] O_2	<5 × 10^{-6}	<70 h
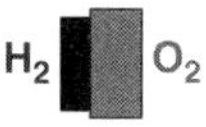 H_2 [catalyst][membrane] O_2	<5 × 10^{-6}	<70 h
N_2 [catalyst][membrane][catalyst] O_2	<2 × 10^{-8}	>17 kh
H_2 [catalyst][membrane][catalyst] N_2	<2 × 10^{-8}	<70 h
 H_2 [catalyst][membrane][catalyst] O_2	<5 × 10^{-6}	<70 h

The *black bar* is the catalyst layer; the *gray bar* is the membrane. The expected lifetime is calculated on the basis of the assumption that membrane fails at 10±5% fluoride inventory loss (Liu et al. 2001).

As can be seen from Table 1 under open circuit voltage (OCV) conditions, H_2 or O_2 cannot directly attack the membrane in the absence of a platinum catalyst. Even in the presence of platinum, significant membrane degradation is not observed when the membrane is exposed to a single type of gas (H_2 or O_2). It is also found that adsorbed oxygen species, in the form of electrochemically generated PtO, do not lead to membrane chemical degradation, even with the presence of H_2 gas. It can be concluded that H_2 and O_2 molecules have to react on a catalytic surface (Pt in this case) to introduce membrane chemical degradation. In the PEMFC configuration, gas crossover (i.e., permeation of reactants through the membrane) and catalyzed chemical reaction of the crossed-over gas are consequently critical for the degradation mechanism.

Early investigation of PEMELC degradation drew similar conclusions for hydrocarbon-based membranes (LaConti et al. 2005). With use of Aclar-*g*-polystyrene sulfonic acid as the membrane material, it was found that H_2, O_2, and platinum black were necessary for accelerated membrane degradation.

It has been widely accepted that hydrogen peroxide (H_2O_2) and related radical species such as hydroperoxyl ($HO_2•$) and hydroxyl (HO•) are generated by the crossed-over gases on the catalytic surface (LaConti et al. 2003). These oxidizing species will attack the ionomer, leading to chain scission, unzipping, and loss of functional groups. Discussion on the detailed mechanism will be presented later.

Although platinum is among the most active catalytic surfaces in modern PEMFCs and PEMELCs, other electrochemically active surfaces, such as carbon blacks, can also facilitate membrane degradation (Endoh et al. 2007). On the basis of findings from rotating ring-disc electrode (RRDE) experiments, carbon black has a higher H_2O_2 yield from electrochemical oxygen reduction than a carbon-supported platinum (Pt/C) catalyst at a potential less than 0.54 V versus the reference hydrogen electrode (RHE) (Antoine and Durand 2000). Detailed discussion on catalyst type and the role of carbon will be presented in a following section.

The role of hydrogen in PFSA chemical degradation can be more complicated. It is believed that PFSA can lose fluorine by H_2 substitution: C–F + H_2 → C–H + HF (Yu et al. 2005b). Although such substitution has been proven to be negligible compared with main membrane degradation, the C–H bond formed is a weak link that renders the membrane more susceptible to peroxyl chemical degradation.

Another possible mechanism for generating membrane-degrading H_2O_2 is the two-electron reduction of O_2 ($O_2 + 2H^+ + 2e^- \rightarrow H_2O_2$), which is independent of the gas crossover phenomenon. This mechanism can generate H_2O_2 under normal PEMFC operating conditions where current is generated with an external load. Compared with other nonnoble metal based catalysts, a platinum-based catalyst is highly efficient for completely reducing O_2 to H_2O. Since it is generally accepted that membrane chemical degradation is most severe under OCV conditions (Mittal et al. 2006a), the two-electron reduction mechanism should not play a significant role. Additionally, it is found that H_2O_2 generation is unfavorable under cathode potentials (Inaba et al. 2004). Thus, the two-electron oxygen reduction mechanism can be considered as a minor pathway in membrane degradation compared with the gas crossover mechanisms when platinum is employed as the catalyst.

To summarize this section, O_2 or H_2 gas crossover and reaction of the crossed-over gas on a catalytic surface are the most fundamental governing mechanisms for membrane chemical degradation. This lays a foundation before experimental parameters, other impacting factors, detailed degradation mechanisms, and potential mitigation methods can be discussed in later sections.

3 Experimental Parameters Affecting Chemical Degradation

From an engineering point of view, deciphering the relationship between stack operating conditions and membrane degradation rate will identify the desired operating conditions to maximize stack lifetime. Scientifically, underlying degradation mechanisms can be elucidated by studying the correlations. The experimental parameters discussed here include external load, relative humidity (RH), gas pressure, and stack temperature.

3.1 Effect of External Load

During normal operation of PEMFCs, an external load is applied to the stack to draw current from the fuel cells to drive other electric devices, such as a motor, on the outer circuit. When demand for current diminishes, such as during a vehicle idle period, the external current is cut down to a minimal level, while the voltage increases. Depending on the vehicle operating strategy, no current may be drawn from the fuel cell stack, which is usually referred to as OCV condition.

For PEMFCs, OCV condition generally leads to higher chemical degradation and shorter lifetime for both PFSA membranes (Mittal et al. 2006a; Aoki et al.

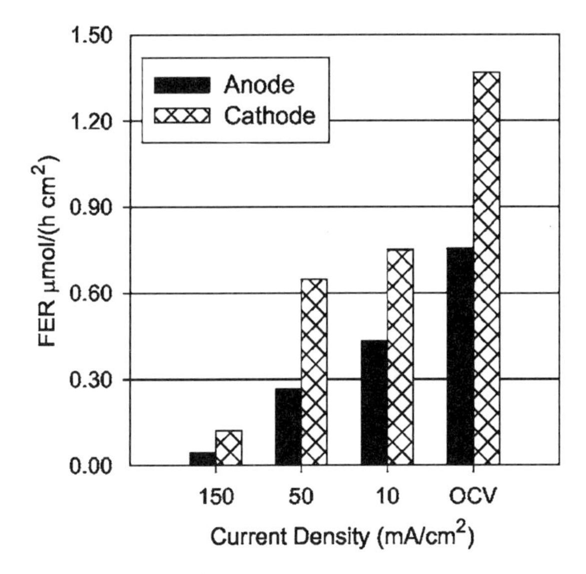

Fig. 1 Relationship between fluoride release rate (FRR; *FER*) and external load from a standard cell with Nafion 117 membrane tested at 80°C, 50% inlet relative humidity (RH), H_2/O_2 at 200 cm^3 min^{-1} (Mittal et al. 2006a)

2005; Knights et al. 2004) and radiation-grafted hydrocarbon membranes (Wang and Capuano 1998; Gubler et al. 2005). Monotonic decrease of the FRR is observed when a higher load is applied to the PEMFC (Fig. 1).

Several mechanisms have been proposed to explain this phenomenon.

3.1.1 Depletion of Reactant Gas

When an external load is applied to the fuel cell, the reactant gases are consumed to generate water and electricity. This hypothesis suggests that the concentration of the reactant gas near the membrane/catalyst interface can be reduced owing to the gas consumption (Aoki et al. 2005; Büchi et al. 1995; Wang and Capuano 1998). Such behavior will both decrease the gas crossover rate and reduce the radical/H_2O_2 generation rate owing to lower reactant concentration.

It is unlikely that the gas concentration is lower at the catalyst/membrane interface for modern catalyst structures. If the reactant concentration is lower, the catalyst utilization will be less, which results in a loss of catalytic efficiency near the interface. Extensive research has been conducted in optimizing the catalyst layer to prevent significant reactant concentration gradients in the electrodes. Thus, the reactant concentration near the membrane/catalyst interface should not be significantly lower than in the bulk under normal operating conditions.

A similar effect, the FRR decrease with current density, can also be observed when a H_2/5% air mixture is used in the anode and H_2 is evolved on the cathode by an external power supply. Thus, it is proposed that, unlike the commonly suggested gas crossover mechanism, the FRR reduction is not due to a lower gas crossover rate under load.

3.1.2 Electrochemical Consumption of Crossed-Over Gas

It is proposed that the crossed-over gas may react electrochemically when current is drawn or applied to the stack. Water is the dominant product of electrochemical reaction. At or near OCV conditions, the crossed-over gas can undergo catalyzed chemical reaction, which potentially leads to significant H_2O_2/radical generation.

3.1.3 External Load and Relative Humidity

The impact of external load can also be the result of RH conditions. Generally, lower RH leads to higher membrane degradation. Detailed discussion of the RH effect is provided in Sect. 3.2.

Water is generated on the cathode side during the normal operation of fuel cells. If the gas is supplied at constant RH regardless of current density, the actual RH will be higher under increased current density (Knights et al. 2004). Since membranes generally sustain less chemical degradation under high RH conditions, lower membrane degradation is observed when an external load is applied to the stack.

In contrast to the observation discussed above, there are reports that higher current density actually can cause faster membrane degradation (Xie et al. 2005; Chen and Fuller 2007). Although a faster oxygen reduction reaction (ORR, i.e., more side reaction to generate H_2O_2) is suspected (Xie et al. 2005), it is unlikely to be the case as discussed above. This supposedly contradictory finding can be due to a water transport induced RH effect. Because of proton electro-osmotic drag of water molecules, high current density can cause dry anode conditions even when a large amount of water is generated on the cathode side (Fuller and Newman 1993; Springer et al. 1991). Such a low RH condition on the anode side will inevitably cause a higher degradation rate, as discussed earlier. Additionally, in situ micro-Raman experiments have revealed that overall water content is lower under load, which may be due to higher membrane temperature (Matic et al. 2005).

By comparison, the OCV condition for PEMELCs is the lowest potential point, while for PEMFCs, OCV shows the highest voltage. Since electrolyzers are usually well controlled under continuous operating conditions, the impact of OCV has not been extensively studied. It is reported that stacks on standby (0.34 mA cm^{-2}) last 50,000 h with 2,300 h of high-current-density operation, while stacks under continuous high-current-density operation (1,000 mA cm^{-2}) failed after 15,000 h (Stucki et al. 1998). The shorter lifetime during operation can be due to mechanism stress induced during gas generation.

3.2 Effect of Relative Humidity

It is generally accepted that lower RH leads to a higher membrane degradation rate (Mittal et al. 2006a; Inaba et al. 2006; Yu et al. 2005a). As can be seen in Fig. 2, the FRR decreases with higher humidification.

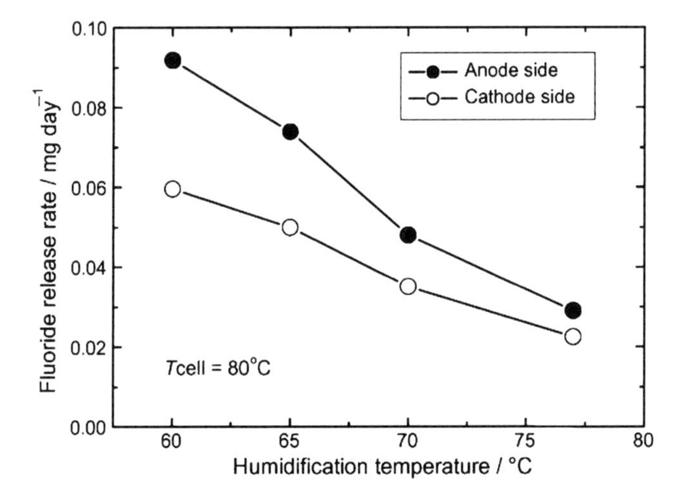

Fig. 2 Effects of humidification on FRR during an open-circuit voltage (OCV) durability test at 80°C, H_2/air 150 cm^3 min^{-1} at atmospheric pressure. Humidifier temperature from 60 to 77.5°C corresponds to RH from 42 to 88% (Inaba et al. 2006)

This observation seemingly contradicts the gas crossover mechanism discussed previously. Under dry conditions, the mechanical property improves compared with a fully humidified condition. The H_2 permeability does not strongly depend on RH at higher temperature (Inaba et al. 2006). It is believed that H_2 permeates through the hydrophobic interfacial backbone region (Ogumi et al. 1984, 1985). The permeability of O_2 is lower compared with that of wet conditions. Thus, the overall gas crossover rate should be lower under dry conditions, which in theory will cause less membrane degradation.

It must be noted that the partial pressure of reactant gases is higher under dry conditions when the total pressure is maintained owing to lower water vapor pressure. As will be discussed in the next section, higher gas pressure will lead to higher membrane degradation. But the chemical degradation is significantly higher than the amount of pressure increase predicts (Fig. 3).

The first mechanism proposed to explain the finding is that the concentration of H_2O_2 is higher under low RH conditions. It is argued that the boiling point of H_2O_2 (150.2°C) is significantly higher than that of H_2O, which leads to higher H_2O_2 concentration under dry conditions (Inaba et al. 2006). In an ex situ experiment employing a high concentration of trifluoromethanesulfonic acid simulating low RH condition, H_2O_2 yield versus the number of water molecules per sulfonic acid group was studied (Zhang and Mukerjee 2006). It found that the H_2O_2 yield increases under dry conditions. Thus, it is possible that the H_2O_2 concentration inside the membrane is higher at lower RH.

To verify the hypothesis, membrane degradation rate versus H_2O_2 concentration was studied without additional iron doping (Liu et al. 2005). The membrane was submerged in liquid, while the H_2O_2 concentration was varied. A semilinear relationship was found for H_2O_2 concentration from 1 to 10 ppm. Further increase of H_2O_2

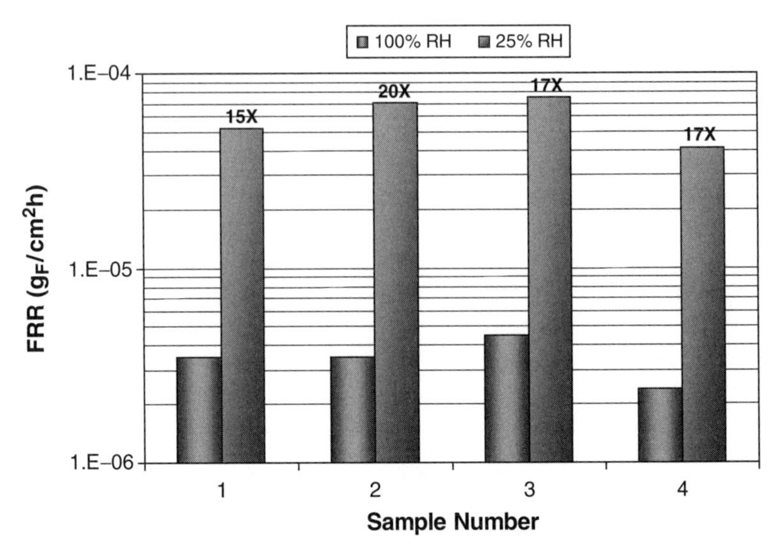

Fig. 3 Impact of RH on the chemical degradation rate of perfluorinated sulfonic acid (PFSA) membranes. All tests employed Nafion® 112 with 0.4 mg Pt cm^{-2} per electrode and were conducted with H_2/O_2 at 95°C, 300 kPa_{abs}, and 525 sccm gas flow (Liu et al. 2006)

concentration up to 10% does not significantly degrade the membrane faster. It is suggested that H_2O_2 self-quenching can convert the potent hydroxyl radicals to the less active hydroperoxyl radicals: $HO\bullet + H_2O_2 \rightarrow HO_2\bullet + H_2O$. Recently, such a mechanism has also been suggested to be the main mechanism for Fenton's reagent (Cipollini 2007). Since the overall membrane degradation rate measured in these tests is significantly less than the values observed in fuel cell testing, it is suggested that simply higher H_2O_2 concentration cannot directly cause the severe membrane chemical degradation under low RH conditions.

By direct measurement of the H_2O_2 concentration in the membrane with a chemical method, it is found that the H_2O_2 concentration in the membrane is in the range of approximately 6 μg cm^{-3} from 100 to 15% RH for full membrane electrode assemblies (MEAs) (Chen and Fuller 2007). This value agrees with findings of earlier studies on H_2O_2 concentration by electrochemical methods (Liu and Zuckerbrod 2005). There is no strong dependence of H_2O_2 concentration on RH conditions. With anode-only half MEAs (i.e., only the H_2 side of the MEA has a catalyst layer), it is found that the H_2O_2 release rate increases to approximately 2.5 times when the inlet RH on the anode side decreases from 100 to 25% RH. It must be noted that the anode/cathode RH may not be well maintained owing to the large membrane area (25 cm^2) and relatively low gas flow rate (500 sccm). The majority of the membrane should be exposed to a balanced RH of approximately 60% owing to water transport.

During in situ electron spin resonance (ESR) experiments, it was also found that the signal is much stronger when dry reactant gas is fed to the miniature fuel cell, compared with wet gases (Panchenko et al. 2004). Unfortunately, it is believed that the signal originated mainly from carbon and the catalyst since the concentration of organic radicals from chemical degradation can be too low to be detected by ESR. Nevertheless, it was demonstrated that more radicals can be present in MEAs when the environment is dry.

It must be stressed that it is the radical, not theH_2O_2, that can attack the membrane, especially when the membrane is in contact with water or under high humidification. The H_2O_2 can generate large amounts of radicals in the presence of certain transition metals, but itself is not able to react with the membrane (LaConti et al. 2003). Thus, larger numbers of radicals can directly cause more membrane degradation, while a higher concentration of H_2O_2 cannot, owing to self-quenching, discussed above.

Although H_2O_2 degrades the as-received membrane very slowly in a liquid (Liu et al. 2005; Mittal et al. 2006b), under dry conditions (12% RH) at elevated temperatures (120°C), severe membrane degradation can occur by exposing the membrane to H_2O_2 without a significant amount of Fe^{2+} (Endoh et al. 2007). Consequently, it is possible that H_2O_2 can attack the membrane more aggressively under lower RH conditions.

On the basis of theoretical calculations, hydroxyl radicals have a tendency to concentrate on the water/air interface owing to a 4.2 kJ mol^{-1} energy difference between the surface-absorbed state and the fully hydrated (bulk-solvated) state. Such an energy difference will cause the hydroxyl radical concentration on the interface to be more than one order of magnitude higher than the bulk concentration (Roeselova et al. 2004). Additionally, hydroxyl radicals can be stabilized by water solvation by approximately 16 kJ mol^{-1} (Autrey et al. 2004). Thus, it was proposed by Panchenko that the membrane was degraded faster under dry operating conditions owing to the loss of water stabilization for the hydroxyl radicals (Panchenko 2006).

Since PEMELCs are typically operated with liquid water flooding the O_2 anode or H_2 cathode side, the effect of RH is generally of less concern. Recently, a study at Giner indicated that local MEA or membrane area dry-out, due to uneven water distribution, may lead to premature stack failure. This is discussed further in the following sections.

3.3 *Effect of Partial Pressure of Reactant Gases*

Since H_2O_2/radicals are generated by reaction of reactant gases on a catalytic surface, the concentration of the reactant, i.e., the partial pressure of the reactant gases, should have a direct impact on the membrane degradation rate.

A linear relationship between the membrane degradation rate and reactant gas partial pressure is observed over a wide pressure range for both H_2 and O_2 gases under accelerated conditions (Fig. 4) (Liu et al. 2006). The FRR reaches a plateau after P_{O_2} exceeds approximately 40 kPa. The detailed mechanism of this behavior is not well understood. A similar finding was reported when the H_2 concentration was reduced by 5 times, from 100 to 20% H_2, the FRR dropped by 5 times, while the FRR increased by 3 times when P_{O_2} increased from 15 to 80 kPa (Mittal et al. 2006b).

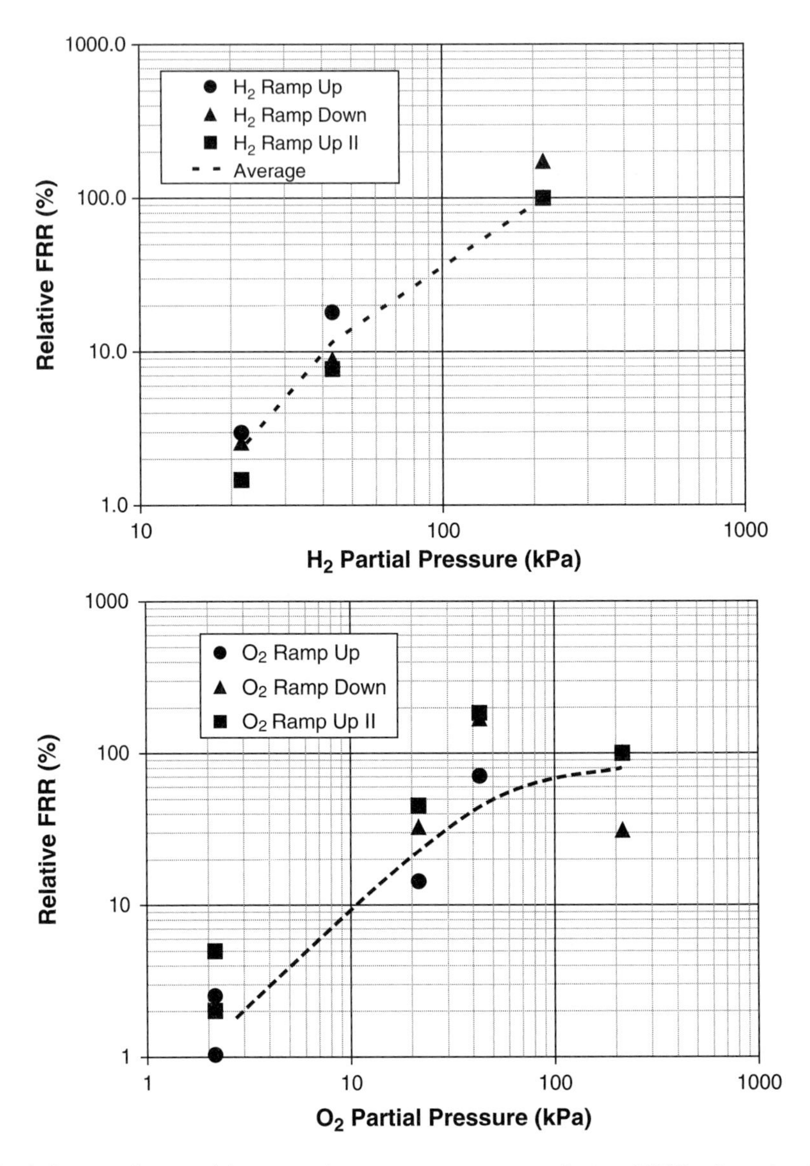

Fig. 4 Impact of gas partial pressure (counter gas pressure was kept at $215\,kPa_{abs}$) on chemical degradation rate of PFSA membranes. The *dashed line* is for reference only. All tests employed Nafion® 112 with $0.4\,mg\,Pt\,cm^{-2}$ per electrode and were conducted at 95°C, $300\,kPa_{abs}$, 100% RH, and 525 sccm gas flow

In a model experiment where a H_2/air mixture was bubbled into an aqueous suspension of Pt/C-Nafion, it was found that high H_2 partial pressure leads to very high FRR in a H_2-rich environment (Aoki et al. 2005). The FRR increases almost linearly with the partial pressure of the abundant gas for both H_2-rich and O_2 environments.

There are two abnormal cases reported regarding the effect of H_2 partial pressure. In one case, lower pressure of H_2 does not reduce the FRR under OCV conditions, while the FRR decreases with lower partial pressure of O_2 (Ohma et al. 2006). Another experiment shows that lower H_2 partial pressure does not reduce the FRR. It is proposed that H_2 can potentially be self-quenching, where radical species react with H_2 to form benign products (Mittal et al. 2006c). During a study employing a H_2/O_2 mixture in wastewater treatment by Pt/TiO_2, it was proposed that excessive H_2 can consume the hydroxyl radicals produced, which is exemplified by the fact that a high H_2 to O_2 ratio leads to less efficacy of waste decomposition (Chen et al. 2005). It is proposed that the adsorbed hydrogen atoms (H_s) can react with hydroxyl radicals and form water molecules. $H_s\text{–}Pt + \bullet OH \rightarrow H_2O$.

PEMELCs are usually designed for significantly higher operating pressure (above 3,450 kPa) than PEMFCs (below 210 kPa). PEMELCs manufactured by Giner can sustain a pressure of 20,000 kPa or higher. Under these high pressures, membrane mechanical failures are generally of more concern than chemical degradation. Membrane-treatment techniques have been developed at Giner to significantly limit the concentration of crossed-over gas (H_2 in O_2 and O_2 in H_2), as well as chemical degradation. Liquid water, O_2 anode feed, and electrolyzer stack lifetimes in excess of 30,000 h without incident have been demonstrated at approximately 1,450 mA cm^{-2}, with voltage decay rates of 2–3 μV h^{-1} and a FRR of less than 2×10^{-8} g cm^{-2} h^{-1} (LaConti et al. 2005).

3.4 *Effect of Operating Temperature*

During fuel cell development, high operating temperature is always preferred. The advantages of high operating temperature are (1) the cooling requirement is reduced, (2) cogeneration of heat and electricity for stationary application becomes more efficient, (3) the contamination problem is lessened, and (4) water management is easier (Knights et al. 2004; Mitov et al. 2006a).

Higher operating temperature will lead to more severe chemical degradation (Healy et al. 2005; Curtin et al. 2004). The kinetics of chemical reactions inevitably increases with temperature. Since membrane chemical degradation is highly irreversible, the degradation rate will increase with operating temperature.

Results from Fenton's tests conducted at different temperatures indicate that higher temperature leads to more severe membrane degradation (Chen et al. 2007a). The FRR increases 2 times when the temperature increases by 10°C.

Projected lifetimes of PEMFC stacks based on different membrane types are shown in Fig. 5. These results are based on early research at General Electric (GE) (LaConti et al. 1977). The FRR is found to increase by 7 times when the operating temperature is increased from 65 to 90°C (Preli 2005).

Similar to PEMFCs, higher temperature also accelerates membrane degradation in PEMELCs (Fig. 6). It must be stressed that the projected lifetime is based on the FRR. In some cases, the actual lifetime may be shortened owing to localized degradation, which will be discussed later.

The typical rate of fluoride release into protonically transported water for PEMELC (50–60°C, 670 kPa, 1.08 A cm^{-2}) is 0.005–0.020 μg h^{-1} cm^{-2}. The accelerating

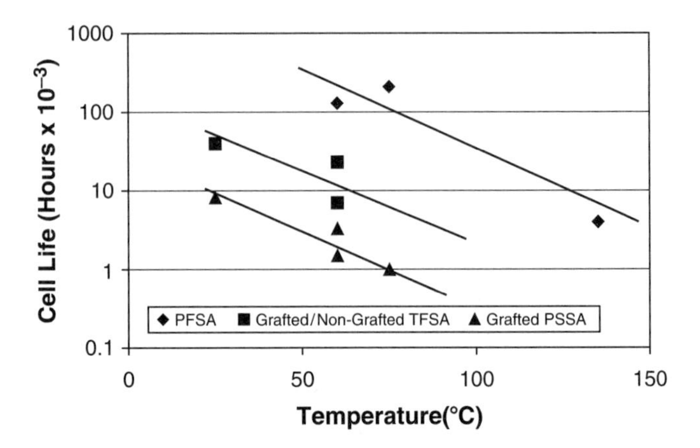

Fig. 5 Projected life capabilities for certain polymer electrolyte membrane (PEMs) in H_2/O_2 fuel cell stacks (210–480 kPa; 50–300 mA cm^{-2}; fully humidified; grafted polystyrene sulfonic acid (PSSA), grafted trifluoromethansulfonic acid (TFSA), and Nafion, (1,250 to 1,100 equivalent weight); nongrafted TFSA (400–600 equivalent weight); 175–300-μm PEM thickness; 4–8 mg Pt cm^{-2} each side) (LaConti et al. 2005)

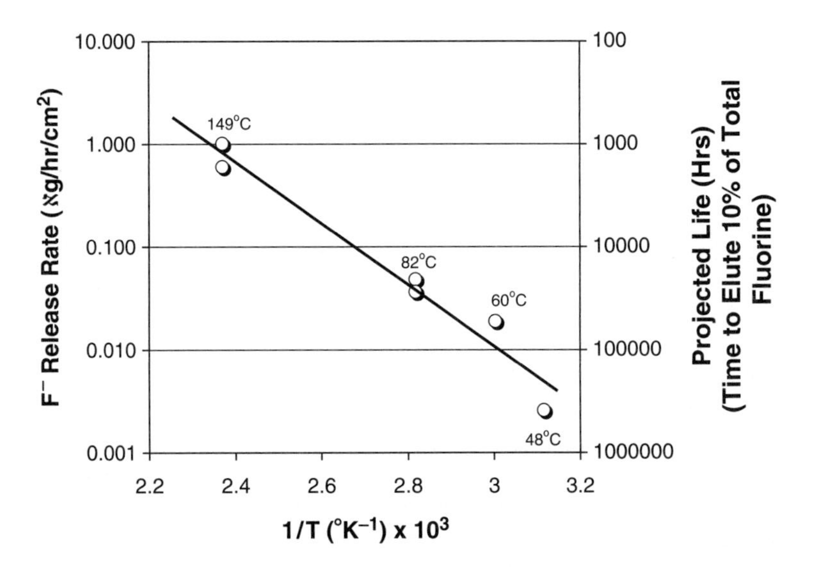

Fig. 6 Accelerating effect of temperature on FRR into protonically transported water (water electrochemically transported from the O_2 anode to the H_2 cathode side) at 1.08 A cm^{-2} for anode water feed H_2/O_2 PEM electrolysis cells with Nafion membrane electrode assemblies (MEAs) and 670-kPa balanced pressure (assumes projected end of useful life is approximately 10% loss of total fluorine) (LaConti et al. 2005)

effect on the PEMELC FRR caused by temperature increase is shown in Fig. 6. The FRR increases approximately by two orders of magnitude as the temperature is increased from 55 to 150°C.

4 Other Parameters Affecting Chemical Degradation

In addition to operating conditions, there are several other factors that have an impact on the membrane chemical degradation. Three major aspects, membrane thickness, catalyst type, and contamination, will be discussed here. In this section, general testing results are summarized. Extensive discussion on the mechanisms will be presented in later sections.

4.1 Membrane Thickness

To maximize performance, efficiency, and lower material cost, thinner membranes have been the focus of recent fuel cell development (Mathias et al. 2005). The impact of membrane thickness on membrane chemical degradation is, thus, of great importance.

When PFSA membranes are exposed to H_2O_2 without additional iron doping, the FRR is proportional to the membrane thickness (Liu et al. 2005). The chemical degradation rate of an Nafion 111 membrane is compared with that for a 10-mm-thick membrane fabricated by thermally laminating ten N111 membranes together. For a given concentration H_2O_2, the FRR is higher for the thicker membrane simply owing to the fact that there is more material to react with the H_2O_2.

In the PEMFC configuration, the FRR is governed by two conflicting factors (1) gas crossover rate will increase with a thinner membrane, which can cause more H_2O_2/radicals and worse degradation and (2) a thinner membrane has less material between anode/cathode structures to degrade, which can reduce the FRR.

Results from a series of degradation tests on grafted polystyrene sulfonic acid (*g*-PSSA) membranes have shown that thinner membranes have higher degradation rates than thicker ones (Gubler et al. 2005). Divinylbenzene cross-linked membranes have higher durability than non-cross-linked membranes owing to their lower gas crossover rate.

The effect of membrane thickness was evaluated in a series of fuel cell OCV tests (Mittal et al. 2006c). Two setups were used: setup A and setup B. The main difference between setup A and setup B is that setup B was a series of well-controlled tests with membrane thickness being the only variable parameter, while the samples in setup A also contained different catalysts. With PFSA membrane thickness increasing from 50 to 150 μm, the FRR also increases correspondingly under similar test conditions (Fig. 7). Further increase of membrane thickness leads to a lower FRR. The concentration of membrane-degrading species is not the only rate-limiting factor; a thicker membrane (i.e., higher reaction volume) also increases the FRR. It must be noted that the overall FRR change is fairly small; the error of sample to sample FRR variation is not clear.

From a practical point of view, the FRR during OCV fuel cell tests is more or less independent of membrane thickness. Therefore, if the membrane degradation rate were normalized by membrane weight, Fig. 7 would show that a 50-μm-thick

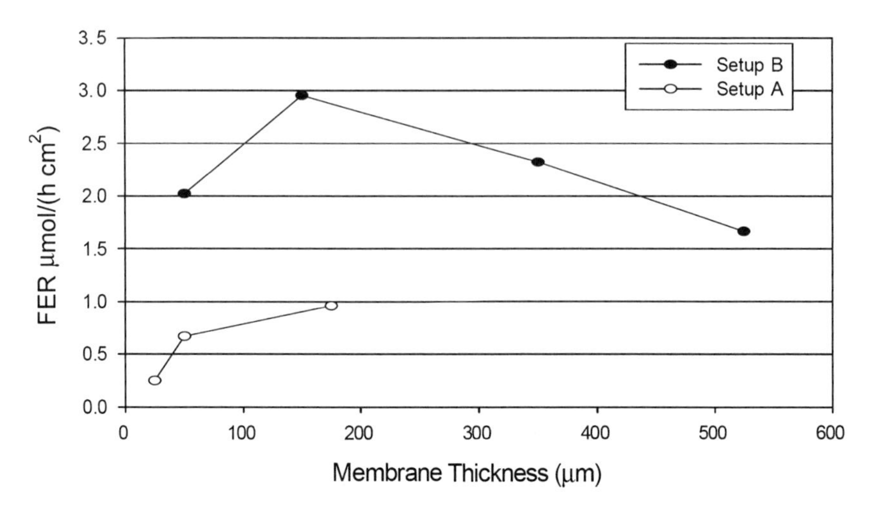

Fig. 7 FRR from Nafion membranes of different thickness in standard cells at 90°C, 30% RH, OCV conditions with H_2/O_2 as the reactant gases (Mittal et al. 2007)

membrane would degrade at approximately ten times the rate of a 500-µm-thick membrane. Consequently, if one considers that the useful membrane life in a fuel cell is limited to a loss of approximately 10% of the polymer in the membrane (Liu et al. 2001), a longer lifetime for thicker membranes can be readily projected.

4.2 Contaminant

All three major components of an MEA, membrane, catalyst, and diffusion medium, can be contaminated by foreign species during PEMFC operation. Contamination of the diffusion medium usually lowers its hydrophobicity, which leads to water transport problems and can potentially create a continuous liquid path from the membrane to the gas flow channel. Foreign ionic contaminants can directly migrate to the membrane via this liquid path. Consequently, faster contamination may occur under this condition. Since diffusion medium contamination does not directly cause more severe membrane degradation, the discussion of the contamination issue will focus on two aspects: membrane contamination and catalyst contamination.

4.2.1 Membrane Contamination

As discussed earlier, H_2O_2 is one of the main reactants for membrane chemical degradation. When the membrane is contaminated by metal ions, these ions can catalyze H_2O_2 decomposition and generate radicals. It is important to evaluate which metal ions can effectively convert H_2O_2 to radicals, i.e., which are Fenton-active.

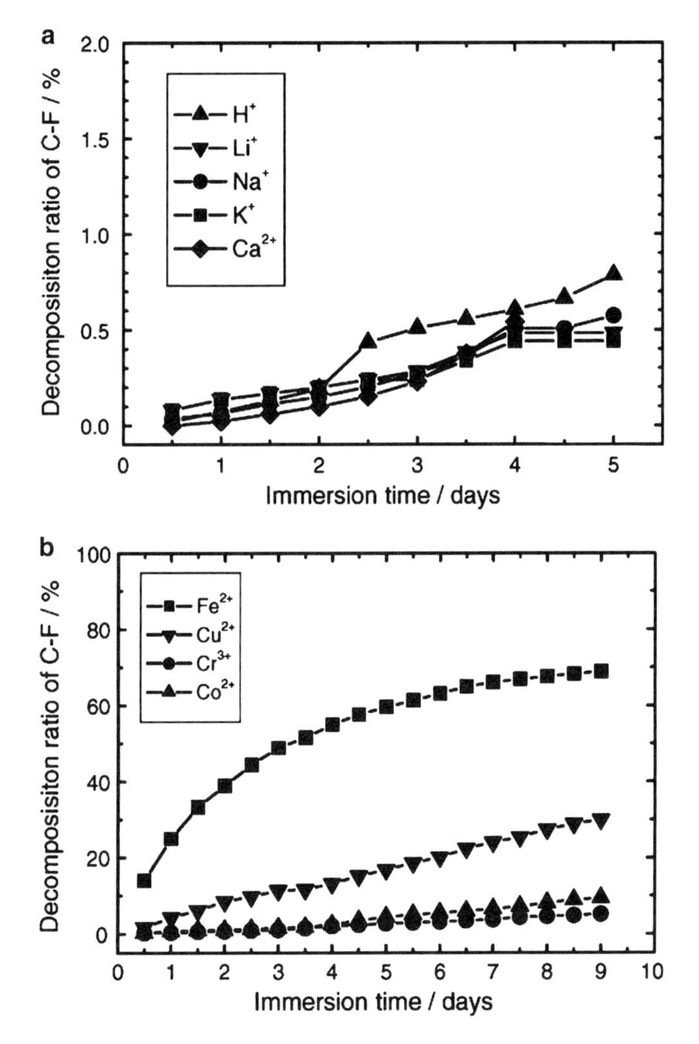

Fig. 8 Fluoride loss of fully cation exchanged Nafion 117 membranes in 30% H_2O_2 at 80°C (Kinumoto et al. 2006)

A series of PFSA membranes were doped with various common metal ions and tested for chemical stability against H_2O_2 solution (Kinumoto et al. 2006). It was found that alkali and alkaline metal ions do not accelerate membrane degradation in ex situ Fenton-style H_2O_2 tests (Fig. 8). The membrane degradation rate resembles that for protonated membranes even when the membranes are fully exchanged with the metal ions. On the other hand, certain transitional metal ions, such as iron and copper, will drastically increase the membrane degradation, while cobalt and chromium ions do not play a significant role. The iron-doped membrane degrades faster than copper-based membranes.

Table 2 Fluoride release rate (*FRR*) from Nafion 117 membranes, in different ionic forms, in a cathode-only cell at 90°C, 30% relative humidity, with H_2/O_2 as reactant gases (Mittal et al. 2007)

Nafion 117 counterion	FRR (μmol h^{-1} cm^{-2})
H^+	2.616
Na^+	0.0114
Cs^+	0.0155
Li^+	0.0074

As shown in Table 2, when the membrane is fully exchanged to non-Fenton-active alkali metals (Li, Na, and Cs), the FRR is drastically reduced (Mittal et al. 2006c). The relationship between percentage ion exchange and the FRR was not investigated. Since these metal ions are not Fenton-active, the protecting behavior will likely diminish when the exchange percentage is significantly less than 100.

Under normal PEMFC operation conditions (<95°C), thermal degradation of membrane ionomer is generally of little concern. Sulfonate groups in PFSA ionomer decompose at approximately 280–300°C (St Pierre et al. 2000). Thermal decomposition of membranes by local combustion can become a potential pathway leading to membrane failures, especially when a pinhole is formed (Aoki et al. 2006a). By exchange of the proton for potassium ion, the thermal stability of Nafion is enhanced (Wilkie et al. 1991). So a heavily contaminated MEA can potentially slow down catastrophic failures. On the other hand, the required contamination level can be so high that the conductivity suffers to an unacceptable level.

Studies have been conducted with PEMELCs to identify the effect on performance and chemical degradation of certain cation contaminants. Generally, the same ions, such as Fe^{2+}, that may accelerate membrane chemical degradation in PEMFCs are also accelerants for PEMELCs. Contaminant ions such as Na^+, Mg^{2+}, Ca^{2+}, and Al^{3+} appear to have minimal effect on accelerating membrane chemical degradation in PEMELCs. Contaminant metal ions such as Fe^{2+}, Fe^{3+}, and Cu^{2+}, during PEMELC operation, are electrochemically transported to the membrane/cathode interface and generally deposit in the cathode structure as metal or metal oxides. However, during standby or inadvertent shutdown periods, they can redissolve into the membrane and accelerate membrane chemical degradation at OCV conditions, with H_2 and O_2 crossover. During PEMELC operation, contaminant ions such as amines are electrochemically transported to the membrane/cathode interface. They can "poison" the cathode catalyst, as well as decrease local membrane or ionomer water content, leading to enhanced peroxide/radical formation and chemical degradation.

Typically, when the electrolyzer MEA is contaminated with a metal ion such as Na^+ and the current (1.08 A cm^{-2}) is applied to the electrolyzer stack, there is a voltage spike that decreases with increasing operating time, depending on the degree of contamination. The pH of the protonically transported water that is discharged on the cathode side is typically alkaline (NaOH). By discarding the transported water or passing it through a deionizer prior to recycling, one can restore the electrolyzer cell performance by operational in situ self-regeneration (electrolysis only) to approximately the same level as originally measured prior to contamination. The voltage spike is attributed to a localized pH, as well as resistance effect. With divalent ions such as Ca^{2+}, there appears to be electrochemical transport and some deposition of calcium species at the membrane/cathode interface and it is difficult to perform an operational in situ self-regeneration. With more complex multivalent ions such as Fe^{2+}/Fe^{3+}, the ions are electrochemically transported to the membrane/cathode interface and reduced iron oxide species are found in the cathode catalyst layer and compartment, making in situ self-regeneration very difficult.

PEMELCs have also been occasionally contaminated by organic species such as degradation products from ion-exchange deionizer beads used to purify the circulating feed water. The amine degradation products from the quaternary ammonium ion exchangers are electrochemically transported to the membrane/cathode interface and are very difficult to remove by in situ self-regeneration. The amines locally decrease membrane and ionomer water content, as well as "poison" the cathode structure. Some success in removing amine contaminants has been achieved by purposely exchanging some Na^+ ion into the MEAs of the amine-contaminated stack and performing an "in situ" self-regeneration. The alkaline NaOH solution generated at the cathode helps remove the amine from the membrane/cathode interface and the contaminants are removed from the system by discarding the cathode water effluent or ion exchange.

The same regeneration concepts, as well as the limitations, as described above, could be applied to a PEMFC stack, providing the PEMFC is operated as an electrochemical hydrogen pump. In this mode of operation, water is circulated over the cathode side. The cell is electrochemically driven, similar to an electrolyzer. Hydrogen is consumed at the anode with production of protons and the protons are electrochemically transported to the cathode to form hydrogen gas. The contaminant positive ions are also electrochemically driven to the cathode for possible removal, as described above for the PEMELC.

4.2.2 Catalyst Contamination

In the hydrogen adsorption region, where the platinum surface is covered by underpotentially deposited hydrogen (H_{upd}), the presence of Cl^- ion leads to a significant amount of oxygen two-electron reduction, i.e., H_2O_2 production (Stamenkovic et al. 2001). At potential less than 0.2 V versus the RHE, the H_2O_2 yield seems to be independent of chloride concentration, indicating similar chloride coverage (Schmidt et al. 2001). Above 0.2 V, the yield increases with higher chloride concentration. High-concentration H_2O_2 can cause higher membrane degradation.

Sulfur and selenium also show a similar effect. It is well known that sulfur is a poison for O_2 reduction on a platinum surface owing to strong adsorption. Sulfur contamination cannot only hinder the ORR, but can also change the oxygen reduction pathway. When a platinum surface is modified by sulfur or selenium, the faradic efficiency of H_2O_2 generation can be as high as 100% (Mo and Scherson 2003).

With less than 10% proton-exchanged Na^+/Ca^{2+}, the moderately contaminated Nafion-coated platinum electrode shows significant reduction of the ORR activity

(Okada et al. 1999). In contrast, 0.1 N H_2SO_4 containing as much as 0.01 N impurity does not have any appreciable impact. When the Nafion-filmed platinum electrode is contaminated, the charge transfer kinetics deteriorates (60–70%) significantly to a level lower than the oxygen transport rate (20–30%) (Okada et al. 2003). It is believed that the ORR degradation is connected to the Nafion/Pt interface, instead of Nafion bulk properties. There is no evidence of changing the oxygen reduction mechanism. Consequently, cation contamination reduces the ORR activity but does not promote H_2O_2 generation.

4.3 Catalyst

Platinum has been widely used for PEMFCs owing to its excellent oxygen reduction catalytic activity. Platinum is usually implemented in the form of Pt/C catalysts because of its significantly higher surface area compared with that of platinum black catalysts. Furthermore, platinum alloy catalysts are also employed to enhance oxygen reduction activities. Pt black, Pt/C, and Pt/C alloys will be discussed in this section.

As mentioned previously, carbon black has a higher H_2O_2 yield than a Pt/C catalyst at a potential below 0.54 V versus the RHE (Antoine and Durand 2000). A so-called *electro-Fenton* process, where cathodic reduction of oxygen on a graphite electrode generates H_2O_2, has been studied for oxidization of various waste species (Da Pozza et al. 2005). The presence of carbon, compared with a pure platinum surface, can potentially have a negative impact on membrane stability. A projection can be made that MEAs based on Pt/C should have a higher degradation rate than MEAs based on platinum black.

One fuel cell study revealed that platinum black actually can cause a higher FRR than Pt/C under 100% RH conditions (Mittal et al. 2006a). At 30% RH, the FRR drops to the same level as for Pt/C. This behavior may be an artifact of catalyst poisoning (Fig. 9). Commercially available platinum black may have some chloride adsorbed on the platinum surface. As discussed earlier, high H_2O_2 yield is observed when platinum is contaminated by chloride. It is found that MEAs based on purified platinum black (50 wt% MeOH aqueous solution at 60°C for 1 h) have a significantly lower FRR than Pt/C, while as-received platinum black causes a much higher FRR (Endoh et al. 2007).

When the carbon functional groups on the Pt/C catalyst are neutralized by Na_2CO_3, NaOH, or EtONa, the FRR drops significantly (Tanuma and Terazono 2006). Such an effect is only observed for an anode catalyst; there is no impact of ion exchange for a cathode catalyst. Since it is found that carbon black has a lower H_2O_2 yield than a Pt/C catalyst at a potential >0.54 V, the property of the carbon does not influence membrane degradation on the cathode side. It is also found that a heat-treated catalyst, which has fewer surface groups than a nontreated catalyst, increases the lifetime of the MEA from 100 to 700 h.

In membrane durability tests, a PtCo/C alloy catalyst significantly improves the lifetime of MEAs (Miyake et al. 2005). Under OCV and 100% RH conditions,

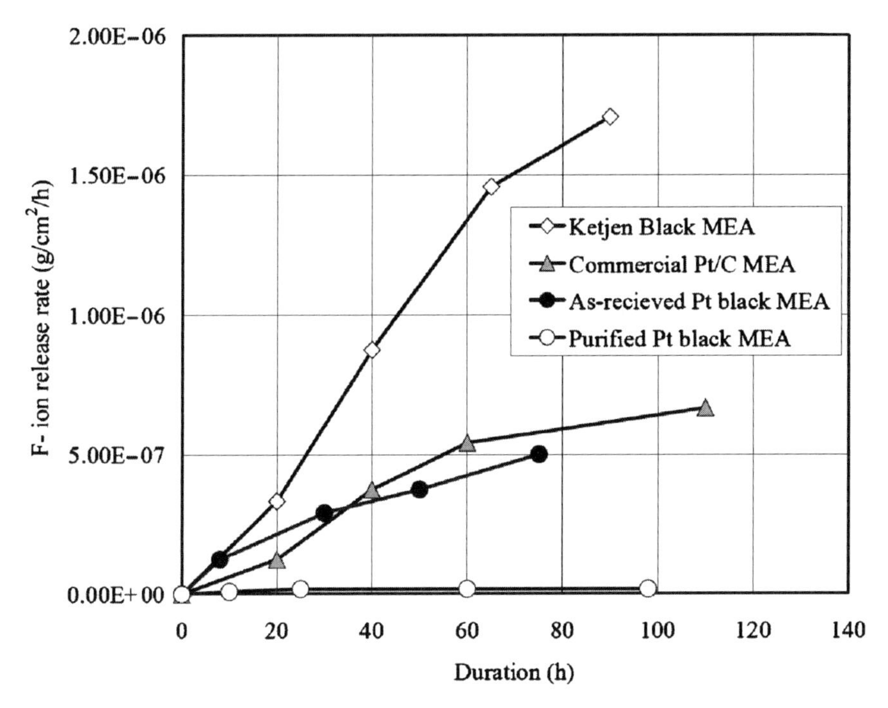

Fig. 9 FRR behavior of MEAs based on different catalysts under OCV conditions. Cell temperature, 80°C; H_2, 100% RH; air, 0% RH; cell pressure, ambient; membrane, Flemion® SH50; platinum black MEA, platinum black loading 6 mg cm^{-2} for both electrodes; commercial Pt/C MEA, platinum loading 0.5 mg cm^{-2} for both electrodes; Ketjenblack MEA, Ketjenblack loading 0.6 mg cm^{-2} for both electrodes (Endoh et al. 2007)

PtCo/C-based MEAs have a FRR comparable to that of Pt/C-based MEAs (Mittal et al. 2006a). When the RH drops to 30%, the FRR from PtCo/C is approximately 50% of that of the Pt/C FRR. The effect of cobalt alloying on membrane degradation is RH-sensitive.

The RH sensitivity of PtCo/C is also demonstrated in RRDE tests. To conduct RRDE experiments under simulated dry conditions, high-concentration trifluoromethanesulfonic acid (CF_3SO_3H) was used. The amount of peroxide formed on PtCo/C in 1 M (50 water molecules per acid) and 6 M (four water molecules per acid) CF_3SO_3H is similar but significantly lower than on Pt/C in 6 M CF_3SO_3H (Murthi et al. 2004). Both Pt/C and PtFe/C show more peroxide formation than PtCo/C in 6 M CF_3SO_3H. Above 0.75 V versus the RHE, no detectable H_2O_2 can be observed (Zhang and Mukerjee 2006).

In an extensive study on various noble catalysts using a RRDE in dilute perchloric acid (0.1 M $HClO_4$), platinum (bulk, polycrystalline), Pt_3Co, Pt/C, and Pt_3Co/C have similar H_2O_2 yield (Paulus et al. 2002). Owing to the dilute concentration of the acid, these experiments simulate H_2O_2 yield under extremely wet conditions.

Although a RRDE can effectively evaluate the electrochemical yield of H_2O_2, it is not a direct analogy to gas crossover reactions, where H_2 and O_2 directly react on a catalytic surface. A gas mixture bubbling test was designed to simulate gas crossover reaction (Endoh 2006). In these tests, air with 1.3% hydrogen mixed gas is bubbled into an aqueous suspension of catalyst or carbon and H_2O_2 concentration is measured. When Ketjenblack is exposed to the gas mixture, H_2O_2 can be detected. On the basis of a similar test configuration, PFSA degradation can be detected when Pt/C is used (Aoki et al. 2005).

5 Localized Degradation

As will be discussed extensively in the following sections, results from numerous studies suggest that membrane chemical degradation can be localized (i.e., PEMFC or PEMELC anode or cathode side, internal membrane, fluid inlets and outlets). Also, PFSA ionomer is typically used as a binder within the catalyst layer and provides a pathway for transporting protons from the catalyst to the membrane. Although the FRR can be used to monitor the membrane degradation rate, there is no direct relationship in PEMFCs between total fluoride loss and membrane lifetime as measured by hydrogen crossover rate (Liu et al. 2001). Thermal decomposition of the membrane by local combustion, especially when a pinhole is formed, is a potential pathway leading to accelerated membrane failures (Aoki et al. 2006a).

Detailed mechanisms of pinhole formation, resulting from both mechanical and chemical degradation, can be very complicated. In this section, the discussion will focus on several aspects of membrane chemical degradation.

5.1 Anode Versus Cathode

In PEMFCs and PEMELCs, anode and cathode are in drastically different chemical environments. In a PEMFC, the H_2 side is highly reductive, while strong oxidative potential is maintained on the cathode side, especially at OCV. Electrochemical potential can also influence H_2O_2 production as discussed before. Extensive study in this area has been conducted since the early development of PEMFCs.

A summary of recent studies on the anode versus cathode degradation for PEMFCs is shown in Table 3. All three scenarios, from (1) anode degradation less than cathode degradation, (2) uniform degradation to (3) dominating anode degradation, have been observed by different studies.

For the two studies that show more severe cathode degradation, the PEMFC was tested under load under fully humidified conditions (Pozio et al. 2003; Cleghorn et al. 2006; Yu et al. 2003). If it can be assumed that H_2O_2 is generated on the anode side, electro-osmotic drag of water can carry the H_2O_2 to the cathode side, which leads to more severe cathode degradation. In addition, impurities inside the membrane can also accumulate near the cathode under external load.

Table 3 Summary of anode versus cathode membrane degradation results

Observation	Membrane	Catalyst	Gas	Flow rate	Pressure	Temperature (°C)	Relative humidity	Current density	Source
Anode degradation less than cathode	35 (Gore)	PtRu/C 0.45 Anode 0.6 Cathode	H_2/air[a]		Ambient	70	100%	800	Cleghorn et al. (2006)
Low sulfur content on cathode	160 (PSSA)	20% Pt/C 0.4	H_2/O_2	4.5/7.5 stoichiometry	200 kPa	80	100%	300	Yu et al. (2003)
Cathode F concentration in affluent water is higher	Nafion 115	24.5% Pt/C, 0.34	H_2/O_2	–	–	25–70	Humidified	10–150	Pozio et al. (2003)
$FRR_A = FRR_C$ $FRR_{Full} < 0.5\ FRR_{Single}$	–	–	H_2/O_2		–		Humidified	OCV	Takeshita et al. (2005)
$R_A = FRR_C$	Nafion 117	46% Pt/C (Tanaka) 0.5	H_2/O_2	200/200 $cm^3\ min^{-1}$	Ambient	90	30%	OCV	Mittal et al. (2006a)
Anode equals cathode	165 µm PVDF-*g*-PSSA	–	–	–	–	–	–	–	Mattsson et al. (2000)
Uniform degradation	Nafion 117	20% Pt/C, 0.2 anode, 0.4 cathode	H_2/air	–	Ambient	60	<4%	–	Luo et al. (2006)
$FRR_A > FRR_C$	30 µm	Pt/C 0.45 anode, 0.4 cathode	H_2/air	150/150 ml min^{-1}	Ambient	80	42–82%	OCV	Inaba et al. (2006)
Anode degrades faster	50 (Aciplex™, SF1101 2 layers)	Pt/C, 0.5	H_2/O_2	50/50 ml s^{-1}	Ambient	76	100%	OCV	Shim et al. (2007)
Anode degrades faster	50 (Flemion® SH50)	Pt/C, 0.5	H_2/air	–	Ambient	90	16%	OCV	Endoh et al. (2007)

OCV open-circuit voltage, *FRR* fluoride release rate, *PVDF* polyvinylidene difluoride, *PSSA* polystyrene sulfonic acid

[a]For a short time the gas composition is 4% air bleed in 43% N_2 17% CO_2 50 ppm CO, and H_2 balance

To further verify that cathode degradation is faster than anode degradation, a Nafion 101 membrane was laminated on the cathode side of the PSSA membrane (Yu et al. 2003). It was found that membrane stability was significantly increased. Similar results were observed when the Nafion 101 membrane was replaced by a 15-μm recast perfluorinated membrane.

In contrast to the previous finding, two investigations concluded that the anode actually degrades faster than the cathode (Inaba et al. 2006; Shim et al. 2007). Compared with the four other studies that show uniform degradation, there is no significant difference in experimental conditions and materials. A possible explanation is that the membranes were thinner in the studies where more severe anode degradation was observed. If anode degradation is highly sensitive to O_2 crossover rate, then a higher O_2 permeation will cause more serious anode degradation. Also, it may be possible that the anode side is operating relatively dry and H_2O_2/radicals and product water may migrate to the anode.

Similar experiments were also conducted in a PEMELC stack where five membranes were stacked together instead of a single membrane. The membranes on the cathode (H_2) side degraded more severely than the membranes on the anode (O_2) side (Scherer 1990). This observation holds true for both the grafted sulfonated polystyrene PTFE-(SPS) membranes and Nafion membranes. Since protons migrate from the O_2 side to the H_2 side in PEMELCs, electro-osmotic drag of water also occurs in the same direction. Similar to the argument above, such water drag or migration will create higher concentration H_2O_2 on the H_2 side. Additionally, the oxidized Ru-Ir catalyst on the O_2 side in the PEMELCs is under very high potential. The surface of the Ru-Ir anode catalyst should be in oxide form, which can suppress H_2/O_2 direct combination reaction.

5.2 Platinum Precipitation Line

Although platinum is very stable in high oxidative conditions even with the presence of strong acid, platinum dissolution and migration have been observed in both PEMFCs and PEMELCs (Ferreira et al. 2005). Theoretical models have been proposed to predict the location of the precipitated platinum (Burlatsky et al. 2005). The location of the platinum has been successfully verified by varying reactant gas partial pressures (Zhang et al. 2007).

Localized membrane degradation has been observed at the location of the platinum line for PEMFCs (Escobedo 2006). As shown in Fig. 10, membrane void, cracks, and delamination can be observed.

On the basis of results from micro-Raman spectroscopy, severe side degradation is found around the platimum redeposition line (Ohma et al. 2007a). The sulfate ion release rate on the cathode side is more than that on the anode side. A similar trend is also observed for the FRR. The results strongly support the role of the platinum line in membrane chemical degradation.

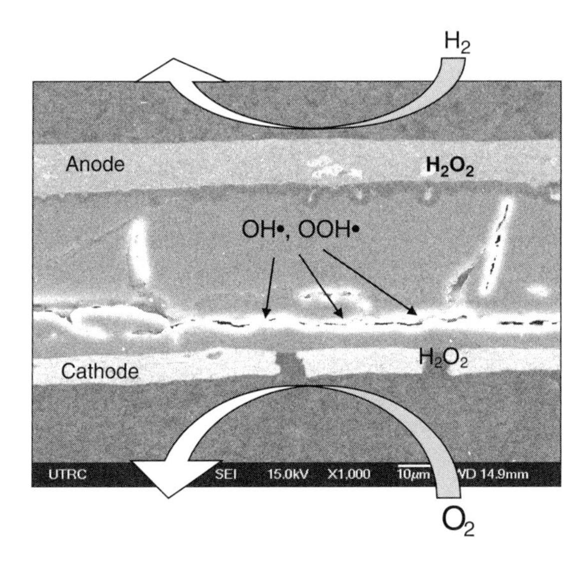

Fig. 10 Localized degradation is observed around the platinum precipitation line(Escobedo 2006)

In similar research, micro-IR results suggest the anode side sustained worse degradation than the cathode side under OCV conditions (Endoh et al. 2007). Since the platinum band is located on the cathode side, it was concluded by the authors that platinum redeposition will not accelerate membrane degradation. An alternative hypothesis can be that the role of the platinum line in chemical degradation does not dominate until the platinum density in the line becomes high. In the initial durability test of the MEAs, degradation species from the anode/cathode electrodes play an important role in membrane chemical degradation. As the test proceeds, more platinum will precipitate along the platinum line; the role of the platinum line becomes a more significant contributor to degradation. Further precipitation of platinum will render the platinum line so dense that all the crossed-over gas will react stoichiometrically on the platinum, which makes the platinum line the main source of degradation.

It was also argued by the authors that when platinum is dispersed evenly throughout the ionomer membrane, the FRR reduces significantly, i.e., platinum inside the membrane should alleviate membrane degradation (Endoh et al. 2007). This "platinized membrane" process has been widely used for PEMELCs to cut down the gas crossover rate and FRR. The benefit of such practice has been well proven. There is a significant difference between the thin platinum precipitation line and a platinized membrane. In case of the platinum line, the H_2O_2/radicals formed on the platinum line can readily diffuse into the bulk nonplatinized membrane. Once the H_2O_2/radicals are in the membrane, there is no quenching agent until they reach either electrode. With the platinized membrane, the platinum nanoparticles are distributed uniformly inside the membrane without changing water content, proton conductivity, or electrochemical transport properties of the membrane. Platinum can act as a source for H_2O_2/radicals, but at the same time it is also an excellent decomposition catalyst. Thus, H_2O_2/radicals formed on a platinum particle will very likely be decomposed by an adjacent platinum particle. From RRDE experiments, it was found that higher loading of platinum leads to a smaller distance between Pt/C agglomerates and lower H_2O_2 yield owing to H_2O_2 decomposition by Pt/C (Inaba et al. 2004).

On the basis of mathematical modeling, it is proposed that platinum oxide can insulate the platinum particle and reduces platinum dissolution (Darling and Meyers 2005). Platinum is more vulnerable to dissolution in the potential region of 0.87–1.2 V, which corresponds to PEMFC OCV conditions.

Platinum particle growth may be also enhanced in the presence of liquids by lowering the activation energy of particle growth (Liu et al. 2004). A similar effect of RH and particle size increase is also observed. It is proposed that platinum line formation will be faster under highly humidified conditions.

5.3 *Inlets and Edges*

During normal operation of PEMFCs, the local conditions, such as RH and gas partial pressure, vary from the inlet to the outlet. Several reports have identified the gas inlet area as more vulnerable to membrane degradation compared with the bulk membrane.

A long-term durability study was conducted on Nafion 112 MEAs, with the following conditions: H_2/air counter flow, 80°C, 65.8% RH, ambient pressure (Yu et al. 2005b). The cell was under a load of 300 mA cm^{-2} for most of the 2,500-h test. The anode inlet region showed the most severe membrane thinning.

With distributed current collection hardware, current distribution during cell operation can be evaluated (Yoshioka et al. 2005). It is found that when the inlet RH is low, the highest current density is observed near the gas outlet, where the humidity is relatively high. Thus, the membrane chemical degradation can be accelerated in the gas inlet region since lower RH usually accelerates degradation.

During normal stack operation, the inlet region is usually drier than the outlet region (Knights et al. 2004). High oxygen concentration in the inlet region, in combination with high temperature owing to less water for heat removal, will accelerate degradation in this area.

An extensive study of localized degradation along catalyst layer edges for PEMFCs was reported recently (Sompalli et al. 2007). In both segmented and unsegmented electrodes, it was found that membrane failures concentrated along the perimeter regions where only one electrode (either cathode or anode) exists. An anode overlap configuration is preferred at 100% RH, while the advantage diminishes under low RH conditions. Anode overlap generally shows high uniform degradation and cathode overlap results in a higher incident of pinholes.

Studies at General Electric with liquid water, anode feed Nafion 117 PEMELC have shown, under certain operating conditions of poor water feed distribution, accelerated membrane chemical degradation can occur, especially at the MEA inlet, outlet, and edge areas (Chludzinski 1982; LaConti et al. 2005). For the General Electric commercial molded graphite electrolyzers, occasionally the molding process was not under control and yielded bipolar plates with deformed fluid ribs and grooves.

The inadvertent poor channel geometry has two adverse effects on the PEMELCs. The first is to lower the overall water flow rate, with resultant increase in operating temperature. The second is to drive the two-phase (O_2/H_2O) flow out of the desirable slug or bubbly region into the annular region. In the annular region, the flow consists of a gas core surrounded by a film of water. Less water is directly available at the porous plate for reaction and for cooling. Heat must be transported through a thin liquid film and a gas core before it reaches the remainder of the water. This could result in higher localized operating temperature of the MEA. Accelerated membrane degradation was found to occur at the catalyst layer edge and fluid port areas of the MEAs, with some loss of material and internal blistering of the membrane near the H_2 cathode side. A substantial amount of fluoride ions was found in the fluid within the blistered area. The blistering of the membrane near the H_2 cathode side at the catalyst layer edge areas is likely caused by migration and concentration of the radical/H_2O_2 species in these areas. This is aggravated by electro-osmotic transport of water, along with radical/H_2O_2 species from the O_2 anode to the H_2 cathode and the relatively dry operational conditions at the catalyst layer edge areas.

At Giner several PEMELC approaches have been developed to maximize fluid flow distribution and avoid dry inlet, outlet, and edge areas, as well as to minimize active catalyst or electrode layer overlap in these critical areas.

5.4 *Ionomer Binder Inside the Catalyst Layer*

The ionomer binder used inside the catalyst layer plays a crucial role in PEMFC performance (Makharia et al. 2005). It forms a pathway for protons to reach the platinum nanoparticles supported on carbon black. Ionomer degradation in the catalyst layer will cause high ionic resistance and poor performance.

It is suggested that the Nafion ionomers inside the catalyst layer may degrade faster than the bulk membrane owing to greater mechanical integrity of the membrane and the fact that the ionomers are located adjacent to the catalytic surface where H_2O_2/radicals are formed (Xie et al. 2005). This hypothesis was discussed by the authors, but there is no strong supporting experimental evidence.

Results from a long-term durability test of a hydrocarbon membrane show behavior opposite to that predicted. The sulfonated polyimide (SPI) membrane demonstrated 5,000-h lifetime when tested at 0.2 A cm^{-2}, 80°C, and ambient pressure (Aoki et al. 2006a). Even if there is Nafion PFSA ionomer in the electrode, the SPI cell does not show any fluoride release, indicating that ionomer degradation in the catalyst layer is minimal. It must be stressed that SPI is 40 times less permeable to

H_2 and 10 times less permeable to O_2 compared with Nafion. The lack of ionomer degradation in the catalyst layer may be due to less gas crossover.

In ex situ experiments where a H_2/air mixture was used to simulate crossed-over gas, it was found that the ionomer decomposition is suppressed greatly by incorporating platinum particles in the ionomer mixture (Aoki et al. 2006c). It was also found that higher RH promotes an increased FRR when other conditions are kept constant. It is suggested that ionomer degradation in the catalyst layer can be more severe under wet conditions. Such behavior is in direct contrast to the membrane degradation during PEMFC operation. The authors suggested that higher RH may enhance permeation of H_2O_2/radicals from the ion cluster region of the PFSA, which is in the vicinity of the platinum, to the hydrophobic region, where no platinum exists.

One possible mechanism that explains the low degradation rate of ionomer inside the catalyst layer is that the crossed-over gas reacts with the abundant gas on the platinum surface. Owing to high reactivity of platinum towards H_2/O_2 recombination, the crossed-over gas can be completely consumed by the catalyst on the membrane/catalyst interface. The ionomer in the bulk of the catalyst layer is thus protected against H_2O_2/radicals. The overall concentration of H_2O_2/radicals inside the catalyst layer can be less than in the bulk of the membrane.

6 Characterization Techniques

A wide range of analytic techniques have been employed to characterize MEAs after a long-term durability test, accelerated degradation test, or ex situ model experiment. Certain characterization techniques, such as viscosity measurement, high-pressure liquid chromatography, and gel permeation chromatography are applicable only to hydrocarbon membranes. These tests are well documented in conventional analytical and polymer textbooks and will be omitted from this discussion. Other widely used routine characterizations, such as membrane thickness, ionic conductivity, water uptake, and dimensional change, will also be skipped. Mass spectroscopy was used to characterize PFSA degraded product (Healy et al. 2005). Further discussion can be found in the referenced paper. This section offers an overview of various characterization techniques for membrane chemical degradation; certain findings based on the characterization, detailed experimental conditions, and analysis can be found in the references therein.

6.1 Fourier Transfer IR Spectroscopy

Fourier transfer IR (FTIR) spectroscopy has been widely used to characterize functional groups in polymers and organic compounds. Membrane chemical degradation, which usually introduces new functional groups, can be readily monitored by FTIR. Furthermore, micro-FTIR can focus the IR beam on a very small area, which offers spatial information in addition to composition information. FTIR has been successfully applied to various membrane types, from PFSA (Kinumoto et al. 2006) and PSSA (Wang and Capuano 1998) to hydrocarbon membranes (Zhang and Mukerjee 2006).

For an FEP-*g*-PSSA membrane (where FEP is fluorinated ethylene propylene), absorption bands at 1,411 and 1,452 cm^{-1} were assigned to aromatic skeleton stretching vibrations. The ratio of the absorption, before and after fuel cell tests, is used to evaluate the remaining degree of grafting (Büchi et al. 1995). The degree of cross-linking can be calculated by the ratio of the 1,510-cm^{-1} band (aromatic skeleton stretching of divinylbenzene only) and the 1,602-cm^{-1} band (aromatic skeleton stretching for all aromatic rings). The peak at 1,039 cm^{-1} is assigned to C–H aromatic in-plane vibrations, while the peak at 982 cm^{-1} is assigned to S=O symmetrical stretching vibration. C–F stretching of FEP CF_3 groups contributes to the peak at 982 cm^{-1}. Dry membrane fully exchanged in potassium form is used to minimize water interference.

In a study on PSSA degradation in fuel cell applications, it was found that peaks at 1,037 cm^{-1} (S=O symmetrical stretching) and 1,008 cm^{-1} (aromatic in-plane vibration) diminish after a durability test, especially on the cathode side (Yu et al. 2003). Both the aromatic rings and the SO_3^- groups are lost after the durability test. On the basis of IR results, the hydroxyl radicals attack the polystyrene via addition to the phenyl groups (Weir 1978).

FTIR has also been used to study the degradation of PFSA (Delaney and Liu 2007). The membrane is in dry potassium form when FTIR experiments are conducted. The potassium exchange is to enhance the intensity of the carboxyl peak since –COOH absorption is very broad and can be easily lost in background. The integral of the CF_2 overtone located at approximately 2,365 cm^{-1} is compared with carbonyl acid C=O stretch at 1,695 cm^{-1}.

In a study of Nafion degraded by Fenton-type tests, several bands were assigned as follows: the 1,148 cm^{-1} band is the antisymmetric vibration of C–F bonds, doublet peaks at 978 and 983 cm^{-1} are C–O–C vibrations, the 1,058 cm^{-1} peak is assigned to SO_3^- on the side chain, and a 1,300-cm^{-1} band is from C–C bonds (Fig. 11; Kinumoto et al. 2006). The ratio is the ratio of C–C/C–O–C peak intensities. Since main chain contributes to the C–C peak and side chain contribute to the C–O–C peak. By comparing there two peaks, we know whether side chains are degraded preferentially. A weak 1,460-cm^{-1} band is assigned to S=O stretching of SO_2F or SO_2–O–SO_2. This band appears after the degradation test, which suggests trace amounts of sulfonic acid groups are converted to SO_2F or SO_2–O–SO_2. Similar results were also found in a long-term degradation test with Nafion in H_2O_2, where a new band located at 1,440 cm^{-1} was discovered (Qiao et al. 2006). When the Nafion is heavily doped with iron, the new band starts to appear within a 24-h period (Chen et al. 2007a).

To study the relative concentration of carbonyl groups, a ratio between the –CF_2-overtone peak (approximately 2,365 cm^{-1}) and the –COOK peak (approximately 1,680 cm^{-1}) is used (Delaney and Liu 2007). For membranes exposed to H_2O_2 vapor, a significant increase of –COOK peak intensity is observed.

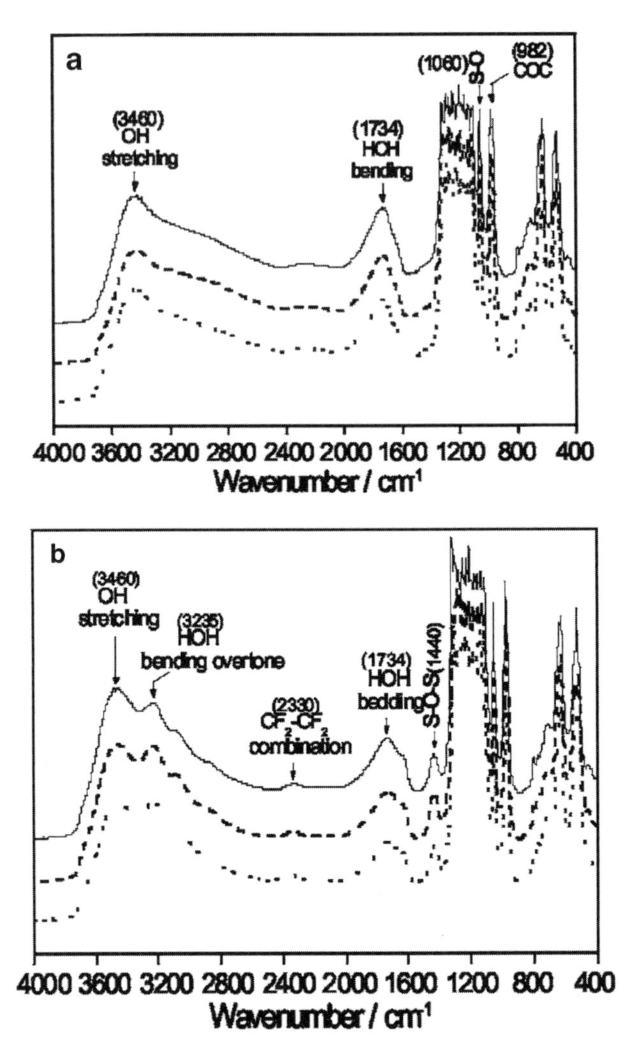

Fig. 11 Fourier transform IR (FTIR) spectra of PFSA membranes with different H_2O_2 (3 wt%) processing times at 80°C: (**a**) as received, (**b**) 22 days. *Top curve* equivalent weight 900, *middle curve* equivalent weight 1,000, *bottom curve* equivalent weight 1,100 (Qiao et al. 2006)

Compared with conventional FTIR, micro-FTIR can offer additional spatial resolution (Endoh et al. 2007). The composition change from the anode to the cathode side can be monitored on membrane cross sections. By monitoring of the carboxylic acid concentration (1,690 cm^{-1}), it was found that anode degradation is more severe than cathode degradation (Fig. 12).

An extensive discussion on end groups in fluoropolymers and their characterization by FTIR, ^{1}H NMR, ^{19}F NMR, and ^{19}F 2D NMR was presented by Pianca et al. (1999). Detailed description will be omitted here.

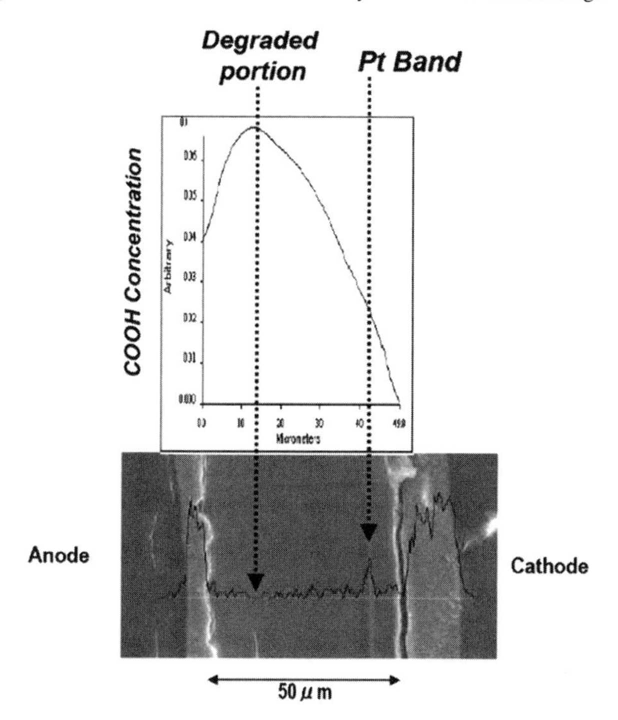

Fig. 12 Cross-sectional energy-dispersive X-ray and micro-FTIR analysis of Pt/C-based MEA following 200 h of testing at OCV (Endoh et al. 2007). OCV conditions were as follows: cell temperature 90°C H_2 18% RH, air 18% RH

6.2 Raman Spectroscopy

Micro-Raman is the main form of Raman spectroscopy used for fuel cell membrane degradations. Micro-Raman can scan the membrane cross section from the anode to the cathode side. Three peaks centered on 1,070, 970, and 810 cm^{-1} are assigned to single-bond stretching of S–O, C–O, and S–C, respectively (Ohma et al. 2007a, b). Since the C–C peak (1,380 cm^{-1}) of Nafion is broader than that of PTFE, a narrower peak near the platinum precipitation line is believed to be due to side chain degradation. The peak ratio between S–O and C–F absorption is also used to evaluate side chain loss and the degree of degradation. It was found that localized degradation is centered on the platinum deposition line (Fig. 13). Quite clearly, this finding is in contrast to that reported in Fig. 12, and it is not clear whether this is due to differences in the aging conditions or whether it is an artifact of one of the two measurement methods.

In addition to characterizing functional groups, in situ micro-Raman experiments were used to monitor the water content inside the membrane. It is found that the anode side membrane has lower water content when a current is applied (Matic et al. 2005). Overall water content is lower under load, which may be due to higher temperature.

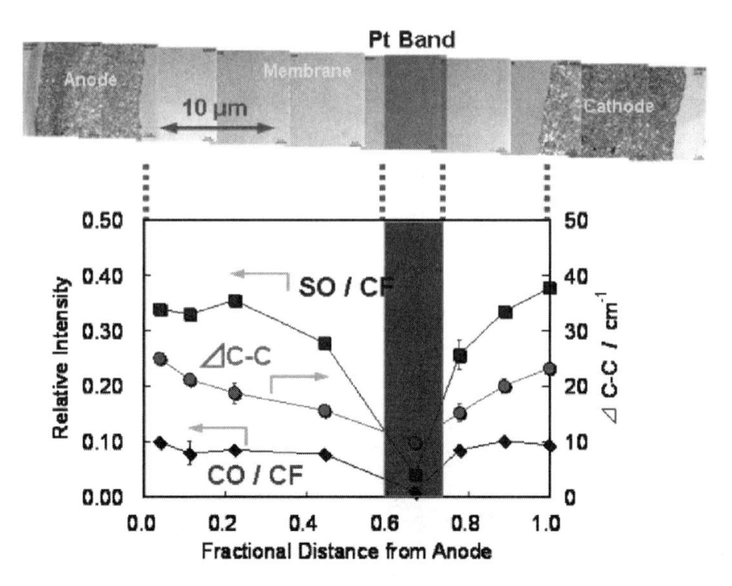

Fig. 13 Relative intensities of S–O and C–O versus C–F, and half bandwidth of C–C as a function of the fractional distance from the anode (Ohma et al. 2007b)

6.3 Electron Spin Resonance Spectroscopy

Among all analytical methods, ESR is capable of directly detecting radicals. The main challenge is the short lifetime of hydroxyl (HO•) radicals. Without spin trap, HO• radicals can be detected only by ESR at low temperatures (below −196°C) owing to their short lifetime (Bosnjakovic and Schlick 2004). HO_2• radicals can be detected by ESR at −73°C. Fluorinated alkyl radicals can be detected after submerging Nafion/Ti^{3+} in H_2O_2 for 14 days at ambient temperature. It seems that the Nafion matrix can stabilize HO_2• radicals.

Owing to the extremely short lifetime of radicals, low temperatures and/or spin trapping have to be used to detect the radicals in ESR experiments (Kadirov et al. 2005). In a UV-assisted Fenton test, it is proposed that side chain C–S bond scission, promoted by the presence of ferric ions, leads to a high membrane degradation rate: $-OCF_2CF_2SO_3^- + Fe^{3+} \rightarrow -OCF_2CF_2SO_3{\bullet} + Fe^{2+}$. The UV irradiation may also play an important role in this mechanism. The radical quintet detected in these tests suggests that the fluorine atom attached to the side-chain-linked carbon atoms can be attacked and leads to possible chain scission (Fig. 14).

To detect the radicals in Fenton's test, 5,5′-dimethyl-1-pyrroline-*N*-oxide (DMPO) was used in a Fenton-type test where a membrane was submerged in 0.3 wt% H_2O_2 at room temperature (Aoki et al. 2006c). 4-Hydroxy-2,2,6,6-teramethylpiperidine-*N*-oxyl was used for spin calibration. The amount of hydroxyl radical is quantified 10 min after H_2O_2 addition. It was found that the number of

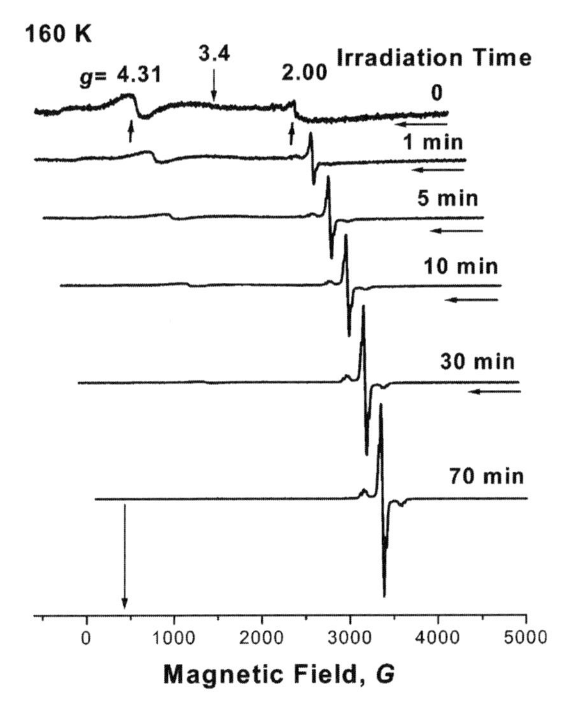

Fig. 14 Effect of UV irradiation time on the intensity of the chain-end radical $ROCF_2CF_2$• measured at −113°C in Nafion/Fe(II)/H_2O_2. Degree of neutralization 40%. Note the progressive horizontal displacement of the spectra to the left, to better visualize the signal from the perfluorinated fragment (Kadirov et al. 2005)

HO• radicals is similar for a membrane in H_2O_2, as well as for H_2O_2 alone. Fe^{2+}-doped PFSA membranes show more than 1 order of magnitude increase of the HO• radicals, while platinum-impregnated membranes show fewer radicals.

For hydrocarbon membranes, the degraded products can be directly measured owing to the stability of the carbon radicals. Fenton-type reaction on hydrocarbon model compounds has been studied (Hubner and Roduner 1999). Possible weak sites for HO• attack have been identified for various hydrocarbon ionomer membranes.

The attempt to detect the radicals directly in model fuel cells has met with limited success. During in situ ESR experiments, it was found that the signal is much stronger when dry reactant gas is fed to the miniature fuel cell, compared with wet gases (Panchenko et al. 2004). On the other hand, it is believed that the signal originated mainly from carbon and the catalyst, since the concentration of organic radicals from chemical degradation is too low to be detected by ESR.

Nonsupported metal black catalysts were then used to eliminate the strong ESR signal from the carbon support. No radicals could be detected for perfluorobutane sulfonate nor for the Nafion ionomer dispersion presumably owing to the short lifetime of the radicals (Vogel et al. 2007). With a DMPO spin trap, HO• adduct can be easily observed in Fenton's test, while a small amount of HO_2• can also be detected.

6.4 Energy Dispersive X-Ray Spectroscopy

Energy dispersive X-ray spectroscopy (EDX) is used predominantly for elemental analysis or chemical characterization. The advantage of this technique is that it can identify the elemental composition for a small area; thus, a profile of element composition can be created across the sample. Since EDX tests are usually conducted in a scanning electron microscope, a composition profile can be directly overlaid onto a secondary scanning electron microscope micrograph. Major elements related to membrane chemical degradation such as fluorine, sulfur, platinum and other metals can be readily detected.

As can be seen in Fig. 15, the relative concentration of the target element can be traced from the anode to the cathode side (Zhang et al. 2007); thus, the chemical degradation can be monitored across the membrane. Similar to the study mentioned above, platinum and sulfur profiles were examined (Aoki et al. 2006a; Yu et al. 2003).

6.5 X-Ray Photoelectron Spectroscopy

X-ray photoelectron spectroscopy (XPS) can determine the chemical composition of the first several atomic layers on a surface. The C1*s* peak with a binding energy of 295 eV is assigned to PTFE, while the peak of carbon black is located at 288 eV

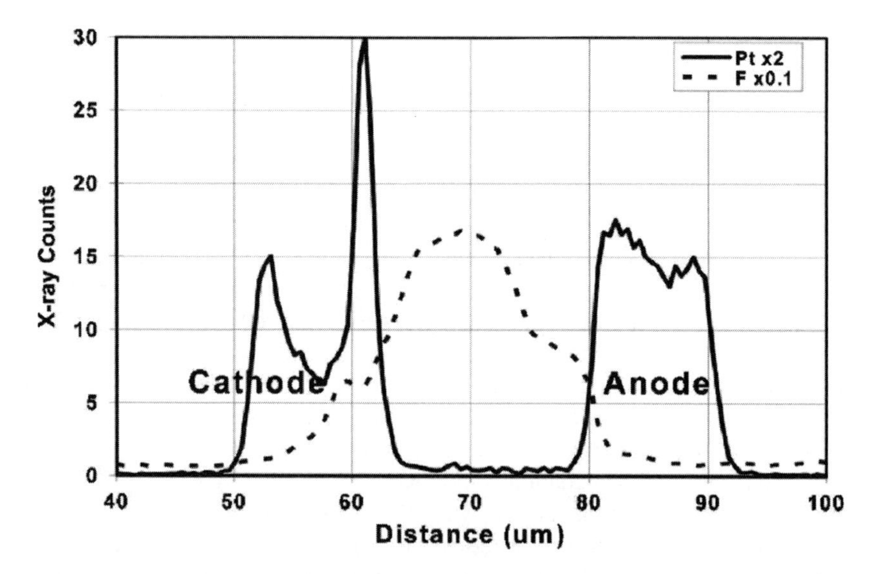

Fig. 15 Cross-sectional EPMA analysis on the elemental profile of platinum and fluorine of an MEA after operation for 2,000 h at OCV conditions with H_2/air [80°C, 100% RH, 150 kPa(abs)] (Zhang et al. 2007)

(Gulzow et al. 2000). It is found that more PTFE is lost on the anode side than on the cathode after a durability test.

XPS of fuel cell degraded MEA indicated destruction of CF_2 groups (Huang et al. 2003). Both the C1*s* peak shift (from 292 to 292.3 eV) and the F1*s* peak shrinkage (689.4 eV) show that the number of CF_2 groups decreases after fuel cell tests, presumably owing to hydrogenation.

The oxygen content of membranes after Fenton's test increases significantly (Chen et al. 2007a). It is believed that peroxide or carbonyl groups are formed during the degradation process.

On the basis of XPS results, it is found that chemical degradation during fuel cell operation occurs mainly on the hydrocarbon fraction of the radiation-grafted ionomer membranes (Nasef and Saidi 2002). It is believed that tertiary hydrogen of the α-carbon is most vulnerable to hydroxyl radical attacks.

6.6 *Nuclear Magnetic Resonance*

^{13}C NMR was used to characterize PTFE-*g*-PSSA membranes (Assink et al. 1991). The experiments were conducted at 50.16 MHz using dioxane (0.3%) as an internal standard set at 67.4 ppm. Resonance peaks at 169, 173, and 178 ppm were assigned to carboxylic acid or ester functional groups. Peaks in the region of 50–80 ppm correspond to carbons next to an alcohol or ether, while peaks in the 15–35-ppm range represent various aliphatic carbons.

^{19}F NMR spectra reveal that degradation species extracted from fuel cell degraded membrane are similar to the compounds found in Fenton's test bath water (Healy et al. 2005). Various peaks have been assigned to different atoms of the side chain (Fig. 16). A similar study was also described in a more recent report (Kinumoto et al. 2006).

As mentioned previously, an extensive discussion on end groups in fluoropolymers and their characterization by FTIR, ^{1}H NMR, ^{19}F NMR, and ^{19}F 2D NMR was

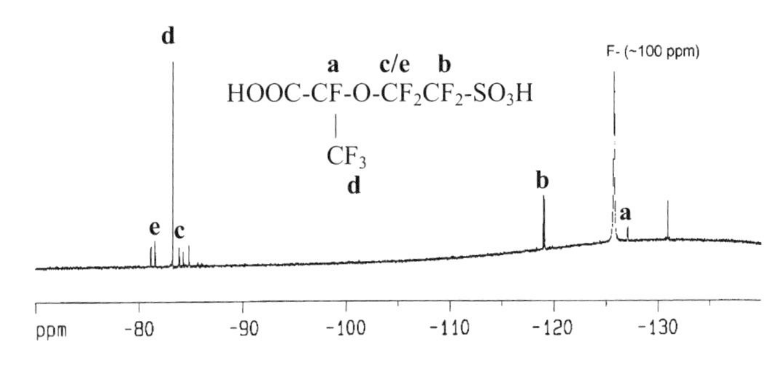

Fig. 16 Band assignment for the ^{19}F NMR spectrum of organic degradation products (Healy et al. 2005)

presented by Pianca et al. (Pianca et al. 1999). Detailed description will be omitted here.

7 Detailed Degradation Mechanisms Discussions

Detailed discussions on the membrane chemical degradation mechanisms will first be divided into three main categories: hydrocarbon membranes, *g*-PSSA membranes, and PFSA membranes. General degradation pathways and correlations between different types of membranes will be discussed at the end of this section.

7.1 Hydrocarbon Membranes

Nonfluorinated hydrocarbon polymer membranes are being developed both as competitive alternatives to commercial perfluorinated membranes and as materials extending the range of operation toward those more generally occupied by phosphoric acid fuel cells (Mitov et al. 2006a; Miyatake and Watanabe 2006). Hydrocarbon polymers usually have polyaromatic or polyheterocyclic repeating units: polysulfone, polyetherketone, polybenzimidazole, polyethersulfone, polyphenylquinoxaline, polyetherimide. Additionally, sulfonated polyetherketone and polysulfone have been used to blend with polybenzimidazole, polyetherimine, polyethylene imine, and poly(4-vinylpyridine). In addition to random copolymers, it is reported that hydrocarbon ionomers in the form of block copolymer have better proton conductivity (Nakano et al. 2005). The advantages of hydrocarbon membranes are lower cost, potentially higher operation temperature, and no environmental pollution by fluoride or disposal (Aoki et al. 2006a). Detailed discussion on synthesis of hydrocarbon ionomers can be found in a recent review (Hickner et al. 2004).

One of the major concerns for hydrocarbon membranes is chemical stability. Typical bond strength in a perfluorinated polymer is C–H < C–C < C–F (Mitov et al. 2006b). C–F bond strengths are generally in the range of 460 kJ mol^{-1}, while C–H bond strengths are approximately 410 kJ mol^{-1} (Pianca et al. 1999; Schiraldi 2006). Generally, fluorinated polymers are considered to have inherently better stability than hydrocarbon membranes.

Long-term durability has been demonstrated on SPI and sulfonated polyarylene ether membranes (Aoki et al. 2006a, b). Both types of membranes successfully sustained 5,000 h of operation under moderate conditions (0.2 A cm^{-2}, 80°C, more than 90% RH, ambient pressure, and no frequent startup/shutdown). Neither membrane showed changes in ion-exchange capacity over time, which indicates that there is no specific cleavage of sulfonic acid groups. It is interesting to note that sulfonated polyarylene ether membranes show no thickness change after the durability test, even after 2,350 h at 60% RH. The molecular weight remains constant for the 90% RH test, but decreases to half of the original value when the membrane is tested under 60% RH. As a result of lower molecular weight, the membranes lose

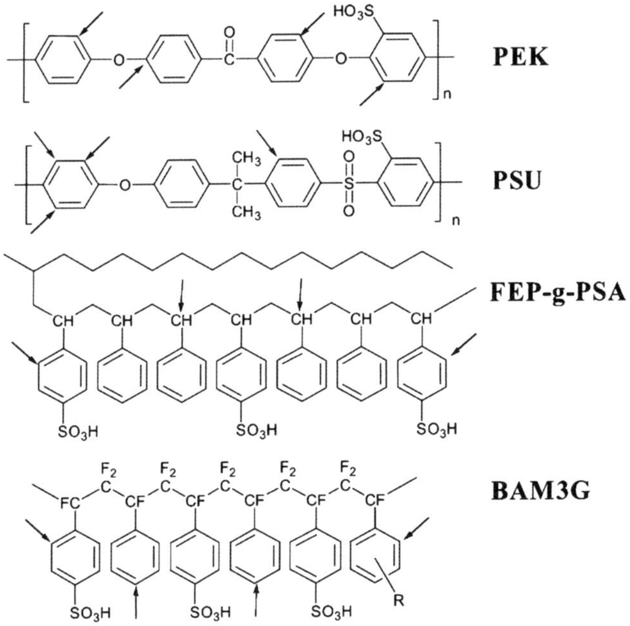

PEK

PSU

FEP-g-PSA

BAM3G

Fig. 17 Sites susceptible to hydroxyl radical attacks on hydrocarbon polymers for PEM fuel cells (Panchenko et al. 2004)

mechanical strength and become brittle. It was proposed by the authors that chain scission is the main degradation mechanism, where the shortened polymer chain segments are still long enough to remain in the solid phase.

On the basis of ESR results from model organic compounds degraded by UV photolysis of H_2O_2, atomic sites that are susceptible to radical attacks were identified (Fig. 17) (Hubner and Roduner 1999). It was also found that HO_2• can be formed by HO• reacting with H_2O_2 during UV photolysis of H_2O_2. Relative to the rate of OH• reaction with hydrocarbon systems, the rate of hydrogen atom transfer from H_2O_2 is rather slow and does not effectively compete. Thus, HO• is believed to be the main damaging species below pH 11.7. Another important finding is that in the absence of UV light, H_2O_2 cannot effectively degrade toluene sulfonic acid and its derivatives. This again proves that H_2O_2 attacks the polymer via radical pathways; H_2O_2 by itself is not potent enough to directly attack even hydrocarbon membranes.

7.2 g-PSSA Membranes

g-PSSA membranes have been used in PEMFCs since the 1960s (LaConti et al. 2005). A comprehensive review of synthesis, properties, and performance of *g*-PSSA membranes has been published recently (Gubler et al. 2005).

As shown in Fig. 17, there are two types of susceptible sites for chemical attack: the phenyl group and the hydrogen on the α-carbon.

On the basis of IR results, the hydroxyl radicals attack the polystyrene via addition to the phenyl groups (Weir 1978). The results were verified by FTIR analysis of PFA-*g*-PSSA (where PFA is perfluoroalkoxy) after fuel cell durability tests (Nasef and Saidi 2002). This pathway is also supported by ESR experiments on model compounds (Hubner and Roduner 1999).

Degradation of the grafted ionomers is believed to be also caused by the reaction between H_2O_2 and the α-hydrogen on the graft chain (Mattsson et al. 2000). When the α-hydrogen is replaced by a methyl group, the grafted membrane with α-methylstyrene is significantly more stable, which proves that the benzylic α-hydrogen atoms are the weak link for chemical degradation (Assink et al. 1991; Hubner and Roduner 1999).

Similar to pure hydrocarbon membranes discussed previously, both the aromatic rings and the SO_3^- groups are lost after the durability test (Yu et al. 2003). No preferential cleavage of sulfonic acid groups was observed. On the basis of XPS results, it also is found that chemical degradation during fuel cell operation occurs mainly on the hydrocarbon fraction of the radiation-grafted ionomer membranes (Nasef and Saidi 2002). Thus, the hydrocarbon fraction has less chemical stability than the fluorinated part.

A long-term durability test with a optimized FEP-25-based membrane showed 4,000 h of continuous operation under moderate conditions: 500 mA cm^{-2} with a cell temperature of 80–85°C, H_2(100% RH)/O_2(dry) under ambient pressure (Gubler et al. 2004). The final lifetime reached 7,900 h, where the membrane degradation was accelerated by several controlled/uncontrolled startup/shutdown cycles.

7.3 PFSA membranes

PFSA membranes, as a commercial product for the chlor-alkali industry, have been widely used in PEMFCs and PEMELCs. Even with its highly stable C–F bonds, the PFSA membrane still sustains appreciable degradation, especially under accelerated conditions such as high temperature, high pressure, and low humidity (Healy et al. 2005). There are three potential degradation pathways: chain unzipping, chain scission, and side-group attack, and these will be discussed separately next.

7.3.1 Chain Unzipping Mechanism

Many polymers synthesized by free-radical polymerization can sustain depolymerization via a chain unzipping mechanism. Thermal degradation of Nafion indicates that chain unzipping is the main mechanism for Nafion degradation (Wilkie et al. 1991).

A possible PFSA degradation mechanism based on end-group peroxyl attack with extraction of HF and formation of CO_2 was proposed (LaConti et al. 2003). Recently, a detailed degradation mechanism through chain-end degradation and chain unzipping was proposed (Curtin et al. 2004). It was pointed out by the authors that PFSA polymer chains are usually terminated by hydrogen-containing end groups, which are susceptible to radical attack. The detailed chain unzipping mechanism is shown below:

$$\text{Rf–CF}_2\text{COOH} + \bullet\text{OH} \rightarrow \text{Rf–CF}_2\bullet + \text{CO}_2 + \text{H}_2\text{O} \quad (1)$$

$$\text{Rf–CF}_2\bullet + \bullet\text{OH} \rightarrow \text{Rf–CF}_2\text{OH} \rightarrow \text{Rf–COF} + \text{HF} \quad (2)$$

$$\text{Rf–COF} + \text{H}_2\text{O} \rightarrow \text{Rf–COOH} + \text{HF} \quad (3)$$

In Fenton-type tests on ionomers with improved end-group perfluorination, the observed lower FRR is consistent with the reduction of the number of reactive end groups. When the C–H bonds in the end groups are converted to C–F bonds, the FRR drastically decreases. In contrast, such a correlation has not been observed in real fuel cell tests. This controversy will be discussed later.

After treatment of the Nafion membrane in Fenton-type degradation tests, both small-angle X-ray scattering and tensile data suggest that there is no major structural change that supports the chain unzipping degradation mechanism (Aieta et al. 2007). On the basis of FTIR study, the ratio of C–C/C–O–C peak intensities does not change over the course of the experiments, indicating that the side chains are not selectively degraded (Kinumoto et al. 2006).

7.3.2 Chain Scission Mechanism

In a UV-induced Fenton test on PFSA membranes, ESR quartets and quintets assigned to $ROCF_2CF_2\bullet$ and $-CF_2C\bullet(OR)CF_2-$ are observed (Kadirov et al. 2005). The results suggested that the fluorine atom attached to the side-chain-linked carbon atoms can be attacked and leads to possible chain scission. Interestingly, a calculation based density function theory (DFT) suggests that HO• radicals should not be able to abstract fluorine from a perfluorinated backbone (Mitov et al. 2005). It is not clear currently if these results can be directly related to fuel cells, where no UV irradiation is present.

When a PFSA membrane is exposed to H_2O_2 vapor at low humidity (approximately 18%) and elevated temperature (120°C), severe degradation can be observed even without doping additional iron (Hommura et al. 2008). The concentration of –COOH functional groups increases with exposure time, while the molecular weight decreases. It is proposed that a significant number of chain scissions, accompanying chain unzipping, occur during the degradation process. The exact chain scission mechanisms were not discussed.

On the basis of a density functional theory study, it was found that the C–F bond dissociation energy required for fluorine abstraction is lowest for tertiary carbon and highest for primary carbon (Mitov et al. 2005). The C–F bond on the side chain α-ether carbon is thus potentially more activated, which creates a weak site for radical attack and chain scission. Even with the less stable C–F bonds, direct HO• attack is unlikely, as discussed above.

7.3.3 Side-Group Attack

The two mechanisms discussed in the previous sections focus on attack on the polymer backbone. Radical attack on the sulfonic side groups can also lead to chain scission. As discussed previously, the equivalent weight of the membrane does not change significantly, which suggests that specific side chain cleavage is not the main mechanism. But even a minor attack of side groups can generate a large number of new chain ends, which can significantly increase the degradation rate.

In a UV-induced Fenton test, it was proposed that side chain C–S bond scission, promoted by the presence of ferric ions, leads to a high membrane degradation rate (Kadirov et al. 2005): $-OCF_2CF_2SO_3^- + Fe^{3+} \rightarrow -OCF_2CF_2SO_3\bullet + Fe^{2+}$. In the presence of UV light, radical generation can occur without H_2O_2. Since there is no UV light in fuel cells, it is not clear whether the mechanism can be directly applied to membrane degradation in PEMFCs.

Solid-state ^{19}F NMR reveals that the fraction of the main chain actually slightly increases with longer decomposition duration, which suggests the side chain has been degraded preferentially (Kinumoto et al. 2006). Such an increase of the fraction is very small compared with the overall degradation.

As discussed previously, when the membrane is fully exchanged with alkali metals (Li, Na, and Cs), the FRR is significantly reduced in OCV fuel cell tests (Mittal et al. 2006c). It was thus proposed by the authors that the radicals can attack side chains in fuel cells presumably owing to the exchange eliminating hydrogen abstraction. It is quite intriguing that the FRR reduction is so drastic (more than 200 times) that the chain end $-COOM^+$ may be more stable than –COOH. Alternatively, it can also be argued that H_2O_2/radical generation can be suppressed by the ion exchange.

One study did find that the equivalent weight of a Nafion membrane increases from 1,100 to 1,221 after a 200-h durability test (Luo et al. 2006). The yield strength of the membrane also increases, which may be the result of the loss of sulfonate groups.

7.4 *Other Aspects of Membrane Chemical Degradation Mechanisms*

There are some issues with membrane chemical degradation mechanisms that are not comprehensively covered in the previous discussions. Instead of being specific to a single membrane type, they are generally more universal from the hydrocarbon to perfluorinated polymers. The discussion will be divided into several topics: source of radicals, end-group stabilization, and stability of hydrocarbon membranes.

7.4.1 Source of Radicals

The early-accepted radical generation pathway for PEMFCs is based on gas crossover (i.e., O_2 permeates to the Pt/H_2 electrode), catalyzed chemical reaction of crossed-over gas, and H_2O_2 decomposition by impurity ions (Hodgdon et al. 1966; LaConti 1988):

Step 1: $H_2 \rightarrow 2H\bullet$ (via a Pt catalyst)
Step 2: $H\bullet + O_2$ (diffused through the PEM) $\rightarrow HO_2\bullet$
Step 3: $HO_2\bullet + H\bullet \rightarrow H_2O_2$ (which can diffuse into the PEM at locus near the degraded front of the PEM)
Step 4: $H_2O_2 + M^{2+}$(Fe^{2+},Cu^{2+}, etc., found in fuel cell MEAs, etc.) $\rightarrow M^{3+} + HO\bullet + OH^-$
Step 5: $HO\bullet + H_2O_2 \rightarrow H_2O + HO_2\bullet$ (H_2O_2 radicals attack the PEM)

Fenton's tests are usually used as an ex situ method to study the decomposition mechanisms. The conventional Fenton's test requires the membrane, usually doped with ferrous/ferric ions, to be submerged in a certain amount of H_2O_2 solution at elevated temperatures. The concentration of H_2O_2 decays over time and the final H_2O_2 concentration at the end of Fenton's test is usually lower than the starting concentration by orders of magnitude. Consequently, the decomposition kinetics and mechanisms are difficult to decipher. A new type of instrument, referred to as H_2O_2 liquid flow cell, is proposed to study the decomposition mechanisms (Liu et al. 2005). The novelty of the experimental setup is that the H_2O_2 is supplied to the membrane as a constant stream, where the H_2O_2 concentration is maintained constant over the entire experiment. It is found that over a wide range of H_2O_2 concentration (1 ppm to 10 wt%), the FRR observed in such tests is significantly lower than that which results from fuel cell tests. The FRR from this ex situ test is compared with fuel cell results only when the membrane is heavily doped with iron (Fig. 18) (Liu et al. 2006). The presence of Pt/C increases the FRR considerably, but a high FRR is observed with a high concentration of H_2O_2. Similar results have also been presented in a more recent study (Mittal et al. 2006b).

By embedding microelectrodes inside the membrane, one can measure the H_2O_2 concentration directly electrochemically (Liu and Zuckerbrod 2005). A thinner membrane has a higher H_2O_2 concentration owing to higher gas crossover rate. The concentration of H_2O_2 inside the membrane ranges from approximately 2 ppm for a 130-μm membrane to approximately 23 ppm for 18-∝m membranes. Severe membrane degradation cannot be expected from such a low concentration of H_2O_2 for noncontaminated membranes. Consequently, direct radical generation, especially on the cathode side, is suggested (Burlatsky et al. 2005; Liu et al. 2005). It is even

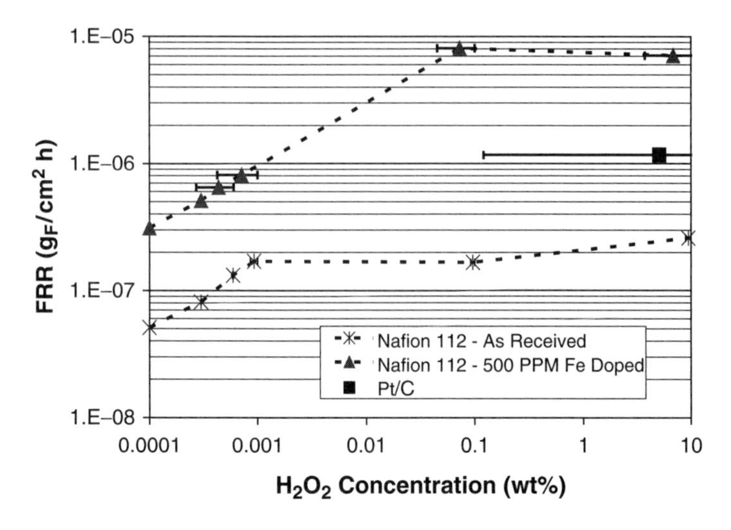

Fig. 18 Effect of H_2O_2 concentration on FRR of as-received, 500 ppm iron-doped, and single-sided Pt/C coated NE112 membranes. The *symbols* represents the average value of the inlet and outlet concentration of the H_2O_2. The *error bars* represent the range of the H_2O_2 concentration from inlet (higher value) to outlet (lower value). This H_2O_2 flow cell study was conducted at 90°C, 0.5 ml H_2O_2 min^{-1}, and 50 sccm N_2 (Liu et al. 2006)

suggested that diamagnetic pathways may play a role in PEMFC membrane degradations (Vogel et al. 2007).

7.4.2 End-Group Stabilization

As discussed earlier, stabilization of PFSA end groups successfully reduced membrane degradation in Fenton's test (Curtin et al. 2004). Unfortunately, such stabilization does not help improve membrane lifetime in actual fuel cell tests.

Conventionally, Fenton's tests are considered as a stability measure for HO• attack. With a DMPO spin trap, HO• adduct can be easily observed in both Fenton's test and UV photolysis of H_2O_2, while only a small amount of HO_2• can also be detected by ESR (Vogel et al. 2007, Hubner and Roduner 1999).

As shown in Table 4, hydrocarbon compounds react with HO• significantly faster than the self-quenching reaction: HO•+H_2O_2 → H_2O + HO_2•. Although the recombination reaction of HO• is also very fast, it should not be considered as a competitive reaction owing to the extremely low concentration of the radicals. Thus, Fenton's test can be considered as a stability test against HO• attack.

The situation can be different for fluorinated compounds owing to their high stability against HO•. On the basis of the rate constant of HO• reactions, it is proposed that hydroxyl radicals can readily react with H_2O_2 to form peroxyl radicals (self-quenching) before it can react with the polymer under Fenton conditions (Liu et al.

Table 4 Rate constants for reactions of HO• (Mitov et al. 2006a)

Reaction	Rate constant (10^9 M^{-1} s^{-1})
HO• + H_2O_2 → H_2O + HO_2•	0.027
HO• + HO• → H_2O_2	5.5
CH_3OH + HO• → CH_3O• + H_2O	0.97
$C_6H_5CH_3$ + HO• → $C_6H_5(OH)CH_3/C_6H_5$•CH_2	3.0
Benzophenone + HO• → adduct	8.8
Benzenesulfonic acid + HO• → adduct	1.6
Benzenesulfonate ion + HO• → adduct	3.0

2005; Cipollini 2007). On the basis of these results, Cippollini proposed that Fenton's test for fluorinated compounds is the peroxyl radical (HO_2•) degradation test, instead of the commonly believed hydroxyl radical test, owing to the high-concentration H_2O_2 employed in these tests (Cipollini 2007). It is further suggested that peroxyl radicals can attack only hydrogen-containing chain ends. In fuel cell tests, the HO• pathway exists owing to the low concentration of H_2O_2. HO• radicals can attack sulfonic acid groups and create chain scissions and more end groups, which leads to a higher degradation rate. Thus, chain-end stabilization can help in Fenton's tests since it eliminates weak hydrogen-containing end groups but it does not help in real fuel cell tests since chain scission generates more weak end groups, which offsets the effect of chain-end stabilization.

In long-term OCV tests, it is found that the starting FRR is low and then approaches a steady value after 24 h. It is proposed that the hydroxyl radicals attack the side chain C–S bonds, which leads to chain scission and additional carboxylic chain ends. When the number of carboxylic chain ends increases, the chain ends effectively react with hydroxyl radicals. Eventually, a steady state is reached (Cipollini 2007).

When an iron-doped stabilized ionomer with “capped” fluorinated end groups is exposed to H_2O_2 vapor, it has similar degradation to nonstabilized ionomer with carboxylated end groups, which is similar to fuel cell test results (Delaney and Liu 2007). Similar to the argument above, it is believed that recombination of hydroxyl radicals and generation of less reactive peroxyl radicals are main causes for this observation. It is proposed that vapor-phase experiments have low H_2O_2 concentration, which leads to high HO• concentration in the membrane owing to less self-quenching. In the liquid-phase tests, abundant H_2O_2 quenches HO• to HO_2•, which can only degrade chain ends but does not cause chain scission. The ratio of the carboxyl and CF_2 overtone peak intensities increases by two orders of magnitude when the membrane is exposed to peroxide vapor, while it is stable after aqueous peroxide tests. Chain scission is the dominating mechanism in vapor-phase degradation tests, while degradation through end groups dominates during liquid-phase degradation.

One controversy regarding this theory is that accelerated decomposition was not observed when the membrane was exposed to low-concentration H_2O_2 solution. If H_2O_2 self-quenching is the main reason for lack of chain scission, low-concentration H_2O_2 tests should generate similar results as the H_2O_2 vapor test. It is thus unclear whether water is playing any role in the degradation process.

7.4.3 Stability of Hydrocarbon Membranes

As discussed in previous sections, the hydrocarbon membranes are generally believed to be intrinsically less stable than perfluorinated membranes owing to lower bond strength of C–H. Results from Fenton's test prove the instability of hydrocarbon membranes to hydroxyl radicals. In contrast, several hydrocarbon membranes show excellent real fuel cell durability.

In an ex situ Fenton's test (80°C, 3% H_2O_2 with 2 ppm $FeSO_4$), an SPI membrane (SPI-5) degraded 190 times faster than Nafion (Aoki et al. 2006a). A sulfonated polystyrene cation-exchange membrane can be severely degraded in Fenton's test, while Nafion sustains less than 1% weight loss (Guo et al. 1999). All radiation-grafted ionomers are completely degraded within 100 h of exposure to Fenton's reagent (60°C 3% H_2O_2) (Chen et al. 2007b). Essentially, the degradation rate of hydrocarbon membranes is orders of magnitude higher than that of PFSA membranes. In contrast, long lifetime has been reported for hydrocarbon and *g*-PSSA membranes (Aoki et al. 2006a, b; Gubler et al. 2004). These results contradict the conclusions based on Fenton's test.

One possible explanation for the long lifetime of hydrocarbon membranes is their lower gas permeability. Since SPI-5 is 40 times less permeable to H_2 and 10 times less permeable to O_2 than Nafion, the reduced gas crossover rate may be responsible for the low degradation rate in fuel cell tests (Aoki et al. 2006a).

In addition to the mechanism described above, Fenton's test can be considered as an HO_2• test for PFSA membranes owing to H_2O_2 self-quenching and slow degradation kinetics, as discussed previously. Fenton's tests for hydrocarbon and *g*-PSSA membranes, in contrast, are HO• challenge tests owing to the fast kinetics of HO• addition to aromatic rings. In Fenton's tests of hydrocarbon membranes, the large amount of HO•, generated by $H_2O_2 + Fe^{2+} + H^+ \rightarrow HO\bullet + Fe^{3+} + H_2O$, immediately reacts with polymer and causes a significant amount of damage. In Fenton's tests of PFSA, the large amount of HO• reacts with H_2O_2 to form the less potent HO•. Thus, the apparently identical Fenton tests actually have different reaction pathways for hydrocarbon membranes and PFSA. In actual fuel cell tests, HO• radicals can be considered as the main degradation reactant. The damage they induce on PFSA membranes can be comparable to that on hydrocarbon membranes, especially under dry conditions where HO• will be more potent due to lack of water stabilization.

8 Mitigation Strategies

Since gas crossover is the fundamental mechanism for membrane chemical degradation, lowering the gas crossover rate can potentially reduce membrane chemical degradations. As discussed earlier, hydrocarbon membranes can effectively reduces gas crossover rate (Aoki et al. 2006a). Further development of membrane/electrode materials and structures can potentially lower the amount of gas crossover and overall degradation rate.

On the basis of degradation mechanisms of hydrocarbon membranes, blocking of the aromatic ring by suitable substitution, such as fluorine atoms, can limit the HO• addition reactions (Mitov et al. 2006a). The SO_3^- groups cannot only promote proton conductivity, but can also reduces the activity of the aromatic ring toward HO• addition.

For the *g*-PSSA membrane, since the α-hydrogen on the graft chain is a weak site (Mattsson et al. 2000), the stability of PTFE-*g*-PSSA is significantly increased when tertiary hydrogen is replaced by a methyl group (Assink et al. 1991). Divinylbenzene cross-linked membranes have higher durability than non-cross-linked membranes owing to their lower gas crossover rate (Gubler et al. 2005).

In alkaline solutions, iron catalyzes H_2O_2 decomposition, forming radicals that promote degradation of organic compounds. In contrast, manganese also promotes H_2O_2 decomposition, but a two-electron transfer mechanism dominates, which leads to no radical generation (Yokoyama et al. 2002). Thus, manganese and other H_2O_2 decomposition catalysts with a two-electron transfer mechanism can potentially be used as additives to the membrane materials.

Membranes based on Asahi Glass's reported results demonstrate FRRs in fuel cell tests which are two orders of magnitude lower than for conventional PFSAs (Endoh 2006; Endoh et al. 2007). Drastic improvements in fuel cell lifetime based on proprietary mitigation strategies (Escobedo 2006; Hicks 2006) have also been reported by DuPont and 3M. The underlying mechanism has not yet been discussed.

9 Summary

The chemical degradation of PEMFCs and PEMELCs is still under active study. The detailed mechanisms are still not well understood and there are contradictory observations that need to be clarified. As discussed previously, there are many correlations between the findings in PEMFCs and PEMELCs. Generally, PEMELCs demonstrate less chemical degradation than PEMFCs. Lower operating temperature, unsupported (pure) platinum catalyst, membrane stabilization, and significantly superior hydration contribute to the long lifetime of PEMELCs.

References

Aieta, N.V., Leisch, J.E., Santos, M.M., Yandrasits, M.A., Hamrock, S.J., Herring, A.M. (2007) Tracking crystallinity changes in PFSA polymers during ex-situ peroxide degradation. ECS Trans. 11, 1157–1164.

Antoine, O., Durand, R. (2000) RRDE study of oxygen reduction on Pt nanoparticles inside Nafion (R): H_2O_2 production in PEMFC cathode conditions. J. Appl. Electrochem. 30, 839–844.

Aoki, M., Uchida, H., Watanabe, M. (2005) Novel evaluation method for degradation rate of polymer electrolytes in fuel cells. Electrochem. Commun. 7, 1434–1438.

Aoki, M., Asano, N., Miyatake, K., Uchida, H., Watanabe, M. (2006a) Durability of sulfonated polyimide membrane evaluated by long-term polymer electrolyte fuel cell operation. J. Electrochem. Soc. 153, A1154–A1158.
Aoki, M., Chikashige, Y., Miyatake, K., Uchida, H., Watanabe, M. (2006b) Durability of novel sulfonated poly(arylene ether) membrane in PEFC operation. Electrochem. Commun. 8, 1412–1416.
Aoki, M., Uchida, H., Watanabe, M. (2006c) Decomposition mechanism of perfluorosulfonic acid electrolyte in polymer electrolyte fuel cells. Electrochem. Commun. 8, 1509–1513.
Asano, N., Aoki, M., Suzuki, S., Miyatake, K., Uchida, H., Watanabe, M. (2006) Aliphatic/aromatic polyimide ionomers as a proton conductive membrane for fuel cell applications. J. Am. Chem. Soc. 128, 1762–1769.
Assink, R.A., Arnold, C., Hollandsworth, R.P. (1991) Preparation of oxidatively stable cation-exchange membranes by the elimination of tertiary hydrogens. J. Membr. Sci. 56, 143–151.
Autrey, T., Brown, A.K., Camaioni, D.M., Dupuis, M., Foster, N.S., Getty, A. (2004) Thermochemistry of aqueous hydroxyl radical from advances in photoacoustic calorimetry and ab initio continuum solvation theory. J. Am. Chem. Soc. 126, 3680–3681.
Baldwin, R., Pham, M., Leonida, A., McElroy, J., Nalette, T. (1990) Hydrogen oxygen proton-exchange membrane fuel-cells and electrolyzers. J. Power Sources 29, 399–412.
Bosnjakovic, A., Schlick, S. (2004) Naflon perfluorinated membranes treated in Fenton media: Radical species detected by ESR spectroscopy. J. Phys. Chem. 108, 4332–4337.
Büchi, F.N., Gupta, B., Haas, O., Scherer, G.G. (1995) Study of radiation-grafted FEP-g-polystyrene membranes as polymer electrolytes in fuel-cells. Electrochim. Acta 40, 345–353.
Burlatsky, S.F., Atrazhev, V., Cipollini, N.E., Condit, D.A., Erikhman, N. (2005) Aspects of PEMFC degradation. ECS Trans. 1, 239–246.
Chen, C., Fuller, T.F. (2007) H_2O_2 Formation under fuel-cell conditions. ECS Trans. 11, 1127–1137.
Chen, Y.L., Li, D.Z., Wang, X.C., Wu, L., Wang, X.X., Fu, X.Z. (2005) Promoting effects of H-2 on photooxidation of volatile organic pollutants over Pt/TiO_2. New J. Chem. 29, 1514–1519.
Chen, C., Levitin, G., Hess, D.W., Fuller, T.F. (2007a) XPS investigation of Nafion (R) membrane degradation. J. Power Sources 169, 288–295.
Chen, J., Septiani, U., Asano, M., Maekawa, Y., Kubota, H., Yoshida, M. (2007b) Comparative study on the preparation and properties of radiation-grafted polymer electrolyte membranes based on fluoropolymer films. J. Appl. Polym. Sci. 103, 1966–1972.
Chludzinski, P.J. (1982) A Mechanistic Model and Proposed Corrections for Solid Polymer Electrolyte (SPE) Degradation in H_2/O_2 Fuel Cells and Water Electrolyzers, GE Direct Energy Conversion Program Internal Report, 1982.
Cipollini, N.E. (2007) Chemical aspects of membrane degradation. ECS Trans. 11, 1071–1082.
Cleghorn, S.J.C., Mayfield, D.K., Moore, D.A., Moore, J.C., Rusch, G., Sherman, T.W., Sisofo, N.T., Beuscher, U. (2006) A polymer electrolyte fuel cell life test: 3 years of continuous operation. J. Power Sources 158, 446–454.
Curtin, D.E., Lousenberg, R.D., Henry, T.J., Tangeman, P.C., Tisack, M.E. (2004) Advanced materials for improved PEMFC performance and life. J. Power Sources 131, 41–48.
Da Pozza, A., Ferrantelli, P., Merli, C., Petrucci, E. (2005) Oxidation efficiency in the electro-Fenton process. J. Appl. Electrochem. 35, 391–398.
Darling, R.M., Meyers, J.P. (2005) Mathematical model of platinum movement in PEM fuel cells. J. Electrochem. Soc. 152, A242–A247.
Delaney, W.E., Liu, W.K. (2007) The use of FTIR to analyze ex-situ and in-situ degradation of perfluorinated fuel cell ionomer. ECS Trans. 11, 1093–1104.
Endoh, E. (2006) Highly durable MEA for PEMFC under high temperature and low humidity conditions. ECS Trans. 3, 9–18.
Endoh, E., Hommura, S., Terazono, S., Widjaja, H., Anzai, J. (2007) Degradation mechanism of the PFSA membrane and influence of deposited Pt in the membrane. ECS Trans. 11, 1083–1091.

Escobedo, G., Enabling commercial PEM fuel cells with breakthrough lifetime improvements. *Department of Energy Hydrogen Program Annual Merit Review Proceedings*, 2006.

Ferreira, P.J., la O, G.J., Shao-Horn, Y., Morgan, D., Makharia, R., Kocha, S., Gasteiger, H.A. (2005) Instability of Pt/C electrocatalysts in proton exchange membrane fuel cells – A mechanistic investigation. J. Electrochem. Soc. 152, A2256–A2271.

Fuller, T.F., Newman, J. (1993) Water and thermal management in solid-polymer-electrolyte fuel-cells. J. Electrochem. Soc. 140, 1218–1225.

Genies, C., Mercier, R., Sillion, B., Petiaud, R., Cornet, N., Gebel, G., Pineri, M. (2001) Stability study of sulfonated phthalic and naphthalenic polyimide structures in aqueous medium. Polymer 42, 5097–5105.

Gubler, L., Kuhn, H., Schmidt, T.J., Scherer, G.G., Brack, H.P., Simbeck, K. (2004) Performance and durability of membrane electrode assemblies based on radiation-grafted FEP-g-polystyrene membranes. Fuel Cells 4, 196–207.

Gubler, L., Gursel, S.A., Scherer, G.G. (2005) Radiation grafted membranes for polymer electrolyte fuel cells. Fuel Cells 5, 317–335.

Gulzow, E., Schulze, M., Wagner, N., Kaz, T., Reissner, R., Steinhilber, G., Schneider, A. (2000) Dry layer preparation and characterisation of polymer electrolyte fuel cell components. J. Power Sources 86, 352–362.

Guo, Q.H., Pintauro, P.N., Tang, H., O'Connor, S. (1999) Sulfonated and crosslinked polyphosphazene-based proton-exchange membranes. J. Membr. Sci. 154, 175–181.

Healy, J., Hayden, C., Xie, T., Olson, K., Waldo, R., Brundage, A., Gasteiger, H., Abbott, J. (2005) Aspects of the chemical degradation of PFSA ionomers used in PEM fuel cells. Fuel Cells 5, 302–308.

Hickner, M.A., Ghassemi, H., Kim, Y.S., Einsla, B.R., McGrath, J.E. (2004) Alternative polymer systems for proton exchange membranes (PEMs). Chem. Rev. 104, 4587–4611.

Hicks, M., MEA & stack durability for pem fuel cells. *Department of Energy Hydrogen Program Annual Merit Review Proceedings*, 2006.

Hodgdon, R.B., Boyack, J.R., LaConti, A.B. (1966) The Degradation of Polystyrene Sulfonic Acid, TIS Report 65DE5, General Electric Company: July 6, 1966.

Hommura, S., Kawahara, K., Shimohira, T., Teraoka, Y. (2008) Development of a method for clarifying the perfluorosulfonated membrane degradation mechanism in a fuel cell environment. J. Electrochem. Soc. 155, A29–A33.

Huang, C.D., Tan, K.S., Lin, H.Y., Tan, K.L. (2003) XRD and XPS analysis of the degradation of the polymer electrolyte in H-2-O-2 fuel cell. Chem. Phys. Lett. 371, 80–85.

Hubner, G., Roduner, E. (1999) EPR investigation of HO. Radical initiated degradation reactions of sulfonated aromatics as model compounds for fuel cell proton conducting membranes. J. Mater. Chem. 9, 409–418.

Inaba, M., Yamada, H., Tokunaga, J., Tasaka, A. (2004) Effect of agglomeration of Pt/C catalyst on hydrogen peroxide formation. Electrochem. Solid State Lett. 7, A474–A476.

Inaba, M., Kinumoto, T., Kiriake, M., Umebayashi, R., Tasaka, A., Ogumi, Z. (2006) Gas crossover and membrane degradation in polymer electrolyte fuel cells. Electrochim. Acta 51, 5746–5753.

Kadirov, M.K., Bosnjakovic, A., Schlick, S. (2005) Membrane-derived fluorinated radicals detected by electron spin resonance in UV-irradiated Nafion and Dow ionomers: Effect of counterions and H_2O_2. J. Phys. Chem. 109, 7664–7670.

Kinumoto, T., Inaba, M., Nakayama, Y., Ogata, K., Umebayashi, R., Tasaka, A., Iriyama, Y., Abe, T., Ogumi, Z. (2006) Durability of perfluorinated ionomer membrane against hydrogen peroxide. J. Power Sources 158, 1222–1228.

Knights, S.D., Colbow, K.M., St-Pierre, J., Wilkinson, D.P. (2004) Aging mechanisms and lifetime of PEFC and DMFC. J. Power Sources 127, 127–134.

LaConti, A.B., McDonald, D.I., Austin, J.F. (1968) Technical Memo, 68–2, General Electric Company.

LaConti, A.B. (1988) Hydrogen and oxygen fuel cell development. *The MIT/Marine Industry Collegium, Power Systems for Small Underwater Vehicles*, Cambridge, MA.

LaConti, A.B., Fragala, A.R., Boyack, J.R. *Proceedings of the Symposium on Electrode Materials and process for Energy Conversion and Storage*, The Electrochemical Society, Los Angels, CA, 1977, p. 354.

LaConti, A.B., Hamdan, M., McDonald, R.C. (2003) Mechanisms of membrane degradation for PEMFCs. In: Vielstich, W., Lamn, A., Gasteiger, H.A. (Eds.), *Handbook of Fuel Cells – Fundamentals, Technology and Applications*, Wiley, New York, NY, vol. 3, p. 647.

LaConti, A.B., Liu, H., Mittelsteadt, C., McDonald, R.C. (2005) Polymer electrolyte membrane degradation mechanisms in fuel cells – Findings over the past 30 years and comparison with electrolyzers. ECS Trans. 1, 199–219.

Liu, W., Zuckerbrod, D. (2005) In situ detection of hydrogen peroxide in PEM fuel cells. J. Electrochem. Soc. 152, A1165–A1170.

Liu, W., Ruth, K., Rusch, G. (2001) Membrane durability in PEM fuel cells. J. New Mater. Electrochem. Syst. 4, 227–232.

Liu, J.G., Zhou, Z.H., Zhao, X.X., Xin, Q., Sun, G.Q., Yi, B.L. (2004) Studies on performance degradation of a direct methanol fuel cell (DMFC) in life test. PCCP 6, 134–137.

Liu, H., Gasteiger, H.A., LaConti, A.B., Zhang, J. (2005) Factors impacting chemical degradation of perfluorinated sulfonic acid ionomers. ECS Trans. 1, 283–293.

Liu, H., Zhang, J., Coms, F., Gu, W., Gasteiger, H.A. (2006) Impact of gas partial pressure on PEMFC chemical degradation. ECS Trans. 3, 493–505.

Luo, Z., Li, D., Tang, H., Pan, M., Ruan, R. (2006) Degradation behavior of membrane-electrode-assembly materials in 10-cell PEMFC stack. Int. J. Hydrogen Energy 31, 1831–1837.

Makharia, R., Mathias, M.F., Baker, D.R. (2005) Measurement of catalyst layer electrolyte resistance in PEFCs using electrochemical impedance spectroscopy. J. Electrochem. Soc. 152, A970–A977.

Mathias, M.F., Makharia, R., Gasteiger, H.A., Conley, J.J., Fuller, T.J., Gittleman, C.J., Kocha, S.S., Miller, D.P., Mittelsteadt, C.K., Xie, T., Yan, S.G., Yu, P.T. (2005) Two Fuel Cell Cars In Every Garage? Electrochem. Soc. Interface 14, 24–35.

Matic, H., Lundblad, A., Lindbergh, G., Jacobsson, P. (2005) In situ micro-Raman on the membrane in a working PEM cell. Electrochem. Solid State Lett. 8, A5–A7.

Mattsson, B., Ericson, H., Torell, L.M., Sundholm, F. (2000) Degradation of a fuel cell membrane, as revealed by micro-Raman spectroscopy. Electrochim. Acta 45, 1405–1408.

Mitov, S., Panchenko, A., Roduner, E. (2005) Comparative DFT study of non-fluorinated and perfluorinated alkyl and alkyl-peroxy radicals. Chem. Phys. Lett. 402, 485–490.

Mitov, S., Delmer, O., Kerres, J., Roduner, E. (2006a) Oxidative and photochemical stability of ionomers for fuel-cell membranes. Helv. Chim. Acta 89, 2354–2370.

Mitov, S., Hubner, G., Brack, H.P., Scherer, G.G., Roduner, E. (2006b) In situ electron spin resonance study of styrene grafting of electron irradiated fluoropolymer films for fuel cell membranes. J. Polym. Sci. Part B Polym. Phys. 44, 3323–3336.

Mittal, V.O., Kunz, H.R., Fenton, J.M. (2006a) Effect of catalyst properties on membrane degradation rate and the underlying degradation mechanism in PEMFCs. J. Electrochem. Soc. 153, A1755–A1759.

Mittal, V.O., Kunz, H.R., Fenton, J.M. (2006b) Is H_2O_2 involved in the membrane degradation mechanism in PEMFC? Electrochem. Solid State Lett. 9, A299–A302.

Mittal, V.O., Kunz, H.R., Fenton, J.M. (2006c) Membrane degradation mechanisms in PEMFCs. ECS Trans. 3, 507–517.

Mittal, V.O., Kunz, H.R., Fenton, J.M. (2007) Membrane degradation mechanisms in PEMFCs. J. Electrochem. Soc. 154, B652–B656.

Miyake, N., Wakizoe, M., Honda, E., Ohta, T., High durability of Asahi kasei aciplex membrane. *Abstracts of 208th Meeting of the Electrochemical Society*, Los Angels, CA, 2005.

Miyatake, K., Watanabe, M. (2006) Emerging membrane materials for high temperature polymer electrolyte fuel cells: Durable hydrocarbon ionomers. J. Mater. Chem. 16, 4465–4467.

Mo, Y.B., Scherson, D.A. (2003) Platinum-based electrocatalysts for generation of hydrogen peroxide in aqueous acidic electrolytes – Rotating ring-disk studies. J. Electrochem. Soc. 150, E39–E46.

Multi-Year Research, Development and Demonstration Plan: Planned program activities for 2004–2015. United States Department of Energy, 2007.

Murthi, V.S., Urian, R.C., Mukerjee, S. (2004) Oxygen reduction kinetics in low and medium temperature acid environment: Correlation of water activation and surface properties in supported Pt and Pt alloy electrocatalysts. J. Phys. Chem. 108, 11011–11023.

Nakano, T., Nagaoka, S., Kawakami, H. (2005) Preparation of novel sulfonated block copolyimides for proton conductivity membranes. Polym. Adv. Technol. 16, 753–757.

Nasef, M.M., Saidi, H. (2002) Post-mortem analysis of radiation grafted fuel cell membrane using X-ray photoelecton spectroscopy. J. New Mater. Electrochem. Syst. 5, 183–189.

Ogumi, Z., Takehara, Z., Yoshizawa, S. (1984) Gas permeation in SPE Method.1. Oxygen permeation through Nafion and neosepta. J. Electrochem. Soc. 131, 769–773.

Ogumi, Z., Kuroe, T., Takehara, Z. (1985) Gas permeation in SPE Method 2. Oxygen and hydrogen permeation through Nafion. J. Electrochem. Soc. 132, 2601–2605.

Ohma, A., Suga, S., Yamamoto, S., Shinohara, K. (2006) Phenomenon analysis of PEFC for automotive use(1) membrane degradation behavior during OCV hold test. ECS Trans. 3, 519–529.

Ohma, A., Suga, S., Yamamoto, S., Shinohara, K. (2007a) Membrane degradation behavior during open-circuit voltage hold test. J. Electrochem. Soc. 154, B757–B760.

Ohma, A., Yamamoto, S., Shinohara, K. (2007b) Analysis of membrane degradation behavior during OCV hold test. ECS Trans. 11, 1181–1192.

Okada, T., Dale, J., Ayato, Y., Asbjornsen, O.A., Yuasa, M., Sekine, I. (1999) Unprecedented effect of impurity cations on the oxygen reduction kinetics at platinum electrodes covered with perfluorinated ionomer. Langmuir 15, 8490–8496.

Okada, T., Satou, H., Yuasa, M. (2003) Effects of additives on oxygen reduction kinetics at the interface between platinum and perfluorinated ionomer. Langmuir 19, 2325–2332.

Panchenko, A., Dilger, H., Moller, E., Sixt, T., Roduner, E. (2004) In situ EPR investigation of polymer electrolyte membrane degradation in fuel cell applications. J. Power Sources 127, 325–330.

Panchenko, A. (2006) DFT investigation of the polymer electrolyte membrane degradation caused by OH radicals in fuel cells. J. Membr. Sci. 278, 269–278.

Paulus, U.A., Wokaun, A., Scherer, G.G., Schmidt, T.J., Stamenkovic, V., Markovic, N.M., Ross, P.N. (2002) Oxygen reduction on high surface area Pt-based alloy catalysts in comparison to well defined smooth bulk alloy electrodes. Electrochim. Acta 47, 3787–3798.

Pianca, M., Barchiesi, E., Esposto, G., Radice, S. (1999) End groups in fluoropolymers. J. Fluorine Chem. 95, 71–84.

Pozio, A., Silva, R.F., De Francesco, M., Giorgi, L. (2003) Nafion degradation in PEFCs from end plate iron contamination. Electrochim. Acta 48, 1543–1549.

Preli, F., Progress in improving durability of pem fuel cells for stationary and transportation applications. *Fourth International Fuel Cell Workshop 2005*, Yamanashi, Japan, 2005.

Qiao, J.L., Saito, M., Hayamizu, K., Okada, T. (2006) Degradation of perfluorinated ionomer membranes for PEM fuel cells during processing with H_2O_2. J. Electrochem. Soc. 153, A967–A974.

Roeselova, M., Vieceli, J., Dang, L.X., Garrett, B.C., Tobias, D.J. (2004) Hydroxyl radical at the air–water interface. J. Am. Chem. Soc. 126, 16308–16309.

Scherer, G.G. (1990) Polymer membranes for fuel-cells. Ber. Bunsen-Ges. Phys. Chem. Chem. Phys. 94, 1008–1014.

Schiraldi, D.A. (2006) Perfluorinated polymer electrolyte membrane durability. Polym. Rev. 46, 315–327.

Schmidt, T.J., Paulus, U.A., Gasteiger, H.A., Behm, R.J. (2001) The oxygen reduction reaction on a Pt/carbon fuel cell catalyst in the presence of chloride anions. J. Electroanal. Chem. 508, 41–47.

Shim, J.Y., Tsushima, S., Hirai, S. (2007) Preferential thinning behaviors of the anode-side of the PEM under durability test. ECS Trans. 11, 1151–1156.

Smitha, B., Sridhar, S., Khan, A.A. (2005) Solid polymer electrolyte membranes for fuel cell applications – a review. J. Membr. Sci. 259, 10–26.

Sompalli, B., Litteer, B.A., Gu, W., Gasteiger, H.A. (2007) Membrane degradation at catalyst layer edges in PEMFC MEAs. J. Electrochem. Soc. 154, B1349–B1357.

Springer, T.E., Zawodzinski, T.A., Gottesfeld, S. (1991) Polymer electrolyte fuel-cell model. J. Electrochem. Soc. 138, 2334–2342.

St Pierre, J., Wilkinson, D.P., Knights, S., Bos, M.L. (2000) Relationships between water management, contamination and lifetime degradation in PEFC. J. New Mater. Electrochem. Syst. 3, 99–106.

Stamenkovic, V., Markovic, N.M., Ross, P.N. (2001) Structure-relationships in electrocatalysis: Oxygen reduction and hydrogen oxidation reactions on Pt(111) and Pt(100) in solutions containing chloride ions. J. Electroanal. Chem. 500, 44–51.

Stucki, S., Scherer, G.G., Schlagowski, S., Fischer, E. (1998) PEM water electrolysers: Evidence for membrane failure in 100 kW demonstration plants. J. Appl. Electrochem. 28, 1041–1049.

Takeshita, T., Miura, F., Morimoto, Y., Abstract 1511. *Abstracts of 207th Electrochemical Society Meeting*, Quebec City, Quebec, Canada, 2005.

Tanuma, T., Terazono, S. (2006) Improving MEA durability by using a catalyst with a small number of functional groups on its surface. Chem. Lett. 35, 1422–1423.

Vogel, B., Aleksandrova, E., Mitov, S., Krafft, M., Dreizler, A., Kerres, J., Hein, M., Roduner, E. (2007) Observation of fuel cell membrane degradation by ex-situ and in-situ electron paramagnetic resonance. ECS Trans. 11, 1105–1114.

Wang, H., Capuano, G.A. (1998) Behavior of Raipore radiation-grafted polymer membranes in H-2/O-2 fuel cells. J. Electrochem. Soc. 145, 780–784.

Weir, N.A. (1978) Reactions of hydroxyl radicals with polystyrene. Eur. Polym. J. 14, 9–14.

Wilkie, C.A., Thomsen, J.R., Mittleman, M.L. (1991) Interaction of poly (methyl-methacrylate) and Nafions. J. Appl. Polym. Sci. 42, 901–909.

Xie, J., Wood, D.L., Wayne, D.M., Zawodzinski, T.A., Atanassov, P., Borup, R.L. (2005) Durability of PEFCs at high humidity conditions. J. Electrochem. Soc. 152, A104–A113.

Yokoyama, T., Matsumoto, Y., Meshitsuka, G. (2002) Enhancement of the reaction between pulp components and hydroxyl radical produced by the decomposition of hydrogen peroxide under alkaline conditions. J. Wood Sci. 48, 191–196.

Yoshioka, S., Yoshimura, A., Fukumoto, H., Hiroi, O., Yoshiyasu, H. (2005) Development of a PEFC under low humidified conditions. J. Power Sources 144, 146–151.

Yu, J.R., Yi, B.L., Xing, D.M., Liu, F.Q., Shao, Z.G., Fu, Y.Z. (2003) Degradation mechanism of polystyrene sulfonic acid membrane and application of its composite membranes in fuel cells. PCCP 5, 611–615.

Yu, J.R., Matsuura, T., Yoshikawa, Y., Islam, M.N., Hori, M. (2005a) In situ analysis of performance degradation of a PEMFC under nonsaturated humidification. Electrochem. Solid State Lett. 8, A156–A158.

Yu, J.R., Matsuura, T., Yoshikawa, Y., Islam, M.N., Hori, M. (2005b) Lifetime behavior of a PEM fuel cell with low humidification of feed stream. PCCP 7, 373–378.

Zhang, L., Mukerjee, S. (2006) Investigation of durability issues of selected nonfluorinated proton exchange membranes for fuel cell application. J. Electrochem. Soc. 153, A1062–A1072.

Zhang, J., Litteer, B.A., Gu, W., Liu, H., Gasteiger, H.A. (2007) Effect of hydrogen and oxygen partial pressure on Pt precipitation within the membrane of PEMFCs. J. Electrochem. Soc. 154, B1006–B1011.

Improvement of Membrane and Membrane Electrode Assembly Durability

Eiji Endoh and Satoru Hommura

Abstract The world's first highly durable perfluorinated polymer-based membrane electrode assembly (MEA) for polymer electrolyte membrane fuel cells, under conditions of high temperature and low humidity, has been developed. The newly developed MEA, which is composed of a new perfluorinated polymer composite membrane, reduces the degradation rate to 1/100th to 1/1,000th of that of the conventional MEA. The new perfluorinated polymer composite MEA can be operated for more than 6,000 h at 120°C and 50% relative humidity.

1 Introduction

In the development of polymer electrolyte membrane fuel cells (PEMFC), the durability of the membrane is one of the most important issues. For automotive utilization of PEMFCs, high-temperature operations between 110 and 120°C and low-humidity conditions are required. For the membranes of PEMFCs, perfluorosulfonic acid (PFSA) polymers are extensively used owing to their considerably higher chemical stability over hydrocarbon polymers. However, conventional PFSA polymers have glass-transition temperatures around 80°C, and are subject to critical breakdown of the membrane at high temperatures. Additionally, conventional PFSA polymers suffer from degradation under low-humidity conditions even at 80°C.

To develop a durable membrane electrode assembly (MEA), it is of the utmost importance that the degradation mechanism is understood. A degradation study of a conventional MEA under low-humidity open circuit voltage (OCV) conditions was conducted (Endoh et al. 2004), and it was confirmed that the hydroxyl radical is the main cause of the MEA degradation. To overcome these challenges, extensive research has been conducted.

The world's first highly durable perfluorinated polymer-based MEA, under conditions of high temperature and low humidity, has been developed (Endoh and

E. Endoh (✉)
Asahi Glass Co. Ltd, 1150 Hazawa-cho, Kanagawa-ku, Yokohama 221-8755, Japan
e-mail: eiji-endoh@agc.co.jp

F.N. Büchi et al. (eds.), *Polymer Electrolyte Fuel Cell Durability*,

Kawazoe 2005). The newly developed MEA, which is composed of a new perfluorinated polymer composite (NPC) membrane, reduces the degradation rate to 1/100th to 1/1,000th of that of the conventional MEA. The NPC MEA can be operated for more than 6,000 h at 120°C and 50% relative humidity (RH). This achievement has opened up a new era of operating PEMFCs at 120°C and low RHs.

2 Experimental

2.1 *Membrane Electrode Assembly*

The electrocatalysts for oxygen reduction and hydrogen oxidation were Pt/C or Pt alloy/C. The cathode platinum loading values and anode platinum loading values were varied. The control MEAs were prepared by hot-pressing the electrodes onto Flemion® SH50 membranes (thickness 50 µm). The NPC MEAs were prepared by hot-pressing the electrodes onto the NPC membranes (thickness 40 or 25 µm). The geometrical area of the electrode was 25 cm^2.

2.2 *Evaluation of Cell Performances*

Current–voltage curves and durability tests of the MEAs were measured by utilizing a single cell with a square-shaped active area (25 cm^2). The cell temperature was kept constant at 120°C. Utilization of hydrogen and air was 50:50. For the measurement of the current–voltage curves, the cell pressures were varied in value from 127 to 280 kPa, and the RH of the inlet gas (both hydrogen and air) was varied from 25 to 100%. The OCV tests were conducted with the same cell.

2.3 *Confirmation of Hydrogen Peroxide Formation*

Air which contained 1.3% hydrogen gas was bubbled into an aqueous dispersion of Ketjenblack or a commercially available Ketjenblack-supported platinum catalyst (Pt/C). The formation of H_2O_2 was examined via titration.

2.4 *Identification of Radical Species*

Thirty milligrams of the Pt/C catalyst or Ketjenblack was dispersed in 1 ml of ultrapure water, which contained 200 µl of 5,5-dimethyl-1-pyrroline-*N*-oxide (DMPO). Next, air which contained 1.3% hydrogen gas was bubbled into the dispersion for 20 min. The filtrate of each dispersion was analyzed with a JEOL JES-TE 300 electron spin resonance (ESR) spectrometer at room temperature.

3 Results and Discussion

3.1 Clarification of the MEA Degradation Mechanism

To develop a durable MEA, it is of the utmost importance that the degradation mechanism is understood. The degradation study of the conventional MEA using a low-humidity OCV test was conducted. Figure 1 indicates the ESR spectra of the degraded MEA at liquid-nitrogen temperature. Large ESR signals at 327 mT were observed within the catalyst layers of the degenerated MEA. The signals observed at 327 mT were assigned to carbon radicals (Endoh et al. 2004). A probable carbon radical formation mechanism was the abstraction of hydrogen, which was bonded directly to carbon atoms, by the hydroxyl or hydroperoxyl radicals.

To identify the radical species, model tests were conducted. First, air which contained 1.3% hydrogen gas was bubbled into an aqueous dispersion of Ketjenblack or the commercially available Pt/C catalyst. The formation of H_2O_2 was confirmed. Next, the radical species which were generated during the decomposition of H_2O_2 were identified. Ketjenblack or the Pt/C catalyst was added into the H_2O_2 solution, which contained DMPO, and the generation of hydroxyl radical was confirmed by ESR measurement. Then, the subsequent experiment of the generation and decomposition of the H_2O_2 was conducted. Figure 2 shows the

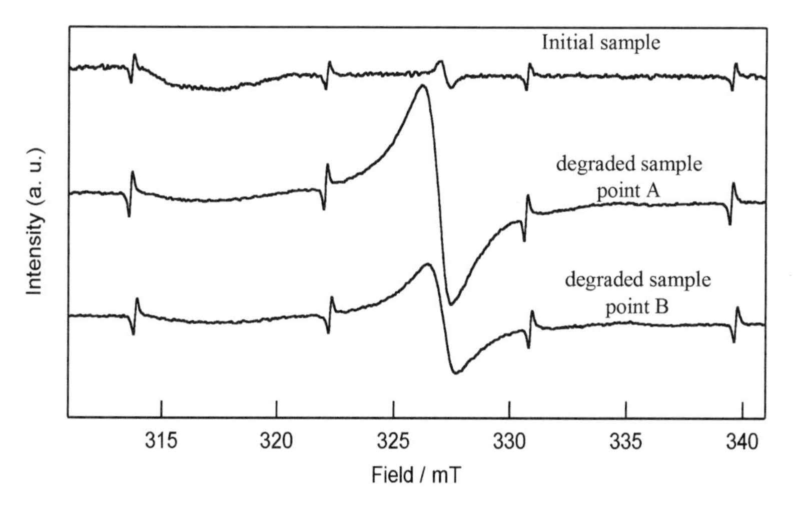

Fig. 1 Electron spin resonance (ESR) spectra of the degraded membrane electrode assembly (MEA) at liquid nitrogen temperature. (Reproduced with permission from Endoh et al. (2004). Copyright 2004, The Electrochemical Society, Inc.)

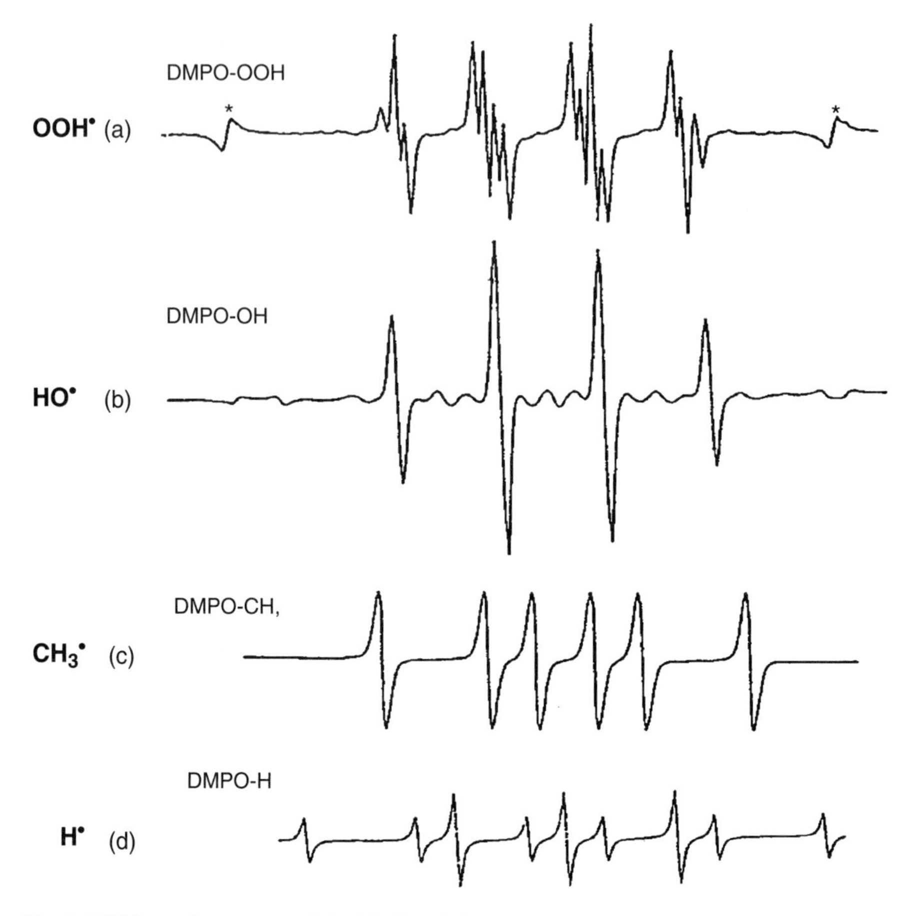

Fig. 2 ESR hyperfine spectra of the 5,5-dimethyl-1-pyrroline-*N*-oxide (DMPO) adducts

ESR hyperfine spectra of the DMPO adducts, which reacted with the respective radicals.

Figure 3 indicates the ESR spectra of the Pt/C catalyst dispersed solution. Figure 4 indicates the ESR spectra of Ketjenblack-dispersed solution. Both spectra indicate the formation of radicals, which corresponded with the hydroxyl radical. From these experimental results, the degradation mechanism of the MEA under the OCV condition can be determined. At the cathode, H_2O_2 is chemically formed, and a hydroxyl radical is generated by the decomposition of the H_2O_2. At the anode, H_2O_2 is chemically or electrochemically generated, and a hydroxyl radical is similarly generated by the decomposition of the H_2O_2. The hydroxyl radical will attack the ionomer, membrane, and carbon support of the catalyst.

Because of these results, it was elucidated that the MEA degradation proceeds via chemical attack by the hydroxyl radical, which is generated in the cathode and anode. The degradation mechanism of MEA under the OCV condition is summarized as follows (Endoh 2006).

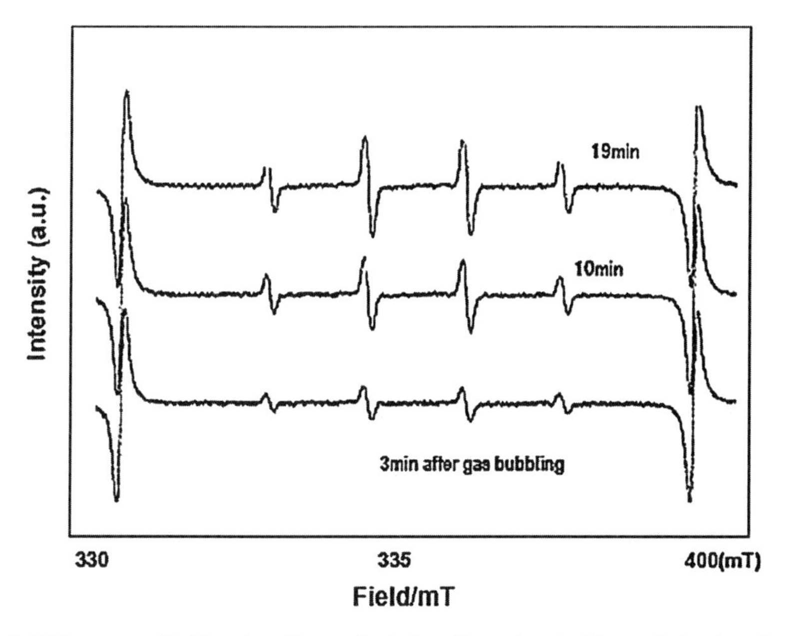

Fig. 3 ESR spectra of Pt/C catalyst dispersed solution. (Reproduced with permission from Endoh (2006). Copyright 2006, The Electrochemical Society, Inc.)

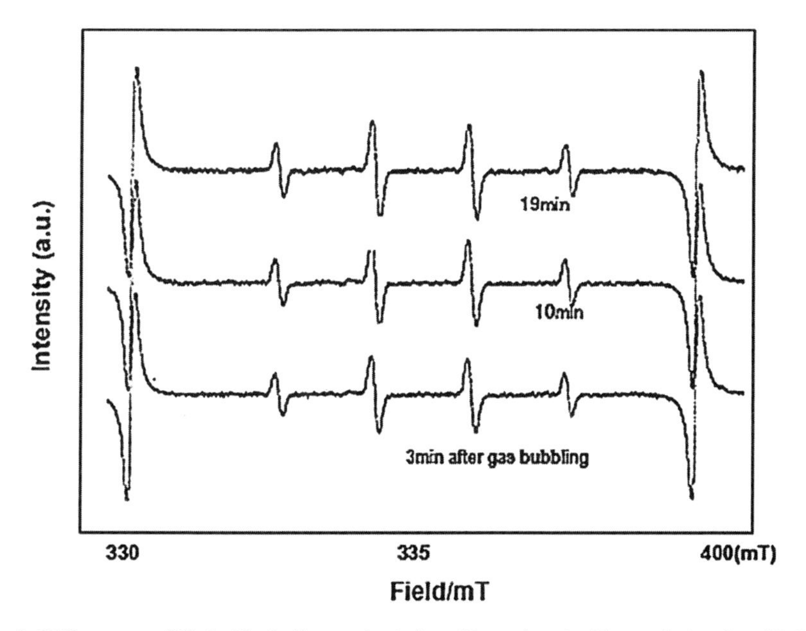

Fig. 4 ESR spectra of Ketjenblack-dispersed solution. (Reproduced with permission from Endoh (2006). Copyright 2006, The Electrochemical Society, Inc.)

At the cathode

$$H_2 + O_2 \rightarrow H_2O_2 \text{ (chemically formed)}$$

$$H_2O_2 \rightarrow 2HO^{\bullet} \text{ (chemically generated)}$$

At the anode

$$H_2 + O_2 \rightarrow H_2O_2 \text{ (chemically formed)}$$

$$2H^+ + O_2 + 2e^- \rightarrow H_2O_2 \text{ (electrochemically formed)}$$

$$H_2O_2 \rightarrow 2HO^{\bullet} \text{ (chemically generated)}$$

3.2 Clarification of the Membrane Degradation Mechanism

Another degradation form of the membrane was discovered. The carboxylic acid concentration profile in the cross section of the OCV-tested membrane was investigated by an infrared imaging method. The carboxylic acid concentration is an indicator of the degradation. It was confirmed that the inside portion of the membrane, especially the anode portion, was significantly degraded. Therefore, the degradation mechanism of the perfluorinated membrane was investigated by a new accelerated method (Hommura et al. 2008).

Table 1 provides information on commonly implemented accelerated tests. The OCV test, which was mentioned before, has an advantage in the fact that its experimental conditions are close to the operational conditions. However, the OCV test can be influenced by impurities in the electrode. In contrast, the Fenton reagent test is easy to implement. However, this test is far from the operational conditions, that is, the test is conducted in an aqueous solution, and contains a large number of ferrous ions. Therefore, a new accelerated degradation method was developed.

The new method simulates operational conditions to which the membrane would be exposed. The method is based on the assumption that H_2O_2 which was formed in the catalyst layer diffuses into the membrane and subsequently attacks it. Additionally, the method takes into account the fact that the degradation is significant under low-humidity conditions.

Figure 5 represents the new accelerated degradation method, in which the membrane is exposed to H_2O_2 gas. In this method, a 30% H_2O_2 solution is heated to 73°C, and the H_2O_2 vapor is fed by nitrogen into a reaction chamber.

The chamber is then heated to 120°C, and the perfluorinated membrane, Flemion® SH50, is placed in the chamber. The RH of the H_2O_2 vapor is approximately 12%. As an indicator of the degradation of the membrane, fluoride ion in the effluent gas is trapped in a KOH solution and analyzed.

Figure 6a indicates the weight loss of the membrane and fluoride ion release rate (μg day^{-1} cm^{-2}). An acceleration of fluoride ion release rate was observed with

Table 1 Commonly implemented accelerated tests

Accelerated tests	Advantage	Disadvantage
OCV Test	Close to operational conditions	Influence of impurities in electrodes
Fenton reagent	Easy to implement	Far from operational conditions (Wet Fe^{2+} ion)

OCV Open-circuit voltage

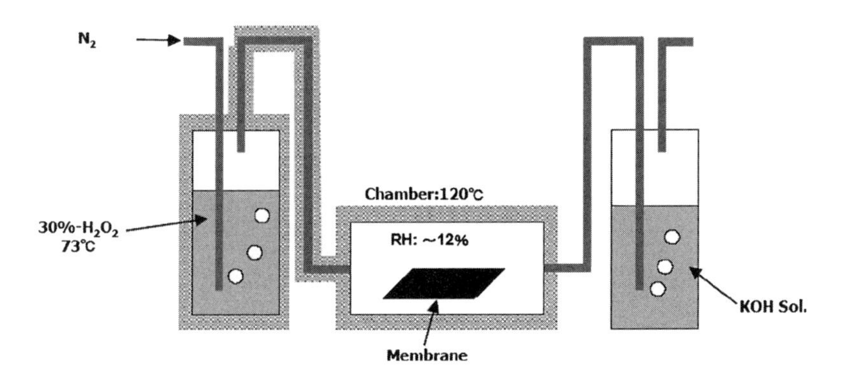

Fig. 5 New accelerated degradation method (H_2O_2 exposure method). RH relative humidity (RH). (Reproduced with permission from Hommura (2008). Copyright 2008, The Electrochemical Society, Inc.)

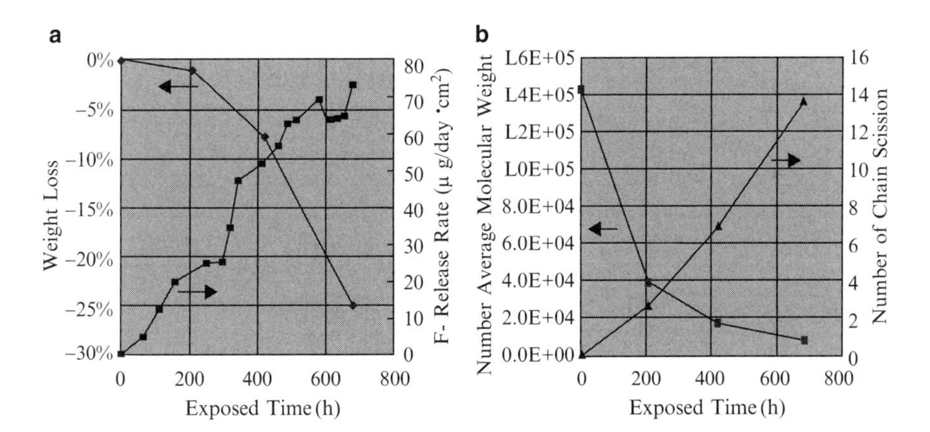

Fig. 6 Variations of weight loss and F^- release rate (**a**), and molecular weight of polymer and number of chain scissions with time in the H_2O_2 exposure test (**b**). (Reproduced with permission from Hommura (2008). Copyright 2008, The Electrochemical Society, Inc.)

time. Since the fluoride ion will be generated by the reaction between the unstable polymer end groups and the hydroxyl radical (Curtin 2004), it is obvious that the number of unstable polymer end groups increased during the exposure of the membrane to H_2O_2.

Molecular weight distributions were measured with size-exclusion chromatography and number-average molecular weights (M_n) were estimated (Fig. 6b). The rapid decrease of M_n is probably due to the scission of the polymer main chains. The number of chain scissions calculated from M_n and the weight loss is also indicated in Fig. 6b. From these results, it was determined that the membranes are decomposed by H_2O_2 alone under low-humidity conditions, that is, the Fe^{2+} ion is not involved. Membrane degradation proceeds via the following two steps:

1. An unzipping reaction at unstable polymer end groups
2. Scission of the main chains

That is, an unzipping reaction at unstable polymer end groups proceeds, and of the main chains occurs, which produces new unstable end groups, and allows the unzipping reaction to continue (Hommura et al. 2008). Figure 7 illustrates the above-mentioned degradation mechanisms.

3.3 Development of a Highly Durable MEA

The MEA must have sufficient chemical stability against the attack by the hydroxyl radical. Therefore, extensive research has been conducted, and a highly durable MEA with a NPC membrane has been developed.

3.3.1 OCV Durability Test at 120°C

Figure 8 indicates the OCV durability test at 120°C. The cell temperature was kept at 120°C, and the RHs of hydrogen and air were 18%. The cell pressure was ambient.

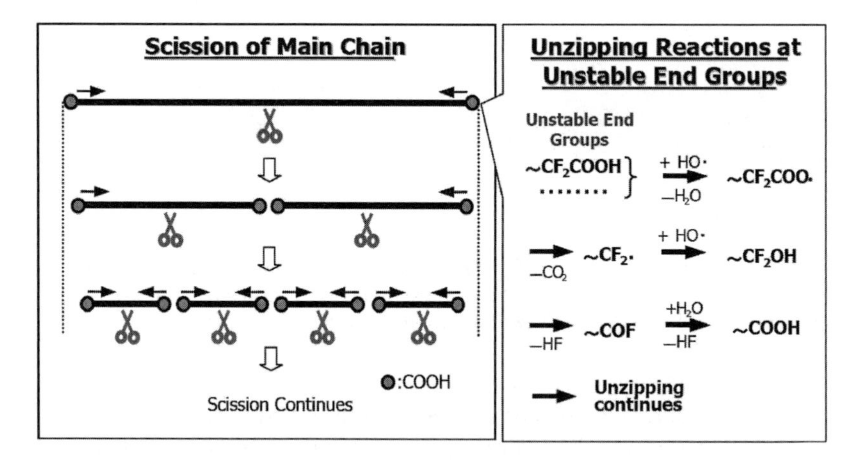

Fig. 7 Degradation mechanisms of the perfluorinated membrane. (Reproduced with permission from Hommura (2008). Copyright 2008, The Electrochemical Society, Inc.)

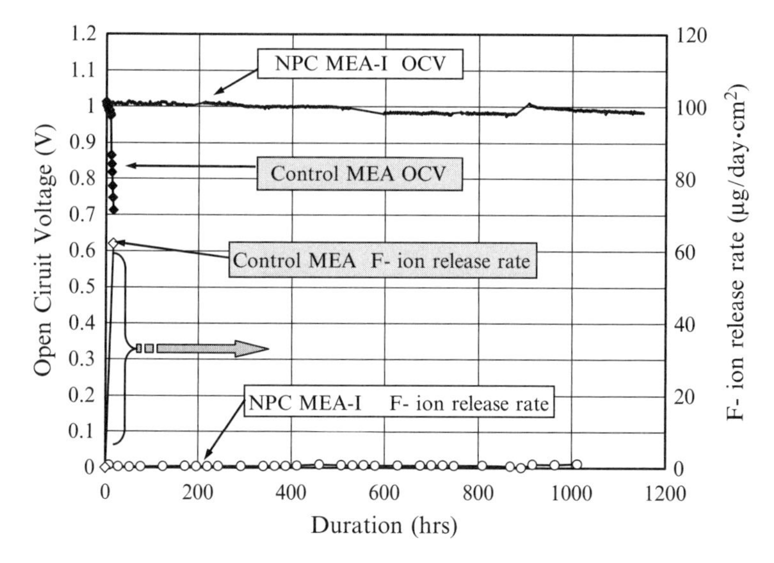

Fig. 8 Open circuit voltage (OCV) durability test at 120°C. Reactant, H/air; pressure, ambient. *NPC* new perfluorinated polymer composite. (Reproduced with permission from Endoh (2006). Copyright 2006, The Electrochemical Society, Inc.)

For the conventional MEA, Flemion® SH50 membrane was used. The Cathode platinum loading was 0.6 mg cm^{-2}, and the anode platinum loading was 0.3 mg cm^{-2}.

For NPC MEA-I, the NPC membrane was applied. The thickness of the membrane was 40 μm. The conventional MEA failed within 10 h of operation, releasing a high amount of fluoride ion. On the other hand, NPC MEA-I showed excellent stability over 1,000 h, and the fluoride ion release rate was less than 1% of that of the conventional MEA. This result indicates the exceptional chemical stability of the NPC MEA against degradation caused by the hydroxyl radical at 120°C and low humidity.

3.3.2 Current–Voltage Curves of NPC MEA-I at 120°C

Figure 9 indicates the current–voltage curves of NPC MEA-I at 120°C, and at various RHs. Both the anode and the cathode were subjected to similar RH. The cell pressure was varied to maintain the oxygen concentration in the cathode gas at 16%. NPC MEA-I showed good performance at 100% RH. The MEA was also able to operate at a lower RH of 25% and 127 kPa.

3.3.3 Constant Current Durability at 120°C and 50% RH

Figure 10 indicates the constant current durability of MEAs at 120°C and 50% RH. The current density was 0.2 A cm^{-2}. The pressure of the cell was kept constant at 200 kPa_{abs}, and the utilization of H_2/air was 50:50. The control MEA failed within

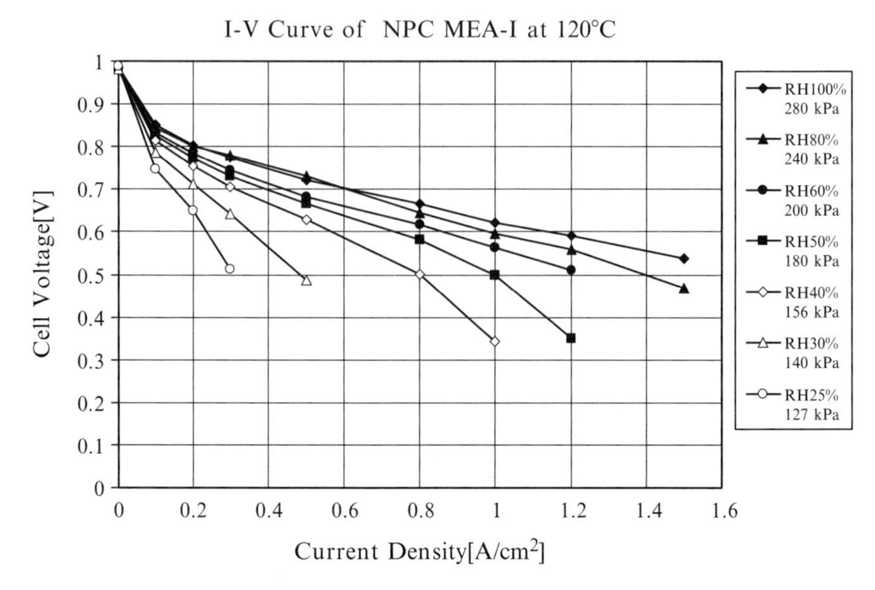

Fig. 9 Current–voltage curves of NPC MEA-I at 120°C. Reactant gases were H_2 and air. Utilization of H_2 and air was 50:50. (Reproduced with permission from Endoh (2006). Copyright 2006, The Electrochemical Society, Inc.)

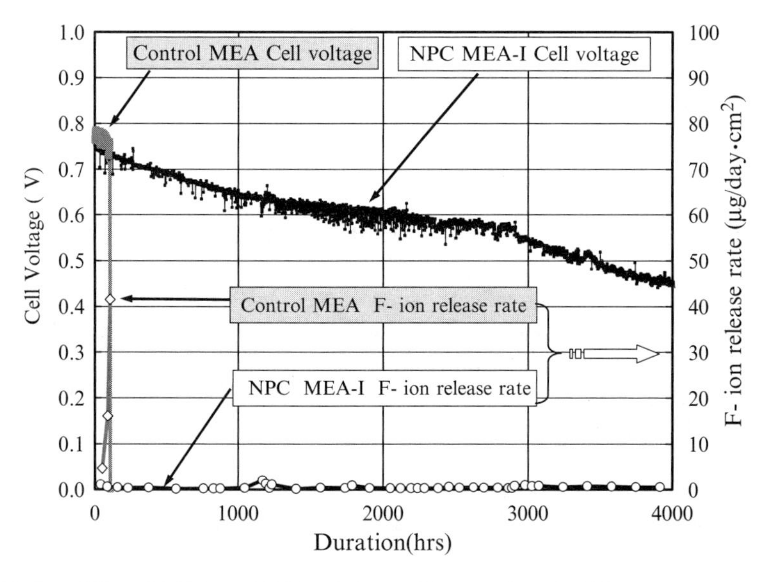

Fig. 10 Durability of MEAs at 120°C, 50% RH, 0.2 A cm^{-2}, and 200 kPa_{abs}. Reactant, H_2/air, utilization:50:50; cathode platinum, 0.6 mg cm^{-2}; anode platinum 0.3 mg cm^{-2}. (Reproduced with permission from Endoh (2006). Copyright 2006, The Electrochemical Society, Inc.)

100 h of operation, releasing a high amount of fluoride ion. NPC MEA-I showed excellent stability and was operated for more than 4,000 h. The fluoride ion release rate of NPC MEA-I was approximately 2×10^{-8} g F^{-} cm^{-2} h^{-1}. The fluoride ion release rate of the conventional MEA operated at 70°C and 100% RH varies from 1×10^{-8} to 10×10^{-8} g F^{-} cm^{-2} h^{-1} (Cleghorn et al. 2003). The fluoride ion release rate of NPC MEA-I which was operated at 120°C and 50% RH was similar to the fluoride ion release rate of the conventional MEA operated at 70°C and 100% RH.

The degradation rate of NPC MEA-I for 4,000 h of operation at 120°C and 50% RH was approximately 75 μV h^{-1}. The degradation rate of NPC MEA-I operated at 70°C and 100% RH was approximately 3 μV h^{-1}. Therefore, the degradation rate of NPC MEA-I at 120°C and 50% RH was approximately 25 times higher than the degradation rate of the MEA at 70°C and 100% RH. To investigate the degradation mechanism, this test was terminated and the MEA was analyzed.

3.3.4 Analyses of the NPC MEA Operated at 120°C for 4,000 h

Figure 11 shows the scanning electron microscope image of the cross section of NPC MEA-I. The thickness of the membrane was 40 μm, and that of the cathode layer was 18 μm. The membrane thickness remained constant after 4,000 h of operation at 120°C; however, the thickness of the cathode catalyst layer decreased to approximately 8 μm. The catalyst layer was further investigated by transmission electron microscopy. The carbon support of the cathode catalyst decreased significantly, and large platinum particles were dispersed in the catalyst coating ionomer. (Fig. 12 shows a schematic illustration of the cross section of the cathode layer.)

This occurred during 4,000 h of operation at 120°C, and resulted in the decrease of the cathode catalyst layer thickness. This may be due to the oxidation of the carbon support by the H_2O_2, which was formed during the oxygen reduction reaction. Substantial crystallite growth of platinum catalyst was observed in that region. X-ray diffraction measurement indicated that the crystallite size of the platinum catalyst increased from 2 to 8 nm. The electrochemical cathode

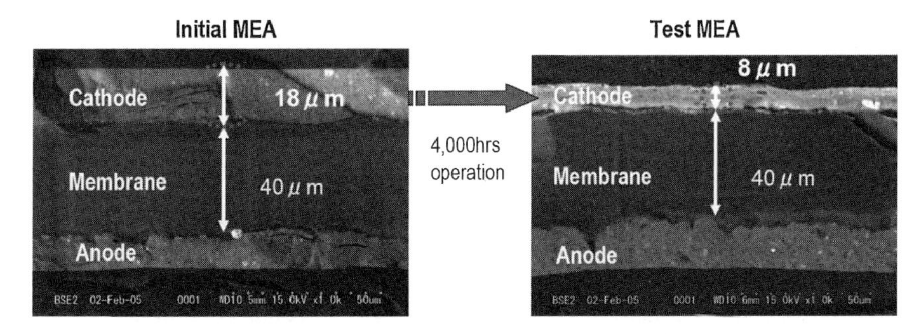

Fig. 11 Scanning electron microscope image of the cross section of NPC MEA-I. (Reproduced with permission from Endoh (2006). Copyright 2006, The Electrochemical Society, Inc.)

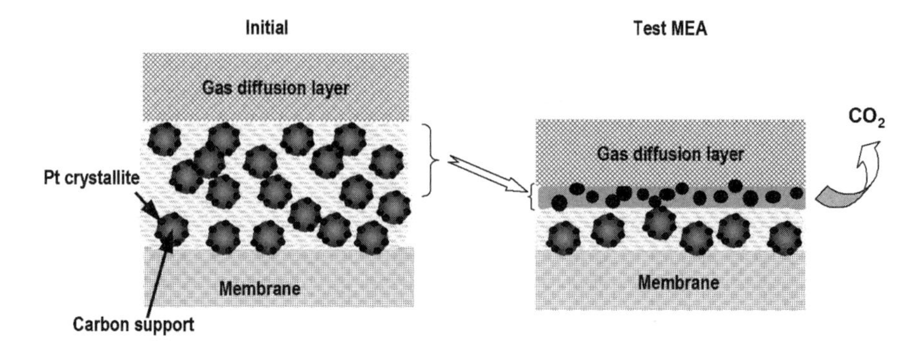

Fig. 12 Cross section of the cathode layer. (Reproduced with permission from Endoh (2006). Copyright 2006, The Electrochemical Society, Inc.)

platinum surface area decreased by 97% after 4,000 h of operation at 120°C. Therefore, the performance loss of the MEA was attributed to the degradation of the cathode catalyst.

Regarding the corrosion mechanism of the carbon support, Hubner and Roduner (1999) reported on the HO• radical initiated degradation reactions of sulfonated aromatics for fuel cell membrane. They reported that HO• radical addition on the aromatic ring occurs, and is followed by the addition of an oxygen molecule. This will eventually lead to a bond-breaking reaction. Figure 13 indicates the degradation reactions of aromatics. From these results, the carbon support may be similarly decomposed to CO_2 by HO• and O_2.

3.3.5 Performance Improvement of the NPC MEAs

Strategies to improve the durability of the cathode catalyst are being considered. They are application of oxidation-resistant carbon support, and application of dissolution-resistant platinum alloy catalyst.

Figure 14 indicates the durability of NPC MEA-VII, which is composed of an originally developed cathode catalyst at 120°C, 50% RH, 200 kPa_{abs}, and 0.2 A cm^{-2}. The cathode platinum loading was 0.2 mg cm^{-2} and the thickness of the NPC membrane was 25 μm. The other experimental conditions were the same as mentioned before.

The MEA showed excellent stability and could be operated for 6,000 h. The degradation rate was approximately 3 μV h^{-1} over 6,000 h of operation, which is a considerably improved value compared with the previous result.

Figure 15 indicates the durability of NPC MEA-VIII. For this MEA, a further improved NPC membrane was applied. The other experimental conditions were the same as mentioned before. The degradation rate was approximately 2 μV h^{-1} over 5,000 h of operation.

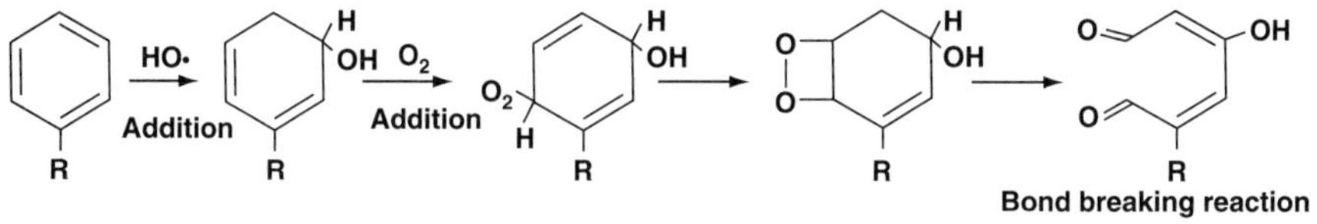

Fig. 13 Degradation reactions of aromatics (Hubner et al. 1999)

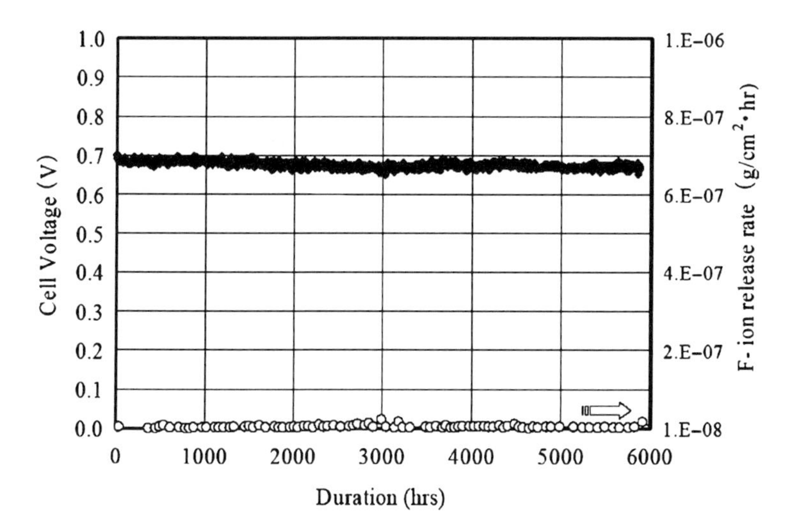

Fig. 14 Durability of NPC MEA-VII at 120°C, 50% RH, 200 kPa_{abs}, 0.2 A cm^{-2}. Reactant, H_2/air, utilization 50:50; cathode platinum, 0.2 mg cm^{-2}; anode platinum, 0.2 mg cm^{-2}. (Reproduced with permission from Endoh (2008). Copyright, 2008, The Electrochemical Society, Inc.)

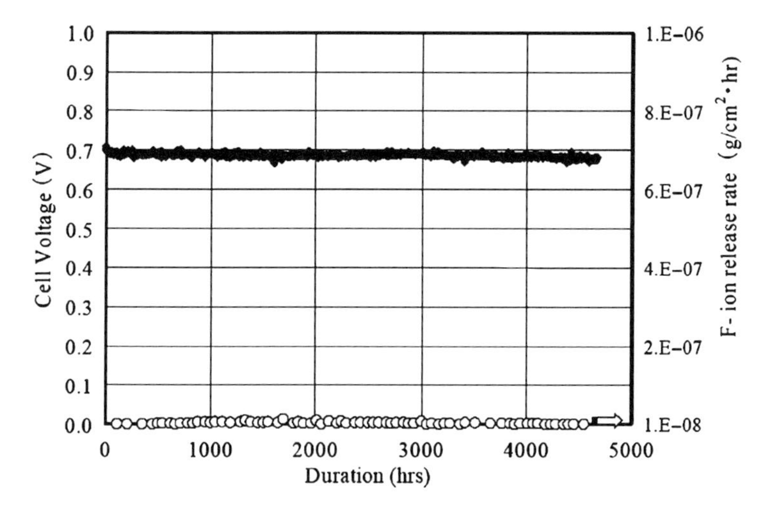

Fig. 15 Durability of NPC MEA-VIII at 120°C, 50% RH, 200 kPa_{abs}, 0.2 A cm^{-2}. Reactant, H_2/air, utilization 50:50; cathode platinum, 0.2 mg cm^{-2}; anode platinum, 0.2 mg cm^{-2}. (Reproduced with permission from Endoh (2008). The Electrochemical Society, Inc.)

4 Conclusions

The world's first highly durable perfluorinated polymer-based MEA for PEMFCs, under high temperature and low humidity conditions, has been developed. The MEA can be operated for more than 6,000 h at 120°C and 50% RH. This achievement has opened up a new era of operating PEMFCs in the range from freezing temperature to 120°C for an extended period. Additionally, this development will solve the degradation problem of the membrane for stationary usage, which will significantly extend the operational life of the PEMFC system.

Acknowledgments The authors would like to acknowledge the contributions of Hisao Kawazoe, Shinji Terazono, Hardiyanto Widjaja, Yasuyuki Takimoto, and Junko Anzai from Asahi Glass Research Center.

References

Cleghorn, J., Kolde, S., and Liu, W. (2003) In: W. Vielstich, A. Lamm and H.A. Gasteiger (Eds.), *Handbook of Fuel Cells – Fundamentals, Technology, and Applications, Chapter 44. Vol. 3*, Wiley, Chichester, UK, pp. 573.

Curtin, D.E. (2004) Journal of Power Sources 131, 41–48.

Endoh, E. (2006) ECS Transactions 3(1), 9–18.

Endoh, E. and Kawazoe, H. (2005) Abstract No.763, *The 207th Meeting Abstracts of the Electrochemical Society*, May 15–20, Quebec City, Canada.

Endoh, E., Terazono, S., Widjaja, H., and Takimoto, Y. (2004) Electrochemical and Solid-State Letters 7(7), A209–A211.

Hommura, S., Kawahara, K., Shimohira, T., and Teraoka, Y. (2008) *J. Electrochem. Soc.*, 155(1), A29.

Hubner, G. and Roduner, E. (1999) Journal of Material Chemistry 9, 409–418.

Endoh, E. (2008) ECS Transactions 12(1), 41–50.

Durability of Radiation-Grafted Fuel Cell Membranes

Lorenz Gubler and Günther G. Scherer

Abstract Partially fluorinated proton exchange membranes prepared via radiation-induced graft copolymerization ("radiation grafting") offer the prospect of cost-effective and tailor-made membranes for the polymer electrolyte fuel cell. The composition and structure of radiation-grafted membranes can be adjusted in a broad range to balance the different requirements of proton transport and mechanical robustness. Styrene, which is readily sulfonated, is predominantly used as grafting monomer. Crosslinking of the structure is a key design parameter, which, if optimized, yields membranes with durability of several thousand hours. Nevertheless, there is potential for improving chemical durability through the use of advanced styrene-derived grafting monomers, such as α-methylstyrene, with enhanced stability against radical attack. *Post mortem* investigations of aged membranes yield important insights into the extent of degradation, in particular locally resolved analysis on the scale of flow field channels and lands. An asset of crosslinked radiation-grafted membranes is their dimensional stability between dry and wet states, which is a key parameter in the context of the mechanical functionality as a separator in the cell. Yet still, the understanding of chemically and mechanically induced degradation, in particular their interplay, is limited, and meaningful accelerated aging test methods have started to be implemented to yield detailed understanding of prevailing degradation mechanisms.

1 Introduction

In the polymer electrolyte fuel cell (PEFC), the electrolyte membrane, allowing transport of protons from anode to cathode, serves at the same time as a separator for electrons and reactant gases. Proton transport is largely determined by the ion-exchange

---

L. Gubler (✉)
Electrochemistry Laboratory, Paul Scherrer Institut, 5232 Villigen PSI, Switzerland
email: lorenz.gubler@psi.ch

F.N. Büchi et al. (eds.), *Polymer Electrolyte Fuel Cell Durability*,

capacity (IEC) and water content of the membrane. Thinner membranes will evidently result in lower ohmic resistance of the membrane electrode assembly (MEA), yet at the expense of mechanical robustness. The membrane is subject to sustained mechanical strain in the presence of the cell compaction force. Sizeable internal stress and stress relaxation via polymer flow may occur within the spatially confined membrane because of changes in temperature and polymer hydration state.

Perfluorinated sulfonic acid (PFSA) type membranes have been used as a solid electrolyte in electrolyzers and fuel cells since the 1960s because of their outstanding chemical inertness and mechanical robustness. However, PFSA membranes, such as Nafion® (Dupont, USA), Flemion® (Asahi Glass, Japan), Aciplex® (Asahi Kasei, Japan), and derivatives thereof, for example the GORE membranes (W.L. Gore, USA), have the disadvantage of being inherently expensive owing to the complex fluorine chemistry involved in their fabrication. The development of alternative membrane materials, which are partially fluorinated or even fluorine-free, is therefore of high interest. Among the promising alternative candidate materials are polyaromatic membranes (for an overview see Hickner et al. 2004) and radiation-grafted membranes (Gubler et al. 2005a).

Membrane aging is associated with the loss of the electrolyte or separator functionality or both. The loss of ion-exchange groups leads to gradual decrease in membrane conductivity, whereas the loss of the mechanical integrity, by pinhole formation or rupture, is perceived as a more "dramatic" event, because it represents catastrophic MEA failure. PEFC membranes undergo chemical degradation through polymer chain scission and loss of functional groups or constituents (blocks, side chains, blend component), caused by HO• and HOO• radicals, which are formed *in situ* through interaction of H_2 and O_2 with the noble metal catalyst on both the anode and the cathode side (Mittal et al. 2006). In addition, hydrolysis may be a cause of chain deterioration for some polymer types with respective susceptible functional units (Meyer et al. 2006). An important factor is the adverse interaction between different aging mechanisms. Under conditions of high chemical exposure, i.e., elevated temperature and low humidity, where the rate of polymer deterioration is high, membranes are likely to fail eventually because of membrane breach (Liu et al. 2001).

2 Design of Radiation-Grafted Membranes

Radiation grafting, short form for "radiation-induced graft copolymerization," offers a versatile method to introduce a functional property, in this case, proton conductivity, into a preformed base polymer film. In the process, the desired component is grafted onto activated centers in the base polymer through radical polymerization, thereby forming a graft copolymer. Base polymers and graft monomers can be selected from a wide range of commercially available commodity products, and all the preparation steps are well-established industrial processes, rendering the final product potentially cost-effective. In view of the application as a solid electrolyte in fuel cells, radiation grafting is a powerful method, as it allows adjustment of the membrane composition and structure over a wide range (Gubler et al. 2005a).

2.1 Base Polymers

Typically used base polymers are fluorinated poly tetrafluoroethylene (PTFE), poly tetrafluoroethylene-co-hexafluoropropylene (FEP), poly tetrafluoroethylene-co-perfluoropropyl vinyl ether (PFA), poly vinylidene fluoride (PVDF), poly ethylene-alt-tetrafluoroethylene (ETFE), films of thickness between 25 and 100 μm. In the most widely used approach, the base film is preirradiated using an electron beam or γ-radiation. The irradiated films can be stored at −70°C for several months without loss of activity. In the next stage, for the graft copolymerization reaction, the activated film is brought into contact with a solution containing monomers amenable to radical polymerization (e.g., styrene; see Fig. 1). The grafted polystyrene side chains are subsequently sulfonated to introduce IEC and thus proton conductivity. As a consequence of introducing sulfonic acid groups into the polymer, the structure becomes capable of absorbing water and swells, leading to dissociation of the acid and enabling proton transport.

Considering desirable base film properties, the polymer is required to yield stable radicals upon irradiation, yet should not undergo chain degradation in the process to a major extent. The grafting kinetics depend on the number of radicals generated in the base film, their accessibility for the grafting monomer(s), and radical lifetime. The partially fluorinated films ETFE and PVDF are advantageous in this respect, because more radicals are generated per unit dose and chain scission is less likely than for perfluorinated films, because breaking of the C–H bond is favored over C–C or C–F bond scission. Structural properties, such as crystallinity and chain mobility in the amorphous state, have a large influence on the radiation chemistry of polymers. For instance, PTFE shows severe loss of mechanical properties upon irradiation at room temperature as a consequence of a large number of

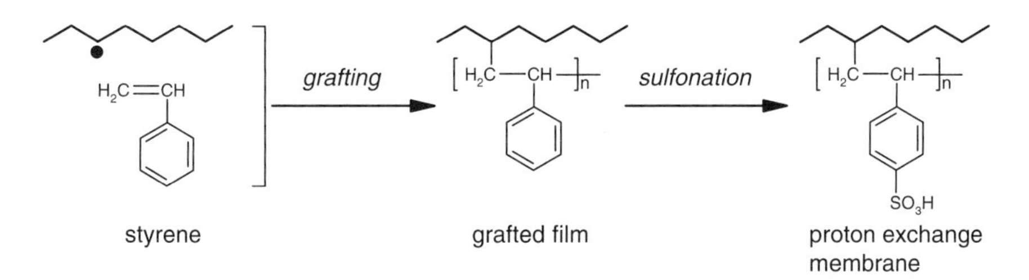

Fig. 1 Preparation of radiation-grafted proton-exchange membranes. Radicals within the base film, created by electron beam irradiation (2.2 MeV, 1.5–25 kGy), act as centers to initiate radical polymerization (here the monomer is styrene). Sulfonation of the grafted film introduces sulfonic acid groups ($-SO_3H$) and thus proton conductivity

chain scission events, owing to the high crystallinity of the polymer. However, if PTFE is preirradiated just above its melting temperature of 327°C, the much higher mobility and flexibility of chain segments carrying radicals lead to the formation of a crosslinked structure with lower crystallinity, which subsequently is much more stable against radiation damage at room temperature (Chen et al. 2005).

When discussing base film properties, we should keep in mind that the chemical stability of the final grafted and sulfonated membranes is mainly determined by the graft component. Yet, the mechanical properties are governed largely by the base film. A high melting point prevents softening of the material, which precludes the use of nonfluorinated materials such as polyethylene or polypropylene. It has to be emphasized at this point that technological film parameters, such as molecular weight (distribution), anisotropy, and the potential presence of additives, should not be left out of the discussion. As a result, the same polymer type from different suppliers may exhibit different properties.

2.2 Grafting Monomers

Styrene has been widely used as a grafting monomer, because it shows favorable radical polymerization kinetics, and the aromatic ring is readily sulfonated to introduce proton conductivity (Fig. 2). The performance and durability of styrene grafted membranes depend largely on the crosslinking of the polymer. A crosslinker is a monomer with two vinyl functions, such as divinylbenzene (DVB) (Fig. 3), which, during graft polymerization, will form links between growing chains, thus yielding a crosslinked structure (Fig. 4). Other crosslinkers used in the grafting process are bis(vinyl phenyl)ethane (BVPE), diisopropenylbenzene (DIPB) m and triallylcyanurate (TAC).

Substituted styrene monomers have been used in various laboratories with the aim of reducing the intrinsic chemical susceptibility of sulfonated polystyrene to graft chain degradation in the PEFC induced by radical species (HO•, HOO•), which are formed *in situ* in the presence of O_2, H_2, and the platinum catalyst (Liu

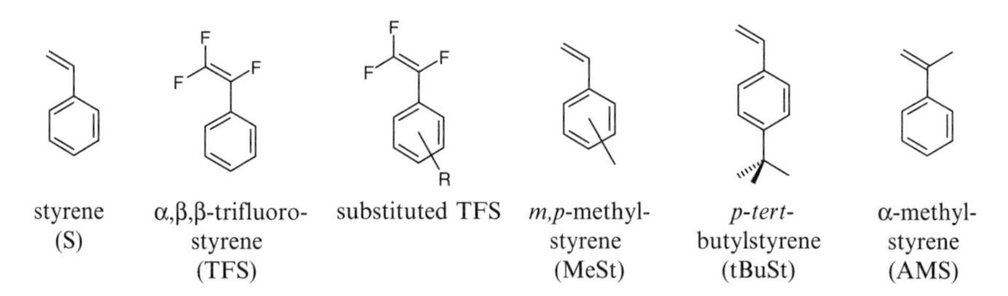

Fig. 2 Monomers of the styrene family used in the preparation of radiation-grafted fuel cell membranes in different laboratories. The aromatic ring can be sulfonated, or the acid functionality may be incorporated via the R substituent

divinylbenzene (DVB) bis(vinyl phenyl) ethane (BVPE) *m*-diisopropenyl-benzene (DIPB) triallylcyanurate (TAC)

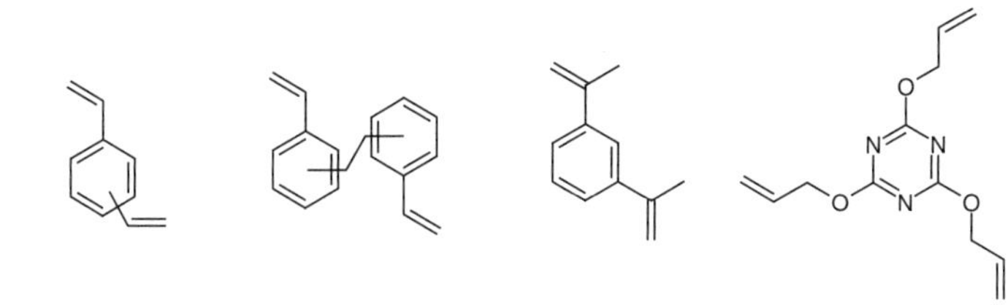

Fig. 3 Crosslinkers, i.e., monomers having two or more double bonds, used as comonomers in the radiation grafting process

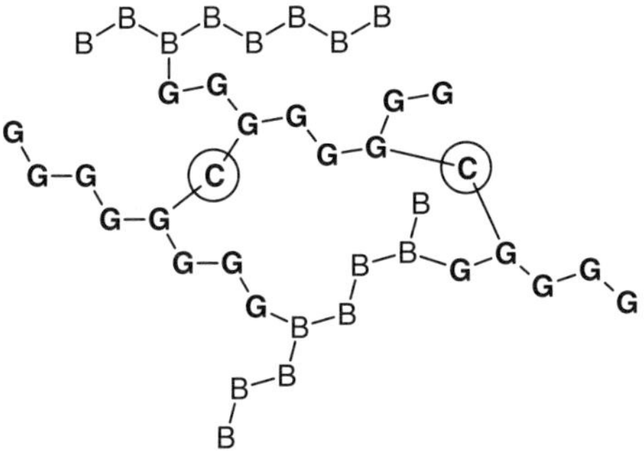

Fig. 4 A crosslinked graft copolymer. *B* Base film units, *G* Graft component, *C* Crosslinker

and Zuckerbrod 2005; Panchenko et al. 2004). The shortcoming of poly(styrene sulfonic acid) (PSSA) appears to be the weakness of the hydrogen at the α-position, which is easily abstracted, leading eventually to chain scission (Hübner and Roduner 1999). The use of fluorinated styrene monomers, such as α,β,β-trifluorostyrene (TFS) or derivatives thereof, has therefore been contemplated. Because of the perfluorinated alkyl chain, poly(α,β,β-trifluorostyrene sulfonic acid) was found to be considerably more stable than PSSA (LaConti et al. 2003).

However, besides the rather moderate polymerization kinetics of TFS and the difficult sulfonation process of the resulting polymer, an evident drawback of TFS is its high cost. Considering the position and chemical environment of the sulfonic acid group, TFS as well as styrene-based monomers where the sulfonic acid group is attached to the aromatic ring via fluoroalkyl units ($-CF_2-SO_3H$) were found to exhibit exceptionally high proton conductivity, probably because of a "superacid" effect (Gürsel et al. 2006; Taniguchi et al. 2001).

The use of readily available styrene-derived monomers of simple hydrocarbon chemistry and low cost also offers the prospect of membranes with chemical stability superior to those based on styrene. At the Japan Atomic Energy Agency (JAEA), work has recently been focused on the ring-substituted monomers methylstyrene (MeSt) and *tert*-butylstyrene (tBuSt) (Chen et al. 2006a–c). *Ex situ* tests in H_2O_2 solution indicated a higher chemical stability of MeSt-based and, in particular, tBuSt-based grafted membranes over styrene-based ones, for which the explanation was put forward that the aromatic ring substituents favorably modify the electronic structure and thus stabilize the α-hydrogen. A concern regarding ring-substituted styrene derivatives is whether the substituents hamper sulfonation owing to steric hindrance. α-methylstyrene (AMS), having the α-position protected by a $-CH_3$ group, yet an unmodified aromatic ring, is a promising monomer in this context. The chemical stability of AMS grafted and sulfonated membranes has been found to be superior to that of styrene-based membranes in an oxidative electrochemical environment (Assink et al. 1991). The major drawback of AMS, however, is its poor radical polymerization kinetics. AMS by itself does not readily graft. Cografting together with an appropriate comonomer is a means to promote, via favorable copolymerization kinetics, incorporation of AMS into the growing chain. Cografting of AMS with styrene has been reported (Li et al. 2006). Increase in the content of AMS versus styrene in the grafted chain improves the chemical stability of the resulting membrane, but the presence of styrene still constitutes a shortcoming. Becker and Schmidt-Naake have reported the use of acrylonitrile (AN) as a comonomer to AMS (Becker and Schmidt-Naake 2001). AN tends towards alternating copolymerization with AMS and greatly improves the effective rate of AMS grafting. We have adopted a similar approach using methacrylonitrile (MAN) as a comonomer (Gubler et al. 2005d). MAN offers potentially higher chemical stability because of the methyl-protected α-position. Both AN and MAN do not undergo sulfonation; the resulting IEC is determined by the AMS graft level. Unwanted side reactions of AN/MAN in the further preparation steps, such as hydrolysis, is not observed (Gubler et al. 2006b).

2.3 Graft Copolymerization

One of the major advantages of radiation grafting is the versatility of the method. The composition of the grafted film can be varied in a large range by choosing appropriate combinations of base polymer type and thickness, grafting monomers, and grafting conditions. The grafting parameter of prime importance is the degree of grafting (or graft level), which is a measure for the amount of polymer introduced through grafting, given by $n_g/n_0 \times 100\%$, where n_g and n_0 are the numbers of repetitive chain units in the grafted component and base polymer, respectively. The degree of grafting is usually determined by weighing the samples before and after grafting. The number of repetitive units is calculated on the basis of the molar weight of the repetitive units of the base polymer and the graft component, respectively. A mass-based degree of grafting may also be defined, but then comparison of membranes with different base polymers and grafting monomers becomes less meaningful. The IEC of the resulting sulfonated membranes, the water uptake, and the conductivity will obviously increase along with the degree of grafting, as the membrane contains more and more graft component (Gubler et al. 2005c).

One might therefore think that higher degrees of grafting will yield "better" membranes. This might be true considering solely the proton conductivity. The mechanical properties and chemical stability of highly grafted membranes, however, were found to be significantly inferior to those of membranes with intermediate graft levels. Rather than trying to maximize the degree of grafting, careful design involves the balancing of conductivity and mechanical properties. In the work carried out at the Paul Scherrer Institut (PSI) on styrene(/DVB) grafted FEP and ETFE (25 μm) based membranes, the optimum graft level was found to be roughly between 14 and 22% (molar basis) (Gubler et al. 2005c). This "window" may of course be at a different position for other combinations of base film and monomer(s), but the qualitative trend is expected to be similar.

The second grafting parameter of utmost importance is the extent of crosslinking (Gupta and Scherer 1994). Crosslinking renders the grafted polymer structure more rigid, resulting in lower water uptake and lower proton conductivity, yet higher dimensional stability (Table 1). Again, the effect of the extent of crosslinking on the membrane and the performance can be discussed by means of a "property map" (Fig. 5b). On the abscissa in Fig. 5b, the crosslinker (DVB) concentration is given with respect to the total monomer (styrene + DVB) content in the grafting solution. The effective crosslinker concentration in the grafted film need not be the same, since the reactivities of styrene and DVB are different. With increasing crosslinker content the conductivity decreases, as expected, yet at the same time the membranes become more brittle, which makes them much more prone to tearing, during either MEA assembly or fuel cell operation (Gubler et al. 2005c). Conversely, membranes with a low degree of crosslinking have not only poor chemical stability, but it was also found that the membrane-electrode interface is of poor quality for FEP-based membranes (Schmidt et al. 2005) as well as for ETFE-based ones (Gubler et al. 2007a). The reason for this phenomenon is not fully understood to date. Possibly, it is related to rapid graft chain degradation at the membrane surface close to the platinum-containing active area. In general, the different nature of the polymers in the membrane and catalyst layer (Nafion® ionomer) yields higher interfacial nonohmic performance losses, compared with an MEA with a Nafion® membrane.

Table 1 Dimensional change of swollen membranes upon drying in a vacuum oven at 60°C for 1 h

Membrane	g-FEP unXL		g-FEP XL		Nafion® 112	
Orientation	MD	TD	MD	TD	MD	TD
Linear shrinkage (%)	13.9	17.6	6.5	9.3	11.1	16.7
Area shrinkage (%)	29.0		15.1		25.9	

Radiation-grafted membranes based on 25 μm grafted FEP (*g-FEP*) with a degree of grafting of approximately 20%. *unXL* Uncrosslinked, *XL* Crosslinked (styrene to divinylbenzene ratio 9:1 v/v in the grafting solution), *MD* Machine direction, *TD* Transverse direction

The quality of the membrane-electrode interface can be improved by impregnating the radiation-grafted membrane with Nafion® solution prior to MEA bonding, which "compatibilizes" the surface of the membrane with that of the electrode (Gubler et al. 2004b). In summary, the property map (Fig. 5a) exhibits a characteristic with an optimum of balanced properties at an intermediate extent of crosslinking. It is noteworthy that the optimum for FEP- and ETFE-based membranes does not emerge at the same crosslinker content (in the grafting solution). This might indicate dissimilar rates of styrene and DVB incorporation into the two types of base film and/or other effects, such as differences in crosslinking between the surface and the interior of the film. An overview of *ex situ* properties for optimized membranes based on 25-μm FEP and ETFE is given in Table 2. At similar

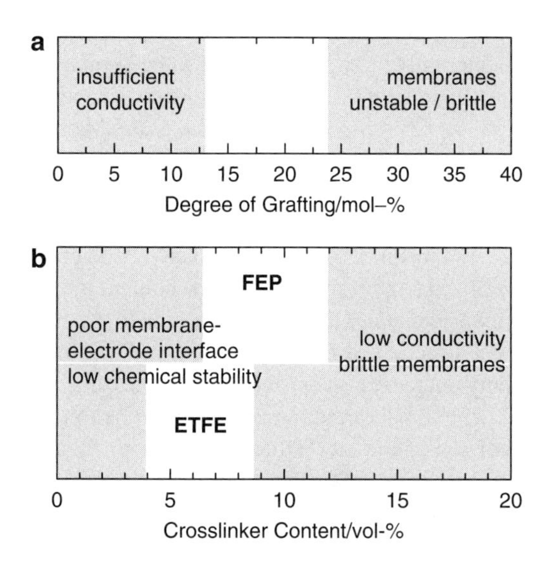

Fig. 5 Property map for styrene/DVB grafted and sulfonated membranes based on 25-μm FEP and ETFE. Influence of the (**a**) degree of grafting and (**b**) extent of crosslinking on global membrane characteristics. The crosslinker content refers to the DVB concentration with respect to total monomer content (styrene + DVB) in the grafting solution. *White areas* near the center denote the range of optimum composition

Table 2 *Ex situ* properties of selected membranes

Membrane type	Degree of grafting (mol%)	Ion-exchange capacity (mmol g^{-1})	Water uptake[a] $[H_2O]/[SO_3H]$	Proton conductivity[a] (mS cm^{-1})
g-ETFE (5% DVB)	16	1.74 ± 0.08	7.0 ± 0.7	62 ± 2
g-FEP (10% DVB)	18	1.36 ± 0.06	6.7 ± 0.7	41 ± 1
Nafion® 112	–	0.91	18.0 ± 0.9	82 ± 6

Base film thickness of FEP and ETFE is 25 μm. Divinylbenzene (*DVB*) crosslinker content is given with respect to the total monomer content in the grafting solution (rest is styrene).
[a]At room temperature in liquid water equilibrated state

molar-based degrees of grafting, the IEC, being a mass-based quantity, of the ETFE membrane is somewhat higher, which is a consequence of the lower density of ETFE compared with FEP. The same argument holds for Nafion® 112. Both crosslinked radiation-grafted membranes have a similar water uptake of around seven H_2O per SO_3H, yet the conductivity of the ETFE-based membrane is somewhat higher.

2.4 Membrane Properties

As the proton-exchange membrane in the PEFC has to provide a physicochemical and mechanical functionality, the following membrane material properties are of primary importance:

- Proton conductivity
- Water uptake and dimensional stability
- Water transport properties (diffusion, hydraulic permeability, and electroosmotic drag)
- Reactant permeability
- Mechanical properties (tensile strength, elongation at break, viscoelastic creep, tear initiation and propagation resistance)

It is evident, but it has to be emphasized nevertheless, that these properties are strongly influenced by temperature and relative humidity. The membrane thickness, being a "component property," not a "material property," should not be left out of the discussion in the context of technological application, as it is the most easily adjustable parameter with straightforward effect: thinner membranes yield a lower ohmic resistance, but at the cost of mechanical resilience, and vice versa. The mechanical properties have received stunningly low attention within the community to date, considering the fact that loss of mechanical integrity leads to catastrophic failure of the MEA. Not only tensile strength and elongation at break values have to be taken into account, but also creep behavior, and resistance to tear initiation and propagation.

The tensile properties of 25-μm FEP-based grafted films have been measured as a function of irradiation dose, grafting solvent, and graft level (Fig. 6). Note that the films are only in the grafted state and not sulfonated, to exclude influences of different water content. An increase in the degree of grafting results in a higher tensile strength on the one hand, and a considerable reduction in the elongation at break on the other hand. In short, the films become more brittle. Similarly, radiation damage at a dose of 30 kGy also leads to an embrittlement of the material. With currently used doses of 3 kGy, however, radiation damage is negligible. Chen et al. (2005) investigated a range of base film materials for their tensile properties and sensitivity to radiation. The partially fluorinated films ETFE and PVDF showed the most favorable properties with respect to tensile strength and elongation at break compared with the perfluorinated films PTFE, crosslinked PTFE, FEP, and

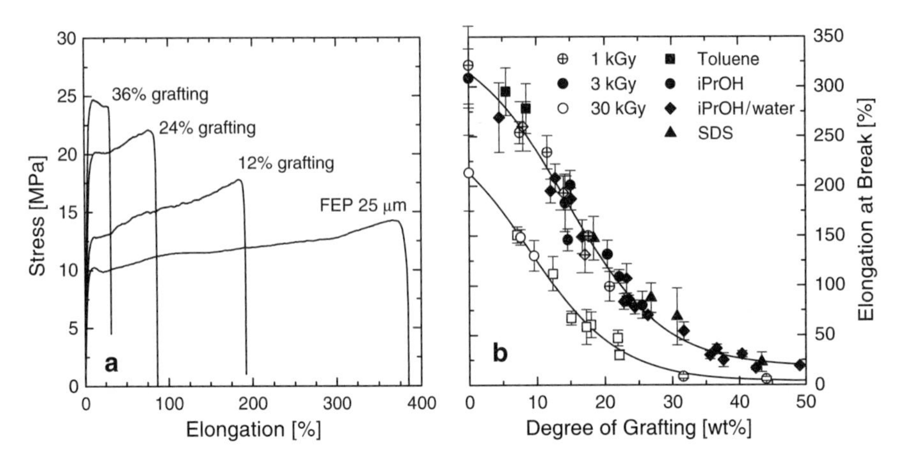

Fig. 6 (**a**) Stress–strain curves for pristine 25-μm FEP and grafted films with different degrees of grafting; (**b**) elongation at break as a function of degree of grafting, preirradiation dose, and type of solvent used. Grafting monomers were styrene and DVB at a ratio of 9:1 v/v (Rager 2003). SDS (Reprinted with the permission of John Wiley & Sons, Inc.)

PFA. Once again, it is emphasized that for a more comprehensive understanding of the mechanical properties of the membrane as a supporting component in the PEFC, more extensive characterization is required, such as tensile tests under relevant temperature and relative humidity conditions, creep tests, and crack initiation and propagation tests.

Water uptake affects both proton conductivity and mechanical properties of the membrane. Whereas the correlation of the former is straightforward, i.e., higher water content yields higher conductivity, the latter requires closer inspection. Water acts as a plasticizer, the material becomes "softer" and more gel-like upon water sorption. Hence, membranes with higher water uptake are more prone to creep under the influence of the compaction pressure in the fuel cell. This may lead to membrane thinning and, eventually, puncturing and crack formation. An effect especially pertaining to swelling of the polymer upon water sorption is a fatigue-type phenomenon when the MEA is subjected to dry–wet cycles, which leads to periodic stress buildup and relaxation in the membrane and, ultimately, to crack formation. This has been observed to be a relevant membrane failure mode (Mathias et al. 2005).

The comparison of proton conductivity and water uptake for a crosslinked and an uncrosslinked radiation-grafted membrane, respectively, based on 25-μm FEP and Nafion® 112 allows us to gain a few insights (Fig. 7). The uncrosslinked membrane (29.5 H_2O per SO_3H) swells substantially more than Nafion® 112 (18 H_2O per SO_3H), yet the conductivities are comparable (72 and 82 mS cm^{-1}, respectively). The crosslinked membrane swells much less (6.7 H_2O per SO_3H), owing to the dense polymer network, with a conductivity (41 mS cm^{-1}) that is roughly half of that of Nafion® 112 and the uncrosslinked FEP membrane. Yet the dimensional stability of the crosslinked FEP grafted membrane is considerably higher (Table 1),

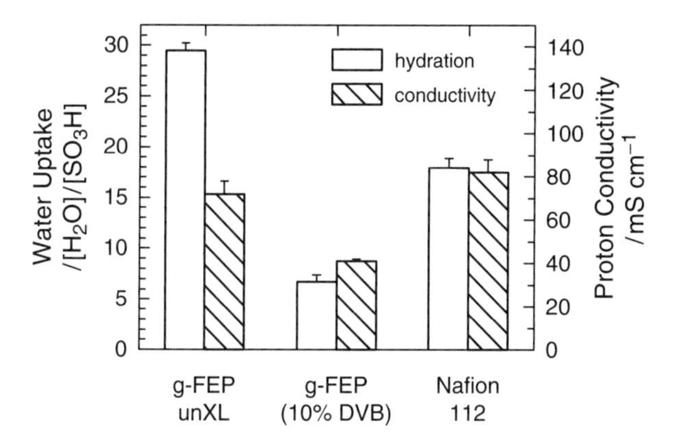

Fig. 7 Water uptake, expressed as the number of water molecules per sulfonic acid site, and proton conductivity at room temperature in the water-swollen state. The uncrosslinked (*unXL*) and crosslinked (10% DVB monomer) radiation-grafted membranes based on 25-μm FEP have a degree of grafting of approximately 20 mol%

which is expected to result in a markedly lower susceptibility of the membrane towards relative humidity cycling fatigue.

The water transport properties of the membrane strongly influence water distribution in the membrane and the MEA. Electroosmotic drag of water with the protons flowing from the anode to the cathode causes drying of the membrane and the ionomer in the catalyst layer on the anode side, and one has to rely on the back-transport of water by diffusion or hydraulic permeation to keep the anode side sufficiently hydrated and reduce flooding of the cathode. Whereas water transport properties of PFSA membranes, such as Nafion®, have been extensively investigated, the data available for radiation-grafted membranes are limited. Evidently, the diffusivity, permeability, and electroosmotic drag coefficient will depend on the exact membrane architecture (base polymer, grafting monomer(s), degree of grafting, extent of crosslinking, and microstructure) and experimental conditions (membrane water content, temperature). For styrene grafted PVDF-based membranes it was found that the water diffusion coefficient increases with the degree of grafting and the water content, as would be expected (Hietala et al. 1999). The electroosmotic drag coefficient, measured in the direct methanol fuel cell at 90°C, in crosslinked membranes based on FEP with different thickness, was determined at 1.7 H_2O per H^+, whereas a value of 5.0 H_2O per H^+ was measured for Nafion® 117 (Gubler et al. 2005b). The difference is most likely a consequence of the tighter polymer structure and lower water uptake in the crosslinked radiation-grafted membranes.

Reactant permeability is an important quantity in the context of durability, since interdiffusing H_2 and O_2 in the PEFC will lead to the formation of aggressive radical species in the catalyst layers. In styrene grafted ETFE (50 μm) based membranes, O_2 permeability was found to increase with increasing graft level and thus water uptake, because oxygen diffusion is faster in the aqueous domains of the membrane

(Chuy et al. 2000). Investigations on the rate of H_2 permeation in the fuel cell indicate that crossover decreases with increasing extent of crosslinking of the membrane (Gubler et al. 2008). In the direct methanol fuel cell, a major challenge is the high methanol permeability of PFSA membranes, leading to major fuel efficiency losses. With use of radiation-grafted membranes with an appropriately chosen extent of crosslinking, the methanol crossover can be reduced by a factor of two, while the performance remains unaffected (Gubler et al. 2005a).

3 Fuel Cell Testing of Radiation-Grafted Membranes

The first experiences with solid polymer electrolytes for fuel cells based on sulfonated polystyrene, crosslinked with DVB, were made in the 1960s at General Electric (LaConti et al. 2003). Significant degradation was observed after 500–1,000 h of operation at 40–60°C. The PSSA broke down into smaller, water-soluble units that were found in the product water. The likely degradation mechanism of PSSA has already been put forward by Hodgdon et al. (1966), involving the attack of the styrene sulfonic acid units by HO•, followed by chain scission owing to the weakness of the benzylic α-hydrogen.

Membranes with PSSA moieties grafted into a base film as a substrate also originated in the 1960s. They were developed for the purpose of membrane separation processes. Interest in radiation-grafted membranes as solid polymer electrolytes in electrochemical conversion devices, such as electrolyzers and fuel cells, grew in the 1980s (Scherer 1990). Owing to the incorporation of the ionomeric groups into a thermally and chemically stable base polymer, some of the drawbacks of the earlier bulk PSSA-based membranes could be alleviated. Since then, the development of radiation-grafted membranes for fuel cells has been pursued in various laboratories around the world, with notable advances in the chemical and structural design, performance, durability, and understanding of limitations of the technology and identification of future prospects (Gubler et al. 2005a).

3.1 Essential Improvements over the Last Decade

The early radiation-grafted membranes prepared at PSI were simultaneously irradiated or pre-irradiated, with FEP base films of 50, 75, or 125 μm thickness, to a dose of 60 kGy or higher using γ-radiation. The grafting was carried out either in bulk monomer, or, in some cases, in mixtures of the monomer(s) and solvents such as benzene or toluene, i.e., good solvents for polystyrene. Yet these membranes showed poor stability in the single cell at 60°C (Fig. 8). In the following years, the membrane preparation as well as the MEA fabrication process and single cell hardware were continuously optimized, which resulted in a markedly enhanced durability, even at a lower membrane thickness and higher operating temperature. The most important improvements to the membrane preparation process were (Gubler et al. 2004a):

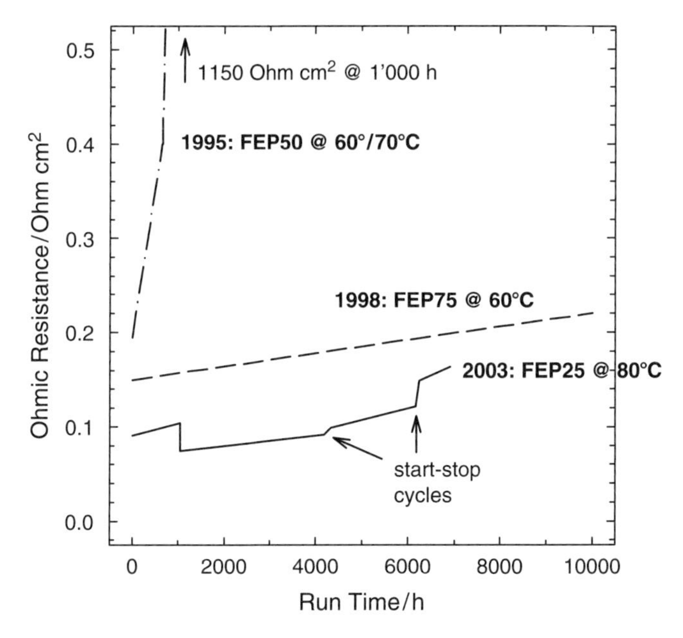

Fig. 8 Evolution of performance and durability of FEP-based radiation-grafted membranes at the Paul Scherrer Institut, expressed through ohmic (i.e., membrane) resistance, recorded on-line using the auxiliary current-pulse method

- Use of electron-beam radiation instead of γ-radiation, resulting in less radiation damage of the base film, owing to the much higher dose rate associated with electron-beam irradiation.
- Instead of using benzene or toluene as a solvent for styrene/DVB in the grafting solution, polar solvents, such as methanol, 2-propanol, and 2-propanol/water mixtures, brought about substantially enhanced grafting kinetics, because these nonsolvents do not swell the grafted polystyrene and thereby result in extended radical lifetimes.
- The improved grafting kinetics resulting from the use of nonsolvents allowed the reduction of the irradiation dose by over one order of magnitude from over 30 kGy down to 3 kGy or less, with concomitant reduced radiation damage of the base film and markedly improved mechanical properties.
- Use of thin base films, 25-μm FEP and ETFE, allowed the use of lower graft levels, yielding membranes with improved mechanical properties, to achieve the desired area resistance.

In addition to improvements to membrane design and preparation, the single cell construction was optimized, for instance, by eliminating stainless steel as the cell housing material. The longest fuel cell test duration achieved to date is 10,000 h, at a cell temperature of 60°C, using a 75-μm FEP-based membrane (Fig. 8). A gradual increase in membrane resistance could be observed, indicating ongoing degradation. With all the improvements pertaining to styrene-based membranes implemented, an MEA lifetime of 7,900 h was achieved at 80°C (Gubler et al. 2004b). Unfortunately, the test run was hampered by adverse start–stop cycles as a consequence of test facility malfunctions.

By all means, at some point one has to reconsider the manner of membrane lifetime testing, as experiments over thousands of hours are time-consuming, resource-intensive, and slow down innovation cycles for the development of improved or novel membranes. Methods to increase sample and testing throughput will be discussed in Sect. 3.4.

3.2 *Influence of Grafting Parameters*

For the discussion of the influence of graft level and extent of crosslinking on the durability of styrene/DVB-based membranes, we shall revisit the property map in Fig. 5. Membranes with a high degree of grafting may have attractive conductivity, but unfortunately they also have poor mechanical properties and a high rate of chemical degradation, because the higher the graft level, the more "PSSA-like" the behavior of the membrane will be, with the well-known shortcomings.

Membranes with moderate extent of crosslinking exhibit the most favorable properties in terms of performance *and* durability (Fig. 5b). Crosslinking leads on the one hand to a reduction in gas (H_2, O_2) transport across the membrane as a consequence of a higher compactness (lower swelling) of the crosslinked polystyrene domains. Hence, the formation rate of HO•/HOO• radicals through interaction of H_2 and O_2 with the platinum catalyst is reduced. On the other hand, even though chain scission events will continue to take place, a crosslinked polymer network is less susceptible to the loss of chain fragments, because a given chain segment is likely to be attached to other chains at more than one point. Excessive crosslinking is detrimental. Not only does the water uptake of the membrane decrease and thus the conductivity, but also the polymer network becomes increasingly dense and renders the membrane brittle, such that MEA and cell assembly is aggravated by crack formation in the membrane.

3.3 *Innovative Monomer and Crosslinker Combinations*

Even though stable operation and encouraging fuel cell lifetime over several thousand hours have been achieved using membranes with optimized composition (Fig. 8), sulfonated polystyrene remains a component intrinsically susceptible to chain degradation induced by radical species and the approach is therefore inherently flawed.

The influence of crosslinker type (DVB, BVPE, TAC) and concentration on the *ex situ* stability of the resulting membranes was studied by Chen et al. (2006a) using 50-μm ETFE as the base film and MeSt as the primary grafting monomer

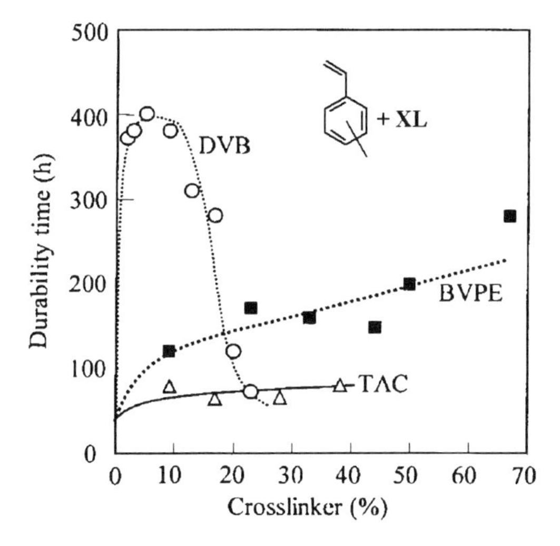

Fig. 9 *Ex situ* chemical stability of (*m,p*-methylstyrene + crosslinker) grafted and sulfonated membranes in 3% H_2O_2 solution at 50°C as a function of the crosslinker content (vol%) in the grafting monomer mixture. "Durability time" is associated with the onset of significant (more than approximately 10%) membrane weight loss. *XL* crosslinker (data reprinted from Chen et al. 2006a. Reprinted with the permission of John Wiley & Sons, Inc.)

(Chen et al.). The membranes were immersed in 3% H_2O_2 solution at 50°C, representing aggressive conditions related to the ones *in situ* for accelerated degradation testing. The decrease of the membrane weight over time is associated with the loss of the graft component. The time after which significant weight loss (more than approximately 10%) is observed was quoted as "durability time" by the authors (Fig. 9). With all three crosslinkers, an increase in durability time, i.e., chemical membrane stability, was observed. The effect of TAC was rather weak; BVPE appeared to be more effective. The influence of DVB on the stability is most pronounced. Interestingly, unlike for BVPE and TAC, the durability rapidly decreases above a concentration of 10% DVB. The authors speculated that owing to the unreacted pendant double bonds, incomplete sulfonation and inhomogeneous distribution of crosslinks in the thickness direction may lead to higher sensitivity to oxidation by H_2O_2. Membranes with the highest stability were obtained at 5–10% DVB, which is in agreement with our own findings for styrene/DVB grafted 25-μm FEP membranes investigated *in situ* in single cells (Schmidt et al. 2005). It should be kept in mind, though, that the crosslinker content given on the abscissa in Fig. 9 is the concentration in the grafting solution. The effective crosslinker content in the grafted film is not necessarily the same owing to the difference in reactivities of the monomer and the crosslinker and their diffusivities in the film. The DVB to styrene ratio in 25-μm ETFE-based grafted films has been determined using IR spectroscopy (Ben youcef et al. 2008). The average composition was determined in transmission mode, whereas the surface region was probed via attenuated total reflectance. On average, the DVB to styrene ratio in the bulk of the film was around

30% lower than that in the grafting solution. This may be a consequence of the lower diffusivity of the larger DVB molecule compared with styrene. Conversely, higher degrees of crosslinking were observed close to the film surface, with DVB to styrene ratios being a factor of 1.5–3 times higher than in the grafting solution. Higher crosslinking at the surface can be explained with the higher reactivity of *para*-DVB versus styrene, resulting in higher incorporation of the crosslinker close to the film surface, where grafting is not limited by diffusion.

The crosslinking resulting from BVPE appears to be less tight, according to Chen et al. (2005), owing to the less rigid structure of the molecule compared with DVB, and is more homogeneous over the thickness of the film, because of the similar reactivities of styrene and BVPE. A further aspect worth considering is the potential influence of the base film properties and structure on the grafting and cografting kinetics in the presence of a crosslinker. With crosslinked PTFE used as the base polymer (see Sect. 2.1), BVPE crosslinked membranes exhibited higher chemical stability in H_2O_2 solution compared with DVB crosslinked ones (Yamaki et al. 2007). Such "double" crosslinked membranes (crosslinked base film and crosslinked graft component) are expected to exhibit encouraging chemical and mechanical stability properties in the fuel cell.

AMS is a promising grafting monomer in the light of absent benzylic hydrogen and associated higher intrinsic chemical stability, yet it is difficult to graft owing to poor radical polymerization kinetics, as discussed in Sect. 2.2, but by using MAN as a comonomer, practical graft level, IEC, and conductivity (approximately 100 mS cm^{-1}) can be obtained (Gubler et al. 2006b). Uncrosslinked AMS/MAN grafted and sulfonated membranes have been tested at 80°C in single cells to characterize the durability and rate of degradation against uncrosslinked styrene grafted membranes (Fig. 10). The MEAs failed due to excessive gas crossover, if not otherwise stated. The lifetime of the AMS/MAN grafted membrane was approximately 500 h, which is an order of magnitude higher than the lifetime of the styrene grafted one (approximately 50 h). To assess the rate of chemical degradation, the residual IEC of the membranes disassembled at the end of test was determined. Thus, with the known time on test and the IEC of the pristine membrane, the rate of IEC loss (in % h^{-1}) was calculated. The rate of IEC loss for an uncrosslinked AMS/MAN grafted membrane was found to be approximately 30 times lower than that of a styrene grafted membrane. The optimized styrene/DVB grafted membrane, as discussed earlier, has an MEA lifetime of several thousand hours, and the associated IEC loss rate is 150 times lower than that of an uncrosslinked styrene grafted membrane. Evaluation of DVB crosslinked AMS/MAN grafted membranes is ongoing; preliminary results indicate a stability even superior to that of styrene/DVB grafted membranes.

3.4 Increasing Sample and Testing Throughput

Extended single cell durability tests over thousands of hours are unattractive for various reasons: the duration of the experiment, the requirement of constant

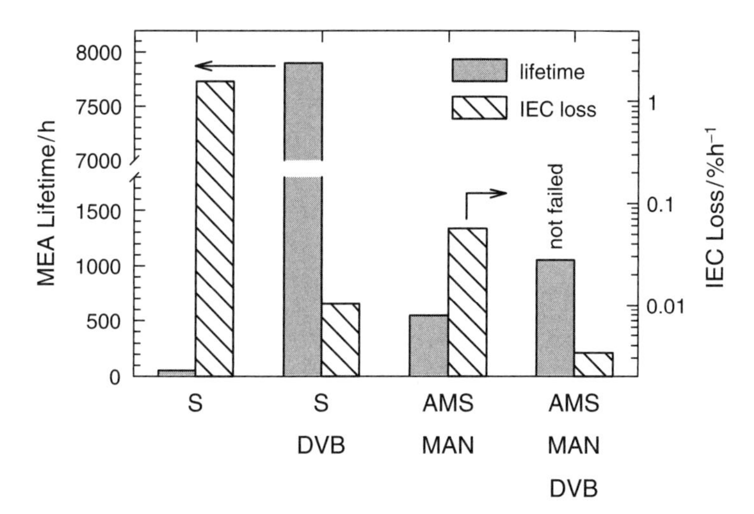

Fig. 10 Comparison of single cell lifetime and rate of ion exchange capacity (IEC) loss for different radiation-grafted membranes. The rate of IEC loss is a measure for the susceptibility of a membrane towards chemical degradation. Cell temperature 80°C. Grafting monomers *S* Styrene, *DVB* Divinylbenzene (crosslinker), *AMS* α-methylstyrene, *MAN* methacrylonitrile (comonomer), *MEA* membrane electrode assembly

supervision, poor result statistics, and risk of test bench malfunction. For the purpose of shortening innovation cycles for novel membranes, the aim has to be to increase sample throughput and improve result statistics, i.e., get more and better information in a shorter time. For this, two approaches have been identified and implemented: parallel testing of several MEAs in a stack, and testing under accelerated aging conditions.

One has to consider the following aspects for parallel testing of several MEAs in a stack (1) upon failure of one cell, the test has to be discontinued and (2) electrical and thermal coupling of cells (Santis et al. 2006). Consequently, the following precautions have been taken to minimize these effects. Stack hardware with thick bipolar plates (6 mm), made from graphite with low resistivity (SGL Carbon Diabon® NS2, 8 mΩ·m), was employed, comprising internal water cooling. Durability tests under faradaic load are not taken to failure, but rather the "health" state of the MEAs was monitored continuously using appropriate diagnostic tools. In the case of open circuit voltage (OCV) hold tests, i.e., in the absence of stack current, electrical coupling is inexistent anyway.

A parallel test of an MEA comprising a styrene/DVB grafted and sulfonated membrane based on 25-μm FEP and a Nafion® 112 based MEA in a two-cell stack has been carried out, with operation in one case under mild temperature (75°C) and cyclic load operation conditions (Gubler and Scherer 2006), and in the other case using an accelerated test protocol involving operation at OCV and a temperature of 90°C (Fig. 11). In the experiment under mild conditions, the performance of the two MEAs did not decrease over 1,130 h, after which the

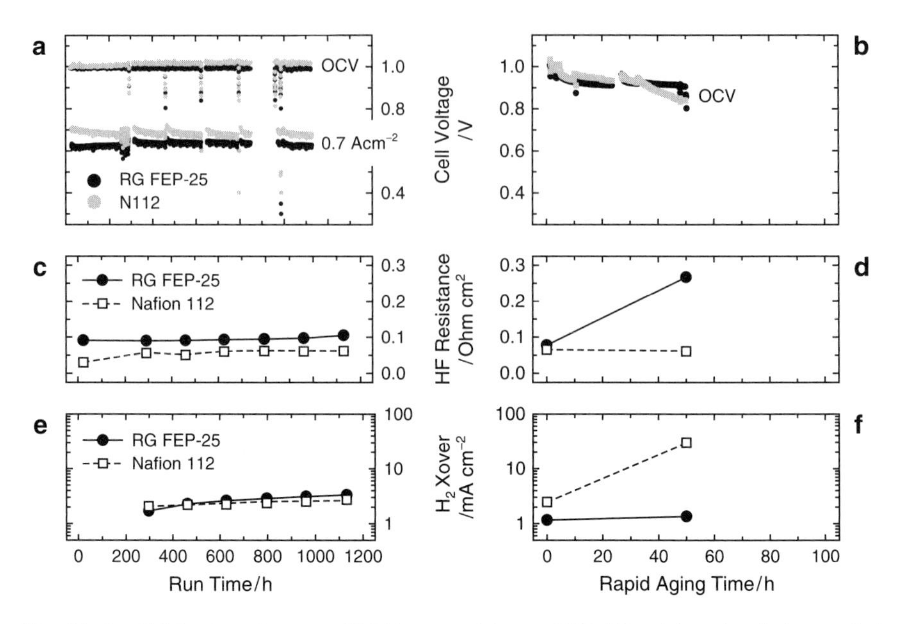

Fig. 11 Parallel *in situ* aging test of MEAs comprising an FEP (25 µm)-based radiation-grafted (RG) membrane (crosslinked) and Nafion® 112 in a two-cell stack of 100-cm^2 active area. Reactants were H_2 and O_2 with a stoichiometry of 1.5 each, pressure 2.5 bar_a. (**a**, **c**, **e**) simulated application profile; temperature 71–76°C, load profile with current density ranging from open circuit voltage (OCV) to 0.7 A cm^{-2}, gas inlet dew points 40°C. (**b**, **d**, **f**) accelerated aging conditions; temperature 90°C, OCV, gas inlet dew points 78°C. (**a**, **b**) cell voltage; (**c**, **d**) high-frequency (HF) resistance measurement at 2 kHz and a stack temperature of 75°C and gas inlet dew points of 40°C as a measure for *chemical* membrane integrity. (**e**, **f**) electrochemical H_2 crossover measurement at a stack temperature of 75°C and gas inlet dew points of 40°C as a measure for *mechanical* membrane integrity

experiment was stopped (Fig. 11, plot a). The OCV values did not indicate any change in membrane state either. In contrast, the accelerated degradation experiment had to be stopped after 50 h owing to failure of the Nafion® 112 MEA (Fig. 11, plot b).

The state of the membranes was analyzed periodically for loss of ion-exchange groups by measuring the high-frequency resistance (Fig. 11, plots c and d), and for mechanical integrity via H_2 crossover measurement (Fig. 11, plots e and f). After an operating time of 1,130 h under mild conditions, the high-frequency resistance had increased only by a small amount for both MEAs, indicating negligible degradation (Fig. 11, plot c). Although a small increase in H_2 crossover was observed, indicating some loss in mechanical membrane integrity and gradual change of membrane morphology, dramatic failure did not occur (Fig. 11, plot e).

During the accelerated aging experiment the resistance of the Nafion® membrane did not change, but that of the grafted membrane increased substantially (Fig. 11, plot d). This suggests that the grafted polymer looses ion-exchange groups under these aggressive conditions. On the other hand, the Nafion® membrane showed mechanical failure, whereas the grafted FEP membrane remained intact. This indicates a different response of the two membrane types to applied stress, in this case OCV operation at elevated temperature (90°C), which may help the experimenter in identifying critical membrane properties and indicate possible approaches to improve membrane composition and design.

3.5 Local Post Mortem *Degradation Analysis*

The understanding of fuel cell membrane degradation on a local scale is important in order to correlate aging mechanisms with local operating conditions (temperature, current density, H_2/O_2 and H_2O concentration, etc.). Furthermore, the results may help in improving cell design to yield more homogeneous operating conditions over the active area. The flow field plate or separator plate in a PEFC comprises channels for reactant gas distribution and water removal, and lands (or "ribs") that establish contact to the electrode for current collection. The widths of the channels and lands, respectively, are on the order of millimeters; therefore, it can be expected that the electrochemically active areas under the lands and channels will not produce the same local current density. The measurement of locally resolved current density distribution in the submillimeter range has been reported recently (Freunberger et al. 2006). Both with air and with oxygen as oxidants the difference in current generated under channels and lands can easily be a factor of 2 or more. In consequence, it is reasonable to assume that degradation of catalyst and membrane material is also inhomogeneous at this length scale. For the purpose of investigating local degradation effects, a styrene/DVB grafted and sulfonated 25-μm FEP-based membrane was artificially aged over 690 h and subsequently analyzed via transmission Fourier transform IR spectroscopy to determine the local extent of degradation by quantifying the intensity of vibrational bands related to the sulfonic acid group (Gubler et al. 2006a). It appears that, on average, degradation in the channel areas is about twice as high as degradation in the land areas (Fig. 12). Different reasons may be put forward as a possible explanation. Stronger degradation in the channel area might be a consequence of higher local H_2/O_2 concentration and thus greater HO• and HOO• radical formation rates, since gas access to land areas is associated with a longer diffusion path. On the other hand, removal of heat is more efficient in the land areas, presumably resulting in higher local temperatures in the channel regions, causing higher membrane degradation rates. The understanding of the differences in degradation associated with channel and land areas of the membrane is still limited, though, and experimental series are under way to identify effects of current density and other parameters.

In the context of localized degradation, one has to address also the through-plane dimension. Degradation may not be homogeneous over the thickness of the membrane. Experience in this regard is limited at this point. Whether degradation is stronger

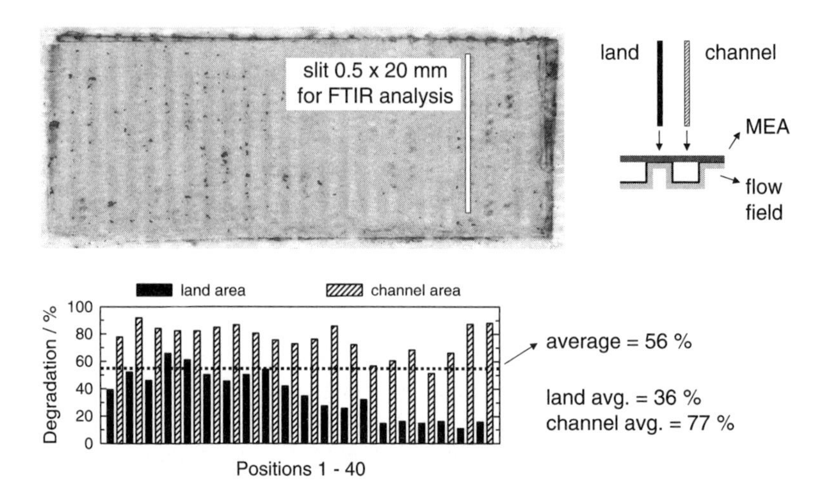

Fig. 12 Local degradation analysis using transmission Fourier transform IR (FTIR) spectroscopy in an FEP (25 μm) radiation-grafted membrane, artificially aged for 690 h using H_2 and O_2 as reactants at various operating conditions: 0–1 A cm^{-2}, 78–95°C, 1–3 bar_a. Local degradation analysis in channel/land areas of the flow field is accomplished via a slit mask for the quantification of the *p*-styrene sulfonic acid ring vibrational band, with the pristine membrane as a reference

on the anode or on the cathode side is still a subject of debate, but obviously it depends on the type of electrodes used and the operating conditions. LaConti et al. (2003) found higher degradation on the anode side for PSSA-based membranes, for which the explanation was put forward that O_2 that diffused from the cathode side reacts with hydrogen adsorbed on the anode catalyst to form H_2O_2, yet its oxidation is inhibited at the hydrogen-covered platinum electrode; therefore, it can diffuse into the membrane and attack the polymer.

4 Concluding Remarks

The properties of proton-exchange membranes for the PEFC have to be carefully balanced to yield satisfactory proton conductivity, chemical stability, and mechanical robustness, in particular in view of long-term operation under practical operating conditions involving rapid load changes and numerous start–stop cycles. Often, these requirements are in conflict with each other. The composition and properties of fuel cell membranes prepared via radiation grafting can be tuned and adjusted in a broad range. In most cases, styrene is grafted onto thermally and chemically stable base polymer films, followed by sulfonation. Research into substituted styrene-derived grafting monomers, such as *m,p*-methylstyrene, *p-tert*-butylstyrene, α-methylstyrene, or trifluorostyrene, has resulted in the preparation of advanced membranes with higher intrinsic chemical stability with reduced susceptibility to radical-induced degradation compared with PSSA.

By far the most substantial improvement in durability is achieved by introducing crosslinking monomers. Optimized crosslinked styrene-based membranes have achieved a lifetime of several thousand hours at 80°C in the single cell. Crosslinking improves the effective chemical stability and reduces water uptake. A lower swelling imparts the membrane with a higher dimensional stability, which reduces internal stress accumulation and membrane damage as a result of dry–wet cycles. Excessive crosslinking, though, yields brittle membranes, which are prone to rupture during cell operation.

Aging tests have to be fast, statistically relevant, and meaningful with respect to the prevailing degradation mechanisms. Methods to increase sample and testing throughput and shorten innovation cycles are being implemented. These involve parallel testing of MEAs in a stack and tests under conditions of accelerated aging (e.g., OCV hold test). It has to be emphasized, though, that the quantitative understanding of the accelerating effect of the various stress factors (temperature, cell voltage, relative humidity, etc.) is not yet well developed, especially the cross-correlation between different degradation modes. *Post mortem* analysis of membranes can yield valuable quantitative information on the extent of degradation, particularly when the analysis is carried out with local resolution on the millimeter scale to resolve differences associated with channel and land areas of the flow field.

Although encouraging performance and lifetimes have been attained with radiation-grafted fuel cell membranes, durability continues to be the major challenge. The potential, however, is not fully exploited yet. Considerable scope for further development exists in the selection of appropriate combinations of advanced grafting monomers, in particular pertaining to the choice of crosslinker type and concentration, with the aim of obtaining membranes with a tight yet flexible polymer network. Moreover, gradients in composition and extent of crosslinking in the thickness direction are largely uncontrolled to date. We found, in our combination of materials and processing steps, that there is more crosslinking close to the surface of the film compared with the bulk. Furthermore, the chain length distribution of the graft component is likely to have a large impact on the morphology and properties of the membrane, yet to date it remains experimentally inaccessible. Control over these parameters would allow the experimenter to design membranes of a next generation, with specific and tailored composition and consciously introduced gradients and varying microstructures over the thickness or area of the membrane.

Acknowledgment The authors gratefully acknowledge long-term funding by the Swiss Federal Office of Energy (http://www.swiss-energy.ch) of the work on radiation-grafted membranes.

References

Assink, R.A., Arnold, C. and Hollandsworth, R.P. (1991) Preparation of oxidatively stable cation-exchange membranes by the elimination of tertiary hydrogen. J. Membr. Sci. 56, 143–151.

Becker, W. and Schmidt-Naake, G. (2001) Properties of polymer exchange membranes from irradiation introduced graft polymerization. Chem. Eng. Technol. 24, 1128–1132.

Ben youcef, H., Alkan Gürsel, S., Wokaun, A. and Scherer, G.G. (2008) The influence of crosslinker on the properties of radiation-grafted films and membranes based on ETFE. J. Membr. Sci. 311, 208–215.

Chen, J., Asano, M., Maekawa, Y. and Yoshida, M. (2005) Suitability of some fluoropolymers used as base films for preparation of polymer electrolyte fuel cell membranes. J. Membr. Sci. 277, 249–257.

Chen, J., Asano, M., Yamaki, T. and Yoshida, M. (2006a) Effect of crosslinkers on the preparation and properties of ETFE-based radiation-grafted polymer electrolyte membranes. J. Appl. Polym. Sci. 100, 4565–4574.

Chen, J., Asano, M., Yamaki, T. and Yoshida, M. (2006b) Improvement of chemical stability of polymer electrolyte fuel cell membranes by grafting of new substituted styrene monomers into ETFE films. J. Mater. Sci. 41, 1289–1292.

Chen, J., Asano, M., Yamaki, T. and Yoshida, M. (2006c) Preparation and characterization of chemically stable polymer electrolyte membranes by radiation-induced graft copolymerization of four monomers into ETFE films. J. Membr. Sci. 269, 194–204.

Chuy, C., Basura, V.I., Simon, E., Holdcroft, S., Horsfall, J. and Lovell, K.V. (2000) Electrochemical characterization of ethylenetetrafluoroethylene-g-polystyrenesulfonic acid solid polymer electrolytes. J. Electrochem. Soc. 147, 4453–4458.

Freunberger, S.A., Reum, M., Evertz, J., Wokaun, A. and Büchi, F.N. (2006) Measuring the current distribution in PEFCs with sub-millimeter resolution. J. Electrochem. Soc. 153, A2158–A2165.

Gubler, L. and Scherer, G.G. (2006) Fuel cell durability of the radiation grafted PSI membrane under high H_2/O_2 pressure and dynamic operating conditions. PSI Electrochemistry Laboratory – Annual Report 2006, 19.

Gubler, L., Gürsel, S.A., Hajbolouri, F., Kramer, D., Beck, N., Reiner, A., Steiger, B., Scherer, G.G., Wokaun, A., Rajesh, B. and Thampi, K.R. (2004a) Materials for polymer electrolyte fuel cells. Chimia 58, 826–836.

Gubler, L., Kuhn, H., Schmidt, T.J., Scherer, G.G., Brack, H.P. and Simbeck, K. (2004b) Performance and durability of membrane electrode assemblies based on radiation-grafted FEP-*g*-polystyrene membranes. Fuel Cells 4, 196–207.

Gubler, L., Gürsel, S.A. and Scherer, G.G. (2005a) Radiation grafted membranes for polymer electrolyte fuel cells. Fuel Cells 5, 317–335.

Gubler, L., Gürsel, S.A., Slaski, M., Geiger, F., Scherer, G.G. and Wokaun, A. (2005b) Radiation-grafted fuel cell membranes: Current state of the art at PSI. *Proceedings of Third European PEFC Forum*, July 17–22, Lucerne, Switzerland, p. 111.

Gubler, L., Prost, N., Alkan Gürsel, S. and Scherer, G.G. (2005c) Proton exchange membranes prepared by radiation grafting of styrene/divinylbenzene onto poly(ethylene-*Alt*-tetrafluoroethylene) for low temperature fuel cells. Solid State Ionics 176, 2849–2860.

Gubler, L., Slaski, M. and Scherer, G.G. (2005d) A method for preparing a radiation grafted fuel cell membrane with enhanced chemical stability and a membrane electrode assembly, Patent Application WO2006084591, Paul Scherrer Institut, Switzerland.

Gubler, L., Müller, R. and Scherer, G.G. (2006a) Local degradation analysis of an aged fuel cell membrane. PSI Electrochemistry Laboratory – Annual Report 2006, 20.

Gubler, L., Slaski, M., Wokaun, A. and Scherer, G.G. (2006b) Advanced monomer combinations for radiation grafted fuel cell membranes. Electrochem. Commun. 8, 1215–1219.

Gubler, L., Ben youcef, H., Alkan Gürsel, S., Wokaun, A. and Scherer, G.G. (2007a) Crosslinker effect on fuel cell performance characteristics of ETFE based radiation grafted membranes. Electrochem. Soc. Trans. 11, 27–34.

Gubler, L., Ben youcef, H., Alkan Gürsel, S., Wokaun, A. and Scherer, G.G. (2008) Crosslinker Effect in ETFE Based Radiation Grafted Proton Conducting Membranes. I. Properties and Fuel Cell Performance Characteristics. J. Electrochem. Soc. 155, B921–B928.

Gupta, B. and Scherer, G.G. (1994) Proton exchange membranes by radiation-induced graft copolymerization of monomers into *Teflon*-FEP films. Chimia 48, 127–137.

Gürsel, S.A., Yang, Z., Choudhury, B., Roelofs, M.G. and Scherer, G.G. (2006) Radiation grafted membranes using a trifluorostyrene derivative. J. Electrochem. Soc. 153, A1964–A1970.

Hickner, M.A., Ghassemi, H., Kim, Y.S., Einsla, B.R. and McGrath, J.E. (2004) Alternative polymer systems for proton exchange membranes (PEMs). Chem. Rev. 104, 4587–4612.

Hietala, S., Maunu, S.L., Sundholm, F., Lehtinen, T. and Sundholm, G. (1999) Water sorption and diffusion coefficients of protons and water in PVDF-g-PSSA polymer electrolyte membranes. J. Polym. Sci. Part B: Polym. Phys. 37, 2893–2900.

Hodgdon, R.B., Boyack, J.R. and LaConti, A.B. (1966) The Degradation of Polystyrene Sulfonic Acid. TIS Report 65DE 5, General Electric Company, USA.

Hübner, G. and Roduner, E. (1999) EPR investigations of $HO^{\bullet}$ radical initiated degradation reactions of sulfonated aromatics as model compounds for fuel cell proton conducting membranes. J. Mater. Chem. 9, 409–418.

LaConti, A.B., Hamdan, H. and McDonald, R.C. (2003) Mechanisms of membrane degradation. In: W. Vielstich, H. A. Gasteiger and A. Lamm (Eds.), *Handbook of Fuel Cells – Fundamentals, Technology and Applications*. Wiley, Chichester, pp. 647–662.

Li, L., Muto, F., Miura, T., Oshima, A., Washio, M., Ikeda, S., Iida, M., Tabata, Y., Matsuura, C. and Katsumura, Y. (2006) Improving the properties of the proton exchange membranes by introducing α-methylstyrene in the pre-irradiation induced graft polymerization. Eur. Polym. J., 1222–1228.

Liu, W. and Zuckerbrod, D. (2005) *In situ* detection of hydrogen peroxide in PEM fuel cells. J. Electrochem. Soc. 152, A1165–A1170.

Liu, W., Ruth, K. and Rush, G. (2001) Membrane durability in PEM fuel cells. J. New Mater. Electrochem. Syst. 4, 227–231.

Mathias, M.F., Makharia, R., Gasteiger, H.A., Conley, J.H., Fuller, T.J., Gittleman, C.J., Kocha, S.S., Miller, D.P., Mittelstaedt, C.K., Xie, T., Yan, S.G. and Yu, P.T. (2005) Two fuel cell cars in every garage ? Electrochem. Soc. Interface 3, 24–35.

Meyer, G., Perrot, C., Gebel, G., Gonon, L., Morlat, S. and Gardette, J.-L. (2006) Ex situ hydrolytic degradation of sulfonated polyimide membranes for fuel cells. Polymer 47, 5003–5011.

Mittal, V.O., Kunz, H.R. and Fenton, J.M. (2006) Membrane degradation mechanisms in PEMFCs. Electrochem. Soc. Trans. 3, 507–517.

Panchenko, A., Dilger, H., Möller, E., Sixt, T. and Roduner, E. (2004) *In situ* EPR investigation of polymer electrolyte membrane degradation in fuel cell applications. J. Power Sources 127, 325–330.

Rager, T. (2003) Pre-irradiation grafting of styrene/divinylbenzene onto poly(tetrafluoroethylene-co-hexafluoropropylene) from non-solvents. Helv. Chim. Acta 86, 1966–1980.

Santis, M., Freunberger, S.A., Papra, M., Wokaun, A. and Büchi, F.N. (2006) Experimental investigation of coupling phenomena in polymer electrolyte fuel cell stacks. J. Power Sources 161, 1076–1083.

Scherer, G.G. (1990) Polymer membranes for fuel cells. Ber. Bunsenges. Phys. Chem. 94, 1008–1014.

Schmidt, T.J., Simbeck, K. and Scherer, G.G. (2005) Influence of cross-linking on the performance of radiation grafted and sulfonated FEP 25 membranes in H_2-O_2 PEFC. J. Electrochem. Soc. 152, A93–A97.

Taniguchi, T., Morimoto, T. and Kawakado, M. (2001) High Temperature Proton Conductive Electrolyte Membrane. Patent JP2001213987A2, Toyota Central Research & Development Laboratory, Inc., Japan.

Yamaki, T., Tsukada, J., Asano, M., Katakai, R. and Yoshida, M. (2007) Preparation of highly stable ion exchange membranes by radiation-induced graft copolymerization of styrene and bis(vinyl phenyl)ethane into crosslinked polytetrafluoroethylene films. J. Fuel Cell Sci. Technol. 4, 56–64.

4
GDL

Durability Aspects of Gas-Diffusion and Microporous Layers

David L. Wood III and Rodney L. Borup

Abstract The polymer electrolyte fuel cell (PEFC) gas-diffusion layer (GDL) is the critical bridging component between the bipolar plate flow-field and electrocatalyst layer. It must participate in all mass-transport processes of a PEFC. These consist primarily of reactant transport and liquid-water handling – either excess water removal to prevent catalyst-layer flooding under humidified conditions or suppression of water removal to prevent membrane dehydration under subsaturated conditions. Other requirements of a GDL include electron collection and transport, and sharing stack compression load with the cell gaskets. To achieve this broad range of functions, state-of-the-art GDLs consist of a complex, porous composite network of graphite fibers, carbon particles, and hydrophobic fluoropolymer. They are manufactured via a series of intricate processing steps, all of which can affect the final properties of the GDL, and may contain several discrete layers in the final form. The most popular configuration is a bilayer structure with the macroporous substrate facing the flow field and a microporous layer (MPL) facing the catalyst layer. All properties of the GDL must be preserved within the PEFC operating environment to ensure required stack lifetimes and power densities. This chapter discusses GDL substrate processing variables, hydrophobic posttreatments, MPL addition, and material selection in the context of their affects on long-term PEFC performance, i.e., loss of hydrophobicity, loss of MPL material, carbon corrosion, increase in mass-transport resistance, etc. Advanced physical property characterization methods are shown and are related to durability data. Finally, considerations for improving GDL durability and extending membrane lifetime under dry operating conditions through novel GDL designs are discussed.

D.L. Wood III (✉)
Materials Physics and Applications, MPA-11, MS D429, P.O. Box 1663,
Los Alamos National Laboratory, Los Alamos, NM 87545, USA
e-mail: dwood@lanl.gov

F.N. Büchi et al. (eds.), *Polymer Electrolyte Fuel Cell Durability*,

1 Introduction

The gas-diffusion layer (GDL) plays a critical role in determining overall polymer electrolyte fuel cell (PEFC) performance in the concentration overpotential region of a H_2/air polarization curve, particularly the cathode GDL. Proper implementation of the cathode GDL allows for operation at high current (power) densities without sacrificing cell voltage, or inducing excessive cathode overpotential. When high cell voltage is preserved, the electrochemical efficiency is preserved, which in turn contributes to higher PEFC system efficiency. Another important operating parameter associated with the concentration overpotential is the limiting current density (i_{max}), which is governed by water removal from the cathode, cell operating conditions, GDL surface chemistry, and pore structures of the substrate and microporous layer (MPL).

For approximately 10–12 years, the state-of-the-art GDL material configuration has been a bilayer structure, and in instances a trilayer structure. Using the bilayer case as an example, it consists of a macroporous, graphite-fiber substrate with the fiber surfaces treated with a fluoropolymer to introduce hydrophobic surfaces. A composite particle layer known as the MPL is coated on one side of the GDL substrate, which consists of a dispersion of carbon black and fluoropolymer. Together this configuration introduces a porous, composite nature to both layers, along with heterogeneous surface chemistry of hydrophobic and hydrophilic domains. When a cell or stack is being assembled, the MPL is placed against the electrocatalyst layers and the uncoated side of the substrate is placed against the bipolar plate for optimum liquid-water management. GDL substrate materials are commercially available from a variety of suppliers, such as SGL Carbon Group, Freudenberg, Toray Composites America, Spectracorp, BASF Fuel Cell, W.L. Gore & Associates, Ballard Material Products, and Mitsubishi Rayon. Not all suppliers include a MPL, and purchasing untreated substrate GDL material may require additional internal processing if required.

Figure 1 shows a schematic of a typical bilayer GDL configuration with the MPL coating on top and the highly porous substrate underneath. Included in Fig. 1 are scanning electron micrographs of the fibrous substrate and fluoropolymer nanoparticles. Scanning electron micrographs s of the particle-based MPL are displayed in Fig. 2, which strongly resemble the microstructure of an electrocatalyst layer. Of particular interest in Figs. 1 and 2 are the length scales, dimensions, and morphology of the various material phases. Structural features occur mainly on the submicron level of MPLs (tens of nanometers to approximately 1 μm), as seen in Fig. 2, and those of the substrate occur mainly in the range from 0.5 to approximately 50 μm, as seen in Fig. 1.

2 Background

The substrate fiber matrix acts to bridge the many structural and functional gaps between the catalyst layers and bipolar plates, as well as ensuring the entire membrane electrode assembly (MEA) active area is utilized. MPLs act as a further transitional

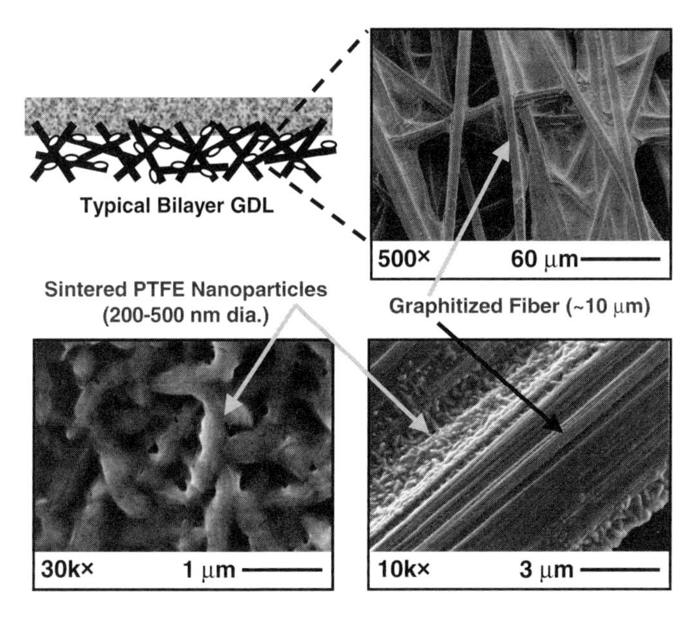

Fig. 1 Schematic and scanning electron micrographs showing gas-diffusion layer (GDL) (Toray TGP-H) graphite-fiber and polytetrafluoroethylene (PTFE) microstructure spanning approximately two orders of magnitude of length scale. Each scale plays a critical role in optimized gas and liquid transport

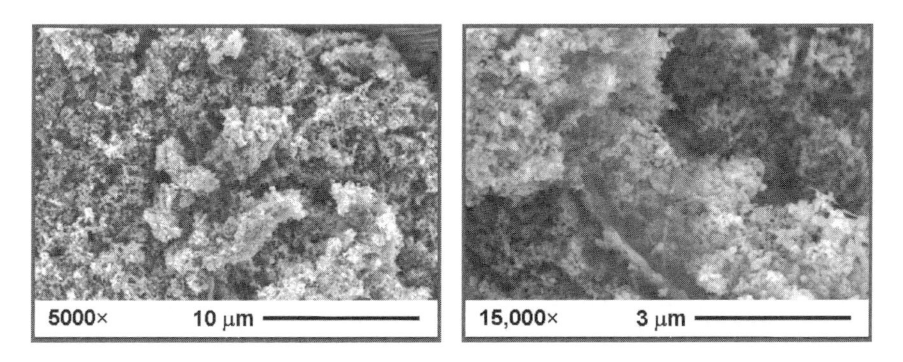

Fig. 2 Scanning electron micrographs s of BASF ELAT® microporous layer (MPL) coating showing the fractal nature of the microstructure and similar morphological characteristics to an electrocatalyst layer

layer by providing a discrete zone of pore sizes in the same range as those between the "secondary" carbon-support particles of the catalyst layer, and they also provide the necessary gradual hydrophobicity gradient between the catalyst layer and the substrate (Wood 2007).

2.1 Conventional Materials of Construction

The three main classes of GDL substrate materials are papers, felts (both "nonwovens"), and cloths ("wovens") and are made from carbon fibers, which have previously been converted into a mat in textile form. Typical textiles used for this process are polyacrylonitrile or cellulose fibers. These fibers are carbonized by heat-treating them to approximately 1,000°C, either before or after the mat is formed. Once the production of the carbonized mat is complete, it is further heat-treated to the "graphitization" range of approximately 1,750–2,700°C (depending on the desired extent of graphite crystallinity), yielding the final graphite-fiber matrix. Phosphoric acid fuel cells use similar GDLs as PEFCs, but they require greater durability owing to the highly corrosive nature of the phosphoric acid fuel cell operating environment. GDLs for phosphoric acid fuel cells must, therefore, be heat-treated to approximately 2,700°C. Owing to the less corrosive operating environment, PEFC GDLs only require heat treatment to the low end of the graphitization range (1,700–2,000°C).

Substrate raw-material fibers undergo a delicate oxidation step in either continuous fibers or roll good, which converts them into a lower molecular weight form of the original polyacrylonitrile polymer. After the polyacrylonitrile-oxidation step, a series of "carbonization" steps are performed at progressively higher temperatures that volatilize the remaining hydrogen, nitrogen, and oxygen atoms bound to the carbon surfaces. Early in the series of carbonization steps, the fiber mat is impregnated with phenolic resin (polymerized phenol–formaldehyde), which serves as a binder network for the individual fibers and improves the mechanical integrity of the mat. A well-known type of phenolic resin called Bakelite® is commonly used in carbon- and graphite-fiber paper-making.

By 1,000–1,200°C, the fibers and resin are 95–99% carbon (depending on the processing variables) and are subsequently sent to the final step of graphitization, which may be completed in one or several steps. Cloths do not contain a binder because the weaving process serves this purpose. Volume 3 of the *Handbook of Fuel Cells* contains a nonproprietary overview of GDL substrate processing that can be consulted for further details (Mathias et al. 2003). The substrate class determines many of the physical properties of the GDL, especially electrical conductivities, mechanical strengths (compression, bending, and tensile), and pore-volume characteristics. In addition, the graphitization temperature is a critical processing variable that governs the crystallinity of the carbon fibers and resin in the finished product. A graphitized material means that it has been heat-treated in an inert atmosphere to at least approximately 1,700–2,000°C, where fully carbonized material begins to take on graphitic character. The graphitic crystallinity continues to increase as the heat-treatment temperature is increased to near the decomposition temperature of approximately 3,000°C (Fleming 2000). This processing variable may be as important to overall GDL durability as the hydrophobic treatments for PEFC water management.

The plain substrate material must be "hydrophobized" with a Teflon® fluoropolymer such as polytetrafluoroethylene (PTFE) or fluorinated ethylene propylene (FEP) copolymer to create porous volume that is nonwetting to water. This leaves that fraction of pore volume open for vapor-phase transport (i.e., reactant supply and water-vapor removal), as well as creating separate hydrophobic and hydrophilic surface domains. State-of-the-art GDLs are hydrophobized by treatment with dispersion forms of the fluoropolymer, which are solvent-diluted (usually deionized water), commercially available products. The most commonly used are DuPont PTFE 30 (60 ± 2 wt% PTFE, approximately 6 wt% surfactant) and FEP 121A (55 ± 1 wt% FEP, approximately 6 wt% surfactant). Application may be carried out by submersion or spraying, and the dilution level is chosen to meet the desired loading of fluoropolymer in the substrate.

The MPL has similar material properties to the GDL fiber matrix, with the main exceptions being the porosity and pore size distribution (PSD) properties and the chemical/geometric properties of the carbon phase. The "wet" MPL (or MPL ink) is essentially an emulsion or suspension of carbon black (such as acetylene black, furnace black, and channel black), commercial fluoropolymer dispersion, additional surfactant for wetting control, and solvent, such as isopropyl alcohol or methanol for viscosity control and fine-tuning of wetting characteristics. The resultant, well-mixed slurry (up to 15–20 wt% solids) is then typically applied to the GDL substrate surface by blade-coating or spraying. The hydrophobized substrate (prior to MPL addition) and MPL-coated substrate both undergo a separate three-step heat-treatment process to evaporate the solvents (approximately 120°C), volatilize the surfactants such as Triton X-100 or Triton N-101 (above 200°C), and sinter the fluoropolymer (approximately 275–285°C for FEP and approximately 350–380°C for PTFE).

2.2 *Evaluated GDL Substrates and MPL Materials*

Cloth substrates from Zoltek (PANEX® 30 PWB3, Zoltek, St. Charles, MO, USA) Gore (CARBEL® CL, W.L. Gore & Associates, Elkton, MD, USA), and BASF (ELAT® versions 2.0 and 2.22, BASF Fuel Cell, Somerset, NJ, USA) were evaluated. Gore and BASF provide the finished GDL (including fluoropolymer impregnation and MPL coating) after the carbon-cloth substrate had been purchased elsewhere. The suppliers of the paper substrates were SGL Carbon Group (SIGRACET® GDL 24, SGL TECHNOLOGIES, Meitingen, Germany), Toray (TGP-H 060, Toray Composites America, Tacoma, WA, USA), and Spectracorp (Spectracarb 2050-C, Spectracorp, Lawrence, MA, USA). The basis for selection of the three paper substrates was that they represent a low, medium, and high degree of graphitization. Many posttreated materials from SGL TECHNOLOGIES (SIGRACET®) were also evaluated.

Two different PTFE contents were chosen for evaluation for both the substrate and the MPL, giving a matrix of four combinations of PTFE loadings for the customized GDLs. Loadings of 5 and 20 wt% were selected for the substrate, and 10 and 23 wt% were selected for the MPL. Neither the type of carbon black nor the MPL thickness was varied, in order to limit the already-sizeable number of parameters. Acetylene black was chosen for most MPLs because of the favorable combination of water management and corrosion resistance (Mändle and Wilde 2001).

ELAT® versions 2.0 and 2.22 have been used by many researchers since near the time of their inception (Mueller et al. 1999) and are considered suitable cloth benchmarks. Version 2.0 contains a double-sided hydrophobic MPL instead of the usual combination of a single-sided MPL plus bulk treatments. One side has a thicker MPL protrusion above the substrate surface that is placed towards the MEA. CARBEL® CL from Gore contains an MPL made from an expanded PTFE membrane filled with carbon black, as opposed to the common carbon black/PTFE-particle dispersions and emulsions. This type of expanded PTFE MPL may have superior long-term performance over conventional PTFE-particle MPLs (Cleghorn et al. 2006; Kolde et al. 2002).

2.3 Overview of GDL and MPL Limitations

By design, these conventional processing steps leave portions of both the graphite-fiber and MPL carbon-particle surfaces uncoated by the fluoropolymer (presumably for enhanced two-phase transport). The fluoropolymer has a second function beyond that of creating coexisting hydrophobic/hydrophilic surface domains. A fluoropolymer layer also protects some of the carbon surface from corrosion due to the operating environment or chemical degradation products from the membrane and catalyst layer. The membrane and catalyst-layer perfluorosulfonic acid ionomer degrade over time and form trace amounts of HF and H_2SO_4 in the process. In addition the carbon support for the platinum nanoparticles also corrodes owing to localized electrochemical driving forces in the immediate vicinity of the Pt/C interface (Wagner 2008; Atanassova 2008). This corrosion process forms CO_2, which may further react to form H_2CO_3. Carbon and graphite corrode (oxidize to CO_2) in the presence of acidic media, especially H_2SO_4, HNO_3, and $HClO_4$ (Kinoshita 1988).

The surface properties of carbon and graphite are further susceptible to gaseous O_2, H_2O, and O_2 dissolved in H_2O, and they may form a variety of surface oxides, carbon oxides, or graphite oxides under many different conditions (Kinoshita 1988). Under most PEFC operating conditions, there is ample liquid water for O_2 dissolution. These possible corrosion and surface-modification processes of the GDL carbon and graphite materials raise concerns for long-term PEFC performance in the high-current-density (concentration overpotential) region of the polarization curve. The question of trade-off is also raised between complete and incomplete surface coverage of the fluoropolymer on the carbon phases (Figs. 2 and 3). Complete coverage might result in performance losses early in the life of the PEFC owing to reduced two-phase mass-transport effectiveness, but it could lead to better corrosion resistance over longer operating times. This improved corrosion resistance of the GDL could, in turn, translate into improved two-phase mass-transport when compared against state-of-the-art GDL fluoropolymer treatments when viewed over the entire life of the PEFC.

3 Hydrophobicity Loss

The comingled hydrophobic fluoropolymers PTFE, FEP, and sometimes ethylene tetrafluoroethylene and perfluoroalkoxy resin yield different hydrophobic character and surface chemistry within the carbon-fiber matrix. This treatment creates a balance

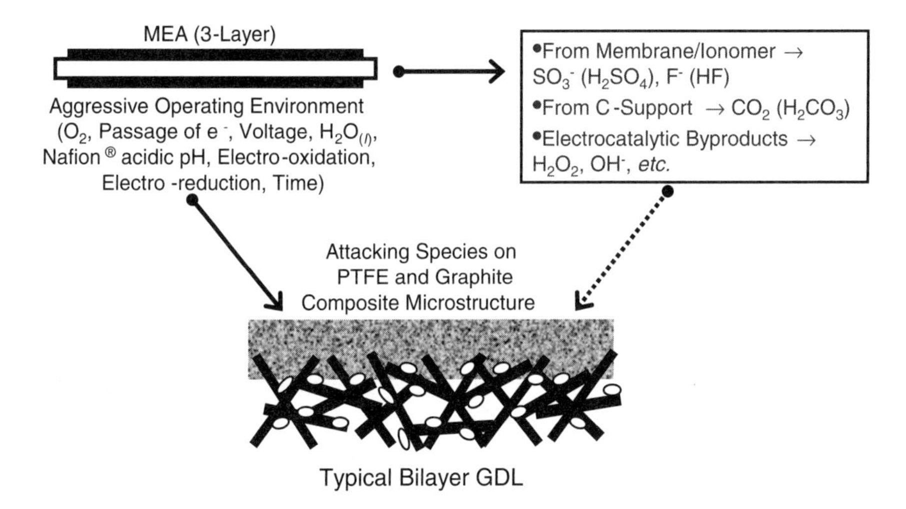

Fig. 3 Proposed pathway of membrane electrode assembly (MEA) and GDL interaction leading to observed hydrophobicity loss of GDL constituent materials

between hydrophobic (predominantly coated with fluoropolymer) and hydrophilic (predominantly exposed carbon surface) pore domains. The fresh GDL exhibits a "net" hydrophobic character, especially when examined with simple contact-angle or water-wetting experiments. Experimental data suggest the overall hydrophobicity deteriorates in a PEFC over long operating times. These data are from aging experiments, post-mortem analysis (from fuel cell, aging, and accelerated tests), and fundamental wetting and surface chemistry studies of GDLs with different fiber chemistry, substrate hydrophobicity, and MPL hydrophobicity.

3.1 GDL and MEA Chemical Interaction During Long-Term Testing

The aggressive PEFC operating environment causes MEA subcomponent degradation, as considered in Sect. 2.3. Corrosion of the carbon support leads to CO_2 and some CO formation, most of which is converted to aqueous H_2CO_3. Furthermore dissolution and migration of platinum particles results in increased H_2O_2 formation, either (1) via hydroxyl and hydroperoxyl intermediates formed by O_2 crossover to the anode or (2) by H_2 crossover to the cathode (LaConti et al.). These processes result in increased corrosion and/or surface modification of the GDL carbon and fluoropolymer through liquid-phase diffusion of the attacking trace species. In fact F^- from ionomer degradation has been measured in outlet water samples at concentrations as high as about 0.1–1.0 mM. When these concentrations were compared with concurrent pH measurements, most of the F^- was found to exist as HF (Healy et al. 2005).

Figure 1 shows a PTFE-treated GDL substrate, emphasizing the random fiber-matrix structure and adhesion of PTFE particles to an individual fiber. The PTFE loading of the GDL is 5wt%, and it can be seen that plenty of the surface of individual fibers remains uncoated. A rough, nonuniform exposed surface of both the carbon fibers and the PTFE particles is also observed. Possible pathways for both primary (operating environment induced – O_2, H_2O, temperature, voltage, etc.) and secondary (MEA degradation product induced) GDL degradation are shown in Fig. 3 and are indicated by a solid and a dashed line, respectively. When considering the porous structure and nanoscale surface features of both carbon and PTFE visible in Figs. 1 and 2, we find that plenty of pathways exist for GDL chemical and physical degradation.

The focus of this work is on the cathode since mass-transport limitations are small to negligible at the anode of H_2/air PEFCs up to approximately 1.0–1.5 A cm^{-2}. Observed performance changes within the mass-transport region during durability testing are, therefore, ascribed to the cathode. Increases in GDL mass-transport resistance (loss of hydrophobicity, loss of MPL material, etc.) due to MEA degradation are likely an interplay between all cathode degradation processes. Performance and characterization data examine how these processes work in concert with each other. For true understanding of this interrelationship, the long-term effects and extents of primary and secondary GDL degradation must be separated. For instance, under certain circumstances, such as operation with a dry cathode inlet stream, F^- formation from ionomer degradation and H_2 crossover may be high, thus resulting in a higher concentration of HF (owing to less liquid water). However, the contact of the liquid phase (with perhaps higher pH) with the GDL porous network would be less. Contact with dissolved O_2 would also be less. These examples are only two of many where individual competing effects must be understood to gain fundamental understanding of GDL durability and long-term surface chemistry.

3.2 Single-Fiber Contact Angle

Single-fiber contact-angle measurements were performed using the Wilhelmy technique, which is identical to the more common Wilhelmy-plate technique except for the sample geometry and measured wetting forces. Figure 4 shows illustratively the difference between the single-fiber Wilhelmy, static sessile drop, and dynamic sessile drop measurements and the breadth of information each provides when analyzed collectively. A drastic difference between the contact angle of an approximately 5-μL water droplet on the surface of Toray TGP-H paper and a single-fiber of the same material penetrating the water droplet interface was observed – about 41° less for the single fiber. These results show that even the most graphitic fibers (TGP-H is graphitized to temperatures approaching 2,700°C) are hydrophilic when considered as a nonporous solid (i.e., the Wilhelmy measurement). However, when a group of many fibers are considered together as a highly porous, nonwoven mat with significant surface roughness

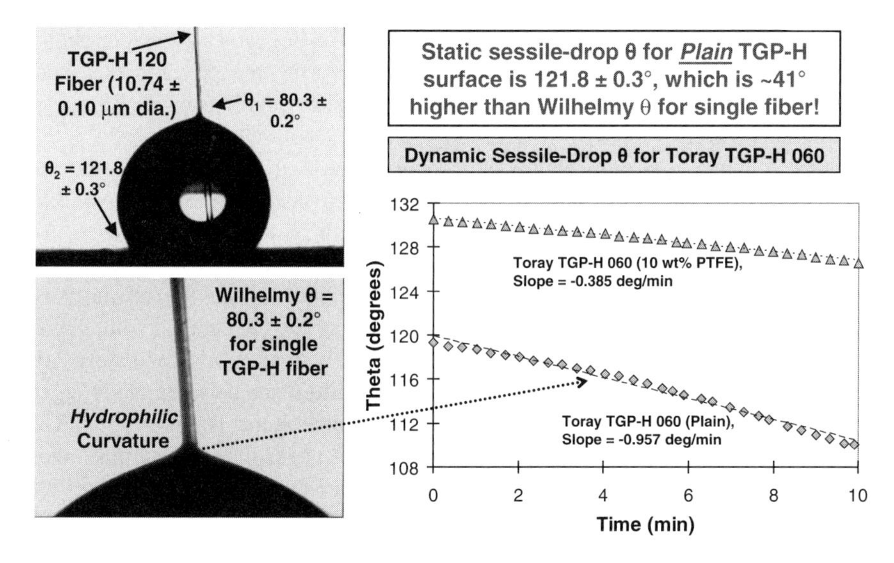

Fig. 4 Comparison between single-fiber Wilhelmy, static sessile drop, and dynamic sessile drop contact angles for plain Toray TGP-H paper. The substrate on which the water droplet is sitting (*top-left corner*) is plain TGP-H paper, with a single approximately 10 µm diameter TGP-H fiber penetrating the water droplet (the fiber was extracted from the paper substrate)

(i.e., the sessile-drop measurement), the geometric surface is quite hydrophobic upon initial exposure to the water droplet.

In the graph in Fig. 4, data for TGP-H paper treated with 10 wt% PTFE are also shown, and a noticeably smaller slope magnitude (40% less) was observed. This result demonstrates the complex nature of the dual hydrophobic and hydrophilic solid-surface domains. The addition of the hydrophobic PTFE causes a higher initial (static) contact angle and a much more gradual slope decline of the dynamic contact angle, but the impartial surface coverage of the fibers allows for some spreading and penetration of the water droplet. This behavior is indicative of the in situ PEFC liquid-transport physics, mostly those of the unknown mechanisms, which are involved with the capillary transport of water through the internal substrate pores during cell operation. The variation of inlet relative humidity (RH), gas pressure, and current density only complicates these phenomena by adding a directional liquid-water flux component.

Single-fiber Wilhelmy measurements were conducted on Toray TGP-H materials aged under different liquid-phase conditions to quantify changes in the GDL-substrate surface chemistry. A matrix of four environments was used (60 or 80°C and N_2 or air sparging gas) with a substrate hydrophobic treatment of 17.0 ± 0.3 wt% FEP (no MPL for any samples). Contact-angle results are shown in Fig. 5, along with actual photographs of the fiber/water interface and liquid meniscus for each sample. Four separate fibers were extracted from all samples with excellent reproducibility of the contact angles, with two different fresh samples being evaluated. The treatment of the substrate with 17 wt% FEP is a high loading, and a correspondingly large increase of 14° in contact angle was measured for the unaged samples. Errors represent one standard deviation of four separate fibers measured from each sample; fiber diameters were 11.1 ± 0.1 μm (see Fig. 5).

Figure 5 shows a strong relationship between the aging environment and the postmortem contact-angle value. The general aggressiveness of the aging environments was found to be the least severe for 60°C with N_2 and the most severe for 80°C with air. The temperature increase over the range investigated had less of an impact on the contact angle and loss in hydrophobicity than the presence of dissolved O_2. The difference between the final values of the GDLs aged with N_2 and those of the GDLs aged with air for each temperature were only approximately 2.0° and approximately 1.5°, respectively. However, the difference between the N_2-aged samples and the air-aged samples, regardless of temperature, is about 4–7°. The samples aged with air were exposed for about 200 h longer than the samples aged with N_2 sparging, but it is unlikely that this difference contributed significantly to the postmortem contact angles. This statement is supported by the inversely proportional relationship of gas (O_2) solubility to temperature: the contact angles of the 80°C/air sample were lower than those for the 60°C/air sample. This finding indicates the lower dissolved O_2 concentration at 80°C had more of an effect on the degradation (likely in conjunction with the increased temperature of the water in contact with the fibers) than the higher dissolved O_2 concentration at 60°C. These results stand in agreement with other extensive published surface-energy, contact-angle, and PSD data (Borup et al. 2007; Wood et al. 2004a, 2006).

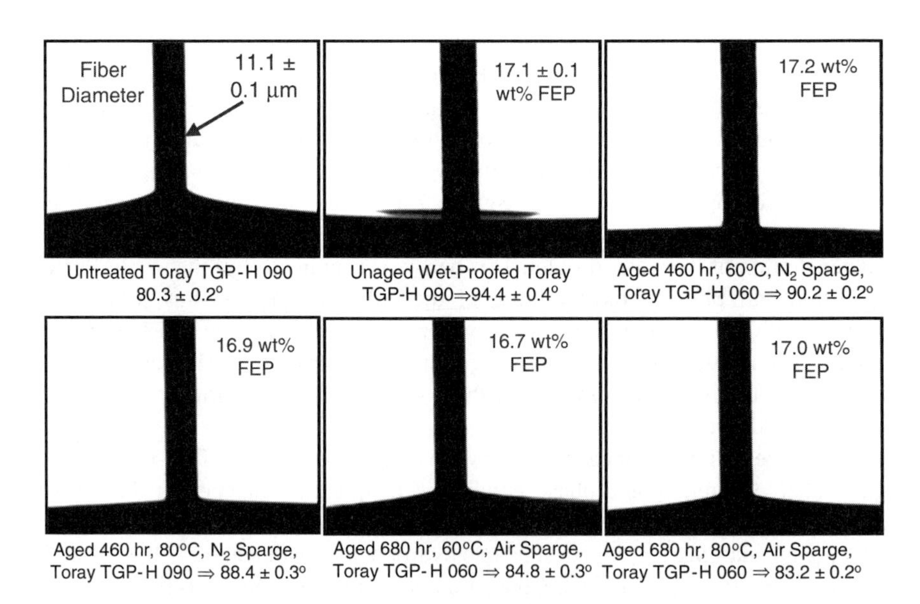

Fig. 5 Single fiber (Wilhelmy) contact angles of Toray TGP-H materials (hydrophobized to 17 wt% fluorinated ethylene propylene, FEP) aged in accelerated fashion in different liquid-water environments

3.3 *Changes in Composite Surface Energy and Dynamic Contact Angle*

In collaboration with W.L. Gore & Associates, GDLs from a 26,300-h (approximately 3 years) life test were characterized. The GDLs were Gore cloth-based CARBEL® CL and the MEA was a PRIMEA® series 5621. The details of the Gore in-house, postmortem investigation are summarized elsewhere (Cleghorn et al. 2006). Zisman surface energies, i.e., a one-component model or "total" surface energy (Zisman 1964), and dynamic sessile drop contact angles were measured for both the anode and the cathode GDLs, and the data are shown in Fig. 6. The surface energy of both the anode and the cathode GDLs increased significantly after 26,300 operating hours. The Zisman surface energy of fresh CARBEL® CL was measured at 19.0 mJ m^{-2}, which increased after the life test to 20.7 and 20.8 mJ m^{-2} for the anode and the cathode, respectively. As shown and discussed in detail elsewhere (Wood et al. 2005, Wood 2007), this magnitude of change in total surface energy (+1.7–1.8 mJ m^{-2}) is substantial and more than enough to induce a change in mass-transport and water-management characteristics. The data were correlated with dynamic sessile drop measurements, also shown in Fig. 6, and excellent agreement was found in terms of increase in slope magnitude.

Of particular interest in Fig. 6 are the changes in surface energy and contact-angle slope relative to those of plain Zoltek PANEX cloth (a material similar to the

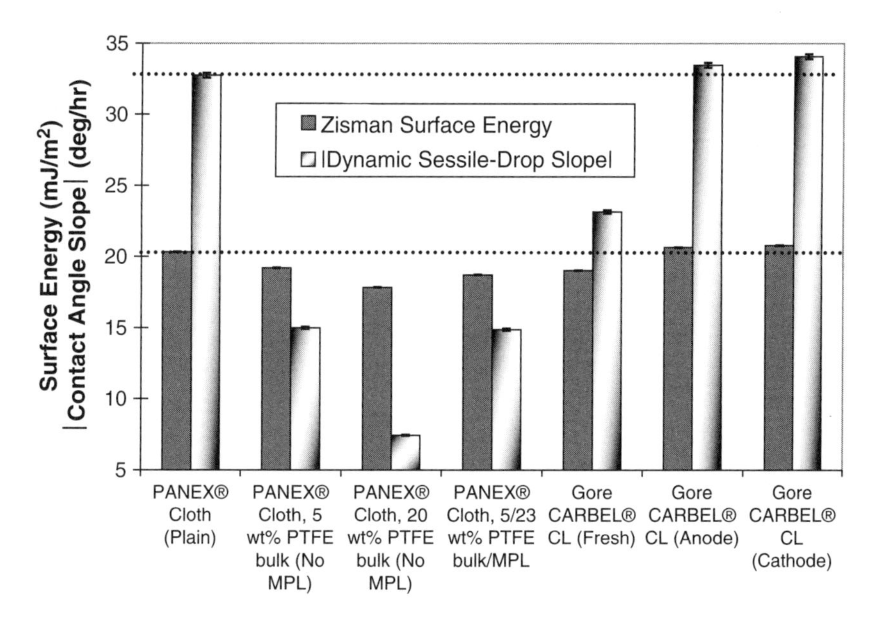

Fig. 6 Surface energy (by the Zisman method) and sessile-drop contact-angle slopes for various fresh and durability-tested cloth GDLs. CARBEL® CL GDL data represent before and after values from a 26,300-h test operated under stationary conditions by W.L. Gore & Associates (Cleghorn et al. 2006, Wood 2007)

CARBEL® CL substrate), which are highlighted by the horizontal dotted lines across the figure. The values of the fresh CARBEL® CL were below those of the plain Zoltek cloth, indicating the effects of the Gore hydrophobic treatments on the substrate. After the durability test run under typical stationary conditions (70/70/70°C anode/cathode/cell and near-ambient pressure), the surface energies of the anode and the cathode GDL became more hydrophilic than that of plain untested cloth. This finding suggests that the collective effects of the bulk hydrophobic treatment and CARBEL® MPL were entirely negated, and the GDL surface was modified to a point that increased the hydrophilic character well beyond that of the initial material.

Electrochemical impedance spectroscopy measurements performed by Gore after the life test revealed substantially increased low-frequency resistance at 0.6 and 1.5 A cm^{-2}, which directly confirmed mass-transport limitations. After a follow-up experiment where the cell was allowed to equilibrate at a substantially reduced cathode RH (dew point below 70°C), the electrochemical impedance spectroscopy measurements were reacquired at 1.5 A cm^{-2} with a notable reduction in total impedance (Cleghorn et al. 2006). This finding was most likely due to alleviation of cathode GDL flooding. It also indicates (independently from the surface chemistry data), that the cathode GDL devolved from a highly hydrophobic material to a much less hydrophobic one. Neither scanning electron microscopy (SEM) observations nor a boiling-water uptake test performed internally by Gore revealed any changes in either GDL, validating the need for the postmortem surface-energy and dynamic sessile drop contact angle measurements.

Surface energies (by the Owens–Wendt method), comprising two components accounting for dispersive (weaker van der Waals forces) and polar (strong dipole and hydrogen-bonding forces) molecular interactions (Good and Girifalco 1960; Owens and Wendt 1969), were measured for two different GDL sets. The single cells are denoted as "G7," which contained a Gore PRIMEA® series 5620 MEA, and "M5," which contained a 3 M Nafion®-based catalyst-coated membrane. Table 1 shows the results for before/after relatively short life testing periods. The surface energy and percentage polarity for both the anode and the cathode GDLs increased, though the testing periods were less than 200 h. There was a larger increase for GDL 24DC10 (20 wt% PTFE substrate/10 wt% MPL) over 122 h, which had a low cathode inlet RH of 25%, than for GDL 24BC (5 wt% PTFE substrate/23 wt% MPL) over 185 h. This small, but statistically significant difference, is attributed to the more aggressive environment imposed by the 25% cathode RH on the membrane and ionomer phases, which increases the backbone end-group degradation, HF formation, and H_2SO_4 formation rates. Trace concentrations of F^- and SO_4^{2-} in the anode and cathode effluent water have been well studied over the past several years. This proposed effect of low cathode RH is also supported by the higher total weight percentage of PTFE of GDL 24DC10 versus that of GDL 24BC (17.2 versus 11.2 wt%, respectively). Cell G7 was also operated for a shorter time, but the changes in percent polarity of the Owens–Wendt surface energy were larger for GDL 24DC10, suggesting that exposed carbon phases were more significantly affected.

Table 1 Surface energies (by the Owens–Wendt method) of two different SIGRACET® gas diffusion layer (GDL) types before and after accelerated durability testing (different catalyst-coated membranes were used in the cells)

Sample	Total surface energy ($mJ\ m^{-2}$)	Polar component ($mJ\ m^{-2}$)	Dispersive component ($mJ\ m^{-2}$)	Polarity (%)
Cell G7: GDL 24DC10; 122 h at 0.90 A cm^{-2}; cell 80°C; 75%/25% relative humidity (anode/cathode); gas pressures 180 kPa (abs); 1.4/2.2 equivalent stoichiometric flow (anode/cathode)				
Fresh	19.5	0.97	18.5	5.0
Anode	19.8	1.19	18.7	6.0
Cathode	19.9	1.21	18.7	6.1
Cell M5: GDL 24BC; 185 h at 1.04 A cm^{-2}; cell 80°C; 75%/75% relative humidity (anode/cathode); gas pressures 180 kPa (abs); 1.5/2.0 equivalent stoichiometric flow (anode/cathode)				
Fresh	20.4	1.47	18.9	7.2
Anode	20.7	1.67	19.1	8.1
Cathode	20.8	1.66	19.2	8.0

3.4 Increases in Cathode Mass-Transport Overpotential

Over the course of life testing, a change in the water-management characteristics of the hydrophobized GDL has been observed by researchers. These changes seem to occur on the micron to submicron level, but they are easily observed on a macroscopic scale with simple postmortem analysis. When a cell or stack is disassembled after durability testing, the GDLs can be sprayed with water and the water adheres to, or wets, the GDL surface. The same experiment performed with an identical fresh GDL will result in spherical water beads bouncing off the surface. The GDL gradually transforms from a generally hydrophobic porous body into a more hydrophilic one when exposed to the PEFC operating environment. Even though many researchers have noticed this effect after a quick water-spraying experiment, almost nothing has been published on this phenomenon (LaConti et al. 2003). Performance effects are typically negative at higher power densities and with high-humidity inlet gas streams (i.e., catalyst-layer flooding and constricted vapor-phase mass transport), although it will be shown later that some of the effects can have a positive impact with dry inlet streams. Mass-transport performance losses (or increased concentration polarization) due to increased flooding of the GDL and/or MPL pores is the main consequence of these surface chemistry changes. This means the liquid-water content within the pore volume is too high and the liquid-water flux through the pores is too low.

To determine the cause of GDL hydrophobicity loss and associated degradation mechanism(s), changes in the exposed carbon phases (graphite fibers and carbon particles), the fluoropolymer coating (PTFE, FEP, etc.) overlaying the carbon surfaces, and the bonding interface between the sintered fluoropolymer particles and carbon surfaces must be studied. All three of these aspects play at least some role in GDL and MPL hydrophobicity losses, and delineation of the relative importance of each surface or interface is a central theme of this work. The relative amounts of fluoropolymer treatment in both the GDL and the MPL, known to be critical to beginning-of-life performance and operating conditions, likely affect the composite surface chemistry of liquid-water interaction within the GDL porous volume, as well as the long-term PEFC performance characteristics. Maintaining the local hydrophobic character of the GDL and MPL pores, known as a "GDL hydrophobicity gradient" (Wood et al. 2002a), is also important for maintaining sufficient mass-transport-region performance over the course of life testing.

There have been several informative studies in recent years regarding the physical property characterization and experimental methods required for this compulsory understanding (Bluemle et al. 2004; Dohle et al. 2003; Ihonen et al. 2004; Jordan et al. 2000; Liu et al. 2004; Weber and Newman 2005; Williams et al. 2004; Wood et al. 2002b), particularly the work of Bluemle et al. and Liu et al. However, all these data and analyses are in the context of fresh (untested) GDLs or novel GDLs evaluated during initial testing. Unfortunately, there is a gap in the literature between GDL physical property characterization and meaningful correlation of the properties to PEFC data (beginning-of-life or otherwise). This chapter will assist in explaining changes in these properties caused by durability and accelerated testing and their relationship to beginning-of-life performance versus end-of-life performance.

A complete polarization analysis of a single-cell denoted as "G2" using methods described elsewhere (Gasteiger et al. 2003; Williams et al. 2005; Wood et al. 2006; Wood 2007) is shown in Fig. 7 as a function of steady-state durability testing time. This cell was similar to G7 in that it contained a Gore series 5620 MEA, with the only major differences being the GDLs (SIGRACET® GDL 24BC) and the cathode RH (75%). A significant increase in cathode GDL mass-transport overpotential ($\eta_{tx,GDL}$) is seen between 536 and 1,014 h in Fig. 7, but remained unchanged through the first approximately 500 h. There was also a noticeable increase in oxygen reduction reaction overpotential (η_{ORR}) with the opposite trend from that of $\eta_{tx,GDL}$ – a larger change was observed in η_{ORR} over the first approximately 500 h than over the second approximately 500 h. As the Ohmic-resistance-corrected cell voltages ($V_{iR\text{-}free}$) in Fig. 7 show, these increases in $\eta_{tx,GDL}$ and η_{ORR} were offset by improvements in the mass-transport overpotential of the cathode catalyst layer ($\eta_{tx,elec}$). Figure 7 shows that reductions in the high-frequency resistance (HFR) in the region below 0.5 A cm^{-2} also contributed to this effect, but the cathode catalyst-layer proton-transport resistance (R_{H+}) remained constant. Based on the way R_{H+} is calculated, it should not change because the cathode catalyst-layer material compositions remained the same, and the stoichiometric flow ratios and current density also remained the same.

Figure 8 shows the GDL mass-transport overpotential expressed in terms of resistivity ($R_{tx,GDL}$), with a more sharply increasing trend observed in the same region where HFR and R_{H+} level out. This finding suggests that the GDL mass-transport losses at the cathode would have continued to worsen indefinitely, in addition to being irreversible. Comparing the trends of the ionomer-phase resistivities and the cathode-GDL resistivity underscores the operating paradox of a PEFC, in that the membrane prefers excess (liquid) water and exhibits minimum resistivity

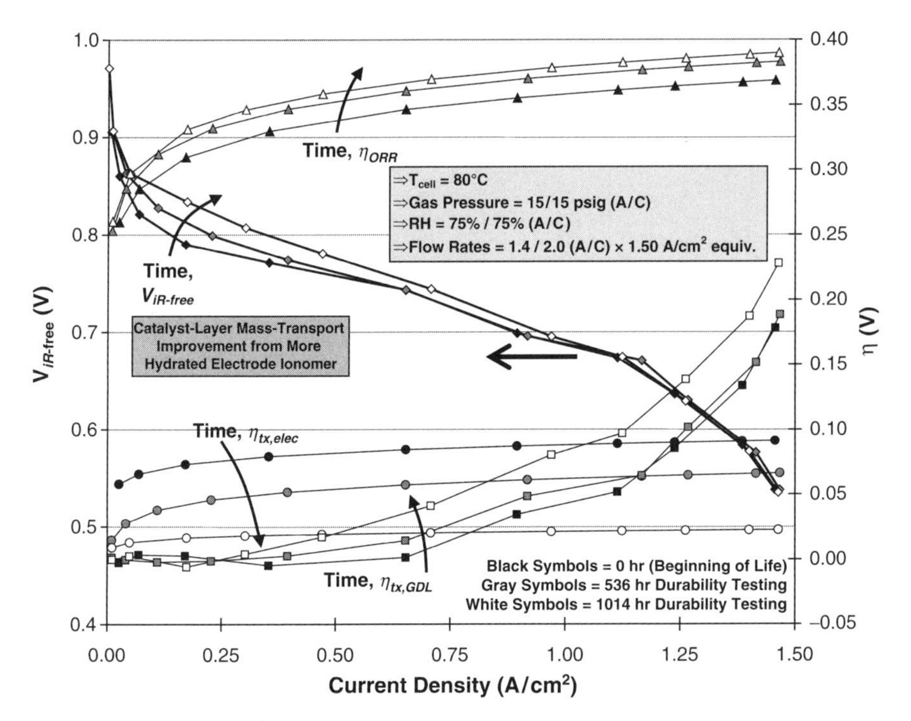

Fig. 7 Cell "G2" cathode catalyst-layer and GDL mass-transport overpotentials, oxygen reduction reaction (ORR) overpotential, and iR-corrected cell voltage as a function of current density at different durability testing times (1,054 total hours at constant 0.90 A cm^{-2}); Gore PRIMEA® series 5620 MEA (50.0-cm^2 active area, 0.40/0.60 mg Pt cm^{-2}, 35-μm Gore SELECT® membrane); SGL SIGRACET® GDL 24BC (5 wt%/23 wt% PTFE substrate/MPL) at both sides. See the Appendix for durability testing conditions. *RH* Relative humidity, *A* Anode, *C* Cathode

in this situation (Fig. 8). However, these operating conditions that lead to minimum ionomer resistivity also increase the mass-transport resistance of the cathode GDL (and perhaps the cathode electrode) owing to liquid-water flooding. The engineering challenge is to extend the Ohmic region of the polarization curve so that flooding is avoided.

On the basis of long-term "US06" drive-cycling and steady-state life-testing results (Borup et al. 2006b; Markel 2004; Wood et al. 2004b), there is evidence that suggests the mass-transport losses associated with drive-cycling experiments are significantly different from those associated with steady-state testing. Two different single cells were exposed to the US06 drive-cycle conditions (Fig. 9). The main difference between the cells was that the durability tests were carried out at different inlet RHs (100%/100% anode/cathode for the cell denoted as "D1" in Fig. 9 and 48%/48% anode/cathode for the cell denoted as "D2"). The polarization data seen in Fig. 9 were acquired intermittently under the same conditions as for the durability testing, with the exception of the inlet dew points. The most important conclu-

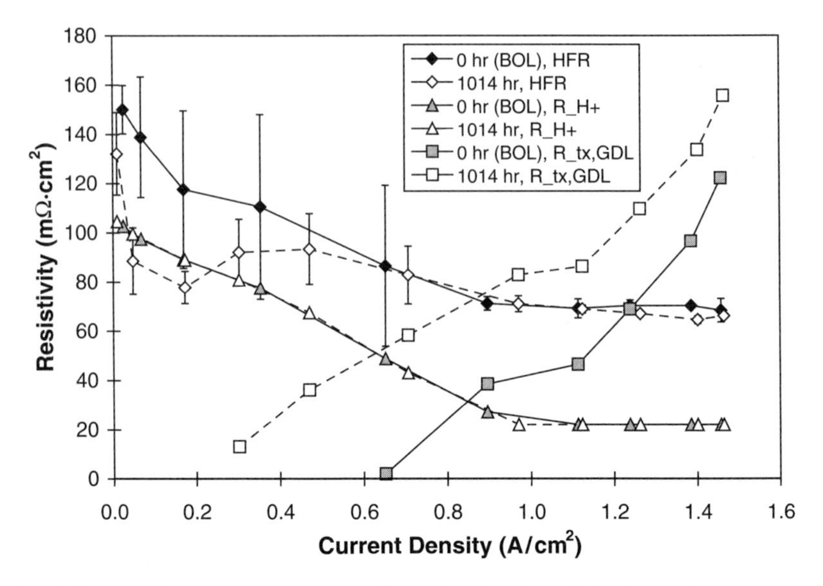

Fig. 8 Various resistivity trends versus durability testing time for cell "G2"; note the sharp increases in $R_{tx,GDL}$ where R_{H+} and high-frequency resistance (HFR) level out. All polarization conditions were the same as those for Fig. 7. See the Appendix for durability testing conditions

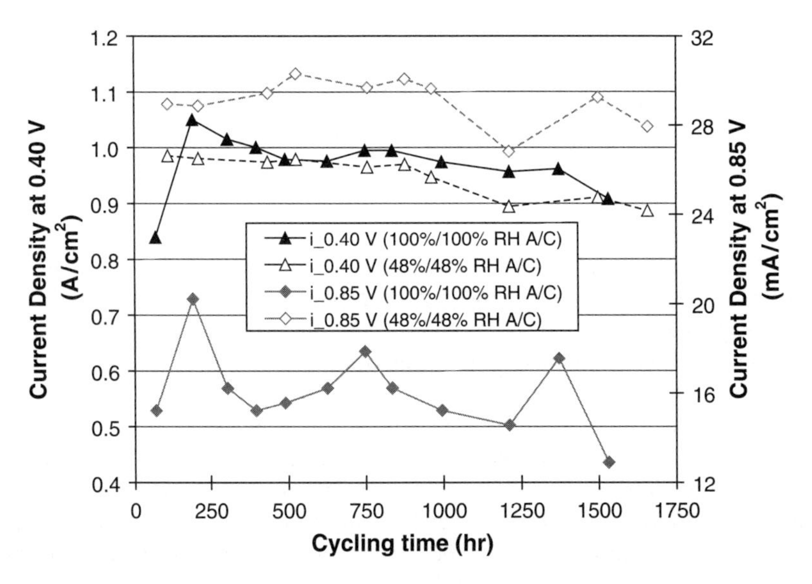

Fig. 9 Comparison between kinetic and mass-transport performance as a function of time for US06 drive – cycle durability – tested cells; cell "D1" was tested at 100%/100% RH for 5,895 cycles (1,965 h); cell "D2" was tested at 48%/48% RH for 7,045 cycles (2,348 h). See the Appendix for durability testing conditions

sions from this data set were that the mass-transport performance for each cell was about the same and that the kinetic performance was significantly higher for cell D2 operated under dry conditions. The performance decline in both regions of operation as a function of time was substantial for each cell, but it transpired at about the same rate. This was a surprising finding since the inlet RHs were significantly different and suggests the load cycling plays a more critical role in long-term mass-transport performance than the RH values themselves. It is also interesting to note that the kinetic performance of cell D2 with much lower RHs was more stable through about 1,600–1,700 operating hours.

4 Carbon Corrosion

The focus of this section is on the corrosion phenomena of the GDL substrate graphite fibers driven by solid–liquid surface chemical interactions and surface energetics. Particular importance is assigned to uncoated areas of fibers by hydrophobic fluoropolymer and interfacial zones between fiber and fluoropolymer. Since Teflon® products such as PTFE and FEP are known to have excellent thermal resistance, electrochemical corrosion stability, chemical durability, and mechanical properties, the GDL and MPL fluoropolymer phase is not considered to be vulnerable in a PEFC operating environment.

4.1 Surface Chemistry and Wettability Changes

An example of the postmortem utility of basic static-sessile-drop contact-angle measurements is shown in Fig. 10. Two pieces of SGL SIGRACET® GDL 24BC were aged in 80°C deionized water with air sparging gas for 1,006 h (in the manner previously described in Sect. 3.2). This accelerated test resulted in a significant reduction in the external contact angle of the MPL side of the GDL 24BC material (i.e., a reduction well below that of the statistical variation of the contact angle of the fresh MPL surface). The higher variability in external contact angle of the back (non-MPL) side of the fresh 24BC material is due to intrinsic, in-plane variability of the hydrophobic treatments, a problem that all GDL manufacturers are currently tackling. Specifically, the MPL ink can penetrate completely through the substrate to the flow-field side via the largest pores and induce varying hydrophobicity on a submillimeter scale. There is also the possibility of variation in the bulk hydrophobic treatment on the same scale, and this length scale represents the same area covered by an approximately 5–8 μL water droplet. These contact-angle measurements were made with a LANL in-house goniometer, which does not have the precision, accuracy, or reproducibility of the KRÜSS DSA100 instrument (as in Figs. 4 and 6).

Liquid water and dissolved O_2, both of which are in abundance at the cathode side of an operating PEFC, are direct causes of hydrophobicity loss of the MPL. This experiment also gives an indication that the intrinsic hydrophobicity of the

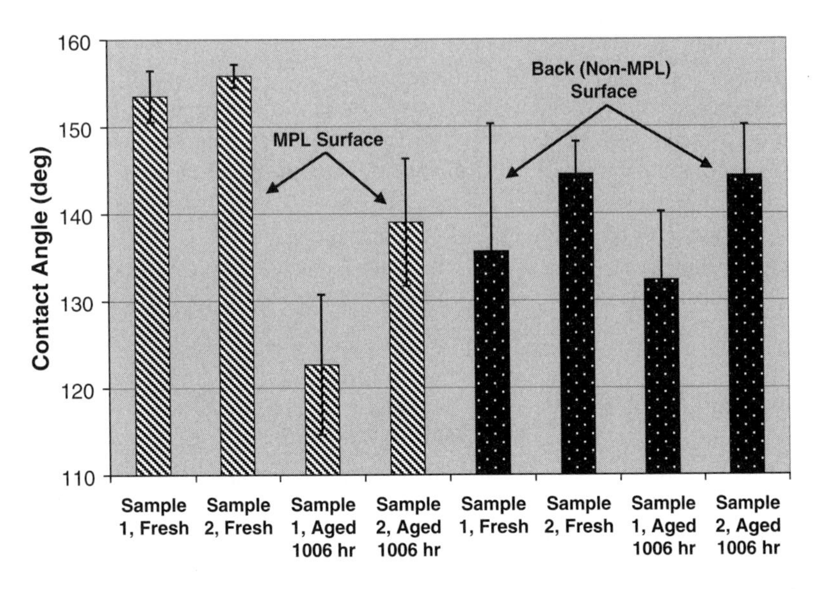

Fig. 10 Before and after comparison of static sessile-drop contact angles of SGL SIGRACET® GDL 24BC aged for 1,006 h in 80°C deionized water with air sparging gas; *Bars on left* represent the MPL surface and *bars on right* represent the non-MPL surface

MPL is more vulnerable to carbon-surface oxidation by liquid water and dissolved O_2 than the graphitized substrate-fiber surfaces (at least for the SIGRACET® GDL 24 substrate type), since the contact-angle reduction of the non-MPL sides was statistically insignificant. However, this might not have been the case for a substrate material that was graphitized to a higher temperature, such as Toray TGP-H.

Figure 11 shows results for seven different Toray TGP-H GDLs. The polar components are almost negligible, even for the plain TGP-H sample. The polarity contribution of plain TGP-H is only 9.0% of the total surface energy. This low extent of surface polarity shows that the wetting of even the nonhydrophobized GDL materials is driven by dispersive molecular interactions between water and graphite. Once PTFE is added to the substrate at a 5-wt% loading, the surface polarity drops to less than 4.5% even with the presence of a MPL, which contained 77–90 wt% acetylene black (Fig. 11). By raising the substrate loading of PTFE to 20 wt%, the surface polarity dropped to 2.7% or less – the surface polarity of the 20-wt% sample with no MPL was only 1.0% and exhibited an overall surface energy of 17.9 mJ m^{-2}. This trend is remarkably consistent with the known surface energetics of pure PTFE, which is highly inert and has no surface polarity owing to its high fluorine content and molecular structure. The overall surface energy of PTFE has been measured at 18.0 mJ m^{-2} for materials provided by a variety of suppliers. The overall surface energy of Durafilm® ethylene tetrafluoroethylene has been measured at 22.4 and 1.14 mJ m^{-2} for the polar component (Rulison 2007). Figure 11 confirms the dominant intermolecular forces driving liquid-phase transport and GDL two-phase performance, with *or* without fluoropolymer treatments and MPL addition, are dispersive even though

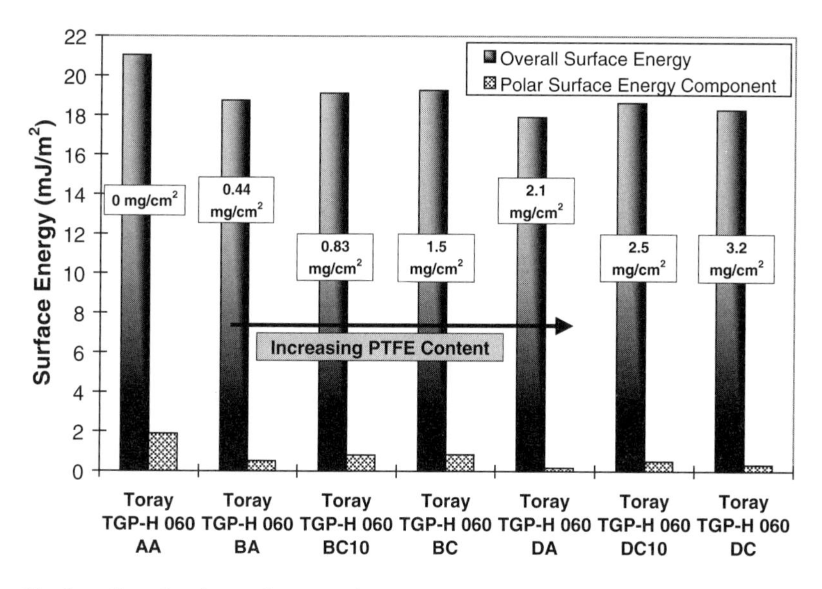

Fig. 11 Overall and polar surface energies measured by the Owens–Wendt technique (with Washburn absorption) for various types of Toray TGP-H paper GDLs. Values in *white boxes* represent the total aerial loading of PTFE for each sample. *AA* plain, *BA* 5 wt% substrate/0 wt% MPL, *BC 10* 5 wt% substrate/10 wt% MPL, *BC* 5 wt% substrate/23 wt% MPL, *DA* 20 wt% substrate/0 wt% MPL, *DC 10* 20 wt% substrate/10 wt% MPL, DC 20 wt% substrate/23 wt% MPL

water has such a high surface polarity. Almost 64% of the total 72.8 mJ m^{-2} of surface tension is due to the polar nature of the water molecule.

An interesting correlation between the data in Fig. 11, the discussion above, and the data in Fig. 10 is that an increase in total surface energy and percentage polarity was measured for the GDL 24BC samples after the 1,006 h of aging. Fresh GDL 24BC has a total surface energy of 20.4 mJ m^{-2} and 7.21% polarity, but after the accelerated aging in liquid water with air sparging these values increased to 21.3 mJ m^{-2} and 9.46%, respectively (Wood 2007). This is a strong indication that corrosion or surface oxidation of the substrate fibers, MPL carbon, or both occurred during aging. Asserting this conclusion is not possible using the one-component Zisman method because it is the carbon phase in PEFC GDLs that represents the surface polarity component of total surface energy.

4.2 Modeling of Heterogeneous Sessile-Drop Contact Angles

To understand the differences between single-fiber contact angles discussed in Sect. 3.2 and sessile-drop contact angles discussed in Sects. 3.3 and 4.1, a more advanced equation is needed than the Young equation to explain the phenomena.

The Cassie equation (Cassie 1948) takes into account both a heterogeneous solid surface (i.e., hydrophobic and hydrophilic solids, etc.) and porosity, so it is well suited for studying GDL substrates and MPLs with respect to wettability. Figure 12 shows modified-Cassie-model calculations as a function of fiber surface coverage for two different fluoropolymers (PTFE and FEP) and three different values of substrate porosity for Toray TGP-H paper, with the details discussed elsewhere (Wood 2007). The original Cassie equation did not include a term for porosity, but it was expanded to include surface porosity (Cubaud and Fermigier 2004). This modification treats the contact angle of any liquid with air as 180°. The results of model calculations shown in Fig. 12 show that both surface coverage of the fluoropolymer and surface porosity have a substantial impact on the observed (i.e., equilibrium) sessile-drop contact angle. The assumption was made that the fraction of fluoropolymer that covers the individual fibers is the same as the fraction of coverage at the geometric surface of the substrate. Geometric substrate surface refers to the apparent material surface (i.e., if the roughness factor were 1). Another assumption was made that the surface porosity is equivalent to the porosity within the bulk of the substrate after treatment. Both of these assumptions are reasonable, but their soundness depends on the differences in and uniformity of the bulk-PTFE treatment processing.

As the fractional coverage of fluoropolymer increases, the curvature of the model lines increases sharply in the $+\theta$ direction, especially for PTFE. This shows that the hydrophobicity of the fluoropolymer (or pure-component contact angle) plays a large role in the equilibrium contact angle. When the porosity of TGP-H is 80%, model calculations suggest only a fractional surface coverage of approximately 0.8 is needed to reach the theoretical maximum of 180°. Figure 12 also shows a substantial difference in equilibrium contact angle between FEP and PTFE at a given fractional surface coverage, as evidenced by the lower degree of curvature of the FEP lines. This phenomenon is due to a pure FEP solid surface having an 8° lower water contact angle than that of PTFE. This explains why PEFC performance and water management can vary extensively as a result of only a change in GDL fluoropolymer type(s). The separation between calculated model lines for a given surface porosity and fluoropolymer also contains useful phenomenological information. There is nearly a linear separation between all three porosity values considered in Fig. 12 for FEP until the surface coverage fraction reaches approximately 0.8. In contrast, an increase in separation between the corresponding constant-porosity lines for PTFE is noticeable at approximately 0.6 fractional surface coverage. This suggests that PTFE-coated fibers at the GDL substrate surface do not require as great an extent of surface coverage as FEP-coated fibers for the surface porosity to have an effect on the equilibrium contact angle. Furthermore, this confirms that a difference of only 8° in water contact angle on the pure fluoropolymer surface and a span of only 10% in substrate void volume has tremendous implications for the solid–liquid interfacial physics within the GDL pores of an operating PEFC.

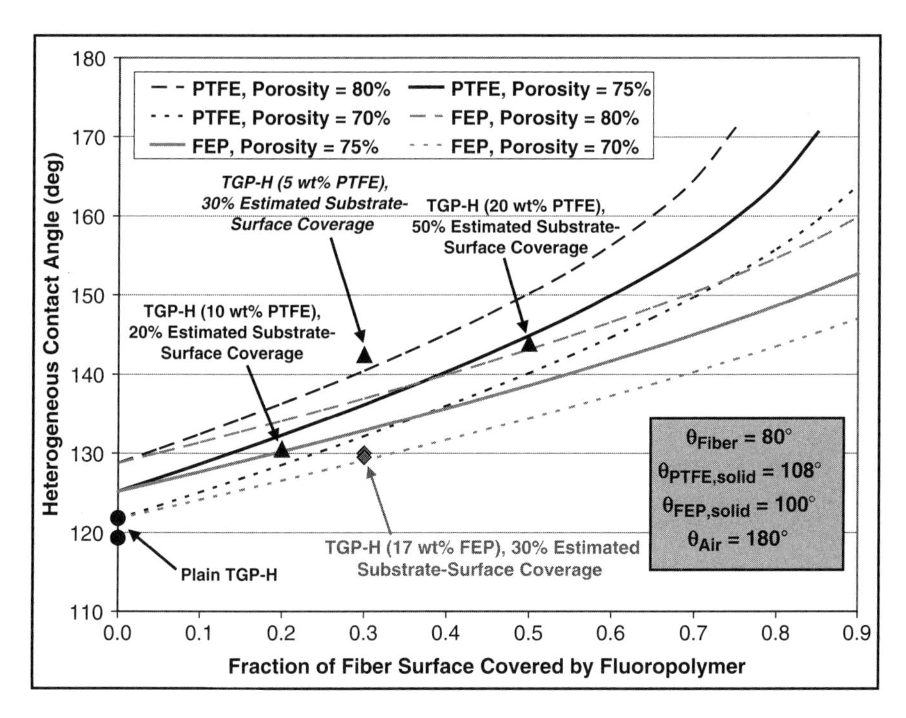

Fig. 12 Cassie model of equilibrium static sessile drop water contact angles as a function of TGP-H fiber surface coverage for FEP (*gray*) and PTFE (*black*) (Wood 2007)

There are several experimental data points plotted in Fig. 12 with the model calculations for various treated TGP-H substrates (plain TGP-H, 5 wt% PTFE, 10 wt% PTFE, 20 wt% PTFE, and 17 wt% FEP). The model lines can be used to predict the approximate fiber surface coverage for a given fluoropolymer loading when the GDL substrate porosity is known. The comparison between the equilibrium contact angles of 5 wt% PTFE and 10 wt% PTFE TGP-H h shows the role the surface porosity plays in yielding a higher contact angle at a lower fluoropolymer loading.

The model discussed in this section and shown in Fig. 12 has been used to predict changes in equilibrium contact angle of 17 wt% FEP TGP-H after aging in 80°C deionized water with air sparging. The sample underwent an approximately 7° reduction from 130.0 ± 0.2° to 123.3° after 680-h aging. The modified Cassie model predicts a change from 130° to 122° in equilibrium contact angle using the concept of gradual formation of graphite oxide on the uncoated surfaces of the substrate fibers (Dawe and Stevens 1960; Kinoshita 1988). This also assumes a 10° loss in plain TGP-H single-fiber contact angle (see Fig. 5) to approximately 70°. Quantitative agreement of this type is a strong indication that carbon corrosion to CO_2 or hydrophilic adsorbed oxygen-containing species must play a role in long-term water-management characteristics of PEFC GDLs.

4.3 *X-Ray Photoelectron Spectroscopy Analysis*

To gain further insight into the mechanisms of carbon corrosion of the GDL and the MPL and associated loss of hydrophobicity, X-ray photoelectron spectroscopy (XPS) was used to evaluate material surfaces of GDLs before and after long-term US06 and DOE start/stop drive-cycle testing (Borup et al. 2006a, b). The XPS data in Figs. 17 and 18 are presented in terms of signal-intensity ratios.

The cathode GDLs used in these drive-cycling experiments were ELAT® version 2.0, which contain a double-sided hydrophobic MPL. One side has a thicker MPL protrusion above the substrate surface that is placed towards the MEA. Both the catalyst-layer and flow-field sides of these GDLs were analyzed after testing and compared with their fresh counterparts. Figure 13 shows ratios of fluorine to graphite intensity; there was little change on either side of the anode GDL after testing with the US06 drive cycle. However, there was about a 10% increase in signal ratio for the flow-field side when start/stop cycles of the cell were included. This is indicative of surface corrosion of the graphite over the approximately 950 start/stop cycles. The behavior of the cathode GDL under drive cycling was quite different, however. The most notable change was an approximately 30% decrease in the fluorine to graphite ratio after the DOE start/stop cycling on the flow-field side of the cathode

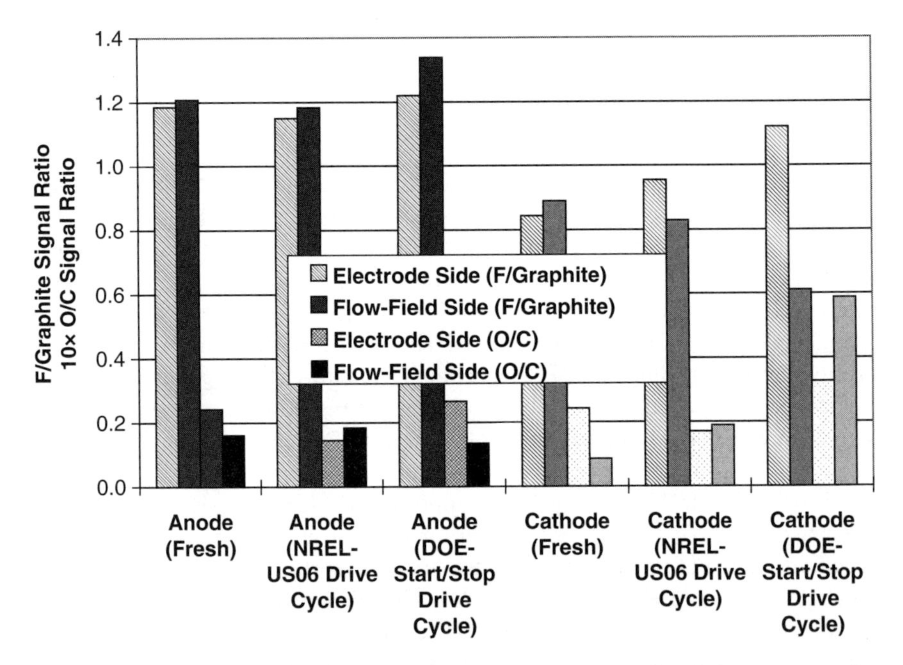

Fig. 13 X-ray photoelectron spectroscopy results of fluorine to graphite and oxygen to carbon signal ratios of ELAT® cloth GDLs for before/after drive-cycle testing (with LANL standard catalyst-coated membranes, CCMs): anode GDL, ELAT® version 2.22; cathode GDL, ELAT® version 2.0; and membrane, Nafion® N1135. DOE start/stop drive-cycle testing lasted 1,000 h (950 start/stop cycles). See the Appendix for durability testing conditions

GDL, which can only indicate a loss of PTFE in this region. The opposite trend was observed on the electrode side of the cathode GDL, with an approximately 15% and an approximately 35% increase in the fluorine to graphite signal ratio for the US06 cycling and start/stop cycling, respectively. These measurements are a strong indication of corrosion of the MPL carbon, as opposed to the case of the fibers of the version 2.22 anode GDL in contact with the flow field.

The oxygen to carbon signal intensity ratios are also shown in Fig. 13, where the numerator represents *total* carbon. There were no significant changes from the fresh sample for either side of the anode GDL, as well as for the electrode side of the version 2.0 cathode GDL. However, flow-field-side oxygen to carbon intensity ratio for the flow-field side doubled after the US06 cycling and increased by about 6 times after the DOE start/stop cycling. These data are in strong agreement with the heterogeneous contact-angle modeling discussed previously and suggest the cathode operating environment plays a significant role in carbon corrosion/oxidation of the GDL materials.

4.4 RH Sensitivity Changes

When a highly hydrophobic MPL and a less hydrophobic MPL are in contact with electrocatalyst layers, a very different response in cell performance is seen when scanning across a range of inlet dew points. This method, developed by Roth and Wood, has been described elsewhere (Mathias et al. 2003; Roth and Wood 1999) and is useful for evaluating new GDL configurations and treatments. Figure 14 shows the comparison of GDLs between identical cells (except for the GDL type). The cell designated as "M5" contained SIGRACET® GDL 24BC and was used for the hydrophobic MPL interface (23 wt% PTFE); the cell designated as "M11" contained GDL 24DC10 and was used for the less hydrophobic interface (10 wt% PTFE). Only the direction of the hydrophobicity gradient and the fraction of carbon in the MPL were different between the two cells. The hydrophobicity for GDL 24DC10 increases in the direction away from the MPL (with 20 wt% PTFE in the substrate), and it decreases for GDL 24BC with respect to the same direction (i.e., 5 wt% PTFE in the substrate). Characteristic trends for these two cases were acquired for both current density and HFR. A hydrophobic interface with lower fluoropolymer content in the substrate exhibits good water management and performance with saturated inlets and inlets with water contents above saturation, but the performance steadily drops off as the inlet streams become drier. This trend was observed for the GDL 24BC case and is indicative of "overremoval" of product water from the cathode catalyst layer. As the anode and cathode dew points were lowered from 100 to 45°C, the electrode ionomer and membrane continued to dry, resulting in an increase of nearly a factor of 5 in HFR and a simultaneous reduction by a factor of 2 in the constant-voltage current density (at 0.70 V). GDL 24DC10 exhibited its highest performance for the inlet dew points of 39 and 45°C (15 and 20% RH, respectively, for a cell temperature of 80°C) and maintained a stable HFR reading. This GDL also showed reasonable performance under wet conditions

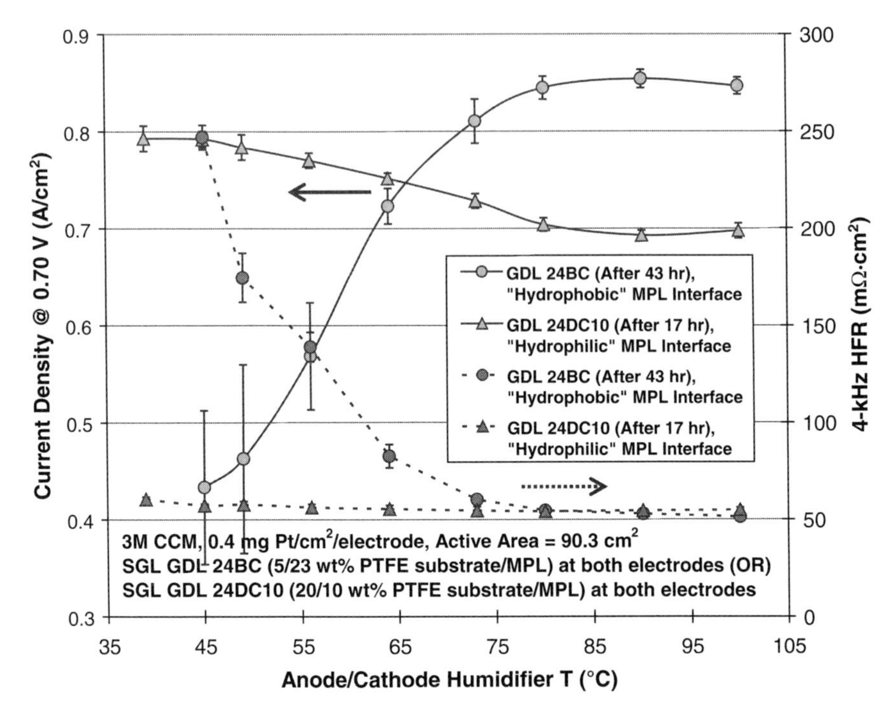

Fig. 14 Comparison of beginning-of-life RH-sensitivity scans for cells with a highly hydrophobic MPL interface (23 wt% PTFE) facing the electrodes (SIGRACET® GDL 24BC in cell "M5") and a less hydrophobic interface (10 wt% PTFE) facing the electrodes (SIGRACET® GDL 24DC10 in cell "M11"). See the Appendix for durability testing conditions

between the 80 and 100°C dew points, which is exactly the set of characteristics an automotive GDL needs to have – good performance under dry conditions and acceptable performance under wet conditions.

The RH-sensitivity method was also used to compare the effects of ex-situ aging on the performance of GDL 24BC. Two identical cells were constructed with 3 M catalyst-coated membranes (30-µm Nafion® membrane and 0.4 mg Pt cm^{-2} per electrode), one with fresh GDL 24BC (see cell M5 data in Fig. 15) and the other with aged GDL 24BC – 80°C deionized water with air sparging for 1,006 h as discussed in Sect. 4.1 (cell designated as "M2"). Cell M5 was scanned after 43 h of testing at constant 1.04 A cm^{-2}, and cell M2 was scanned after 153 h at constant 1.00 A cm^{-2}, in addition to the ex situ aging.

Since the RH-sensitivity scan for the GDL 24BC material was taken after only 153 h of constant-current operation, it is primarily representative of the ex situ aging effects. Figure 15 shows the aged GDL 24BC took on the characteristics of a *hydrophilic* MPL interface and altered surface-energy gradient. Both the current-density and HFR trends as a function of inlet dew points were nearly identical to those of the fresh GDL 24DC10, as shown in Fig. 14. This indicates that the aging environment attacked the exposed carbon phases of the GDL (carbon oxidation to CO_2,

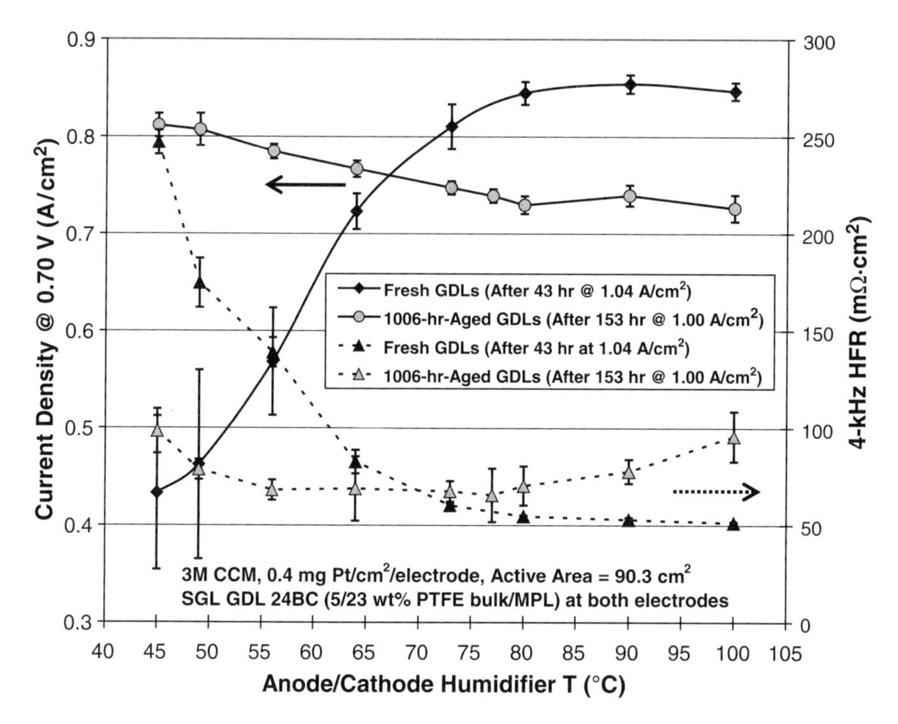

Fig. 15 Comparison of RH-sensitivity scans for cells built with fresh SIGRACET® GDL 24BC (cell "M5") and GDL 24BC aged ex situ for 1006 h (cell "M2"). Aging conditions were 80°C deionized water with air sparging. See the Appendix for durability testing conditions

adsorbed oxygen groups, etc.) and is in excellent agreement with other data sets in this chapter, in particular the single-fiber Wilhelmy contact-angle data in Fig. 5 and the Zisman surface-energy data in Fig. 6. The increases in HFR for the aged GDL 24BC in Fig. A at the dry and wet extremes (inlet dew points approaching 45 and 100°C) are attributed to localized dry portions of the active area at either electrode caused by nonuniform water transport (liquid or vapor) from the accelerated testing.

5 MPL Degradation

Since the bilayer structure of the substrate/MPL configuration is important to optimal mass transport and PEFC performance, its properties must be preserved throughout the course of the system lifetime. The surface chemistry of the MPL is critical to long-term liquid-water transport, and the surface chemistry is tied closely to the PSD. The performance of the entire GDL will suffer over the PEFC lifetime if the structural integrity of the MPL is compromised by gradual corrosion of the carbon phase or detachment of carbon particles from the PTFE phase, which serves as the MPL binder.

5.1 Loss of MPL Material and Air Permeability Increase

Capillary flow porometry (CFP) measurements, a technique for measuring the porous volume and PSD in a solid medium which contributes to mass flux through the medium, were taken on the aged SIGRACET® GDL 24BC samples discussed in Sect. 4.1 after they had been subjected to 664 h of durability testing. The principles of operation of the capillary flow porometer and associated PSD measurements have been summarized elsewhere (Jena and Gupta 2002a, b). Jena and Gupta have successfully extended the CFP method to in-plane PSDs with and without a compression force, as applied to battery separators (Jena and Gupta 1999). This method was further extended to PEFC GDLs to simulate the through-plane and in-plane transport of the GDL material under stack compression force (Wood and Lehnert 2000).

CFP data are shown in Fig. 16 for the anode and cathode GDLs from the single-cell designated as "M2" after the 1,006-h aging and 664-h testing periods, as well as for two separate samples of fresh GDL 24BC. The pore sizes for the flow-based PSDs decreased slightly for the anode GDL over the entire range and increased for the cathode GDL over most of the smaller-pore range. The most significant increases for the cathode GDL occurred within the Q_{dry}/Q_{wet} region of 50–80%, which corresponds to the substrate/MPL transition layer and larger-pore domain of the MPL. Here Q_{dry} is the volumetric flow rate of air through the unfilled/nonwetted sample (or air permeability curve), and Q_{wet} is the volumetric flow rate of air through the sample as the Galden® HT 230 (Solvay Solexis, Alessandria, Italy) wetting fluid is being evacuated from the pores. This low-surface-tension, heat-transfer fluid is assumed to be perfectly wetting for most solids. The observed change in cathode-GDL pore size in Fig. 16 corresponds to what must be a loss of particulate matter that comprises the MPL and penetrates into the substrate fiber matrix. It is unlikely that oxidation of graphite fibers or MPL carbon to CO_2 could fully account for the observed changes in CFP PSD, especially a PSD based on convection and flow-through rates. This suggests that the bilayer structure of GDL 24BC was compromised by the harsh combination of aging followed by durability testing, and eventually led to at least partial washing away of the MPL material. Since the PTFE phase within the MPL is mostly continuous, it is likely that the carbon particles dislodged from the binding matrix and were swept away with the effluent water. It is no surprise this was observed at the cathode where the product water is generated during operation. The reduction in anode PSD in Fig. 16 is attributed to the compression set that all GDLs experience (nonwoven GDLs more so than woven GDLs) once assembled into a cell or stack. Had the cathode GDL not lost a portion of its MPL material, a similar reduction would have been observed because symmetric gasketing was used (i.e., the same material was used for sealing at each side).

The CFP data in Fig. 16 were correlated with Darcy air permeability data extracted from the same measurement set and are shown in Fig. 17. Permeabilities as a function of pressure differential for the two fresh GDL 24BC samples were close to one another (about 10–20% difference), and the variance corresponds to the CFP error bars for the fresh GDL 24BC data points in Fig. 16. The anode GDL permeability dropped by approximately 40% over the measured range of pressure

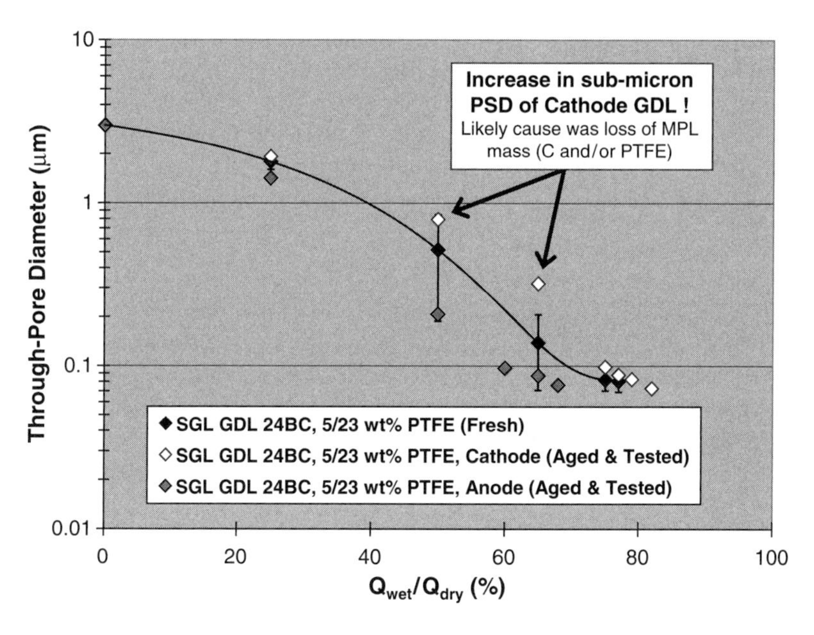

Fig. 16 Before and after comparison of capillary flow porometry (CFP) pore size distributions (*PSDs*) for cell M2 – GDLs aged in 80°C deionized water with air sparging gas for 1,006 h, then durability-tested (with fresh 3 M CCM) for 664 h at 1.00 A cm^{-2}. See the Appendix for durability testing conditions

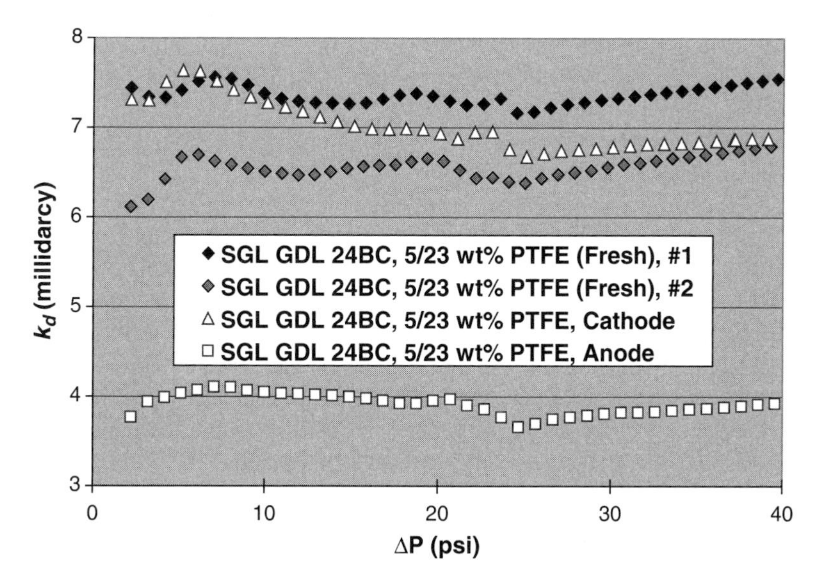

Fig. 17 Before and after comparison of Darcy air permeability from CFP for cell M2 – GDLs aged in 80°C deionized water with air sparging gas for 1,006 h, then durability-tested (with fresh 3 M CCM) for 664 h at 1.00 A cm^{-2}. See the Appendix for durability testing conditions

differentials, which partially explains the overall reduction in CFP PSD. However, the cathode GDL was measured at the same permeability levels as for the fresh GDL 24BC, although it did exhibit a slightly decreasing trend with increasing pressure differential. This trend was close to that observed for the anode GDL and confirms the effect of cell compression on the cathode GDL as well. The cathode-GDL Darcy permeabilities are in agreement with the observed changes in flow-based PSD, and it is concluded on the basis of the supporting evidence in Fig. 17 that only a loss in cathode MPL material can explain the extent of these differences. This may not be the sole explanation given the other contributing factors to the aggressive cathode operating environment.

5.2 Total and Hydrophobic PSD Changes

The CFP and Darcy air-permeability data discussed in Sect. 5.1 were correlated with mercury porosimetry (total PSD) and water porosimetry (hydrophobic PSD) before and after the consecutive aging/durability-testing experiments for cell M2. Mercury porosimetry can be effectively used to measure the total porosity and PSD of a GDL. This technique measures all porosity that exists (including constricted or dead-ended pores). The mercury intrusion volume also represents the hydrophobic plus hydrophilic surface domains because mercury is nonwetting for both types of pores.

The technique of water (intrusion) porosimetry can be effectively used to characterize the number of hydrophobic pores; however, it must be used in conjunction with mercury porosimetry to separate the hydrophobic pore volume from the total pore volume to quantify the *hydrophilic* pore volume. This "dual-intrusion" porosimetry method as applied to PEFC GDLs has been described previously (Gupta and Jena 2003).

A change in intrusion-based mercury porosimetry was found for both the anode and the cathode GDLs after aging and durability testing, as shown in Fig. 18. A small increase in the transition-layer and MPL pore volumes (below approximately 2 μm) over those for the fresh GDL 24BC material was measured for both durability-tested GDLs. There was little change in the anode GDL pore volume above approximately 4–5 μm, except for the largest pore sizes (more than 100 μm). An increase in pore volume was measured in this region for both durability-tested GDLs (above approximately 140 μm for the cathode GDL). There was one noticeable difference between the anode and the cathode GDL, which was the height and area of the mode pore-size peaks (approximately 11–80 μm). Both the height and the area for the cathode GDL were smaller than for the anode GDL, which suggests that the cathode GDL was compressed to a slightly greater extent than the anode. The main conclusion drawn from the mercury porosimetry data is that they suggests only a modest difference between the anode and the cathode GDLs after aging/durability testing – the total pore volumes were 2.05 and 1.77 $cm^3\ g^{-1}$, respectively. These data indicate that on the basis of the intrusion volume, the anode suffered to a greater extent from these experiments. Since most of the 0.28 $cm^3\ g^{-1}$ difference

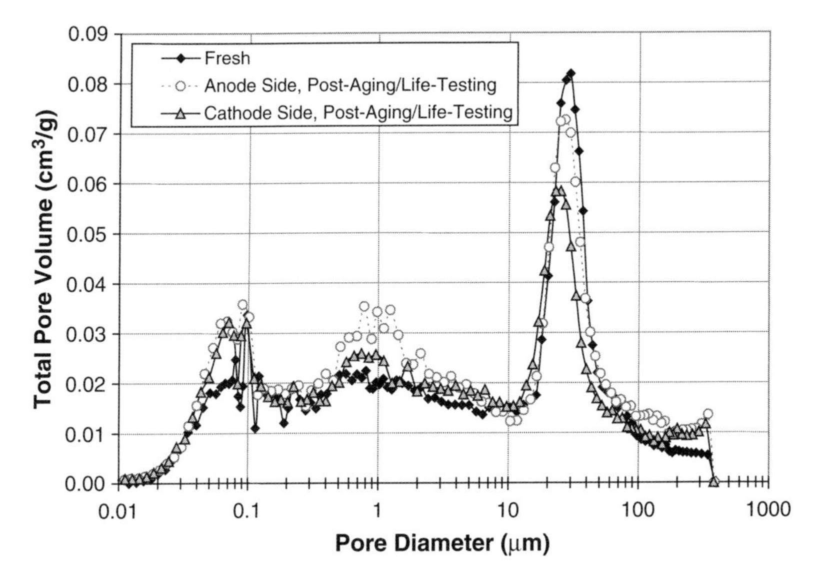

Fig. 18 Before/after mercury porosimetry data for cell M2 showing changes in total specific pore volume after aging/life testing; GDLs aged in 80°C deionized water with air sparging for 1,006 h, then durability-tested with fresh 3 M CCM (30-μm Nafion® membrane, 0.4 mg Pt cm^{-2} per electrode) for 664 h at 1.00 A cm^{-2}. See the Appendix for durability testing conditions

occurred within the large-pore region, it would not directly contribute to changes observed in the smaller-pore region of a flow-based PSD (i.e., CFP). This is because the smaller pores act as a flow restrictor for the removal of the wetting fluid (see Sect. 5.1), which also limits the air-permeability values as well. Therefore, the intrusion-based findings of the mercury porosimetry do not necessarily contradict those of the CFP measurements.

Results for before and after *water* porosimetry PSDs are seen in Fig. 19, and show that the fresh and cathode-GDL hydrophobic PSDs were nearly identical within the material and experimental errors. As for the anode GDL, the hydrophobic pore volume in the larger-pore region (more than 200 nm) was much greater than for the fresh GDL 24BC material for most pore sizes. The cathode GDL pore volume over the same range of pore sizes was either the same or slightly smaller than for the fresh material, which supports the statement that the cathode GDL was slightly more compressed than the anode GDL during the constant-current durability testing. The opposite trend from the mercury porosimetry data was found for the anode and cathode mode pore-size peaks (as compared with water porosimetry data), which directly corresponded to the MPL pore sizes for the hydrophobic PSDs (see the region between about 10 and 100 nm in Fig. 19). This observation is in agreement with the CFP and air-permeability data (in Figs. 16, 17, respectively), which suggest the occurrence of more than a basic physical loss of MPL material. The surface chemistry also likely changed (carbon corrosion, adsorbed hydrophilic species on the carbon, degradation of the carbon/PTFE interface, etc.). Figure 19

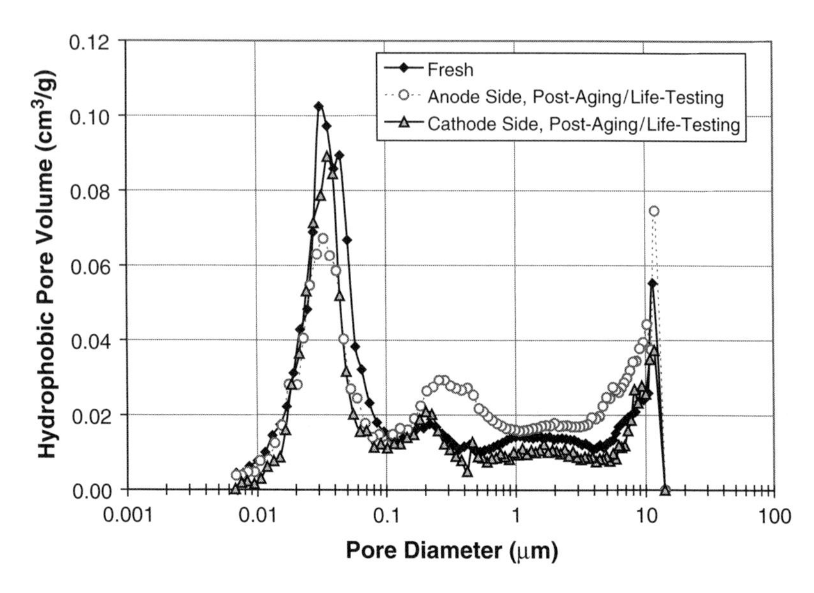

Fig. 19 Before/after water porosimetry data for cell M2 showing changes in hydrophobic specific pore volume after aging/life testing; GDLs aged in 80°C deionized water with air sparging for 1,006 h, then durability-tested with fresh 3 M CCM (30-μm Nafion® membrane, 0.4 mg Pt cm^{-2} per electrode) for 664 h at 1.00 A cm^{-2}. See the Appendix for durability testing conditions

further suggests that the cathode GDL experienced an increase in the amount of compressed hydrophobic pore volume within the domain of the MPL, since the mode pore-size peak is closer to the size of that of the fresh GDL 24BC than that of the anode GDL.

5.3 MPL Carbon Corrosion

Only one study is known that has attempted to address the issue of carbon corrosion of the MPL at the cathode-electrode/MPL interface (Owejan et al. 2007). This corrosion mechanism occurs in the same manner as for the cathode catalyst carbon support, by the formation of localized "air–air" cells during startup of a PEFC when H_2 mixes with air after a system shutdown (Meyers and Darling 2006; Reiser et al. 2005). Owejan et al. implemented a technique of comparing voltage differences of two MPL types. This technique is in contrast to measuring carbon-corrosion current and CO_2 evolution under potentiostatic conditions at cell voltages of approximately 1.2–1.5 V and equating the CO_2 formation rate to carbon-support weight loss. An important point which validates the findings is that the membrane, catalyst layers, GDL substrate, and MPL ink constituents were the same – the only material variation was the MPL carbon.

The experimental methods involved comparing performance loss (H_2/air polarization curves) after periodic holds at cell voltages of 1.2 V for MPLs containing a highly graphitized carbon and acetylene black. In addition, comparisons in voltage loss at 0.2 and 1.2 A cm^{-2} for the same two MPL carbon types were made after 1,000 startup/shutdown cycles. These cycles induce localized air–air cell(s) where the local cell voltage can approach 1.5 V. For both experiments and variations in conditions, the highly graphitized carbon was more resistant to corrosion than the acetylene black. Since the performance losses due to cathode-electrode carbon corrosion and MPL carbon corrosion cannot be separated, total cell polarization losses were compared. Major findings included a 63% current-density loss at 0.6 V under H_2/air operation for the acetylene black MPL after 25 h of potentiostatic hold at 1.2 V. This performance loss was substantially reduced by more than 50% when switching to the highly graphitized carbon MPL. The results of the startup/shutdown cycling were similar. The highly graphitized carbon MPL exhibited approximately 75 mV higher performance at 0.2 A cm^{-2} than its acetylene black counterpart after 1,000 cycles. At 1.2 A cm^{-2}, approximately 110 mV higher performance was observed (Owejan et al.), suggesting a phenomenological effect of increased mass-transport overpotential.

6 Effects of Compression Nonuniformity

Perhaps the most important mechanical function of the GDL is to evenly distribute stack compression force within each cell in harmony with the sealing material and thickness. As with many other MEA materials and subcomponents, this function becomes increasingly important over operating lifetimes. Consequences of GDL compression maldistribution include the following: (1) substrate fibers may puncture thin ionomer membranes, causing gas crossover; (2) local thermal nonuniformities may develop across the active area of the MPL/electrocatalyst interfaces; (3) regions of over- or undercompression could result in nonuniformities of PSDs; (4) maldistribution of PSDs could cause inhomogeneous reactant and product transport of both the liquid and the vapor phases; (5) designed anisotropic functionalities (i.e., in-plane vs. through-plane) such as electrical, thermal, and PSD properties could be compromised; and (6) undesirable changes in the GDL substrate and MPL surface chemistry owing to localized changes in mass transport and PSD. With these many possible negative consequences of compression nonuniformity, it is important to adequately consider optimal compatibility of GDL and gasket thickness tolerances with system considerations such as thermal cycling, power-density cycling, and stack cooling/thermal management concepts

6.1 *Substrate Fiber Puncturing of Membrane*

Although fiber puncture of the membrane has become less of an issue over the last 3–4 years owing to substantial improvements in perfluorosulfonic acid membrane mechanical durability and properties and improvements in sealing thickness tolerances, it cannot be entirely neglected. Excessive thermal cycling or thickness nonuniformities in the MPLs or catalyst layers could lead to unforeseen membrane puncturing owing to quality control issues in even the most well thought out stack designs.

This issue was studied primarily using postmortem SEM of MEA cross sections (Stanic and Hoberecht 2004), which has become a popular technique for diagnosing MEA failure. The SEM results revealed that there can be a multistep pathway to fiber puncturing of the ionomer membrane when operating under dry (low-RH) reactant conditions. Stanic and Hoberecht found that RH cycling combined with dry gases exacerbates membrane creep and brittleness. This eventually leads to localized membrane thinning and when combined with thermal and mechanical-load cycling may assist GDL substrate fibers in puncturing the MPL, catalyst layer, and membrane. If substrate fibers are broken or cracking develops in the MPLs or catalyst layers, the puncturing process would be further facilitated.

6.2 Effects of Electrical and Thermal Maldistribution

Several studies have been done over the last 5–7 years on the effects of inhomogeneous electrical and thermal gradients in operating cells or stacks as a result of poor GDL compression distribution (Eckl et al. 2006; Hwang 2006).

One of the studies that focused primarily on experimental measurements related to these phenomena systematically covered the effects of inhomogeneous compression on both GDL physical properties and GDL long-term performance (Nitta 2008). These results were also combined with a mathematical model. The emphasis of this research was to provide fundamental understanding of the inherent inhomogeneous compression effects of the conventional land/channel configuration on GDL physical properties and long-term performance. They correlated GDL thickness and their mathematical model results with various properties, including mechanical properties, gas permeability, in-plane and through-plane electrical conductivities, electrical contact resistances, bulk thermal conductivity, and thermal contact resistance (Nitta 2008). One of the most significant findings was that bulk electrical conductivity increased with increasing compression force, which agrees with prior work (Wood et al. 2002b), but the bulk thermal conductivity was unaffected by compression force. Total electrical and thermal contact resistances, as well as gas permeability, were found to decrease systematically and logarithmically with increasing compression force. These findings also agree with previous findings of efforts to correlate GDL structural and physical property information with stack/cell compression. However, this work carried the study further in an attempt to understand these effects on current and temperature distribution across the active area. Using the characterization data with the mathematical model produced the findings that GDL compression inhomogeneity had a substantial effect on current distribution, but had less effect on temperature distribution than many previous studies of this topic had predicted (Nitta 2008).

This research highlights the importance of thorough and systematic component characterization for input into mathematical models. Even the most sophisticated models with comprehensive, accurate governing and boundary equations are of minimal use without properly measured component parameters. In addition, the major findings suggest that the GDL does a better job of thermal management than previously thought. Hot spots in the cathode catalyst layer due to inhomogeneities in current distribution may play a more critical role in long-term MEA durability.

6.3 GDL/MPL Mass-Transport Effects

A study investigating the effects of liquid-water transport through hydrophobized substrates recently revealed an important finding of preferential pathways of water flow during application of different levels of compression force (Bazylak et al. 2007). The experimental method involved placing samples of TGP-H 060 paper between transparent clamping plates and applying pressure. An optical microscope was used to obtain fluorescence images, which were captured with a charge-coupled-device (CCD) camera. These images were used to investigate water breakthrough pressures (sometimes referred to as hydrohead pressures) and localized through-plane flow phenomena. The study found that the water breakthrough pressures decreased with higher clamping forces, in contrast to expected behavior. When hydrophobic pores are made smaller (i.e. via compression) liquid-water permeability should decrease.,

SEM was also used to study the GDL substrate macrostructure after application of different compression forces, in conjunction with the fluorescence and CCD imaging. Many broken TGP-H fibers were observed together with tears in the PTFE domains coating the fibers. These findings have several critical implications for long-term PEFC performance (1) a change in effective substrate PSD that would result in the preferential water-flow pathways observed; (2) an alteration in the substrate surface chemistry balance (hydrophobic to hydrophilic pore-volume ratio); and (3) an increased likelihood of damage to the MPLs, catalyst layers, or membrane owing to excessive broken fibers.

7 Summary

The PEFC GDL consists of the relatively simple materials of carbon, graphite, fluoropolymer (PTFE, FEP, etc.), and a substantial amount of porosity. However, the combination of these constituents creates a complex composite, heterogeneous medium, which operates in a complex PEFC environment, especially at the cathode side. Gaining fundamental understanding of liquid-water transport phenomena in this environment is critical to understanding the degradation mechanisms at work within the GDL and MPL structure. The data presented and studies discussed in this chapter shed light on the physical properties required to understand GDL durability and long-term performance for next-generation GDL components. The information presented further creates a platform of fundamental science for facilitating technology transfer generally related to mass-transport phenomena in PEFCs.

Appendix

Durability test conditions for cell G2 (Figs. 7, 8): Cell T = 80°C, humidifier T = 73/73°C, RHs 75%/75%, gas pressures 180/180 kPa (abs), constant gas flows (anode/cathode) 1.4/2.2 × 0.90 A cm^{-2} equiv (450/1,630 sccm H_2/air). Polarization-curve conditions were the same except for the gas flow rates, which were constant 740/2,430 sccm H_2/air (1.4/2.0 anode/cathode stoichiometric flow × 1.50 A cm^{-2} equiv.

Durability test conditions for cells D1 and D2 (Figs. 9, 13): Cell T = 80°C; humidifier T = 80/80°C for cell D1 and 63/63°C for cell D2; RHs 100%/100% for cell D1 and 48%/48% for cell D2; gas pressures 214/214 kPa (abs), constant gas flows (anode/cathode) 1.0/1.7 × 1.00 A cm^{-2} equiv (350/1,390 sccm H_2/air). All polarization conditions were the same except for the humidifier temperatures, which were 105/80°C (255%/100% RH anode/cathode).

Durability test conditions for start/stop cells (Fig. 13): Cell T = 80°C, humidifier T = 64/64°C (anode/cathode), RHs (anode/cathode) 50%/50% (anode/cathode), gas pressures 214 kPa (abs), constant stoichiometric flows 1.1/2.0 (anode/cathode). Conditions for US06 drive-cycle testing were the same as those for cell D1, except that the RHs were 100%/100%.

Durability test conditions for cells M5 and M11 (Fig. 14), cells M2 and M5 (Fig. 15), and cell M2 (Figs. 16–19): Cell T = 80°C, M2/M5 humidifier T = 73/73°C, M2/M5 RHs 75%/75%, M11 humidifier T = 73/49°C (anode/cathode), M11 RHs 75%/25% (anode/cathode), gas pressures 180/180 kPa (abs), constant gas flows (anode/cathode) 1.5/2.0 × 1.00 A cm^{2} equiv (940/3,060 sccm H_2/air). RH-sensitivity conditions were the same except for the load (constant 0.70 V) and the gas flow rates, which were constant 940/2,320 sccm H_2/air (1.5/1.6 anode/cathode stoichiometric flow × 1.00 A cm^{-2} equiv).

Acknowledgments Special thanks is extended to the US Department of Energy Office of Hydrogen, Fuel Cells, and Infrastructure Technology (program manager Nancy Garland) for the financial support of and commitment to this research. Sincere appreciation is extended to the LANL MPA-11 team for the many helpful and intellectually stimulating discussions that assisted in this research. A debt of gratitude is also owed to Trung Nguyen and Plamen Atanassov for steadfast encouragement in the pursuit of this research topic long before it was fashionable in the PEFC research community. Supporting component/materials characterization and fuel-cell measurements from the following colleagues are greatly appreciated: Chris Rulison (comprehensive contact-angle and surface-energy measurements of GDLs and catalyst-coated membranes), Josh Powers (CFP and compression porometry), Terri Murray (mercury and water porosimetry), Rong Chen (Toray GDL-fiber SEM), John Davey (static sessile-drop contact angles of fresh/aged GDL 24BC samples and single-cell data acquisition), and Susan Pacheco (single-cell data acquisition). Special acknowledgement is given to Simon Cleghorn of W.L. Gore and Associates, Inc. for providing the 3-year durability-tested CARBEL® GDLs for surface chemistry analysis. Peter Wilde and Michael Mändle of SGL TECHNOLOGIES GmbH provided experimental and specially processed GDLs for use in this research.

References

Atanassova, P. (2008) Cabot Corporation, Private communication.

Bazylak, A., Sinton, D., Liu, Z.-S. and Djilali, N. (2007) Effect of compression on liquid water transport and microstructure of PEMFC gas diffusion layers. J. Power Sources 163, 784–792.

Bluemle, M.J., Gurau, V., Mann, J.A. Jr., Zawodzinski, T.A. Jr., De Castro, E.S. and Tsou, Y.-M. (2004) Characterization of transport properties in gas diffusion layers for PEMFCs. Presented at the 2004 Electrochemical Society Joint International Meeting, Honolulu, HI, Abstract No. 1932, October 3–8, 2004.

Borup, R., Wood, D., Davey, J., Welch, P. and Garzon, F. (2006a) PEM fuel cell durability. Presented at the 2006 DOE Hydrogen Program Annual Merit Review, Arlington, Virginia, May 16–19, 2006.

Borup, R.L., Davey, J.R., Garzon, F.H., Wood, D.L., Welch, P.M. and More, K. (2006b) PEM fuel cell durability with transportation transient operation. ECS Trans. 3 (1), 879–886.

Borup, R., Meyers, J., Pivovar, B., Kim, Y.S., Garland, N., Myers, D., Mukundan, R., Wilson, M., Garzon, F., Wood, D., Zelenay, P., More, K., Zawodzinski, T., Boncella, J., McGrath, J.E., Inaba, M., Miyatake, K., Hori, M., Ota, K., Ogumi, Z., Miyata, S., Nishikata, A., Siroma, Z., Uchimoto, Y. and Yasuda, K. (2007) Scientific aspects of polymer electrolyte fuel cell durability and degradation. Chem. Rev. 107, 3904–3951.

Cassie, A.B.D. (1948) Contact angles. Trans. Faraday Soc. 44 (3), 11–16.

Cleghorn, S.J.C., Mayfield, D.K., Moore, D.A., Moore, J.C., Rusch, G., Sherman, T.W., Sisofo, N.T. and Beuscher, U. (2006) A polymer electrolyte fuel cell life test: 3 years of continuous operation. J. Power Sources 158, 446–454.

Cubaud, T. and Fermigier, M. (2004) Advancing contact lines on chemically patterned surfaces. J. Colloid Interface Sci. 269, 171–177.

Dawe, H.J. and Stevens, R.F. (1960) Proceedings of the fourth conference on carbon. Pergamon Press, New York, p. 17.

Dohle, H., Jung, R., Kimiaie, N., Mergel, J. and Müller, M. (2003) Interaction between the diffusion layer and the flow field of polymer electrolyte fuel cells – experiments and simulation studies. J. Power Sources 124, 371–384.

Eckl, R., Grinzinger, R. and Lehnert, W. (2006) Current distribution mapping in polymer electrolyte fuel cells – a finite element analysis of measurement uncertainty imposed by lateral currents. J. Power Sources 154, 171–179.

Fleming, G. (2000) Spectracorp, Inc., Private communication.

Gasteiger, H.A., Gu, W., Makharia, R., Mathias, M.F., and Sompalli, B. (2003) Beginning-of life MEA performance–efficiency loss contributions. In: W. Vielstich, A. Lamm and H.A. Gasteiger (Eds.), *Handbook of Fuel Cells: Fundamentals, Technology, and Applications*. Vol. 3, Part 1, Wiley, New York, NY, pp. 593–610.

Good, R.J. and Girifalco, L.A. (1960) Estimation of surface energy of solids from contact angle data. J. Phys. Chem. 64, 561–565.

Gupta, K. and Jena, A. (2003) Techniques for pore structure characterization of fuel cell components containing hydrophobic and hydrophilic pores. Presented at the 2003 Fuel Cell Seminar, Miami Beach, FL, Abstract pp. 723–726, November 3–7, 2003.

Healy, J., Hayden, C., Xie, T., Olson, K., Waldo, R., Brundage, M., Gasteiger, H. and Abbott, J. (2005) Aspects of the chemical degradation of PFSA ionomers used in PEM fuel cells. Fuel Cells 5, 302–308.

Hwang, J.J. (2006) Thermal-electrochemical modeling of a proton exchange membrane fuel cell. J. Electrochem. Soc. 153, A216–A224.

Ihonen, J., Mikkola, M. and Lindbergh, G. (2004) Flooding of gas diffusion backing in PEFCs. J. Electrochem. Soc. 151, A1152–A1161.

Jena, A.K. and Gupta, K.M. (1999) In-plane compression porometry of battery separators. J. Power Sources 80, 46–52.

Jena, A.K. and Gupta, K.M. (2002a) Characterization of pore structure of filtration media containing hydrophobic and hydrophilic pores. Fluid/Part. Sep. J. 14, 1–6.

Jena, A.K. and Gupta, K.M. (2002b) Analyse der porendurchmesser von mehrschichtfiltermitteln. Filtrieren und Separieren 16, 13.

Jordan, L.R., Shukla, A.K., Behrsing, T., Avery, N.R., Muddle, B.C. and Forsyth, M. (2000) Diffusion layer parameters influencing optimal fuel cell performance. J. Power Sources 86, 250–254.

Kinoshita, K. (1988) Carbon – Electrochemical and Physicochemical Properties. Wiley, New York, NY.

Kolde, J., Lane, D. and Mongan, J. (2002) Addressing the needs of the PEM fuel cell market through innovation. Presented at the 202nd Meeting of The Electrochemical Society, Salt Lake City, UT, Abstract No. 802, October 20–24, 2002.

LaConti, A.B., Hamdan, M. and McDonald, R.C. (2003) Mechanisms of membrane degradation for PEMFCs. In: W. Vielstich, A. Lamm and H.A. Gasteiger (Eds.), *Handbook of Fuel Cells: Fundamentals, Technology, and Applications*. Vol. 3, Part 1, Wiley, New York, NY, pp. 647–662.

Liu, W., Moore, D. and Murthy, M. (2004) Using AC impedance to characterize gas diffusion media in PEM fuel cells. Presented at the 2004 Electrochemical Society Joint International Meeting, Honolulu, HI, Abstract No. 1930, October 3–8, 2004.

Mändle, M. and Wilde, P., SGL Carbon Group, SGL TECHNOLOGIES GmbH, Private communication, 2001.

Markel, T. (2004) National Renewable Energy Laboratory, Private communication.

Mathias, M.F., Roth, J., Fleming, J. and Lehnert, W. (2003) Diffusion media materials and characterization. In: W. Vielstich, A. Lamm and H.A. Gasteiger (Eds.), *Handbook of Fuel Cells: Fundamentals, Technology, and Applications*. Vol. 3, Part 1, Wiley, New York, NY, pp. 517–537.

Meyers, J.P. and Darling, R.M. (2006) Model of carbon corrosion in PEM fuel cells. J. Electrochem. Soc. 153, A1432–A1442.

Mueller, B., Zawodzinski, T., Bauman, J., Uribe, F., Gottesfeld, S., De Castro, E. and De Marinis, M. (1999) Title. In: S. Gottesfeld and T.F. Fuller (Eds.), *Proton Conducting Membrane Fuel Cells II*. PV 98–27, The Electrochemical Society Proceedings Series, Pennington, NJ, pp. 1–9.

Nitta, I. (2008) Inhomogeneous compression of PEMFC gas diffusion layers. PhD Dissertation, Helsinki University of Technology, Espoo, Finland.

Owejan, J.E., Yu, P.T. and Makharia, R. (2007) Mitigation of carbon corrosion in microporous layers in PEM fuel cells. ECS Trans. 11, 1049–1057.

Owens, D.K. and Wendt, R.C. (1969) Estimation of the surface free energy of polymers. J. Appl. Polym. Sci. 13, 1741–1747.

Reiser, C.A., Bregoli, L., Patterson, T.W., Yi, J.S., Yang, J.D., Perry, M.L. and Jarvi, T.D. (2005) A reverse-current decay mechanism for fuel cells. Electrochem. Solid-State Lett. 8, A273–A276.

Roth, J. and Wood, D.L. (1999) General Motors Corporation, Unpublished Data.

Rulison, C. (2007) Augustine Scientific, LLC, Unpublished Data.

Stanic, V. and Hoberecht, V. (2004) MEA failure mechanisms in PEM fuel cells operated on hydrogen and oxygen. 2004 Fuel Cell Seminar Abstracts, San Antonio, Texas, November 1–5, pp. 85–88.

Wagner, F. (2008) General Motors Corporation, Private communication.

Weber, A.Z. and Newman, J. (2005) Effects of microporous layers in polymer electrolyte fuel cells. J. Electrochem. Soc. 152, A677–A688.

Williams, M.V., Begg, E., Bonville, L., Kunz, H.R. and Fenton, J.M. (2004) Characterization of gas diffusion layers for PEMFC. J. Electrochem. Soc. 151, A1173–A1180.

Wood, D.L. (2007) Fundamental material degradation studies during long-term operation of hydrogen/air PEMFCs. PhD Dissertation, University of New Mexico, Albuquerque, NM.

Wood, D.L. and Lehnert, W.K. (2000) General Motors Corporation, Unpublished Data.

Wood, D.L., Grot, S.A. and Fly, G. (2002a) Composite gas distribution structure for fuel cell. US Patent No. 6,350,539, General Motors Corporation.

Wood, D.L., Wilde, P.M., Mändle, M. and Murata, M. (2002b) Correlation of gas diffusion layer physical properties and PEMFC performance. 2002 Fuel Cell Seminar Abstracts, pp. 41–44.

Wood, D., Davey, J., Garzon, F., Atanassov, P. and Borup, R. (2004a) Effects of long-term PEMFC operation on gas diffusion layer and membrane electrode assembly physical properties. Presented at the 206th Meeting of The Electrochemical Society, Honolulu, HI, Abstract No. 1881, October 3–8, 2004.

Wood, D.L., Xie, J., Pacheco, S.D., Davey, J.R., Borup, R.L., Garzon, F.H. and Atanassov, P. (2004b) Durability issues of the PEMFC GDL & MEA under steady-state and drive-cycle

operating conditions. Presented at the 2004 Fuel Cell Seminar, San Antonio, TX, Abstract No. 24, November 1–5, 2004.
Wood, D., Davey, J., Garzon, F., Atanassov, P. and Borup, R. (2005) Characterization of gas diffusion layers and membrane electrode assemblies for long-term operation. Presented at the 208th Meeting of The Electrochemical Society, Los Angeles, CA, Abstract No. 1010, October 16–21, 2005.
Wood, D., Davey, J., Atanassov, P. and Borup, R. (2006) PEMFC component characterization and its relationship to mass-transport overpotentials during long-term testing. ECS Trans. 3 (1), 753–763.
Zisman, W.A. (1964) Relation of equilibrium contact angle to liquid and solid constitution. ACS Adv. Chem. 43, 1–51.

5
MEAs

High-Temperature Polymer Electrolyte Fuel Cells: Durability Insights

Thomas J. Schmidt

Abstract BASF Fuel Cell (formerly PEMEAS) produces polybenzimidazole-based high-temperature membrane electrode assemblies (MEAs). These Celtec®-P MEAs operate at temperatures between 120 and 180°C, and, therefore, are especially suitable for use in reformed-hydrogen-based polymer electrolyte fuel cells. Owing to these high operating temperatures, CO tolerances up to 3% can be achieved. Additional fuel gas impurities (inorganic or organic) can be tolerated to a much higher concentration than in low-temperature fuel cells. From a fuel cell system perspective, waste heat can be effectively used which increases the overall system efficiency. However, besides the distinct advantages over low-temperature polymer electrolyte fuel cells, some challenges have to be overcome. Especially on the catalyst level, there are several requirements which have to be met. In detail these are (1) anode catalyst activity for the oxidation of CO in the presence of hydrogen, (2) cathode catalyst activity in the presence of an adsorbing electrolyte such as phosphoric acid, and (3) high corrosion stability of the catalyst metal and catalyst support, especially under transient operation conditions such as start/stop or local fuel starvation. Especially the last point is important since for successful commercialization of MEAs, durability, reliability, and robustness are critical factors. That is, all materials used in MEAs have to be highly durable even under nonideal daily life conditions outside the laboratory. This contribution gives insight into the degradation mechanism during start/stop operation. Several tests are presented giving a better understanding of corrosion effects in high-temperature MEAs.

T.J. Schmidt
BASF Fuel Cell GmbH, Industrial Park Hoechst, G865,
65926 Frankfurt am Main, Germany
e-mail: thomas.justus.schmidt@basf.com

F.N. Büchi et al. (eds.), *Polymer Electrolyte Fuel Cell Durability*,

1 Introduction

BASF Fuel Cell produces polybenzimidazole (PBI) based high-temperature membrane electrode assemblies (MEAs) commercialized and sold under the brand name Celtec®. These MEAs operate at temperatures between 120 and 180°C, and, therefore, are especially suitable in reformed-hydrogen-based polymer electrolyte fuel cells (PEFCs). Owing to the high operating temperatures, CO tolerances up to 3% can be achieved. Additional fuel gas impurities [inorganic or organic, such as H_2S (Schmidt 2006b) or methanol] can be tolerated up to much higher concentrations than in low-temperature fuel cells based on perfluorinated sulfonic acid (PFSA) type or hydrocarbon membranes. That is, in a fuel cell system using a fuel processor for reforming hydrocarbon-based fuels, the gas purification can be simplified or sometimes completely avoided. Additionally, from a system engineering point of view, owing to the high stack operating temperatures of up to 180°C, fuel processors can easily be thermally integrated within the system. For example, from a stack using PBI-based MEAs, waste heat can be effectively used to provide heat for a methanol steam reformer when both the stack and the fuel processor are appropriately thermally packaged (Allen et al. 2006).

Owing to the inherent differences in the proton conductivity mechanism in the Celtec® membranes (Grotthus mechanism) compared with that in PFSA-type membranes (water-assisted shuttle mechanism) (Kreuer et al. 2004), in high-temperature PEFCs, no external humidification of the gases is necessary, which leads to a significant reduction of complexity and cost in high-temperature fuel cell systems. Additional issues due to membrane decomposition from peroxide can be avoided with Celtec® membranes owing to the inherent higher stability of PBI compared with PFSA membranes and the high operating temperature, which results in the survival time of hydrogen peroxide being minimized.

Finally, from a fuel cell system perspective, waste heat can be effectively used (either cogeneration of heat and power, CHP systems, or even trigeneration systems, where waste heat can be also used for cooling purposes). This results in a drastic increase of the overall efficiency of the systems compared with the efficiency of low-temperature fuel cell systems.

Although several distinct advantages of high-temperature stack operation can be outlined, care has to be taken when selecting the materials for high-temperature MEAs. The catalyst materials must be highly active for the oxidation of realistic reformates and the oxygen reduction reaction (ORR), but in addition, high stability toward corrosion is needed to ensure long fuel cell lifetimes especially when the stack is operated under realistic conditions. For example, realistic fuel cell operation conditions not only mean operation with real reformates derived from a fuel processor, but also typically involve temperature and start/stop cycling. Especially start/stop cycling produces high stress on the cathode owing to local excursions to high potentials above 1.3 V, where significant corrosion of the cathode carbon materials occurs. Details on the mechanism of cathode potential excursions during start/stop phases can be found in Reiser et al. (2005).

In this contribution, the impact of start/stop cycling of high-temperature Celtec®-P 1000 MEAs on the cathode stability will be demonstrated. Results from several tests will be shown to gain insight into the mechanism of carbon corrosion during start/stop cycling.

2 High-Temperature PBI Membranes

PBIs are a class of well-known polymers that have applications as thermally stable and nonflammable textile fibers, high-temperature matrix resins, adhesives, and foams. The entirely aromatic PBIs were developed for high-performance fiber applications in the early 1960s by the US Air Force Materials Laboratory in conjunction with DuPont and the Celanese Research Company (Vogel and Marvel 1961). Poly(2,2′-*m*-phenylene-5,5′-bibenzimidazole) (*m*-PBI) was produced using a solid-state polymerization route creating PBIs with rather low molecular weight (inherent viscosity of 0.5–0.8 dL g^{-1}). Fibers and textiles made from *m*-PBI (Fig. 1) have excellent properties, such as high-temperature stability, nonflammability, and high chemical resistance. Traditionally, PBI fiber has been used in firehter turnout coats, astronaut space suits, and gloves for use in metalworking industries. Details of the synthesis of *m*-PBI and production of fibers can be found in a recent review paper by Mader et al. (in press).

Although it had already been tested in the early days of phosphoric acid fuel cells as an electrolyte matrix (Breault 2003), the first intensive study of acid-doped PBI was done at Case Western University (Savinell et al. 1994; Wainright et al. 2003; Wang et al. 1996a–c). An *m*-PBI film was put into concentrated phosphoric acid and soaked to produce an imbibed film. This phosphoric acid containing film enabled sufficient proton conductivity, with the phosphoric acid content being in the range 6–10 mol phosphoric acid per mole of polymer repeat unit. This approach was followed later by other groups (Li et al. 2004a, b) using also different PBI structures, e.g., poly(2,5-benzimidazole).

Another approach for producing PBI membranes for fuel cell applications is through a sol–gel process (Xiao et al. 2005a, b). This synthesis route was developed by researchers from BASF Fuel Cell and from Rensselaer Polytechnic Institute, respectively, in a collaborative effort. Figure 2 sketches the production pathway of the now-called Celtec® membranes. The polymerization of the two monomers, viz.,

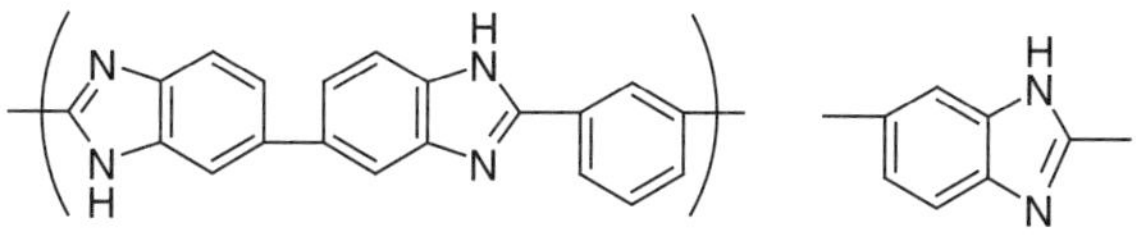

Fig. 1 Poly(2,2′-*m*-phenylene-5,5′-bibenzimidazole) (*m*-PBI) and poly(2,5-benzimidazole) (AB-PBI)

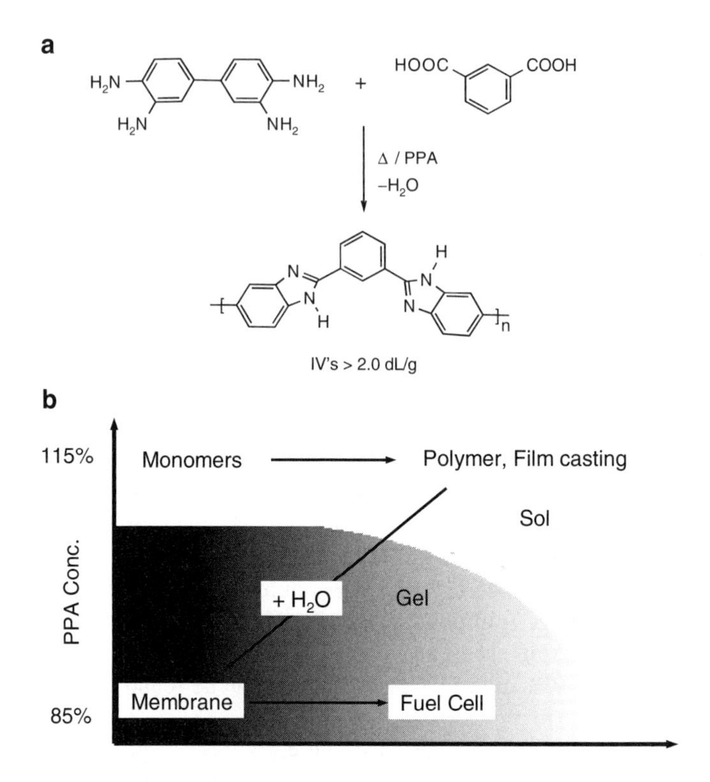

Fig. 2 (**a**) Polybenzimidazole (PBI) polymerization reaction in polyphosphoric acid (PPA). (**b**) State diagram for the Celtec® PBI membrane

tetraaminobiphenyl and an aromatic dicarboxylic acid (e.g., isophtalic acid), occurs in polyphosphoric acid as a solvent, forming PBI in situ.

From the PBI/polyphosphoric acid sol the membrane is cast and further hydrolyzed. During the hydrolysis, the membrane undergoes a sol–gel transition which results in the final membrane used in Celtec® MEAs (Fig. 2b). This production pathway not only produces high molecular weight PBIs with inherent viscosities of up to 7, but the final membrane also has a phosphoric acid content of more than 95 wt%, or up to 70 phosphoric acid molecules per PBI repeat unit.

3 Typical Degradation Mechanisms

Table 1 summarizes the most important degradation modes observed in high-temperature MEAs operating at up to 200°C. It must be noted, however, that except for acid evaporation modes, which are unique to liquid electrolyte-based fuel cells, all other degradation modes are also observed in low-temperature PEFCs.

Table 1 Main degradation modes in high-temperature membrane electrode assemblies

Cause	Primary effect	Secondary effect	Mechanism
Membrane			
Pinhole formation	H_2 crossover	Fuel loss	Creep, *f*(compression)
Membrane thinning	Shortcuts		Fibers, *f*(compression)
PA evaporation from membrane	Proton conductivity	Increase of IR drop	Evaporation, *f*(*T*, *p*)
Electrodes			
Platinum particle growth	Loss of ECSA	Decrease of reaction kinetics	Migration (surface diffusion), *f*(*T*) Dissolution/recrystralization, *f*(*T*, *E*)
Platinum dissolution/ alloy dissolution	Loss of ECSA	Decrease of reaction kinetics	EC dissolution, *f*(*T*, *E*)
PA evaporation from catalyst layer	Loss of ECSA	Decrease of reaction kinetics	Evaporation, *f*(*T*, *p*), *f*(porosity changes)
Carbon corrosion	Loss of ECSA Flooding	Decrease of reaction kinetics, increase of mtx resistances, increase of IR drop	EC oxidation, *f*(*T*, *E*, *p*)
Gas diffusion layer corrosion	Loss of structural integrity Flooding	Increase of mass-transport resistances, increase of IR drop	EC oxidation, *f*(*T*, *E*, *p*)

EC Electrochemical, *ECSA* Electrochemical surface area, *PA* Phosphoric acid

3.1 Membrane Degradation Modes

Typical membrane degradation modes are (1) the possibility of pinhole formation due to thinning of the membrane, which leads to increased fuel crossover and loss of fuel efficiency and (2) acid evaporation. By choosing the appropriate gasket material, one can minimize the former effect even over extended operation times. Most importantly, hard gasket materials which do not deform at the applied pressure between the flow-field plates and at the operating temperature of an actual fuel cell are most suitable. Using these gaskets, one can effectively reduce the pressure on the MEA itself.

The latter effect, that is, acid evaporation due to the operating temperature range of 120–200°C, was found to be of no concern owing to the unique properties of the membrane. Figure 3a illustrates the phosphoric acid evaporation rates at the beginning of life of Celtec® MEAs (determined in the first 100 h of operation) as a function of operating temperature. Typically, at the beginning of life the measured acid evaporation rates are higher than after longer operation since the MEAs are slowly approaching their equilibrium, and first some excess acid is evaporated. As shown in Fig. 3a, at the typical operating temperatures of the Celtec® MEAs of up to 180°C, the maximum acid evaporation rate is around 2 μg $m^{-2}s^{-1}$ and only at higher

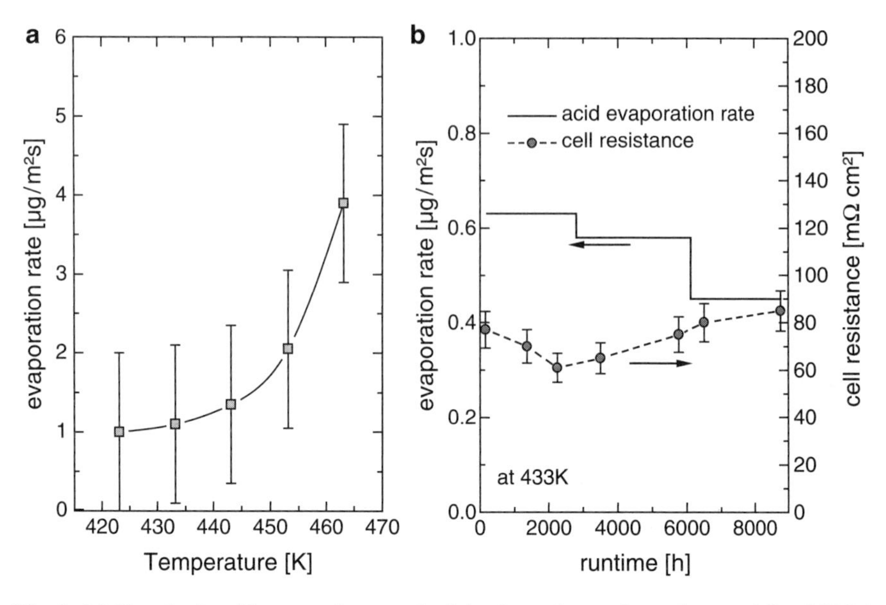

Fig. 3 (**a**) Phosphoric acid evaporation rates in Celtec® membrane electrode assemblies (MEAs) at the beginning of life as a function of temperature. (**b**) Phosphoric acid evaporation rate in a Celtec® MEA operated at 160°C over a period of 1 year (*left axis*) and the cell resistances measured by 1-kHz AC impedance spectroscopy (*right axis*)

temperatures do the rates increase significantly. To gather some experimental values of acid evaporation rates for longer MEA lifetimes, Fig. 3b demonstrates the rates from a 1-year test at 160°C. As is obvious, the acid evaporation rates decrease during the test (see above) and the overall value can be calculated to be approximately 0.5 μg $m^{-2}s^{-1}$. Given the amount of acid in the initial membrane, the experimentally determined acid evaporation rate, and defining the end of life of the MEA when 10% of the initial acid in the membrane is evaporated, one obtains a calculated lifetime on the order of 50,000 h. Additionally, during this test of over 8,800 h of operation, the Ohmic cell resistance changes by only ±15 mΩ cm^2, demonstrating that there is no significant impact of acid evaporation on cell lifetime.

3.2 Electrode Degradation Modes

The typical electrode degradation modes are (1) corrosion of the catalyst metal (both particle growth and dissolution) and (2) corrosion of the carbon materials in electrodes (catalyst support and gas diffusion layer materials).

These two modes are both strong functions of the electrochemical potential and the operating temperatures. Platinum particle growth can occur through Ostwald ripening in a simple surface diffusion process or through a dissolution/recrystallization process (Ross 1987). Platinum dissolution concomitant with an irreversible loss of the active metal phase can become a severe issue when the fuel cell is operated for extended periods of time at potentials above 0.8 V (Ross 1987). Both lead to reduction of the electrochemical active surface area and manifest themselves mainly as a decrease of the cathode kinetics.

Electrochemical corrosion of carbon supports was widely studied in the context of phosphoric acid fuel cell development (Antonucci et al. 1988; Kinoshita 1988), but recently also the low-temperature fuel cell community paid more attention to this process (Kangasniemi et al. 2004; Roen et al. 2004). Carbon corrosion in fuel cell cathodes in the form of surface oxidation leads to functionalization of the carbon surface (e.g., quinones, lactones, carboxylic acids, etc.), with a concomitant change in the surface properties, which clearly results in changes of the hydrophobicity of the catalyst layer. Additionally, and even more severe, total oxidation of the carbon with the overall reaction

$$C + 2H_2O \rightarrow CO_2 + 4H^+ + 4e^- \quad (1)$$

leads to a substantial loss of the catalyst layer itself and can also result in losses of the electrical connection of the platinum particles in the electrode. It has to be noted that carbon itself is thermodynamically not stable in a fuel cell environment ($E^0 = 0.206$ V; Kinoshita 1988) and can only be used owing to kinetic hindrance of its oxidation reaction. Carbon corrosion currents at operating temperatures of 160°C or higher can be quite large (e.g., for Vulcan XC 72 at 180°C and 1.0 V, corrosion currents of 1–10 mA mg_{carbon}^{-1} are observed; Antonucci et al. 1988; Kinoshita, 1988; Schmidt 2006a, b; Stonehart 1984). These currents are high enough to completely oxidize the cathode carbon in a rather short period of time. The main effect of carbon corrosion is an increase in the hydophilicity of the cathode catalyst layer concomitant with electrolyte flooding which, in turn, leads to increase of the cathode mass-transport overpotentials. Concomitant with the corrosion of the carbon, thinning of the catalyst layer and loss of its void volume is observed. In a worst-case scenario, complete deterioration of the carbon finally would affect the structural integrity of the cathode. It should be noted at this point that carbon corrosion is a strong function of temperature, potential, and water partial pressure (Antonucci et al. 1988; Stonehart 1984).

3.3 Electrode Degradation During Start/Stop Operation

During start/stop cycling the two predominant mechanisms which can affect MEA durability are (1) corrosion of the cathode owing to potential excursions to $E >$ 1.2 V (Makharia et al. 2006; Reiser et al. 2005) and (2) electrolyte redistribution owing to volume expansions/contractions. The latter mode is unique to fuel cell operation with at least partly liquid electrolytes, whereas the first is a more general mode observed also in low temperature PEFCs.

Cathode carbon corrosion during start/stop cycling occurs when the cathode electrolyte potential moves to high values when H_2/air (air/H_2) fronts pass through

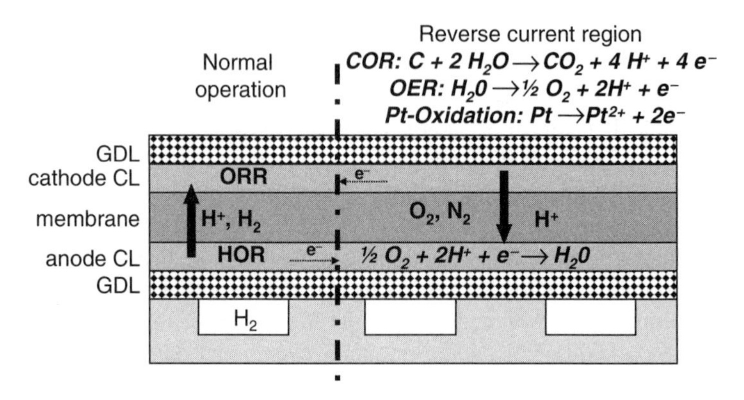

Fig. 4 Principle of the reverse current region. For details see the text. *COR* carbon oxidation reaction, *OER* oxygen evolution reaction, *ORR* oxygen reduction reaction, *HOR* hydrogen oxidation reaction, *GDL* gas-diffusion layer, *CL* catalyst layer

the anode while the cathode compartment is filled with air (Fig. 4). Since the anode potential is basically defined by the hydrogen potential, the oxygen present in the anode will be instantaneously reduced to water.

During the ORR at the anode side, the protons used are generated in an oxidation reaction at the cathode, by carbon oxidation, oxygen evolution, or oxidation of platinum (Fig. 4, right side). This is called the reverse current region (Reiser et al. 2005).

Simultaneously, in the normal operation region (Fig. 4, left side), hydrogen and oxygen are oxidized and reduced, respectively, as usual. Both regions are electrically connected through the gas diffusion layer materials. Basically, the fuel cell is internally short-circuited. One can consider the normal operation region as a power source and the reverse current region as a power sink. Locally around the H_2/air (air/H_2) front the potential of the cathode can be as high as 1.5 V (Reiser et al. 2005). Owing to these high potentials, severe oxidation of the cathode can take place if no precautions are taken.

3.4 Dual-Cell Setup

The dual-cell setup illustrated in Fig. 5a can be used to measure the electrochemical potentials when H_2/air (air/H_2) fronts are passing through the anode while the cathode is filled with air during the start/stop phases described in the previous section. The idea of the dual-cell setup is to (electrically and physically) separate the normal operation region and the reverse current region described in Fig. 4. Therefore, the setup consists of a fuel cell (*the normal operation cell*) where the anode and cathode are filled with H_2 and air, respectively, and a second fuel cell (*the reverse current cell, RCC*) where both the anode and the cathode are filled with air.

When both cells are electrically connected (cathodes C and D, anodes A and B), a continuously standing H_2/air front is created. Figure 5b shows the potential measurement in the dual-cell setup at 160°C using Celtec®-P 1000 MEAs. Initially, in

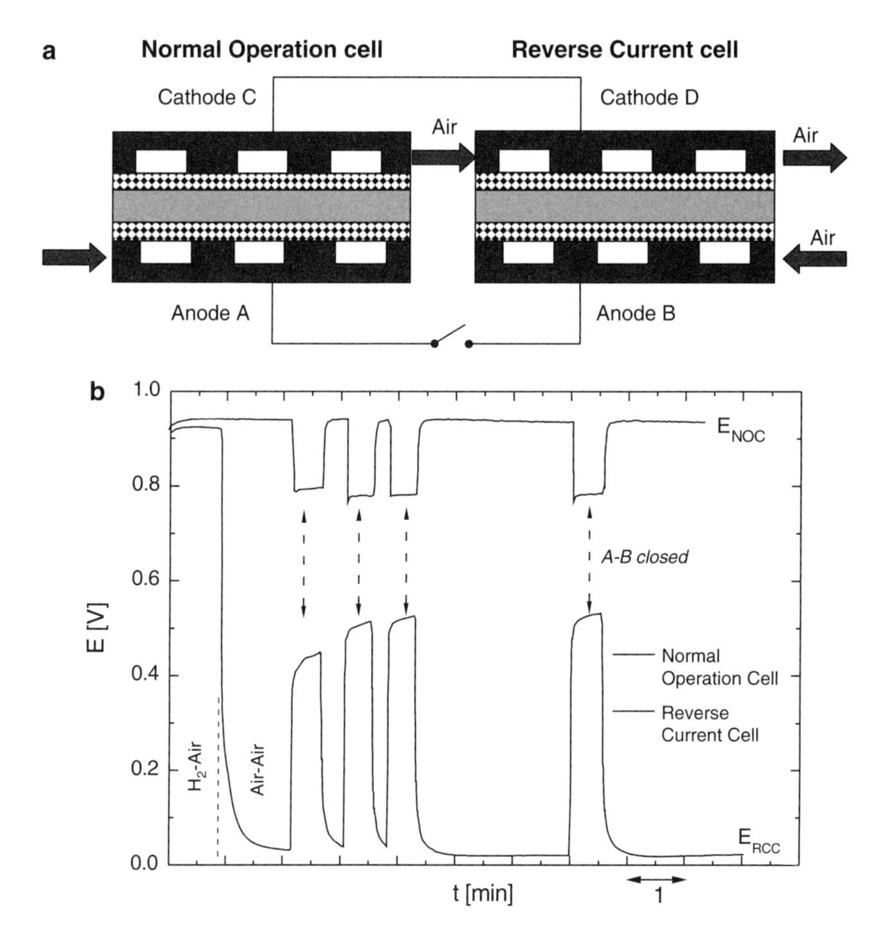

Fig. 5 (**a**) Dual-cell setup for measurement of the cathode potential during start/stop phases. (**b**) Potential measurements of the cathode in the reverse current region at 160°C using the dual-cell setup shown in (**a**). Note that since the test was carried out using dry gases, the changes of the individual potentials are related to changes of the water partial pressures when the cells are polarized

both fuel cells – the normal operation cell and the reverse current cell – the anodes are filled with H_2, the cathodes with air, and both cells are at open circuit potential. Both cells are not yet electrically connected between anode A and cathode B. As shown in Fig. 5b, subsequently in the reverse current cell the anode is filled with air, which can be followed by a drop of its potential E_{RCC} to almost zero.

When anodes A and B (arrows in Fig. 5b) are connected, the potential E_{NOC} decreases to approximately 780–800 mV since the normal operation cell is polarized (i.e., H_2 is oxidized in anode A and oxygen is reduced in cathode C). In the reverse current cell, the potential E_{RCC} increases to 450–500 mV (vs. anode B). Hence, the potential of cathode D in the reverse current cell can be calculated as the sum of E_{NOC} and E_{RCC}, viz., approximately 1.3 V versus the reversible hydrogen electrode That is, having a single cell with a H_2/air front passing through its anode, the cathode potential locally increases to values of about 1.3 V, which certainly represents severe corrosive conditions.

The same dual-cell setup can be used to measure the current flowing when H_2/air and air/H_2 fronts are passing through the anodes. Figure 6a illustrates the experimental setup schematically. Generally, the setup is very similar to that described previously. In the case of Fig. 6a, the gas outlets of anode A and cathode C are connected to the gas inlets of anode B and cathode D, respectively. Additionally, the anodes A and B are electrically connected through an ampere meter to measure the current flowing between the two cells. To simulate the passage of a H_2/air front, both gases are exchanged and the resulting current between anode A and anode B is measured. Generally, two situations have to be considered:

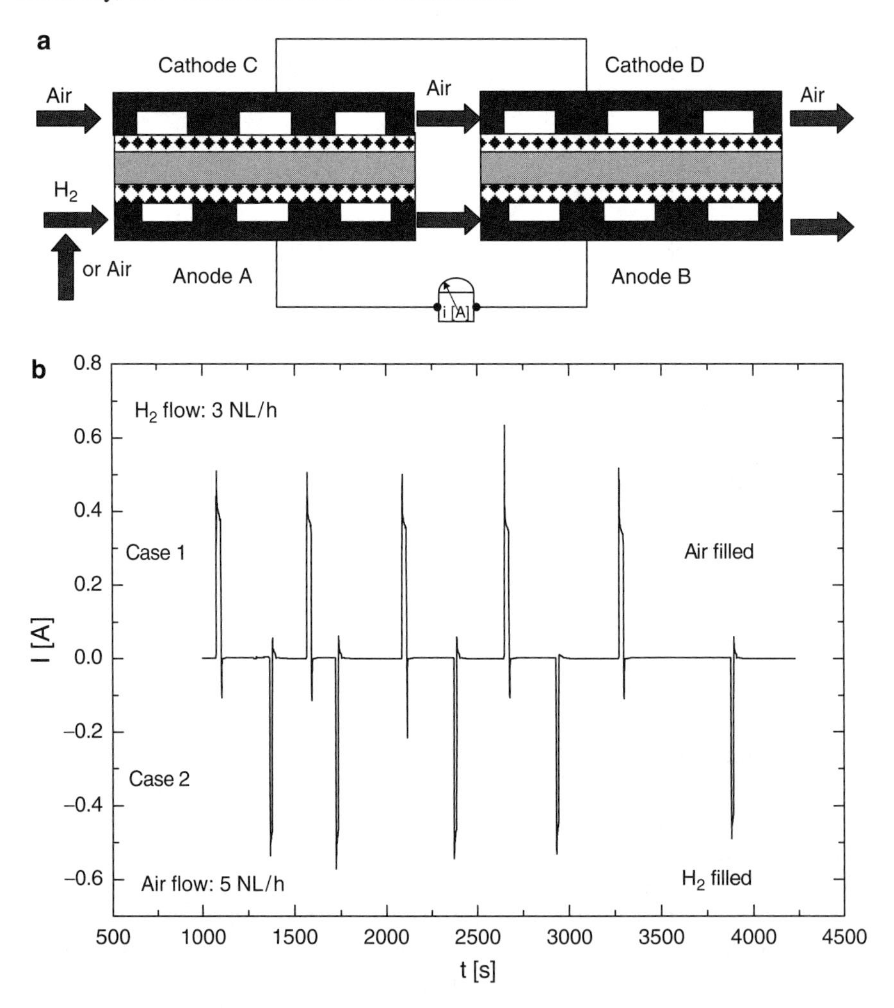

Fig. 6 (**a**) Dual-cell setup for simulation of H_2/air (air/H_2) fronts passing through the anode. (**b**) Current flow measurement during passage of H_2/air (air/H_2) fronts through the dual-cell setup at 160°C

Case 1. Startup situation: Both anodes are filled with air and then exchanged with H_2, resulting in positive currents. The residence time of the H_2/air front was approximately 4.2 s in our experiments.

Case 2. Shutdown situation: Both anodes are filled with H_2 and then exchanged with air, resulting in negative currents. The residence time of the air/H_2 front was approximately 2.5 s in our experiments.

The results of current flowing in the two cases are plotted in Fig. 6b. Note that owing to the differences in the H_2 and air flow (3 vs. 5 L/h) the width of the current peaks in Case 2 is smaller than in Case 1, i.e., the charge flow is strongly dependent on the residence time of the H_2/air (air/H_2) front (see also the charge flow analysis in Fig. 7), which directly implies that quick purging during startup and shutdown is essential to suppress carbon corrosion as much as possible. It has to be noted that in the dual-cell setup only the currents flowing between the two cells can be measured. Any current flowing internally directly through the gas diffusion layer/electrode material is not accounted for. On the basis of an estimate in recent work (Tang et al. 2006), the current flowing within the electrode is roughly one-third of the measured current between the two cells. However, independent of the real ratios of the two currents, in both startup and shutdown situations (cases 1 and 2, respectively), currents up to 0.6 A are flowing between the two anodes. These measured currents are representative of the processes occurring on the cathode of the two cells, viz., oxidation of carbon (to form CO_2), platinum (to form Pt^{2+}), or water (to produce O_2). The extents of the individual reactions can only be assumed; however, as will be shown in Sect. 3.5, platinum oxidation plays only a minor role.

It can be also assumed that water oxidation under the conditions applied most likely is only a secondary side reaction owing to only small amounts of water present in the highly concentrated phosphoric acid used. Therefore, the main reaction occurring may indeed be carbon corrosion. To determine the charge flows during the passage of H_2/air (air/H_2) fronts as a function of temperature, we performed the aforementioned experiments in the temperature range between 60 and 160°C. The results are shown in Fig. 7, with the gray symbols representing the charge flowing under the conditions chosen for a complete simulated start/stop cycle.

On the right axis in Fig. 7, the percentage of carbon consumption during one complete simulated start/stop cycle is plotted. As is obvious, e.g., at 160°C, one complete start/stop cycle under the experimental conditions applied results in corrosion of approximately 0.4% of the carbon present in the catalyst layer, allowing an estimation of approximately 250 cycles for the complete oxidation of the cathode carbon under the experimental conditions applied.

3.5 *Analysis of a Start/Stop-Cycled Cell*

To study the start/stop behavior of Celtec®-P 1000 MEAs under more realistic conditions and to make a correlation with constant cell operation, we compare two fuel cells: one cell is operated in start/stop cycling mode (12-h operation followed by 12-h

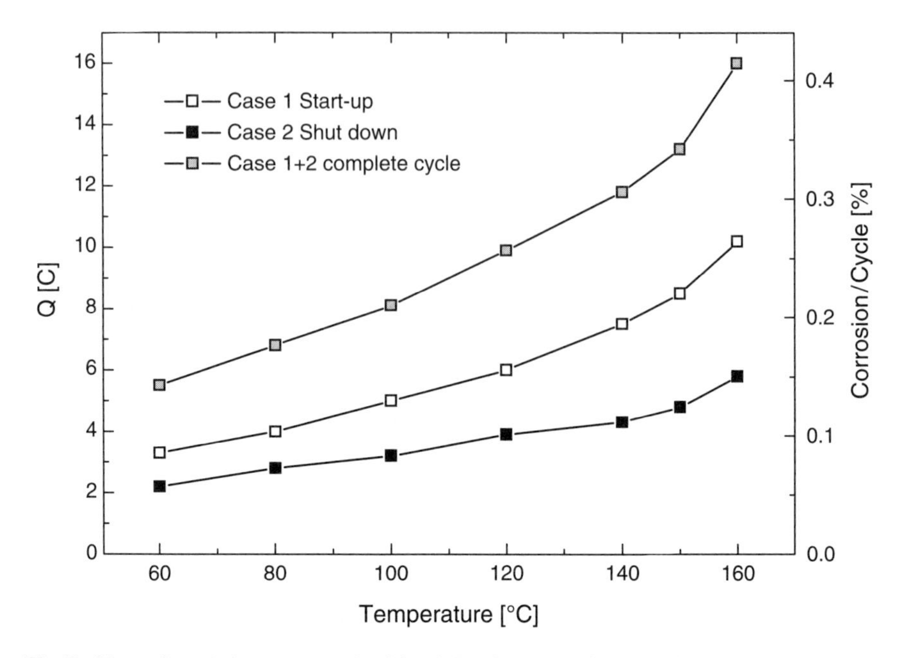

Fig. 7 Charge flow during passage of H_2/air (air/H_2) fronts as a function of temperature. The *right axis* shows the percentage of carbon corroded during each cycle

shutdown with cooling to room temperature, in a total 270 start/stop cycles) and another cell is operated at constant load of 0.2 A cm^{-2}. Both single cells are operated for approximately 7,000 h. Figure 8 illustrates the cell potential at 0.2 A cm^{-2} for both MEAs as a function MEA operation time. To analyze the degradation behavior of the two MEAs (Japanese Industrial Standard 2003), formally, three regions can be defined (note the definition of the region is rather arbitrary and mainly related to the changes in the slopes of the voltage–time plots):

1. Region I in the operation time range from 0 to 1,000 h
2. Region II in the operation time range from 1,000 to 5,000 h
3. Region III in the operation time range from 5,000 to 7,000 h

For purposes of this analysis, regions II and III are certainly the most interesting to understand, whereas region I, within the first several hundred hours of operation, is difficult to interpret owing to complicated cell break-in processes and the limited time resolution of the diagnostic data. Therefore, the analysis will focus on regions II and III. Especially the transition from region II to region III in the start/stop-cycled cell is quite obvious.

Since typically changes in the degradation mode/mechanism can be observed when degradation lines are changing their slope (Japanese Industrial Standard 2003), the last 2,000 h of operation of the start/stop cell is interesting owing to apparent changes in the degradation mode. Following the Tafel slope analysis described elsewhere in more detail (Gasteiger et al. 2003; Neyerlin et al. 2005;

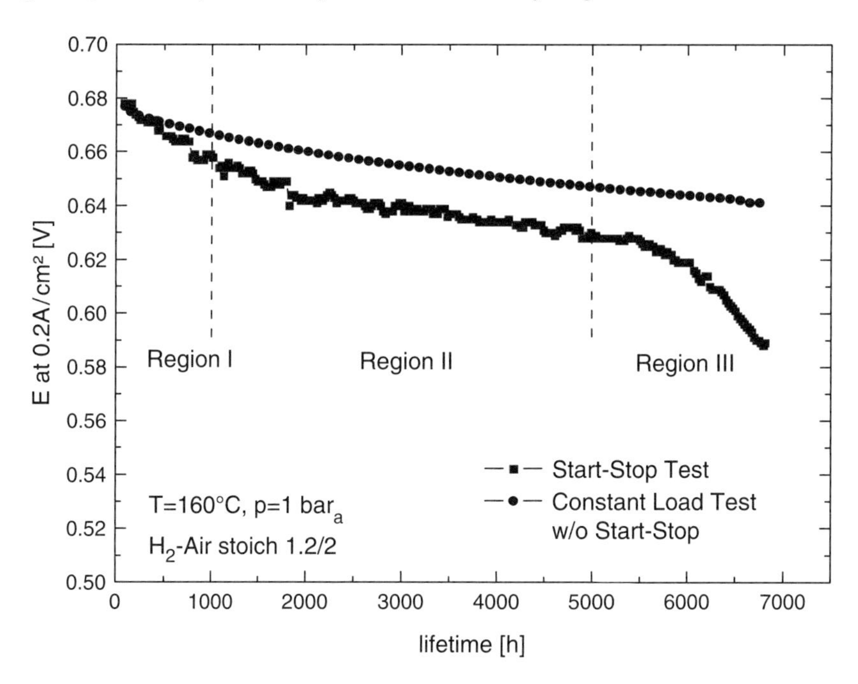

Fig. 8 Comparison of a cell operated at 160°C under constant load at 0.2 A cm^{-2} and a cell operated in start/stop cycling mode at 0.2 A cm^{-2}. The total operation time of the tests was 7,000 h

Schmidt 2006a, b; Schmidt and Baurmeister 2008), we try to separate different loss terms to identify the main degradation modes and their implications for cell performance during these two tests. The strategy for calculating the different loss terms in the MEA is summarized in the Appendix.

3.5.1 Cell Impedance

The Ohmic cell impedance, measured by 1-kHz AC impedance, as a function of lifetime for both cells under investigation is illustrated in Fig. 9a. In regions I and II, no significant difference is observed for both cells. Only at lifetimes above 5,000 h in region III does the start/stop-cycled cell show a slightly higher impedance of approximately 15 mΩ cm^2, which most likely is due to increased contact resistances. However, the overall difference of the cell impedance especially in region III cannot account for the difference in performance in Fig. 8. That is, it appears that changes in impedance have only a minor effect during start/stop cycling.

3.5.2 Cathode (Oxygen Reduction) Kinetics

To get insight into the oxygen reduction kinetics, H_2–O_2 polarization curves were recorded at the beginning of life and throughout the course of the tests, followed by

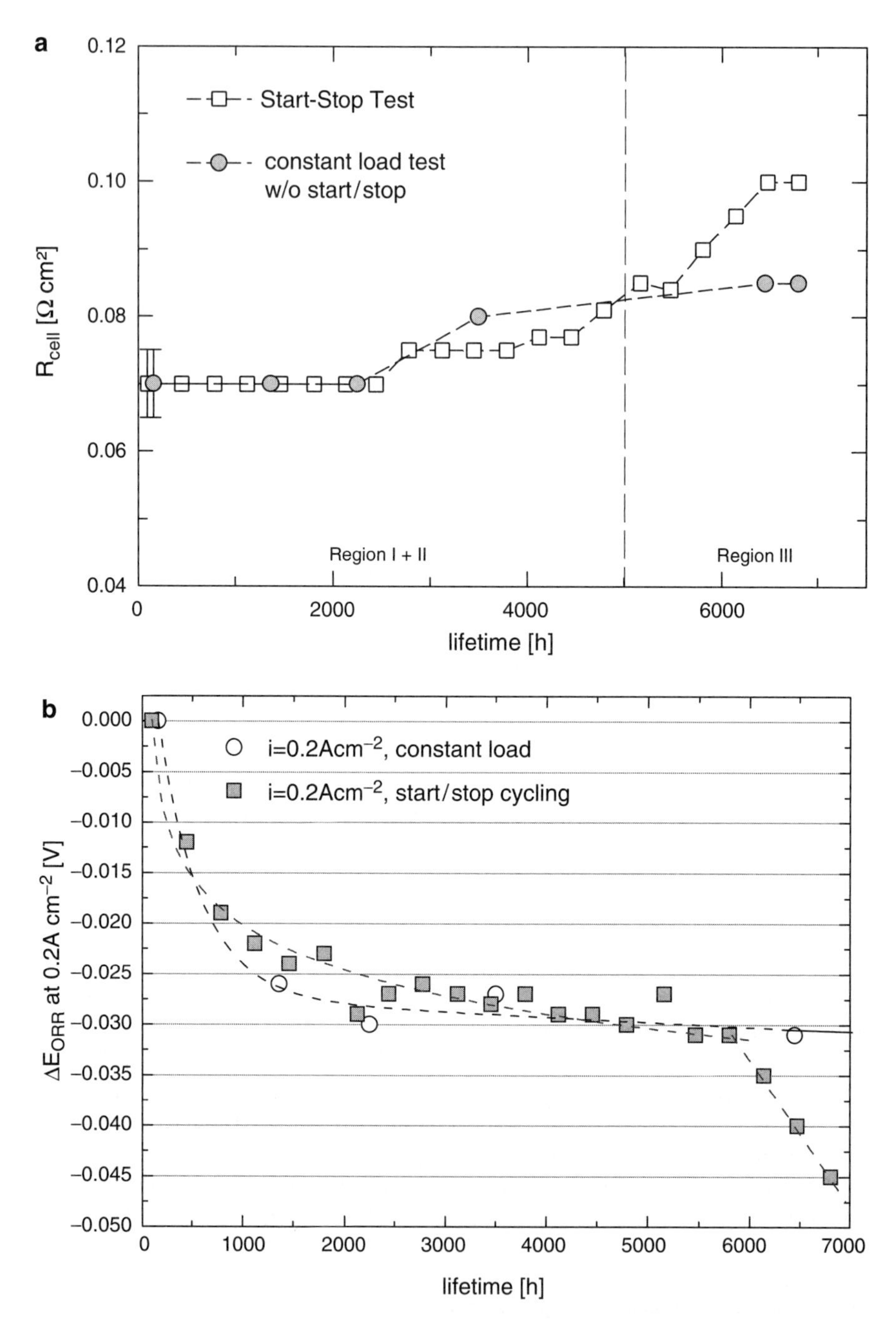

Fig. 9 (**a**) Cell impedance measurements (1-kHz high-frequency resistance) of the continuously operated cell and the start/stop-cycled cell. (**b**) Changes in oxygen reduction overpotentials versus lifetime for a MEA in constant operation (*circles*) and start/stop cycling mode (*squares*). (**c**) Changes in mass-transport overpotentials in H_2/air operation versus lifetime for a MEA in constant operation (*circles*) and start/stop cycling mode (*squares*) at 0.2 and 0.8 A cm^{-2}, respectively

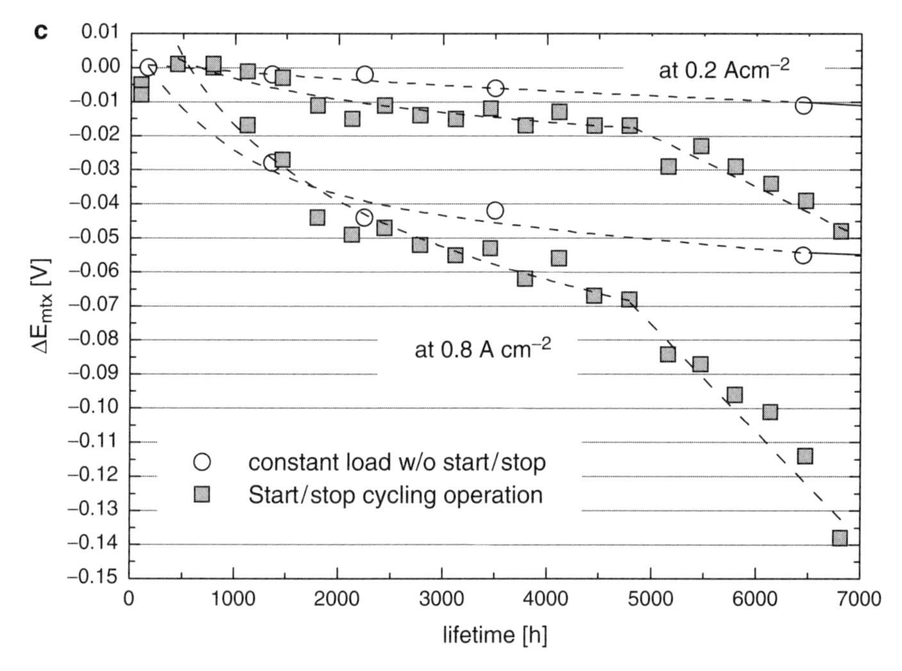

Fig. 9 (continued)

calculation of the oxygen reduction overpotentials (Schmidt and Baurmeister 2008). The changes of the oxygen reduction overpotentials as function of lifetime for both continuously operated and start/stop-cycled cells, respectively, gives information on the impact of start/stop cycling on cathode kinetics (Fig. 9b). Quite obviously, both MEAs exhibit very similar logarithmic behavior with an initial fast increase in overpotential in the first approximately 1,000 h of operation followed by stable values for longer operation times. Based on a Tafel slope of approximately 90 mV dec^{-1}, an increase in oxygen overpotentials of approximately 30 mV translates into a reduction of the electrochemical active surface area by approximately 50%.

This value was confirmed by 53% reduction of the H_{upd} charges in cyclic voltammetry measurements at the beginning versus end of life of the continuously operated MEA. The reduction is mainly related to catalyst particle growth during operation in the first 500–1,000 h. To give an impression, reduction of the platinum surface area by 50% is obtained when catalyst nanoparticles grow from an initial mean diameter of 3 nm to a mean particle diameter of 6 nm (Stonehart 1992), a realistic scenario in fuel cell cathodes (Tada 2003). Interestingly, in the first 5,000 h of operation, both fuel cells show similar behavior. That is, within this operation time (end of region II) the similar kinetic behavior of the two different cathodes studied here points to the fact that start/stop cycling does not adversely affect the cathode catalyst kinetics. Additionally, it also proves that platinum dissolution into the electrolyte can only play a very minor role during start/stop phases with potential excursions to 1.3 V.

In region III, however, the cathode kinetics change significantly for the start/stop-cycled MEA, and at the end of life, the electrochemical surface area was reduced to roughly 30% of its initial value. This behavior is discussed further in Sect. 3.5.3.

3.5.3 Cathode Mass-Transport Overpotentials

As we know from the experiments with the dual-cell setup carbon corrosion during start/stop cycling is the main degradation factor, which typically becomes visible when the cathode mass-transport properties are changing during cell operation. Figure 9c illustrates the changes in mass-transport overpotentials for both a continuously and a start/stop-operated cell at 0.2 and 0.8 A cm^{-2}, respectively. As expected from the underlying carbon corrosion process, the mass-transport properties of the start/stop-cycled MEA significantly differ from those of the continuously operated cell. Interestingly, at short operation times below 1,000h of operation, in the start/stop-cycled cell, the mass transport slightly improves, which may be interpreted by beneficial redistribution of the electrolyte within the MEA. After this initial improvement, however, mass-transport overpotentials increase significantly. At 5,000h of operation, e.g., mass-transport overpotentials for the cycled cell at 0.8 A cm^{-2} are about 30mV higher than those of the continuously operated MEA. This behavior can be related to increased carbon corrosion of the cathode catalyst support during start/stop cycling as described in Sect. 3.5.2. As a result, the cathode becomes more and more hydrophilic and, as a consequence, increasingly flooded with phosphoric acid. This, in turn, leads to increased mass-transport resistances (Kinoshita 1992; Kunz and Gruver 1975; Perry et al. 1998). In region III, mass-transport overpotentials increase further significantly.

One should keep in mind that carbon corrosion, especially in a later stage of the tests performed here, also may change the porous structure of the cathode catalyst layer concomitant with a void volume loss (Patterson and Darling 2006; Yu et al. 2006) and also with complete deterioration of the catalyst layer, which was confirmed by scanning electron microscopy postmortem analysis of the start-stop-cycled MEA.

3.5.4 Summary of Underlying Processes

To summarize and conclude the analysis of start/stop operation, the main processes and effects are collected in Table 2. In region II up to 5,000h of operation, basically no changes in the start/stop cell and in the continuously operated cell are observed. That is, start/stop cycling during this operation time window does not affect the cell impedance. A very similar conclusion can be drawn when the kinetic behaviors of the two cells investigated are compared, since both cells exhibit roughly 30-mV increase of the ORR overpotential at the end of region II. The only significant difference in the two cells within the first 5,000h of operation (region II) can be found in the increase of the cathode mass-transport overpotentials due to flooding of the cathode catalyst layer induced by carbon corrosion. The increase of mass-transport overpotentials is even more pronounced in region III. At end of life and

Table 2 Summary of the processes during the start/stop cycling test

	Region II (1,000–5,000 h)	Region III (5,000–7,000 h)
Impedance	Both cells very similar	Increase in impedance in start/stop cell
	No specific influence of start8stop cycling	Increase in contact resistance due to thickness reduction of electrode
Cathode kinetics	Both cells very similar	Increase in ORR overpotential in start/stop cell
	No specific influence of start/stop cycling	Reduction of ECSA due to loss of structural integrity of cathode catalyst layer
	Platinum dissolution during start/stop cycling insignificant	
Cathode mass transport	Higher mass-transport overpotentials in start/stop-cycled cell	Significant increase of mass-transport overpotentials
	Electrolyte flooding due to carbon corrosion and resulting changes in hydrophobicity	Severe corrosion of carbon results in loss of structural integrity of cathode catalyst layer and void volume loss

ORR Oxygen reduction reaction

0.8 A cm^{-2}, the mass-transport overpotential of the start/stop-cycled call is almost 90 mV higher than that observed in the continuously operated MEA. These drastic increases, which eventually lead to the stopping of the test, are caused by the severe oxidation of the cathode carbon during start/stop cycling concomitant with the loss of void volume and structural integrity of the catalyst layer. That is, drastically expressed, a large part of the cathode carbon is oxidized to CO_2, not only leading to reduction of the electrode thickness and increase in contact resistances, but also to the electrical disconnection of platinum particles in the catalyst layer which is observed in the significant increase of the ORR overpotentials in region III. As a conclusion, the main process during start/stop cycling in these tests is definitively carbon corrosion, which in the later state of the test (region III) also leads to secondary effects on kinetics and cell impedance.

3.6 Comparison of Low- and High-Temperature PEFCs

Finally, in this last section, a comparison of the start/stop behavior of high-temperature and low-temperature PEFCs is given. Figure 10 shows the decrease in cell voltage at 0.2 and 0.8 A cm^{-2}, respectively, for 200 start/stop cycles performed with a high-temperature and a low-temperature MEA. The data for the low-temperature MEA were extracted from a recent paper by Yu et al. (2006).

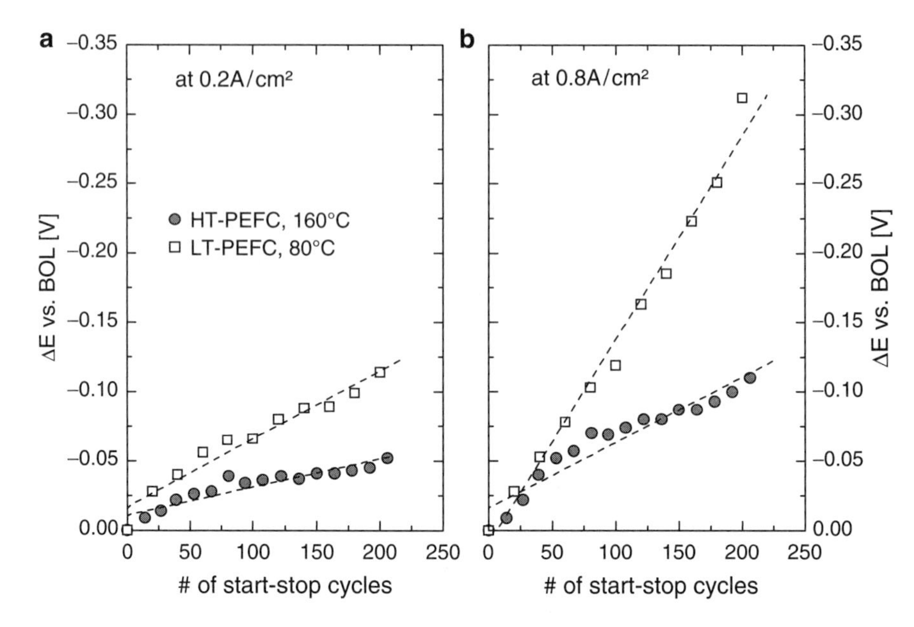

Fig. 10 Comparison of voltage losses of low- and high-temperature fuel cells at 0.2 A cm^{-2} (**a**) and 0.8 A cm^{-2} (**b**). Experimental conditions for the low-temperature polymer electrolyte fuel cell (LT-PEFC) were as follows: Pt/Ketjenblack, 0.4 mg_{Pt} cm^{-2}, 80°C, 150 kPa, 66% relative humidity (inlet), residence time 1.3 s. Experimental conditions for the high-temperature polymer electrolyte fuel cell (HT-PEFC) were as follows: platinum alloy/Vulcan XC72, 0.7 mg_{Pt} cm^{-2}, 160°C, 100 kPa, dry gases, residence time 2.5 s. (The data for the LT-PEFC were extracted from Yu et al. 2006)

At first sight, quite a difference between the low-temperature and the high-temperature MEA is observed. Analyzing the slopes of the linear regression lines results in a difference of a factor of 2 between the low-temperature and the high-temperature case, i.e., the high-temperature MEA appears to be twice as stable as the low-temperature MEA. However, the experimental environment and conditions have to be taken into account to make a fair comparison. The two critical factors for the serious analysis of the both data sets are:

1. The carbon loading in the cathodes
2. The residence time of the H_2/air (air/H_2) boundaries, as corrosion charges are proportional and residence time (Fig. 7).

In the case of the high-temperature MEA, carbon loading in the cathode is approximately 3 mg cm^{-2}. Although the real carbon loading for the low-temperature MEA is not given (only the platinum loading of 0.4 mg cm^{-2}), we can safely assume the maximum metal-to-carbon ratio is 1:1 (i.e., a 50 wt% Pt/C catalyst was used in the experiments of Yu et al.), which would result in a carbon loading of 0.8 mg cm^{-2}, roughly a factor of 4 less than in the high-temperature case. Secondly, the residence time of the H_2/air (air/H_2) boundaries in the low-temperature case is given as 1.3 s, approximately a factor 2 smaller than the 2.5 s used in the high-temperature cell. Considering these two factors, the difference in the slopes of the regression lines in Fig. 10 can be easily explained, and it has to be concluded that the degradation of the low-temperature and that of the high-temperature MEA is similar. This is a quite surprising result, since typically one would expect the 80°C difference in operation temperature would account for significantly higher degradation rates for the high-temperature fuel cell.

A simple explanation for the apparent similarity of properties with respect to start/stop cycling can be found in the differences of the relative humidity. The low-temperature fuel cell is operated with humidified gases (66% relative humidity), which drives the carbon oxidation reaction to the product (CO_2) side (1). Interestingly, we recently found very similar results by a comparison of low- and high-temperature MEAs operated continuously at around 0.8 V (Schmidt 2006a, b).

4 Outlook

Having identified the main degradation modes during start/stop cycling, one can implement mitigation strategies to increase the stability of MEAs operated under conditions where corrosion issues are predominant. One feasible way is external potential control of the MEA (Bekkedahl et al. 2005; Condit and Breault 2003). This system solution simply avoids potential excursions of the cathode during start/stop cycling, and therefore carbon corrosion issues are suppressed. Besides system solutions (for a useful summary, see Perry et al. 2006), also the intrinsic stability of the materials used in the MEA with respect to corrosion can be improved, especially the use of more stable catalyst supports is key to improve the stability of a MEA. Besides the typical improvements of carbon itself (e.g., through graphitization; Kinoshita 1988), also other possibilities have to be considered, such as the adaptation of ceramics or other high-temperature stable electronically conductive materials, e.g., tungsten carbides or metal oxides. However, owing to the difficulty of achieving combined temperature stability and chemical and electrochemical stability in an acidic environment together with electronic conductivity, material sciences of support materials remains one of the most important topics in fuel cell R&D.

Appendix: Calculation of Individual Overpotentials

To determine the individual overpotentials of a fuel cell cathode, we followed the Tafel slope analysis given elsewhere in more detail (Gasteiger et al. 2003; Neyerlin et al. 2005; Schmidt 2006a, b; Schmidt and Baurmeister 2008). However, for better understanding of the analyses presented in this chapter, a summary of the most important equations will be given.

The cell potential corrected for Ohmic contribution can be given by (2):

$$E_{\mathrm{cell,IR\text{-}free}} = E^{0}(p_{\mathrm{H_2/O_2/H_2O}}, T) - \eta_{\mathrm{ORR}} - \eta_{\mathrm{cathode,mtx}} - \eta_{\mathrm{HOR}} - \eta_{\mathrm{anode,mtx}}, \qquad (2)$$

where $E^0(p_x,T)$ represents the temperature and H_2, O_2, and H_2O partial pressure dependent equilibrium potential, η_{ORR} and η_{HOR} the kinetic overpotentials for the ORR and the hydrogen oxidation reaction, and $\eta_{x,\,mtx}$ the mass-transport overpotentials on the cathode and anode, respectively. The temperature and partial pressure dependent equilibrium potential $E^0(p_x,\mathrm{T})$ is calculated according to (3):

$$E^0(p_{H_2/O_2/H_2O},T) = E^{0,*} + \frac{\partial E^{0,*}}{\partial T}(T-298) + \left(\frac{2.303RT}{2F}\right)\log\left(\frac{\bar{p}_{H_2}\bar{p}_{O_2}^{\;0.5}}{\bar{p}_{H_2O}}\right), \tag{3}$$

where the first term, $E^{0,*}$, is the equilibrium potential at 298 K and 101,325 Pa calculated from ΔG values for gas-phase water (CRC Handbook 1995), the second term is used to correct $E^{0,*}$ for the temperature, calculated using ΔS values for gas-phase water (CRC Handbook 1995), the third term corrects $E^{0,*}$ for the individual partial pressures, and $\bar{\mathrm{P}}_{H2}$, $\bar{\mathrm{P}}_{O2}$, and $\bar{\mathrm{P}}_{H2O}$ are the mean partial pressures under operation conditions of hydrogen, oxygen, and water, respectively, normalized to 101,325 Pa.

Given that consensus is emerging regarding the fact of negligible anodic polarization and the pure kinetic control of the oxygen polarization curve, (2) reduces to

$$E_{\text{cell, IR-free}} = E^0(p_{H_2/O_2/H_2O},T) - \eta_{ORR}. \tag{4}$$

By proper calculation of $E^*(p_x,T)$, the oxygen reduction overpotential can be determined using (4) from the measured oxygen polarization curve, which follows an oxygen partial pressure dependent Butler–Volmer expression. The theoretical Tafel line for air polarization should just be shifted to lower cell potentials, by $\Delta E_{O2\,-\,\mathrm{air}}$ given by

$$\Delta E_{O_2-\mathrm{air}} = \Delta E^0 + b\gamma\,\log\left(\frac{p_{O_2}}{p_{\mathrm{air}}}\right) \tag{5}$$

where ΔE^0 is the difference between the equilibrium potentials for pure oxygen and air, b and γ are the Tafel slope ($b = 2.3RT/F$) and the kinetic reaction order ($\gamma \approx 0.6$; defined as the logarithmic change of the current density with the oxygen partial pressure at constant temperature and overpotential), and p_x is the partial pressures for pure oxygen and air. Deviations from the theoretical Tafel line for air polarization to lower cell potentials can be considered to be cathode mass-transport overpotentials.

The changes in oxygen overpotentials plotted in Fig. 9b are calculated using (6):

$$\Delta E_{ORR} = \eta_{ORR,BOL} - \eta_{ORR,t}, \tag{6}$$

with $\eta_{ORR,t}$ and $\eta_{ORR,BOL}$ being the overpotential at a given operation time and at the beginning of life, respectively. The changes in mass-transport overpotentials plotted in Fig. 9c are calculated using (7)

$$\Delta E_{\text{mtx}} = \eta_{\text{mtx,BOL}} - \eta_{\text{mtx},t}, \tag{7}$$

with $\eta_{\text{mtx},t}$ and $\eta_{\text{mtx,BOL}}$ being the overpotential at a given operation time and at the beginning of life, respectively.

Acknowledgments D. Ott, F. Rat, M. Ferreira, and M. Jantos are greatly acknowledged for performing parts of the experiments. I also want to express my sincere thanks to J. Baurmeister for continuous and helpful discussions.

References

Allen, J. N., Sifer, N., and Bostic, E. (2006) The XX25: A 25-watt portable fuel cell for soldier power. *Proceedings of the 42nd Power Sources Conference, Philadelphia*, pp. 403–406.

Antonucci, P. L., Romeo, F., Minutoli, M., Alderucci, E., and Giordano, N. (1988) Electrochemical corrosion behavior of carbon black on phosphoric acid. Carbon 26(2), 197–203.

Bekkedahl, T. A., Bregoli, L. J., Breault, R. D., Dykeman, E. A., Meyers, J. P., Patterson, T. W., Skiba, T., Vargas, C., Yang, D., and Yi, J. S. (2005) Reducing Fuel Cell Cathode Potential During Startup and Shutdown. US Patent 6913845.

Breault, R. D. (2003) Stack materials and stack design. In: Vielstich, W., Lamm, A., and Gasteiger, H. A. (Eds.), *Handbook of Fuel Cells. From Fundamentals, Technology and Applications. Part 3*, Wiley, New York, NY, pp. 797–810.

Condit, D. A. and Breault, R. D. (2003) Shut-Down Procedure for Hydrogen-Air Fuel Cell System. US Patent 6635370

CRC Handbook of Chemistry and Physics (1995), Lide, D. R. (Ed.), 76th Edition, CRC, Boca Raton, FL.

Gasteiger, H. A., Gu, W., Makharia, M., Mathias, M. F., and Sompalli, B. (2003) Beginning of life MEA performance – Efficiency loss contributions. In: Vielstich, W., Gasteiger, H. A., and Lamm, A. (Eds.), *Handbook of Fuel Cells – Fundamentals, Technology and Applications. Volume 3: Fuel Cell Technology and Applications, Part 1*, Wiley, New York, NY, pp. 593–610.

Japanese Industrial Standard (2003) Accelerated Life Test Methods for Phosphoric Acid Fuel Cell. Japanese Industrial Standard JIS C 8802:2003.

Kangasniemi, K. H., Condit, D. H., and Jarvi, T. D. (2004) Characterization of vulcan electrochemically oxidized under simulated PEM fuel cell conditions. Journal of the Electrochemical Society 151(4), E125–E132.

Kinoshita, K. (1988) *Carbon. Electrochemical and Physicochemical Properties*, Wiley, New York, NY

Kinoshita, K. (1992) *Electrochemical Oxygen Technology*, Wiley, New York, NY.

Kreuer, K. D., Paddison, S. J., Spohr, E., and Schuster, M. (2004) Transport in proton conductors for fuel cell applications: Simulations, elementary reactions, and phenomenology. Chemical Reviews 104, 4678.

Kunz, H. R. and Gruver, G. A. (1975) The catalytic activity of platinum supported on carbon for electrochemical oxygen reduction in phosphoric acid. Journal of the Electrochemical Society 122(10), 1279–1287.

Li, Q., He, R., Berg, R. W., Hjuler, H. A., and Bjerrum, N. J. (2004a) Water uptake and acid doping of polybenzimidazoles as electrolyte membranes for fuel cells. Solid State Ionics 168, 177–185.

Li, Q., He, R., Jensen, J. O., and Bjerrum, N. J. (2004b) PBI-based polymer membranes for high temperature fuel cells – Preparation, characterization and fuel cell demonstration. Fuel Cells 4(3), 147–159.

Mader, J., Xiao, L., Schmidt, T. J., and Benicewicz, B. (2008) Polymer-acid blends as high-temperature membranes. In: Scherer, G. G. (Ed.), *Advances in Polymer Science*, Springer, New York, NY.

Makharia, R., Kocha, S. S., Yu, P. T., Sweikart, M. A., Gu, W., Wagner, F. T, and Gasteiger, H. A. (2006) PEM fuel cells electrode materials: Requirements and benchmarking technologies. ECS Transactions 1(8), 3–18.

Neyerlin, K. C., Gasteiger, H. A., Mittelstaedt, C. K., Jorne, J., and Gu, W. (2005) Effect of relative humidity on oxygen reduction kinetics in a PEMFC. Journal of the Electrochemical Society 152(6), A1073–A1080.

Patterson, T. W. and Darling, R. M. (2006) Damage to the cathode catalyst of a PEM fuel cell caused by localized fuel starvation. Electrochemical and Solid-State Letters 9(4), A183–A185.

Perry, M. L., Newman, J., and Cairns, E. J. (1998) Mass transport in gas diffusion electrodes: A diagnostic tool for fuel cell cathodes. Journal of the Electrochemical Society 145(1), 5–15.

Perry, M. L., Patterson, T. W., and Reiser, C. A. (2006) System strategies to mitigate carbon corrosion in fuel cells. ECS Transactions 3(1), 783–795.

Reiser, C. A., Bregoli, L. J., Patterson, T. W., Yi, J. S., Yang, J. D., Perry, M. L., and Jarvi, T. D. (2005) A reverse-current decay mechanism for fuel cells. Electrochemical and Solid-State Letters 8(6), A273–A276.

Roen, L. M., Paik, C. H., and Jarvi, T. D. (2004) Electrocatalytic corrosion of carbon support in PEMFC cathode. Electrochemical and Solid-State Letters 7(1), A19–A22.

Ross Jr., P. N. (1987) Deactivation and poisoning of fuel cell catalysts. In: Petersen, E. E. and Bell, A. T. (Eds.), *Catalyst Deactivation*, Marcel Dekker, New York, NY, pp. 167–187.

Savinell, R., Yeager, E., Tryk, D., Landau, U., Wainright, J., Weng, D., Lux, K., Litt, M., and Rogers, C. (1994) A polymer electrolyte for operation at temperatures up to 200°C. Journal of the Electrochemical Society 141(4), L46–L48.

Schmidt, T. J. (2006a) Durability and degradation in high-temperature polymer electrolyte fuel cells. ECS Transactions 1(8), 19–31.

Schmidt, T. J. (2006b) Durability and reliability in high-temperature reformed hydrogen PEFCs. ECS Transactions 3(1) 861–869.

Schmidt, T. J. and Baurmeister, J. (2008) Properties of high temperature PEFC Celtec P1000 MEAs in start/stop operation mode. Journal of Power Sources 176, 428–434.

Stonehart, P. (1984) Carbon substrates for phosphoric acid fuel cell cathodes. Carbon 22(4/5), 423–431.

Stonehart, P. (1992) Development of alloy electrocatalysts for phosphoric acid fuel cells (PAFC). Journal of Applied Electrochemistry 22, 995–1001.

Tada, T. (2003) High dispersion catalysts including novel carbon supports. In: Vielstich, W., Gasteiger, H. A., and Lamm, A. (Eds.), *Handbook of Fuel Cells – Fundamentals, Technology and Applications. Volume 3: Fuel Cell Technology and Applications, Part 1*, Wiley, New York, NY, pp. 481–488.

Tang, H., Qi, Z., Ramani, M., and Elter, J. F. (2006) PEM fuel cell cathode carbon corrosion due to the formation of air/fuel boundary at the anode. Journal of Power Sources 158, 1306–1312.

Vogel, H. and Marvel, C. S. (1961) Polybenzimidazole, new thermally stable polymers. Journal of Polymer Science 50, 511–539.

Wainright, J. S., Litt, M. H., and Savinell, R. F. (2003) High-temperature membranes. In: Vielstich, W., Lamm, A., and Gasteiger, H. A. (Eds.), *Handbook of Fuel Cells. Fundamentals, Technology and Applications. Volume 3*, Wiley, New York, NY, pp. 436–446.

Wang, J.-T., Savinell, R. F., Wainright, J., Litt, M., and Yu, H. (1996a) A H2/O2 fuel cell using acid doped polybenzimidazole as polymer electrolyte. Electrochimica Acta 41(2), 193–197.

Wang, J.-T., Wainright, J. S., Savinell, R. F., and Litt, M. (1996b) A direct methanol fuel cell using acid-doped polybenzimidazole as polymer electrolyte. Journal of Applied Electrochemistry 26, 751–756.

Wang, J.-T., Wasmus, S., and Savinell, R. F. (1996c) Real-time mass spectrometric study of the methanol crossover in a direct methanol fuel cell. Journal of the Electrochemical Society 143(4), 1233–1239.

Xiao, L., Zhang, H., Scanlon, E., Chen, R., Choe, E.-W., Ramanathan, L. S., Yu, S., and Benicewicz, B. (2005a) Synthesis and characterization of pyridine-based polybenzimidazoles for high-temperature polymer electrolyte fuel cell applications. Fuel Cells 5(2), 287–295.

Xiao, L., Zhang, H., Scanlon, E., Ramanathan, L. S., Choe, E.-W., Rogers, D., Apple, T., and Benicewicz, B. (2005b) High-temperature polybenzimidazole fuel cell membranes via a sol-gel process. Chemical Materials 17, 5328–5333.

Yu, P. T., Gu, W., Makharia, M., Wagner, F. T, and Gasteiger, H. A. (2006) The impact of carbon stability on PEM fuel cell startup and shutdown voltage degradation. ECS Transactions 3(1), 797–809.

Direct Methanol Fuel Cell Durability

Yu Seung Kim and Piotr Zelenay

Abstract This chapter provides an overview of performance durability issues typically occurring in the direct methanol fuel cell (DMFC), in both single cells and short DMFC stacks. The focus of this chapter is on those sources of performance degradation that have been recognized as impacting DMFC operation in a major way (1) the loss of cathode activity due to surface oxide (hydroxide) formation, (2) ruthenium crossover from the anode to the cathode through the proton-conducting membrane, and (3) membrane–electrode interface degradation. Much attention is devoted to the interpretation of performance losses observed during extended operation of DMFCs under "realistic" DMFC operating conditions, including high-voltage cell operation. A separation of the anode and cathode performance losses is attempted whenever possible. Also addressed in this chapter are various methods of mitigating DMFC performance losses, either at the stage of membrane–electrode assembly design and fabrication or in an operating fuel cell.

1 Introduction

The direct methanol fuel cell (DMFC) has been extensively investigated in the past 15 years as a potential new source of electrical power for future generations. The high specific energy, easy fuel storage and delivery, and the convenience of use of a liquid fuel make the DMFC an attractive alternative to batteries in portable electronics and other low-power devices for civilian and military applications. DMFC systems are expected to outperform their primary market competitors, lithium-ion batteries in particular, at sufficiently long operating times. For example, a "reference" 20-W DMFC system, operating with already achievable

P. Zelenay (✉)
Materials Physics and Applications Division, Los Alamos National Laboratory
Los Alamos, New Mexico 87545, USA
e-mail: zelenay@lanl.gov

F.N. Büchi et al. (eds.), *Polymer Electrolyte Fuel Cell Durability*,

efficiency, is supposed to offer an energy-density/specific-energy advantage over the batteries after approximately 10 h of continuous operation.

Together with often complex aspects of DMFC system design (applicable to both "passive" and "active" DMFC system concepts), durable operation of the fuel cell stack and the ability to deliver the required power for several thousand hours are generally considered crucial to the ultimate large-scale commercial success of DMFCs. Other key requirements include cell (stack) operation at a highest achievable voltage and cathode operation at ambient pressure, low airflow, and natural air humidification.

An illustration of the impact that gradual loss in performance of a hypothetical DMFC has on cell operation is shown in Fig. 1. The cell is supposed to deliver 55 mW cm^{-2} in power density, which initially can be achieved by generating 0.10 A cm^{-2} in current density at 0.55 V (0%-loss polarization plot). To ensure constant power delivery, any loss in the cell performance, expressed as a percentage of the initial cell voltage, has to be compensated with higher current. As shown in Fig. 1, cell current density has to be increased to 0.11, 0.13, 0.17, and 0.20 A cm^{-2} for 10, 20, 27, and 30% voltage loss, respectively (see the squares on the "required-power line"). Polarization plots that correspond to different assumed voltage loss values indicate that the requited power density of 55 mW cm^{-2} can be generated up to and including a case of a 27% voltage loss. However, once the voltage performance loss is assumed to reach 30%, the system fails to generate the power needed, which is indicated by the lowest square on the "required-power line" staying above the 30%-loss polarization plot.

In addition to ultimately preventing the DMFC system from generating the required power, performance degradation always leads to a decrease in the total efficiency of the fuel cell, most typically via a decrease in voltage efficiency. In the example shown in Fig. 1, the assumed performance degradation leads to a decrease in voltage efficiency of the "55 mW cm^{-2} cell" from the initial 45% in the no-loss case to 22% following a 27% loss in the cell operating voltage.

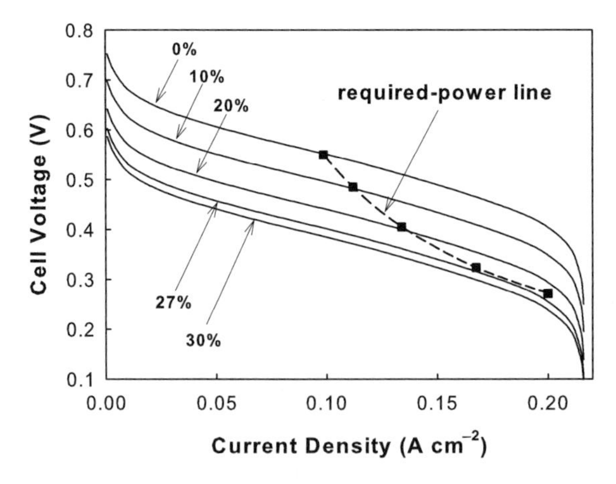

Fig. 1 Impact of voltage loss on current density in a hypothetical direct methanol fuel cell (DMFC) system designed to deliver 55 mW cm^{-2}. Voltage losses assumed are 10, 20, 27, and 30%

Although performance degradation of DMFC systems was realized long ago and often discussed in the past decade, it has still received relatively little attention compared with the degradation of H_2–air fuel cell systems. This may have to do with the fact that lifetime requirements for DMFC systems are generally perceived as less stringent than for H_2–air systems, those for automotive applications in particular. As importantly, lifetime requirements for DMFC systems, which predominantly target the ever-changing portable-electronics market, have not been defined as precisely as for H_2–air fuel cell systems, designed to replace the internal combustion engine in the very mature automotive market.

Although the routes of performance degradation in DMFCs bear some similarity to those found in H_2–air systems, the major causes of DMFC performance loss are often different owing to the use of different materials (e.g., anode catalyst) and cell operating conditions (liquid rather than gaseous fuel, generally lower fuel cell power, significant fuel crossover, etc.). However, like in the case of H_2–air fuel cells, understanding the performance degradation mechanism and development of methods that would either prevent or reverse (fully or in part) performance losses in an operating DMFC system are crucial to both the demonstration of practical viability and the commercialization potential of DMFC power systems.

Steady-state performance of a DMFC system is usually monitored in either constant-current or constant-voltage mode of operation. Constant-current mode better simulates operating conditions of a practical DMFC system, among others allowing for constant generation of water in the oxygen reduction reaction (ORR), which is suitable for studying performance degradation processes related to the formation of water at the cathode. Constant-voltage mode is generally more convenient in studies of those degradation processes that depend on the electrode potential, such as anode and cathode electrocatalysis. In some cases, DMFC life-testing is carried out under the conditions that are more extreme than the expected operating conditions of a practical system to shorten the time needed for certain degradation processes to take place and manifest themselves. Such "accelerated testing" is most commonly achieved by carrying out electrode reactions at high overpotential, elevated temperature, exceedingly low or high humidification level, and under voltage-cycling conditions.

Certain performance losses incurred by the cell during steady-state operation can be recovered, fully or in part, by stopping and then restarting a life test. Such "recoverable" performance losses are usually associated with reversible phenomena occurring in the fuel cell, for example, cathode catalyst surface oxidation, cell dehydration, and incomplete water removal from the catalyst layer and/or gas-diffusion layer (GDL).

DMFC performance losses that cannot be reversed to a significant degree in an operating cell are referred to as "unrecoverable" performance losses. They are usually caused by irreversible changes to the properties of cell components, e.g., decrease in the electrochemical surface area (ECSA) of catalysts, cathode contamination by crossover of ruthenium, and delamination of catalyst layer(s). Other causes of unrecoverable DMFC performance loss include membrane/ionomer degradation and irreversible hydrophobicity loss of the cathode GDL. The magnitude of unrecoverable performance loss can be determined by subtracting the current density (or cell voltage) measured after every cell performance recovery, usually requiring a break in the life test, from the current density (cell voltage) measured at the beginning of the life test.

In this chapter, we review several major routes of long-term performance losses affecting the catalysts, membrane–electrode assembly (MEA), and fuel cell stack, with special emphasis placed on (1) the loss of cathode activity due to surface oxide formation, (2) ruthenium crossover from the anode to the cathode through the electrolyte membrane and its impact on cathode electrocatalysis, and (3) membrane–electrode interface degradation. We also propose techniques for identifying sources of DMFC performance degradation (some in an operating fuel cell) and, if available, methods for mitigating various performance losses.

2 Catalyst Degradation

2.1 Loss of Cathode Activity Due to Surface Oxide Formation

To maintain maximum overall fuel conversion efficiency, the required power (current) should be delivered at the highest possible voltage (Fig. 1). This is a challenging requirement for DMFCs, which, unlike H_2–air fuel cells, suffer from significant anode overpotential. For as long as no major improvements to methanol-oxidation electrocatalysis are in sight, special care must be taken of the cathode to keep the voltage efficiency of the DMFC at the maximum achievable level. At the present state of DMFC technology that level is at approximately 45%, limited by impractically low cell current density at voltages higher than 0.55 V.

The high-voltage requirement translates into an operating potential of the DMFC cathode of at least 0.80–0.90 V versus the reversible hydrogen electrode. At that potential, surface hydroxide and/or oxide is slowly formed at the platinum catalyst, leading to a gradual decrease in cathode catalyst activity in the ORR. A typical performance loss due to the surface oxide formation at the platinum black cathode in a six-cell DMFC stack is shown in Fig. 2. The ensuing cell voltage drop occurs within the first approximately 30 h of the life test, up to 60 mV per cell. In a typical way for a recoverable performance loss in DMFCs, cell voltage can be brought back to the initial level following a break in the life test. The increase in cell voltage shown in Fig. 2 after approximately 100 h is caused by an interruption in cell operation and reduction of the cathode catalyst.

The process of oxide-layer growth on platinum has been thoroughly investigated for smooth platinum surfaces in aqueous electrolytes and in the gas phase (Angerstein-Kozlowska et al. 1973; Conway et al. 1990; Conway and Jerkiewicz 1992; Harrington 1997). While Conway et al. (1990) proposed rapid diffusion of oxide species followed by a slow oxide turnover process, Harrington (1997) opted for slow formation of the oxide species followed by rapid diffusion of oxide species across the surface. The kinetics of surface oxide formation on fuel-cell-type platinum catalysts has also been studied. Paik et al. (2004) observed that surface oxide formation on a platinum electrode occurs rapidly under realistic operating conditions of

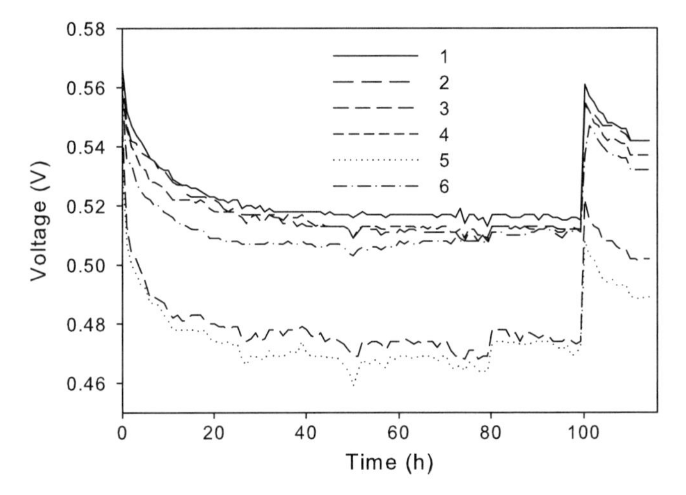

Fig. 2 Voltage change of six cells in a DMFC stack during 110h of a constant-current life test (80mA cm^{-2}). (Reproduced from Eickes et al. 2006)

the polymer electrolyte fuel cell, with much larger quantities of the oxide being generated in the presence of gas-phase molecular oxygen than in an oxygen-free atmosphere.

Eickes et al. (2006) suggested that the kinetics of surface oxide formation on nanocrystalline platinum in the absence of gas-phase oxygen follows the direct logarithmic growth law, which was established earlier for smooth platinum surfaces. On the basis of the measurements of oxide-stripping charges, the process of oxide formation was found to be complete after approximately 2h of oxidation, with restructuring of the oxide layer occurring at longer times. Although no new oxide was detected during the restructuring, the changes occurring in the cathode catalyst layers were found to impact the oxygen reduction rate via changes in the electronic properties of the catalyst rather than by mere blocking of the active catalyst sites.

DMFC performance loss due to the surface oxide formation of the cathode catalyst can be reversed via surface reduction. In the case of H_2–air fuel cells, the required decrease in the cathode potential can be realized by lowering the cell voltage with the help of an external voltage source (Uribe and Zawodzinski 2003). In the DMFC case, the use of an external power source is neither effective nor necessary. Because of intrinsic sluggishness of the anode process, an attempt to lower cell voltage predominantly leads to an increase in the anode potential (overpotential). However, thanks to methanol crossover, the potential of the DMFC cathode can be effectively reduced in an operating cell, even without an external power supply, using air break. In this method, air delivery to the cathode is interrupted, preferably in the constant-current mode of cell operation, which allows for faster consumption of all the remaining cathode oxygen than in the constant-voltage mode. Once all cathode oxygen has been used up, the potential is lowered as result of methanol crossover from the anode side of the fuel cell, within seconds reaching a value of 0.4V, needed for complete reduction of the surface platinum oxide. An effect of air break on the average power output of a single 22-cm^2 DMFC in a 140-h life test is shown in Fig. 3.

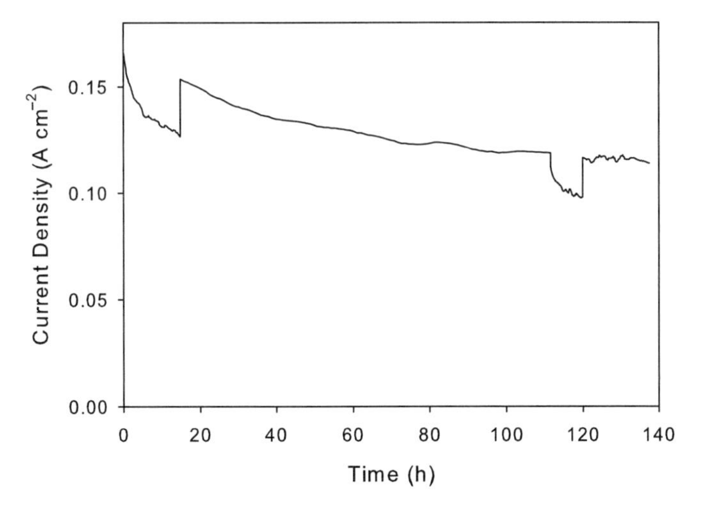

Fig. 3 Long-term performance of a 22-cm^2 air-pulsed DMFC (single cell). Cell voltage 0.5 V; cell temperature 80°C; anode 1.0 M CH_3OH, 3 mL min^{-1}; cathode air humidified at the dew point of 90°C; 200 mL min^{-1}, 0.7 atm. back pressure. "Air break" performed for 4 s every 100 s. (Reproduced from Eickes et al. 2006)

2.2 *Ruthenium Crossover*

Nanocrystalline Pt–Ru alloy with bulk atomic platinum-to-ruthenium ratio close to 1:1 has been widely perceived as the most active catalyst for electrochemical oxidation of methanol at elevated cell temperatures (Eickes et al. 2005). However, Pt–Ru catalysts are likely to suffer from the rather limited stability of the ruthenium in the alloy and the nonalloyed part of the catalyst (often present in large quantities), leading to possible loss of ruthenium from the anode. The ensuing crossover of ruthenium from the anode to the cathode and its impact on long-term performance of DMFCs were first reported by Piela et al. These authors observed that CO stripping scans from the DMFC cathode after a 600-h life test became characteristic of a Pt–Ru alloy rather than of pure platinum (Piela et al. 2004).

Figure 4 shows CO stripping scans recorded with a clean ruthenium-free cathode (dotted line) and after 24 and 600 h of DMFC operation. The peak potential of CO stripping from the cathode catalyst after prolonged cell operation shifts negatively by approximately 150 mV relative to CO stripping from a clean platinum surface. On the basis of the findings of Gasteiger et al. (1994), a negative shift of CO stripping peak from a binary Pt–Ru surface indicates an increase in the surface ruthenium content. The presence of ruthenium at the DMFC cathode was also evidenced by X-ray fluorescence (Piela et al. 2004), energy-dispersive X-ray spectroscopy (Gancs et al. 2007; Chen et al. 2006), and X-ray absorption spectroscopy (Sarma et al. 2007).

While not impacting the performance of the anode, at least for up to 3,000 h in cell operation (Harmon et al. 2006), leaching of ruthenium from the anode catalyst, followed by its accumulation at the cathode, has a negative effect on long-term DMFC performance. Fuel cell performance loss mainly results from a lower rate of the ORR on Pt–Ru alloys relative to pure platinum. Increasing the charge-transfer resistance of the ruthenium-contaminated cathode (Chen et al. 2006) and decreasing the cathode catalyst activity at higher ruthenium surface coverage by ruthenium were observed by AC impedance, rotating ring disk and H_2–air fuel cell experiments (Johnston et al. 2006). The total contribution of ruthenium crossover to unrecoverable performance loss in the DMFC has been measured from approximately 40 to 200 mV, depending on the degree of cathode contamination by ruthenium (Zelenay and Kim 2005). In addition to causing directly measurable ORR performance loss, the presence of ruthenium at the cathode promotes faster oxidation of the catalyst. The time by which the DMFC performance drops by 30% is noticeably shorter for a ruthenium-contaminated cathode than for a ruthenium-free platinum catalyst (Choi et al. 2005).

Two routes for ruthenium crossover have been proposed (1) currentless and (2) current-assisted. The currentless crossover mechanism is evidenced by a negative shift of CO stripping charge found to take place with even newly made MEAs. This route of contamination is responsible for a significant change of CO stripping charge after only 24 h of cell operation (Fig. 4). DMFC performance loss due to the currentless ruthenium crossover cannot be accurately estimated during normal DMFC operation because of the difficulties involved in decoupling the effect caused by ruthenium from catalyst activation (the latter having the opposite, i.e., positive, impact on DMFC performance). According to Piela et al. (2004), the currentless migration of ruthenium species can happen by pure diffusion during the MEA fabrication process. This mechanism is especially likely to occur with

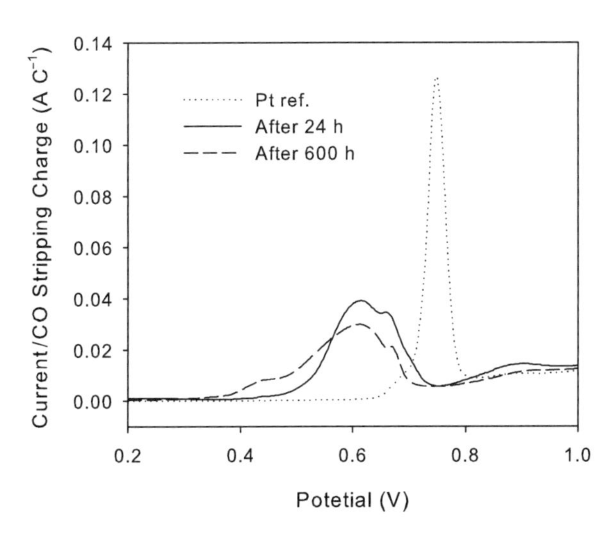

Fig. 4 Carbon monoxide stripping scan of a DMFC cathode at different times in a life test. The reference shown is for a pure platinum black electrode. (Reproduced from Choi et al. 2005)

Pt–Ru black catalysts that contain considerable amounts of nonalloyed ruthenium, the hydrous RuO_2 phase in particular (Dinh et al. 2000). RuO_2 was found to be in form of particles small enough (approximately 1 nm in size) to cross the membrane through hydrated Nafion® channels (approximately 5 nm in diameter) (Dmowski et al. 2002). Another currentless mechanism, proposed by Choi et al. (2005), involves ionic ruthenium species capable of migrating at the CO stripping voltage. This concept is supported by the absence of particles in field-emission scanning electron microscope images of the membranes and the potential dependency of ruthenium crossover rate. Ruthenium contamination of the cathode in a newly prepared MEA was found to be faster at a low anode potential of approximately 0.1 V than approximately 0.35 V (a more typical potential of DMFC anode operation).

Unlike the currentless crossover, the current-assisted mode of cathode contamination by ruthenium occurs under current and is enhanced by higher anode operating potential. In general, a gradual negative shift of the CO stripping potential as well as splitting of the initial single stripping peak into three peaks over the time of the life test are both indicative of the current-assisted ruthenium crossover. Although the Pourbaix diagram of ruthenium indicates that no $Ru(OH)_3$ should form below 0.7 V, the actual formation of ruthenium oxide in the Pt–Ru catalyst has been observed at potentials as low as 0.4 V, possibly owing to the alloying effects with platinum and the high operating temperature of the fuel cell (Jeon et al. 2007).

Several approaches have been used to limit ruthenium crossover in DMFCs, including (1) high-temperature cure of the anode catalyst by hot-pressing (Piela et al. 2004), (2) preleach of the "loose" ruthenium phase in an aqueous solution of inorganic acid (Piela et al. 2004), (3) use of low-permeability membranes (Choi et al. 2005), and (4) electrochemical removal of mobile ruthenium (Choi et al. 2005). Although not entirely preventing ruthenium contamination of the cathode, all these approaches lead to incremental lowering of currentless ruthenium contamination.

Comparison of the 140-h performance of two MEAs, one untreated and the other with mobile ruthenium removed in an electrochemical treatment before the life test, is shown in Fig. 5. The performance of the treated MEA is significantly better than that of the untreated one, yielding less than 1 mA cm^{-2} in unrecoverable performance loss and 67 mA cm^{-2} in recoverable performance loss. By comparison, the recoverable and unrecoverable performance losses of the untreated MEA are 17 and 81 mA cm^{-2}, respectively. It is worth remembering, however, that the treatment used in the case depicted in Fig. 5 is capable of limiting losses caused mostly by the currentless ruthenium contamination. Methods for limiting current-assisted contamination in DMFC systems with a Pt–Ru anode still need to be developed.

2.3 *Electrochemical Surface Area Loss*

Active ECSA loss due to the sintering of electrocatalyst particles has been commonly observed in H_2–air fuel cells and DMFCs (Jiang et al. 2005; Chen et al.

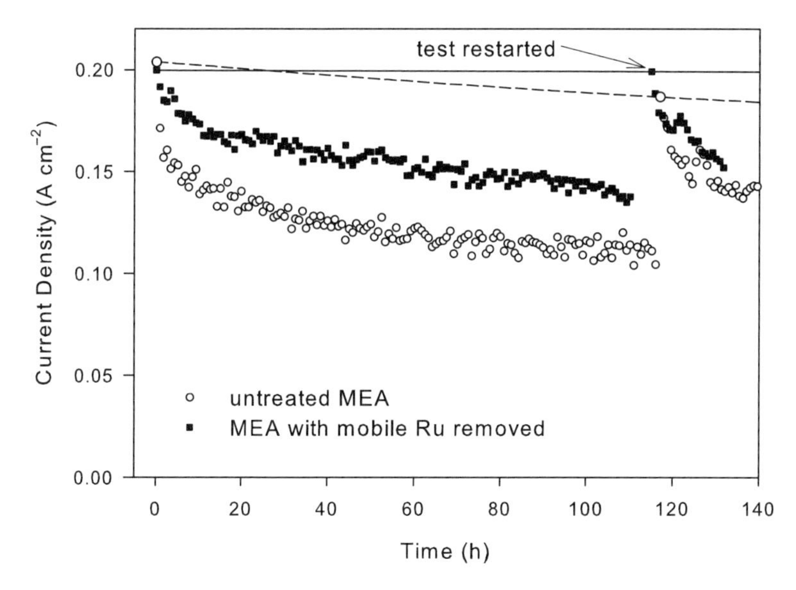

Fig. 5 Performance of an untreated and a treated membrane/electrode assembly (MEA) during a 140-h DMFC life test. Mobile ruthenium removal in the treated MEA was carried out by electrochemical means. (Reproduced from Choi et al. 2005)

2006). A noticeable sintering of DMFC catalysts has been identified not only in the cathode catalyst layer but also in the anode, where catalyst ripening is facilitated by the presence of liquid water (Liu and Wang 2007). Particle size growth in the anode and cathode catalysts during 500 h of DMFC operation, resulting in an ECSA loss of approximately 15%, is shown in Fig. 6.

Three sintering mechanisms have been proposed to explain ECSA loss of fuel cell electrocatalysts (1) dissolution/reprecipitation (Tseung and Dhara 1975; Watanabe et al. 1994; Antolini 2003; Ferreira et al. 2005), (2) migration of platinum particles Bett et al. 1976; Blurton et al. 1978; Wilson et al. 1993; Ferreira et al. 2005; Borup et al. 2006), and (3) carbon corrosion (Roen et al. 2004; Cai et al. 2006; Guilminot et al. 2007). Details of these mechanisms can be found in the chapter "High-Temperature Polymer Electrolyte Fuel Cells: Durability Insights."

In reality, the effect of ECSA change on DMFC durability may have limited direct impact on fuel cell performance thanks to the specific properties of many DMFC systems. These properties are (1) increased specific activity of larger catalyst nanoparticles, possibly compensating for the loss in the catalyst surface area (Kinoshita et al. 1973), (2) relatively high loading of DMFC catalysts making catalysts less "sensitive" to surface area losses, and (3) redeposition of dissolved platinum (ruthenium) species on the surface of catalyst nanoparticles, reducing the rate of catalyst loss (Yasuda et al. 2006).

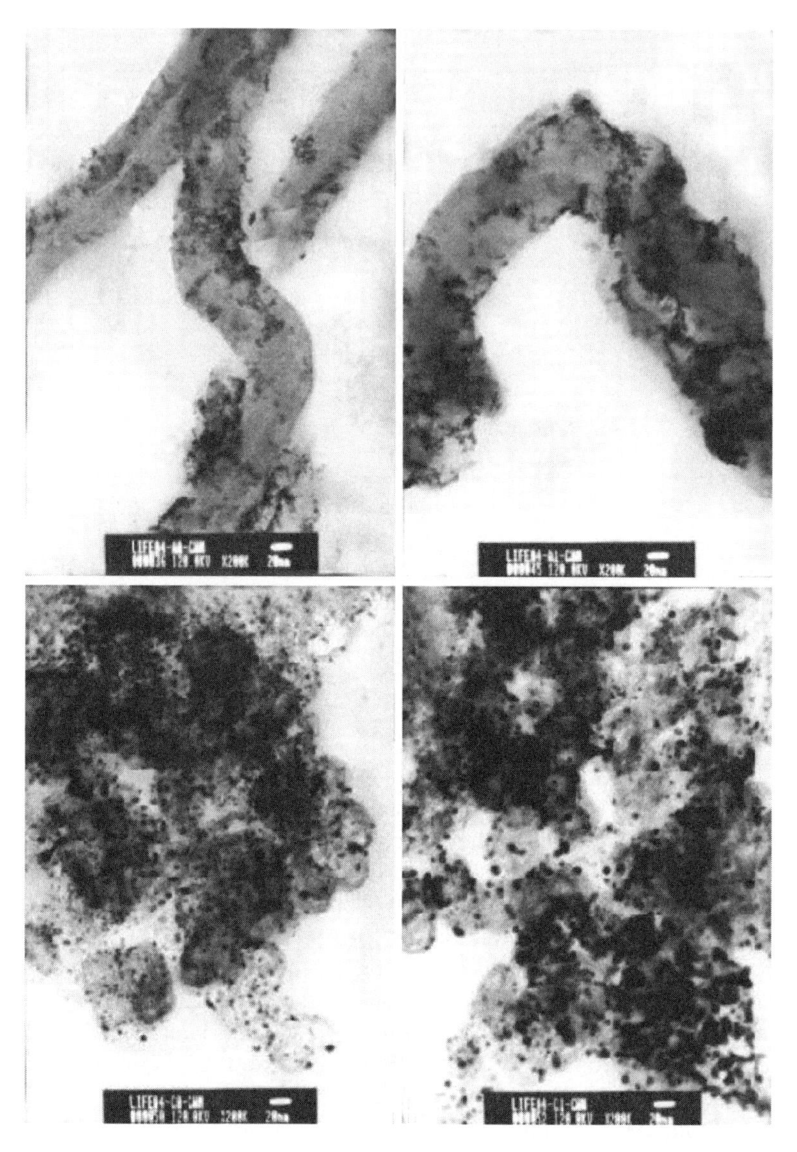

Fig. 6 Transmission electron microscope images of DMFC catalysts: (**a**) as-prepared Pt–Ru/carbon nanotube (CNT), (**b**) 50% Pt–Ru/CNT after a 500-h DMFC life test; (**c**) as-prepared Pt/C; (**d**) Pt/C after a 500-h DMFC life test. The *scale bar* is 20 nm. (Reproduced from Chen et al. 2006)

3 Membrane–Electrode Assembly Degradation

3.1 Membrane Degradation

Most commonly used membranes generally show acceptable performance durability under steady-state DMFC operation. Good membrane performance over thousands of hours under such conditions has been reported with perfluorinated sulfonic acid-based MEAs (Savadogo 1998; Hamon et al. 2006), poly(tetrafluoroethylene-*co*-hexafluoropropylene)-based radiation-grafted polystyrenes (Gubler et al. 2004), and sulfonated poly(arylene ether)s (Roziere and Jones 2003; Kim et al. 2004). However, membrane durability under dynamic operating conditions is still the subject of investigation. Membrane and/or ionomer dissolution in aqueous methanol solution is likely to lead to durability issues at high concentration of methanol in the anode feed stream. Dissolution of a significant fraction of Nafion® was shown to take place in 20 mol% methanol solutions within merely 1 week of DMFC operation (Siroma et al. 2004). Heat treatment of recast Nafion®, e.g., at 120°C for 1 h, was demonstrated to improve the resistance of the Nafion® ionomer to dissolution in methanol solutions. Such treatment induces structural changes on the ionomer, resulting in the growth of ion clusters and enhanced crystallinity of the electrolyte (Lee et al. 2004).

Any major membrane failures in polymer electrolyte fuel cells, including DMFCs, tend to lead to abrupt performance deterioration owing to massive reactant crossover. Such spectacular failures have not been commonly observed in DMFCs.

3.2 *Membrane–Electrode Interface Degradation*

Because of significant differences in the physicochemical properties of the membrane and the electrode, elevated temperature of fuel cell operation, and high flux of charged particles (together with ensuing potential difference) the membrane–electrode interface in all polymer electrolyte fuel cells is subject to degradation that often precedes performance loss of other MEA components. In the case of DMFCs, degradation has been observed predominantly at the anode–membrane interface. This is in a stark contrast to H_2–air fuel cells, which seldom suffer from interfacial failures, but when they do, the failures usually occur at the cathode–membrane interface. Jiang et al. (2007) observed interfacial delamination between Nafion® and a Nafion®-bonded anode after 5,000 h of DMFC operation that resulted in a gradual increase of the interfacial resistance and discharge performance degradation.

Interfacial degradation is especially common in DMFC systems operating with membranes that are alternatives to Nafion®. The main reason for the enhanced probability of an interfacial failure in such systems is poor compatibility between the non-Nafion polymer in the membrane and Nafion®-bonded electrodes (Kim et al. 2004).

A comparison of the long-term DMFC performance between MEAs with good and poor interfaces between the membrane and the electrode(s) is shown in Fig. 7. The cell with a good membrane–electrode interface shows lower unrecoverable performance loss than the cell with a poor interface, 42 versus 70 mA cm^{-2} after 700 h of operation. A greater performance loss of the cell with the poor interface is accompanied by an increase in the high-frequency resistance of the cell (Fig. 7). However, the increased cell resistance is only in part responsible for the total performance loss caused by interfacial degradation. *iR*-corrected H_2–air polarization

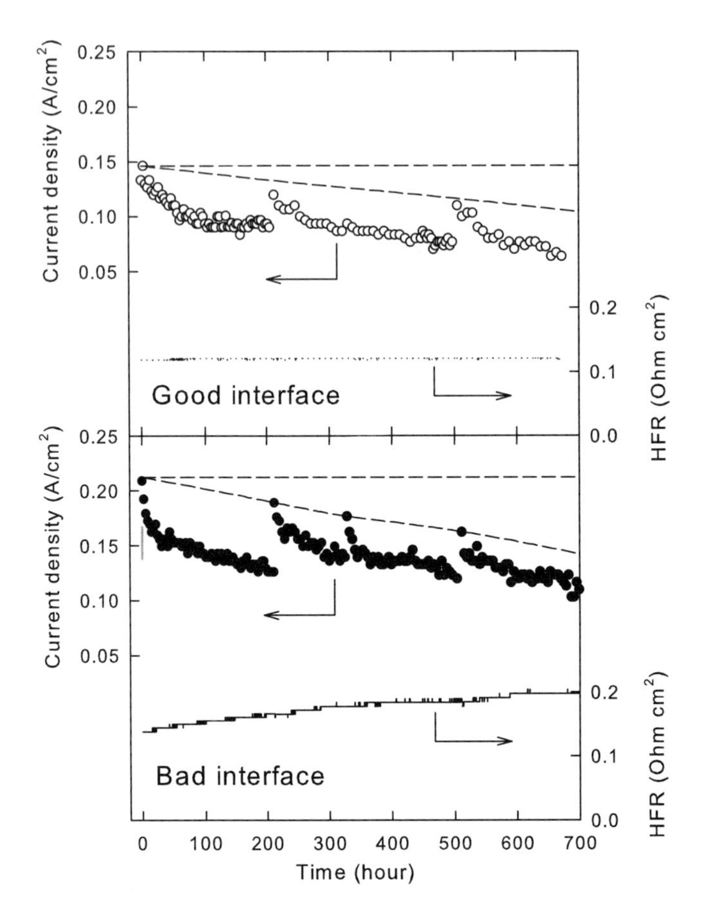

Fig. 7 Long-term performance of DMFC MEAs with good and poor membrane–electrode interface. Cell temperature 80°C; 0.5 M CH_3OH; fully humidified cathode. *HFR* High-frequency resistance

data, which provide methanol-crossover-free information on the cell performance, indicate that cells with a poor interface also suffer from decreased electrode performance, possibly owing to nonuniform current distribution in the catalyst layer (Pivovar and Kim 2007). Postmortem analysis indicates that MEAs with a good interface maintain interfacial integrity during the life test, whereas MEAs with a poor interface are susceptible to interfacial delamination.

Membrane–electrode interfacial compatibility is reflected by the interfacial resistance value. The measurement requires the interfacial resistance to be extracted from the total resistance of the fuel cell, especially that of the membrane. One way of determining the "nonmembrane" part of the cell resistance is from the intercept of a cell resistance plot as a function of membrane thickness (Fig. 8). The difference between the observed nonmembrane resistance and the electronic resistance of a single cell is the interfacial resistance. From the data shown in Fig. 8, it follows that

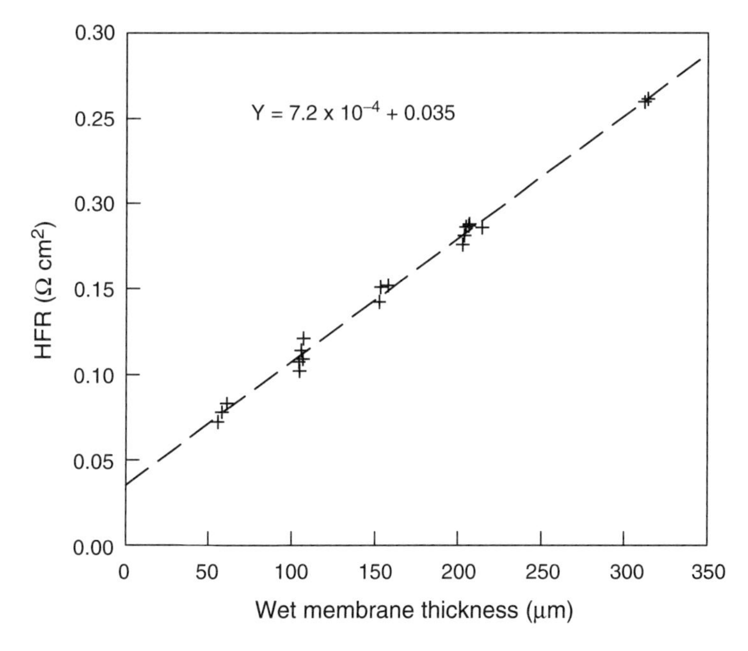

Fig. 8 High-frequency resistance (HFR) of a single DMFC operating with 1,100 equivalent weight Nafion® membranes as a function of membrane thickness. DMFC operation, 80°C. (Reproduced from Pivovar et al. 2007)

interfacial resistance of Nafion® with a Nafion®-bonded electrode is 8 mΩ cm². Similar experiments carried out with alternative membranes reveal much higher interfacial resistance, 17–43 mΩ cm². Interestingly, the resistance of a membrane–electrode interface involving a membrane made of recast Nafion® is 57 mΩ cm², significantly higher than that obtained with a commercial Nafion® membrane and Nafion®-bonded electrodes. This observation clearly points to a major role of membrane processing in membrane–electrode compatibility. In general, systems with an initial interfacial resistance lower than 20 mΩ cm² tend to show stable DMFC performance over thousands of hours.

AC impedance studies have indicated that DMFC performance loss due to interfacial failure can be linked to (1) increased ohmic resistance of the cell, (2) enhanced electrode overpotential, and (3) electrode flooding. Ohmic loss due to the interfacial resistance buildup can contribute several tens of milliohm square centimeters to the total cell resistance. An increase in the overpotential is caused by the loss of contact between the membrane and the catalyst, which renders part of the catalyst layer unusable (a possible major performance loss). Electrode flooding can be attributed to the nonuniform current distribution, which is more significant when membrane–cathode delamination occurs.

Although the exact mechanism of interfacial failure is yet to be determined, the adhesion/wetting properties and membrane water swelling are viewed as two main factors initiating the failure. Poor adhesion/wetting is largely related to the MEA fabrication process. The low temperature of thermal degradation of Nafion® on one hand and the high glass-transition temperature of polyaromatic copolymers on the other hand often limit the usefulness of the hot-pressing process, which consequently may fail to induce proper adhesion between the membrane and the electrode (Cho et al. 2003).

Interfacial failure during DMFC operation seems to be closely related to water swelling of the ionomer. Typically, the dimensional change of the polymer electrolyte membrane (PEM) under hydration is greater than that of the electrode. As a result, mechanical stress at the membrane–electrode interface is likely to initiate local delamination, which then expands further over the time of DMFC operation. Good correlation between membrane water uptake and the gain in cell resistance was demonstrated (Kim and Pivovar 2005).

Several different approaches were described in the literature to enhance the adhesion/wetting properties between the membrane and the electrode (1) electrophoretic deposition process, (2) plasma treatment, and (3) wet-glue process. In the electrophoretic deposition process, carbon particles and Nafion® ionomer are transferred along the potential gradient and deposited on the PEM (Morikawa et al. 2002). In the plasma treatment, plasma activation of the PEM is used to improve adhesion/wetting properties (Feichtinger et al. 2002; Bae et al. 2006). In the wet-glue method, a Nafion® precursor is introduced between the electrode and the PEM as an effective "binding agent" (Saha et al. 2004; Liang et al. 2006).

Reduction in water swelling of the membrane was achieved by incorporating metal oxide/phosphate particles (Silva et al. 2005) or by introducing chemical/physical cross-links into a PEM matrix (Kerres et al. 2004). Since these approaches to reducing membrane swelling also affect proton conductivity and transport of water and methanol in the membrane, balancing the effects brought about by the additives is critically important to the fuel cell performance.

4 Stack Performance Degradation

Uniform distribution of reactants to all cells in a stack is crucial for ensuring uniform stack performance. Nonuniformities in the reactant flow can result in fuel and/or air starvation in individual cells, leading to rapid increase in reaction overpotential, drop in the cell voltage, and possible cell reversal. In this regard, flow-field design could be critical to mass-transfer stability (Arico et al. 2000).

An important factor for stable stack performance is proper water management. Insufficient water supply to the cathode is likely to cause catalyst dry-out, resulting in an increase in the ORR overpotential. While severe membrane dehydration is unlikely to occur in liquid-feed DMFC systems, oversupply of water to the cathode is very common. The number of water molecules per proton (electro-osmotic drag coefficient) is much higher in DMFCs than in H_2–air fuel cells. In addition to the three H_2O molecules produced in oxygen reduction, as many as 18–20 molecules of H_2O are electro-osmotically dragged through the membrane to the cathode per methanol molecule oxidized at the anode. Incomplete water removal from the cathode results in the flooding of the catalyst and/or GDL, possible decrease in the active area of the cell, and reduction in the performance. Flooding of individual cell cathodes is the most common primary reason for the failure of DMFC stacks.

Although the importance of water management for DMFCs has been recognized (Scott et al. 1998; Oedegaard 2006), long-term performance loss due to poor water management is largely ignored in the open DMFC literature.

5 Summary

DMFCs experience recoverable (reversible) and unrecoverable (irreversible) performance degradation over their lifetime. The most detrimental performance losses originate in the MEA, especially in the catalyst layers and at the electrode–membrane interface.

Of the two electrodes, the anode, although made of a less stable Pt–Ru alloy, suffers lower performance loss over time than the cathode. The anode is, however, the source of mobile ruthenium species, either neutral or ionic, capable of permeating the proton-conducting membrane (ruthenium crossover) and depositing at the cathode, leading to a decrease in ORR activity of the platinum catalyst.

Cathode catalyst oxidation is another source of a major DMFC performance loss that is associated with electrocatalytic properties of platinum in aqueous media. Unlike the performance loss caused by ruthenium crossover, the loss due to the surface oxide (hydroxide) formation can be easily reversed via reduction, for example, by breaking the flow of air to the cathode.

Similarly to the catalysts in H_2–air fuel cells, DMFC catalysts lose their ECSA over time, predominantly owing to catalyst particle agglomeration. Once again, the performance loss is more pronounced at the platinum cathode, which tends to undergo frequent oxidation–reduction cycles, than at the anode (possible protective role of oxophilic ruthenium).

An important DMFC performance loss, which involves the membrane, is electrode delamination. The phenomenon, which is rooted in markedly different properties of the two layers, purely polymeric membrane and composite electrode, results in an increase in the cell resistance (enhanced ohmic loss), loss in the MEA area, and interfacial flooding. Expectably, membrane–electrode delamination is more likely in systems with different polymers in the membrane and catalyst layer.

In most cases, the end of the cell life is neither related to catalysis nor to the state of the membrane, but is caused by a major hydrophobicity loss of the cathode GDL (backing). Such loss leads to catastrophic flooding of the cathode catalyst and prevents oxygen from reaching catalytic sites.

In addition to suffering from all the single-cell performance losses, DMFC stacks incur performance degradation caused by poor reagent distribution to individual cells. As a consequence, mass-transfer limitation of the cathode performance (less than of the anode performance) takes place far more often in the stacks than in individual cells. By the same token, cathode flooding and carbon dioxide blinding of the anode are far more common in the stacks than in single cells.

Several methods have been developed over the years to mitigate performance losses in operating DMFCs. Some of the methods, such as cathode catalyst reduction to regain active sites lost owing to platinum oxidation, or several methods aimed at minimizing ruthenium crossover (including acid leach of "mobile" ruthenium species) have been quite successful.

References

Angerstein-Kozlowska, H., MacDougall, B. and Conway, B.E. (1973) Origin of activation effects of acetonitrile and mercury in electrocatalytic oxidation of formic acid. J. Electrochem. Soc. 120, 756–766.

Antolini, E. (2003) Formation, microstructural characteristics and stability of carbon supported platinum catalysts for low temperature fuel cells. J. Mater. Sci. 38, 2995–3005.

Arico, A.S., Creti, P., Baglio, V., Modica, E. and Antonucci, V. (2000) Influence of flow field design on the performance of a direct methanol fuel cell. J. Power Sources 91, 202–209.

Bae, B., Kim, D., Kim, H.J., Lim, T.H., Oh, I.H. and Ha, H.Y. (2006) Surface characterization of argon-plasma-modified perfluorosulfonic acid membranes. J. Phys. Chem. B 110, 4240–4246.

Bett, J.A.S., Kinoshita, K. and Stonehart, P. (1976) Crystallite growth of platinum dispersed on graphitized carbon-black. 2. Effect of liquid environment. J. Catal. 41, 124–133.

Blurton, K.F., Kunz, H.R. and Rutt, D.R. (1978) Surface area loss of platinum supported on graphite. Electrochim. Acta 23, 183–190.

Borup, R.L., Davey, J.R., Garzon, F.H., Wood, D.L. and Inbody, M.A. (2006) PEM fuel cell electrocatalyst durability measurements. J. Power Sources 163, 76–81.

Cai, M., Ruthkosky, M.S., Merzougui, B., Swathirajan, S., Balogh, M.P. and Oh, S.H. (2006) Investigation of thermal and electrochemical degradation of fuel cell catalysts. J. Power Sources 16, 977–986.

Chen, W.M., Sun, G.Q., Guo, J.S., Zhao, X.S., Yan, S.Y., Tian, J., Tang, S.H., Zhou, Z.H. and Xin, Q. (2006) Test on the degradation of direct methanol fuel cell. Electrochim. Acta 51, 2391–2399.

Cho, E.A., Ko, J.J., Ha, H.Y., Hong, S.A., Lee, K.Y., Lim, T.W. and Oh, I.H. (2003) Characteristics of the PEMFC repetitively brought to temperatures below 0 degree C. J. Electrochem. Soc. 150, A1667–A1670.

Choi, J.H., Kim, Y.S., Bashyam, R. and Zelenay, P. (2005) Ruthenium crossover in DMFCs operating with different proton conducting membranes. ECS Trans. 1, 437–445.

Conway, B.E. and Jerkiewicz, G. (1992) Surface orientation dependence of oxide film growth at platinum single-crystals. J. Electroanal. Chem. 339, 123–146.

Conway, B.E., Barnett, B., Angersteinkozlowska, H. and Tilak, B.V. (1990) A surface-electrochemical basis for the direct logarithmic growth law for initial stages of extension of anodic oxide films formed at noble metals. J. Chem. Phys. 93, 8361–8373.

Dinh, H.N., Ren, X.M., Garzon, F.H., Zelenay, P. and Gottesfeld, S. (2000) Electrocatalysis in direct methanol fuel cells: In-situ probing of Pt–Ru anode catalyst surfaces. J. Electroanal. Chem. 491, 222–233.

Dmowski, W., Egami, T., Swider-Lyons, K.E., Love, C.T. and Rolison, D.R. (2002) Local atomic structure and conduction mechanism of nanocrystalline hydrous RuO_2 from X-ray scattering. J. Phys. Chem. B. 106, 12677–12683.

Eickes, C., Brosha, E., Garzon, F., Purdy, G., Zelenay, P., Monta, T. and Thompsett, D. (2005) Electrochemical and XRD characterization of Pt–Ru blacks for DMFC anodes. *Electrochemical Society Series*, vol. 2002, pp. 450–467.

Eickes, C., Piela, P., Davey, J. and Zelenay, P. (2006) Recoverable cathode performance loss in direct methanol fuel cells. J. Electrochem. Soc. 153, A171–A178.

Feichtinger, J., Kerres, J., Schulz, A., Walker, M. and Schumacher, U. (2002) Plasma modifications of membranes for PEM fuel cells. J. New Mater. Electrochem. Syst. 5, 155–162.

Ferreira, P.J., La O', G.J., Shao-Horn, Y., Morgan, D., Makharia, R., Kocha, S. and Gasteiger, H.A. (2005) Instability of Pt/C electrocatalysts in proton exchange membrane fuel cells. J. Electrochem. Soc. 152, A2256–A2271.

Gancs, L., Hult, B.N., Hakim, N. and Mukerjee, S. (2007) The impact of Ru contamination of a Pt/C electrocatalyst on its oxygen-reducing activity. Electrochem. Solid-State Lett. 10, 15–154.

Gasteiger, H.A., Markovic, N., Ross, P.N. and Cairns, E.J. (1994) CO electrooxidation on well-characterized Pt–Ru alloys. J. Phys. Chem. 98, 617–625.

Gubler, L., Kuhn, H., Schmidt, T.J., Scherer, G.G., Brack, H.P. and Simbeck, K. (2004) Performance and durability of membrane electrode assemblies based on radiation-grafted FEP-*g*-polystyrene membranes. Fuel Cells 4, 196–207.

Guilminot, E., Corcella, A., Charlot, F., Maillard, F. and Chatenet, M. (2007) Detection of Pt^{Z+} ions and Pt nanoparticles inside the membrane of a used PEMFC. J. Electrochem. Soc. 154, B96–B105.

Hamon, C., Purdy, G., Kim, Y.S., Pivovar, B. and Zelenay, P. (2006) Novel process for improved long-term stability of DMFC membrane-electrode assemblies. *Proceedings – Electrochemical Society*, vol. P2004–21, pp. 352–362.

Harrington, D.A. (1997) Simulation of anodic Pt oxide growth. J. Electroanal. Chem. 420, 101–109.

Jeon, M.K., Won, J.Y. and Woo, S.I. (2007) Improved performance of direct methanol fuel cells by anodic treatment. Electrochem. Solid-State Lett. 10, B23–B25.

Jiang, L.H., Sun, G.Q., Wang, S.L., Wang, G.X., Xin, Q., Zhou, Z.H. and Zhou, B. (2005) Electrode catalysts behavior during direct ethanol fuel cell life-time test. Electrochem. Commun. 7, 663–668.

Jiang, R.Z., Rong, C. and Chu, D. (2007) Fuel crossover and energy conversion in lifetime operation of direct methanol fuel cells. J. Electrochem. Soc. 154, B13–B19.

Johnston, C.M., Choi, J., Kim, Y.S. and Zelenay, P. (2006) Towards understanding ruthenium crossover effects: the oxygen reduction reaction on Ru-modified platinum surfaces. *209th Electrochemical Society meeting, Denver, Colorado, May 07–May 12*, Abs. no. 1123.

Kerres, J., Ullrich, A., Hein, M., Gogel, V., Friedrich, K.A. and Jörissen, L. (2004) Cross-linked polyaryl blend membranes for polymer electrolyte fuel cells. Fuel Cells, 4, 105–112.

Kim, Y.S. and Pivovar, B. (2005) Durability of membrane–electrode interface under DMFC operating conditions. ECS Trans. 1, 457–467.

Kim, Y.S., Harrison, W.L., McGrath, J.E. and Pivovar, B.S. (2004) Effect of interfacial resistance on long term performance of direct methanol fuel cells. *205th Electrochemical Society meeting, San Antonio, Texas, May 9–13*, Abs. no. 334.

Kinoshita, K., Routsis, K., Bett, J.A.S. and Brooks, C.S. (1973) Changes in morphology of platinum agglomerates during sintering. Electrochim. Acta 18, 953–961.

Lee, K., Ishihara, A., Mitsushima, S., Kamiya, N. and Ota, K. (2004) Effect of recast temperature of diffusion and dissolution of oxygen and morphological properties in recast Nafion. J. Electrochem. Soc. 151, A639–A645.

Liang, Z.X., Zhao, T.S. and Prabhuram, J. (2006) A glue method for fabricating membrane electrode assemblies for direct methanol fuel cells. Electrochim. Acta 51, 6412–6418.

Liu, W.P. and Wang, C.Y. (2007) Three-dimensional simulations of liquid feed direct methanol fuel cells. J. Electrochem. Soc. 154, B352–B361.

Morikawa, H., Mitsui, T., Hamagami, J. and Kanamura, K. (2002) Fabrication of membrane electrode assembly for micro fuel cell by using electrophoretic deposition process. Electrochemistry 70, 937–939.

Oedegaard, A. (2006) Characterization of direct methanol fuel cells under near-ambient conditions. J. Power Sources 157, 244–252.
Paik, C.H., Jarvi, T.D. and O'Grady, W.E. (2004) Extent of PEMFC cathode surface oxidation by oxygen and water measured by CV. Electrochem. Solid-State Lett. 7, A82–A84.
Piela, P., Eickes, C., Brosha, E., Garzon, F. and Zelenay, P. (2004) Ruthenium crossover in direct methanol fuel cell with Pt–Ru black anode. J. Electrochem. Soc. 151, A2053–A2059.
Pivovar, B. and Kim, Y.S. (2007) The membrane–electrode interface in PEFCs: I. A method for quantifying membrane–electrode interfacial resistance. J. Electrochem. Soc. 154, B739–B744.
Roen, L.M., Paik, C.H. and Jarvi, T.D. (2004) Electrocatalytic corrosion of carbon support in PEMFC cathodes. Electrochem. Solid-State Lett. 7, A19–A22.
Roziere, J. and Jones, D.J. (2003) Non-fluorinated polymer materials for proton exchange membrane fuel cells. Annu. Rev. Mater. Res. 33, 503–555.
Saha, M.S., Kimoto, K., Nishiki, Y. and Furuta, T. (2004) A fabrication method for MEAs for PEFCs using Nafion precursor. Electrochem. Solid-State Lett. 7, A429–A431.
Sarma, L.S., Chen, C.H., Wang, G.R., Hsueh, K.L., Huang, C.P., Sheu, H.S., Liu, D.G., Lee, J.F. and Hwang, B.J. (2007) Investigations of direct methanol fuel cell (DMFC) fading mechanisms. J. Power Sources 167, 358–365.
Savadogo, O. (1998) Emerging membrane for electrochemical systems: (I) solid polymer electrolyte membranes for fuel cell systems. J. New Mater. Electrochem. Syst. 1, 47–66.
Scott, K., Taama, W. and Crulickshank, J. (1998) Performance of a direct methanol fuel cell. J. Appl. Electrochem. 28, 289–297.
Silva, V.S., Ruffmann, B., Silva, H., Gallego, Y.A., Mendes, A., Madeira, L.M. and Nunes, S.P. (2005) Proton electrolyte membrane properties and direct methanol fuel cell performance – I. Characterization of hybrid sulfonated poly(ether ether ketone)/zirconium oxide membranes. J. Power Sources 140, 34–40.
Siroma, Z., Fujiwara, N., Ioroi, T., Yamazaki, S., Yasuda, K. and Miyazaki, Y. (2004) Dissolution of Nafion membrane and recast Nafion film in mixtures of methanol and water. J. Power Sources 125, 41–45.
Tseung, A.C.C. and Dhara, S.C. (1975) Loss of surface-area by platinum and supported platinum black electrocatalyst. Electrochim. Acta 20, 681–683.
Uribe, F.A. and Zawodzinski, T.A. (2003) *Method for Improving Fuel Cell Performance*, US Patent # 6,635,369.
Watanabe, M., Tsurumi, K., Mizukami, T., Nakamura, T. and Stonehart, P. (1994) Activity and stability of ordered and disordered Co–Pt alloys for phosphoric acid fuel cells. J. Electrochem. Soc. 141, 2659–2668.
Wilson, M.S., Garzon, F.H., Sickafus, K.E. and Gottesfeld, S. (1993) Surface area loss of supported platinum in polymer electrolyte fuel cells. J. Electrochem. Soc. 140, 2872–2877.
Yasuda, D.A., Taniguchi, A., Akita, T., Ioroi, T. and Siroma, Z. (2006) Characteristics of a platinum black catalyst layer with regard to platinum dissolution phenomena in a membrane electrode assembly. J. Electrochem. Soc. 153, A1599–1603.
Zelenay, P. and Kim, Y.S. (2005) Performance degradation of DMFC MEAs and methods of improving their longevity. *Fuel Cells Durability – Stationary, Automotive, and Portable*. Knowledge Foundation, Washington DC, Dec. 8–9.

6
Bipolar Plates

Influence of Metallic Bipolar Plates on the Durability of Polymer Electrolyte Fuel Cells

Joachim Scherer, Daniel Münter, and Raimund Ströbel

Abstract This chapter describes the behavior and stability of metallic bipolar plates in polymer electrolyte fuel cell application. Fundamental aspects of metallic bipolar plate materials in relation to suitability, performance and cell degradation in polymer electrolyte fuel cells are presented. Comparing their intrinsic functional properties with those of carbon composite bipolar plates, we discuss different degradation modes and causes. Furthermore, the influence and possible improvement of the materials used in bipolar plate manufacturing are described.

1 Introduction

In most fuel cell setups, bipolar plates serve to separate individual electrochemical cells, which are consecutively attached together to from higher-voltage stacks. The main function of the bipolar plate is the reliable separation of the reaction gases (fuel vs. oxidant) and the most often required cooling medium. Intermixing of any of the fuel cell medium would generally lead to severe damage of the cells. Thus, secure and stable sealing within an individual cell as well as between adjacent cells is evident for operation and the long-term stability of the fuel cell stack (see Part 1, Chapter 7, paragraph 7). Beside their sealing function, bipolar plates also comprise channel structures on their outer faces, the so-called flow field, which are necessary to guide and distribute the reaction gases over the active area of the electrochemical cell. In the stack setup, the raised features of the flow field (land area) are in mechanical contact with the gas diffusion medium of the fuel cell. Via this contact, the bipolar plate conducts the electrons involved in the electrochemical reaction from and between the cells. To minimize power loss in the stack, the conductivity of the whole electronic pathway through the bipolar plate has to be optimized. Evident for the long-term stability and the main topic of this chapter is the fundamental stability

J. Scherer (✉)
European Fuel Cell Support Center, DANA Holding Corporation Sealing Products, Reinz-Dichtungs-GmbH, P.O. Box 1909, Neu-Ulm 89209, Germany
e-mail: joachim.scherer@dana.com

F.N. Büchi et al. (eds.), *Polymer Electrolyte Fuel Cell Durability*,

of the bipolar plate in a fuel cell environment, the absence of which will result in increased degradation of the fuel cell or even failure of the whole stack.

Metallic bipolar plates are regarded as the primary pathway to reduce overall stack cost, especially in automotive applications, in the transition to mass manufacturing. With established production technologies, such as stamping and joining, metallic bipolar plates can take advantage of today's high-volume, low-cost production of, e.g., automotive sealing parts. With common forming techniques, high-quality and high-precision parts can be produced at very high output rates. In addition, metallic bipolar plates offer a further benefit in terms of lower plate weight and thickness, thus leading to higher stack power density by weight and volume. In the following, the major counterargument against metallic bipolar plates, the intrinsic stronger tendency to corrosion, will be addressed and discussed.

2 Fuel Cell Operating Environment and Corrosion

Polymer electrolyte fuel cells (PEFC) commonly incorporate a chemically stable polymer membrane (e.g., perfluorinated hydrocarbons), which is coated with thin platinum-containing electrode films on both sides, to from the so-called membrane–electrode assembly (MEA). The polymer electrolyte is able to conduct protons by its intrinsic ability to absorb water owing to the acidic side groups attached to the polymer backbone. With the use of today's most common Nafion-type polymers, the proton conductivity of the membrane is directly linked to its water content. At elevated operating temperatures, the performance-determining conductivity of the polymer electrolyte thus has to be maintained by high water content in the fuel gases fed to the cell. In addition, with the electrochemical reaction, an excess of water is produced on the cathode of the MEA. Hence, the bipolar plate primarily has to deal with wet to sometimes condensing conditions in the gas compartments. Being adjacent to the acidic polymer membrane, the environment close to the MEA usually has a pH below 7, resulting in further corrosive attack on the bipolar plates.

During operation, as hydrogen is fed to the anode of the MEA, the electrochemical potential on this side of the bipolar plate is forced to 0 V versus the reversible hydrogen electrode (RHE). For most stainless steels, this potential is below the region of active corrosion (Fig. 1). At the air-fed side of the bipolar plate, the electrode potential of the cathode comes close to 1 V, which is, with respect to stainless steel, the beginning of the transpassive region. Thus, if the cell is at open cell voltage, or idling at low current densities, significant corrosive stress is applied to the metallic bipolar plate. On the other hand, if the fuel cell is operated under load, the anode potential could rise to 100 mV, approaching the region of active corrosion, while the cathode potential drops to values much lower than 1 V, which correspond to the region of intrinsic passivity. Under unconventional operating conditions, also during startup and shutdown of the stack, the electrodes can be polarized to extreme potentials, which pull the metallic bipolar plate into the regions of active or transpassive oxidation (Shores and Deluga 2003).

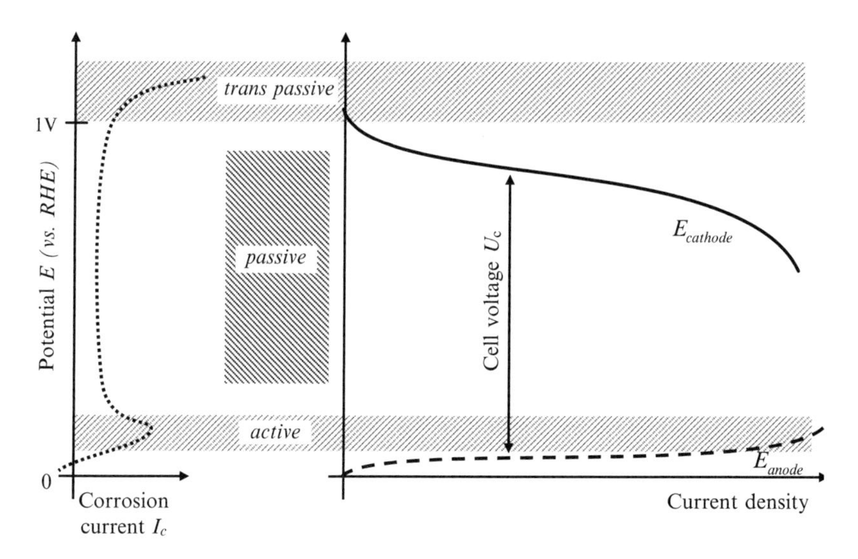

Fig. 1 Comparison of the current density–potential curve of stainless steel (*left*) and the anode and cathode polarization curve of a polymer electrolyte fuel cell (*right*). *RHE* reversible hydrogen electrode

Beside this, corrosive attack to the metallic bipolar plate is further supported by the common PEFC operating temperatures of 80–100°C. Elevated temperatures in particular increase the chemical attack on the polymer membrane. Especially if the membrane tends to dry out, the increasing formation of peroxides results in increasing attack on the membrane polymer (Cleghorn and Kolde 2007; Liu et al. 2001; Endoh et al. 2004). In the case of fluorinated polymers, membrane degradation results in continuous release of fluoride ions, which also support corrosive attack on the stainless steel (Ningshen and Kamachi Mudali 2002).

3 Base Materials for Metallic Bipolar Plates

Many different metallic materials have been proposed and discussed in the scientific and patent literature. Because of their low density, typical lightweight metals such as aluminum and titanium are often used to manufacture metallic bipolar plates (Mepsted and Moore 2003). But owing to difficulties in stamping thin aluminum and titanium foils, most of the known designs use chemically etched plates. In addition, these metals and most of their alloys exhibit only poor stability in acidic conditions, as well as a strong tendency to form very thick and nonconductive oxide layers. Thus, the use of aluminum or titanium in PEFCs requires electrochemically dense and corrosion-resistant coatings.

The aim of achieving very high bulk conductivity would favor copper or nickel as base materials. In both cases, the lack of electrochemical stability precludes the use of these metals without completely dense protective coatings. If only parts of the base material consisting of copper, nickel, or some nickel-based alloys are exposed to the acidic conditions in a PEFC, direct corrosion is likely to proceed, especially if the potential at the bipolar plate exceeds 500–700 mV (vs. RHE), in which case the electrochemical oxidation of the metals causes continuous dissolution of the metal. To stay within the scope of this chapter, we will concentrate in the following on the discussion of stainless steel as a base material for metallic bipolar plates. To some extent, most of the principles presented can also be transferred to other metals, taking into account the individual electrochemical properties of the material.

To withstand the conditions in a PEFC, it is necessary to choose a material which has a wide range of electrochemical passivity in the current density–potential curve for all operating conditions of the fuel cell (Fig. 2), otherwise, the material needs a strong self-healing mechanism for local overloads during operation. Another critical point for the use of metals in this corrosive atmosphere is to avoid a material mixture with different electrochemical potentials, which would cause contact corrosion. This effect is not only a problem for mixtures with a strong electrochemical potential gradient, as known, for example, in the anodic protection between zinc and steel, but also appears in the contact of two stainless steel materials such as AISI301 and AISI316L.

However, high corrosion resistance of the material is not the sole requirement for the use of stainless steel in fuel cells. Furthermore it is important to achieve a low contact resistance, good formability, and, as one of the most important aspects for the automotive industry, low cost. But some of these properties are contradictory, owing to their individual dependence on the composition of the alloy. For a stainless material it is necessary to alloy the steel with a minimum of 12–13 wt% chromium. This element is able to form and maintain a very thin oxide layer of about 3–4 nm on the surface of the steel. However, the layer can grow under oxidizing conditions up to 15 nm (George and Shaikh 2002). The layer mostly consists of a complex composition of chromium, oxygen, and few percent of other alloying elements such as nickel. The minimum content of chromium is not enough for the

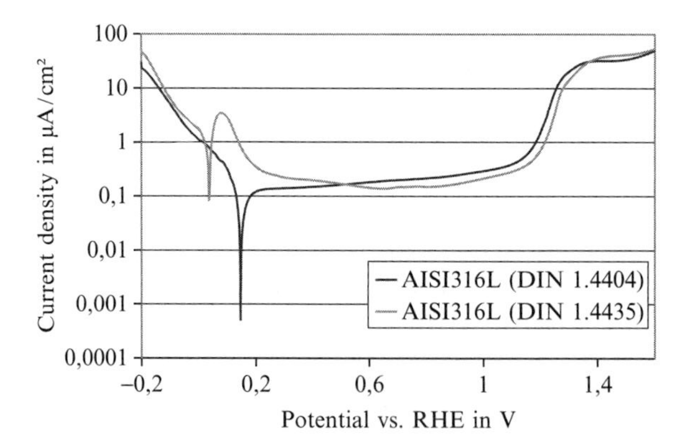

Fig. 2 Current density–potential curve of two 316L alloys in H_2SO_4 pH 2. Dynamic scan, 50 μV s^{-1}, resolution 2 mV, 25°C

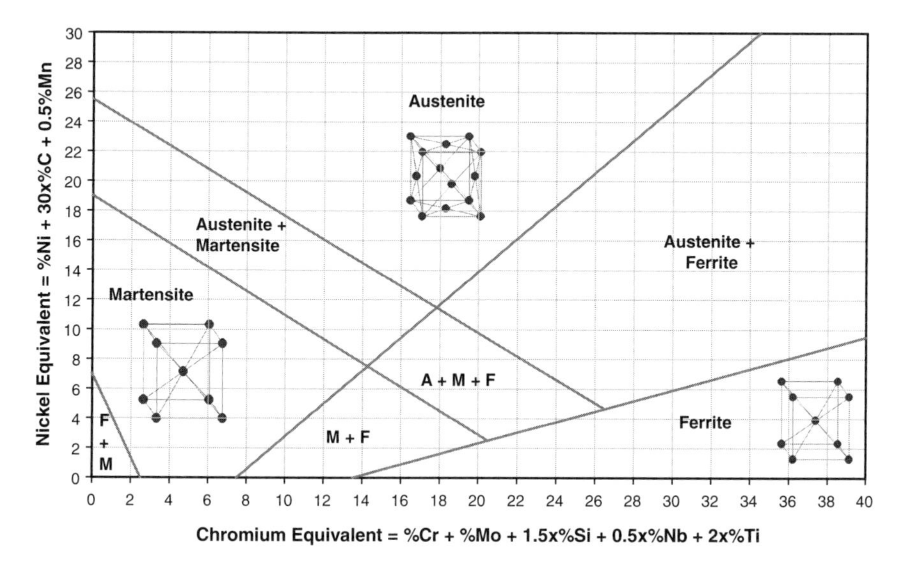

Fig. 3 Schaeffler diagram: relationship between alloying element contents and the crystallite structure of the resulting steel

necessary stability under fuel cell conditions. For that reason, it is essential to increase the content of alloying elements such as chromium, nickel and molybdenum and partially of nitrogen. Nevertheless, the increase of corrosion resistance thereby achieved leads to a disadvantageous increase in contact resistance by the thicker insulating oxide layer. Thus, it is obligatory to increase and stabilize the conductivity of the surface without reducing the corrosion stability, to achieve good performance and durability.

Beside the above-mentioned properties, good formability is another important requirement for suitable bipolar plate materials, and can only be provided by the cubic face-centered structure of austenite. This is due to the denser hexagonal close packing of the austenitic structure in comparison with the cubic base center structure of the ferrites. Through this closer packing it is possible to activate more sliding systems, which are essential for greater elongation and formability. To produce a stable austenitic structure, the steel must be alloyed with elements such as nickel and manganese. The correlation between the different alloying elements is shown in a Schaeffler diagram (Fig. 3).

4 Degradation of PEFC with Metallic Bipolar Plates

Degradation of fuel cell performance is caused by various influences on the members of the cell. Owing to the close proximity of a variety of materials and their electrical, chemical, and electrolytic connection, the degradation of the fuel cell is most often a complex interaction of many components in the PEFC. High and stable performance of the cell is particularly related to high and stable ionic and electrical conductivity of all electrochemically active components. Also the contact resistances between the numerous layered materials, from the bipolar plate via the gas-diffusion layer to the electrode and the membrane, directly contribute to parasitic losses within the cell setup. As mentioned before, all passivating metals such as stainless steel or titanium protect themselves from continuous corrosion by forming a stable and well-adhering surface oxide layer if they are in contact with oxygen. If those metals are used in an electrochemical setup, such as a fuel cell, the advantageous corrosion protection is associated with increasing contact resistance owing to the semiconductive surface oxides. To overcome increasing contact resistance between the metallic bipolar plate and the adjacent gas-diffusion medium, different measures are known to lower and stabilize the surface conductivity of the bipolar plate (see paragraph 6).

The direct degradation of the electrochemically active MEA can be enhanced by contaminants, which are released from surrounding parts such as the bipolar plates, the gaskets, or even materials within the medium loops close to the stack. In the case of metallic bipolar plates, the release of metal ions is contemplated as a material-specific degradation factor. As described before, most of the base metals used contain iron, nickel, chromium, and sometimes aluminum as predominant alloy elements. From ex situ experiments it was shown that even small amounts of metallic cations, deposited in the electrolyte membrane, cause a reduced number of vacant charge carriers (St-Pierre et al. 2000; Kelly et al. 2005; Pupkevich et al. 2007). By displacement of protons from the acidic sites in the polymer membrane and strong coordination to the sulfonic acid groups, a significant decrease in membrane conductivity is induced. Furthermore, some metal ions can also enhance the polymer degradation of the membrane (Pozio et al. 2003). Owing to this, the migration of dissolved metal ions from the bipolar plate or from proximate cell parts into the electrolyte membrane has to be avoided or at least minimized. But not only the amount of dissolved metal influences the quantity of cations in the membrane, also the availability of liquid water in the cell, forming electrolyte bridges from metal parts to the membrane, has a tremendous effect on the migration of cations into the electrolyte. By good control of humidification and other operating conditions, the formation of liquid water in the cell is limited and thus migration pathways of metal ions to the membrane are reduced. It was also shown that continuous discharge of dissolved ions from the cell, e.g., by exhaust water, has a great influence on minimizing the incorporation of cations into the membrane. Using corrosion-stable base materials, as well as controlling the corrosion influences on the bipolar plate, one can significantly reduce the release of metal ions.

5 Influence of Corrosion Initiators

There are many reports on high performance and stable operation of fuel cells with metallic bipolar plates under various operating conditions (Wind et al. 2003). Although even standard operation implies low pH, the presence of fluoride ions, and constant

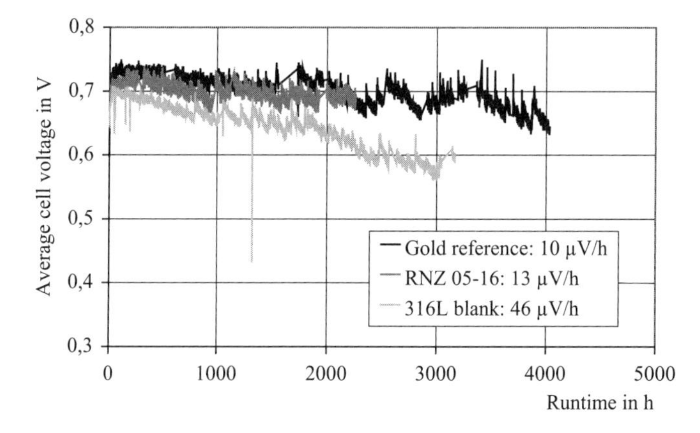

Fig. 4 Long-term operation of identical metallic bipolar plates: uncoated 316L, low-cost coating and gold plated. DANA standard conditions were as follows: constant current operation at 250 m cm^{-2}, 80°C, fully humidified, anode stoichiometry 2, cathode stoichiometry 5

wet to condensing conditions, common highly alloyed austenitic steels exhibit sufficient corrosion resistance in PEFC cells in situ. It was shown by various groups that even uncoated stainless bipolar plates could be operated for several thousand hours without severe corrosion (Fig. 4).

It is noteworthy that the performance of an uncoated metallic bipolar plate significantly decreases with increasing operation time; the main cause by far is the increasing contact resistance between the gas-diffusion layer and bipolar plate. The corrosion resistance of austenitic steels dramatically drops as soon as corrosion catalysts are present in the wet fuel cell environment. From many other, typical applications for stainless steel, it is well known in industry that corrosion initiators such as halides, which are ubiquitous, start and propagate steel corrosion. Even small traces of chloride can lead to local breakdown of the protecting chromium oxide layer of stainless alloys and result in severe pitting corrosion. Continuous contact with halide ions can cause persistent dissolution of metal ions from the steel into the contacting liquid water. The influence of chloride traces as a contaminant in the cell on the corrosion behavior of metallic bipolar plates was detected by the following model experiment.

In a comparison of two stack runs under identical operating conditions, the influence of halide traces in the gas compartments on the degradation of metallic bipolar plates becomes obvious (Fig. 5). As a baseline, one stack, using uncoated 316L metallic bipolar plates, was operated for 2,500 h using standard clean conditions. A typical degradation rate of 38 $\mu V\ h^{-1}$ was achieved. During the operation of the second stack with identical, uncoated 316L bipolar plates, every 100 h, approximately 30 ∝mol chloride was injected into both gas streams fed to the stack. After 690 h the stack reached the defined end-of-life criteria, showing an overall degradation of 65 $\mu V\ h^{-1}$.

After disassembly of the stacks, the effect on the bipolar plates was analyzed. If standard operating conditions are used, commonly no to only minor signs of corrosion

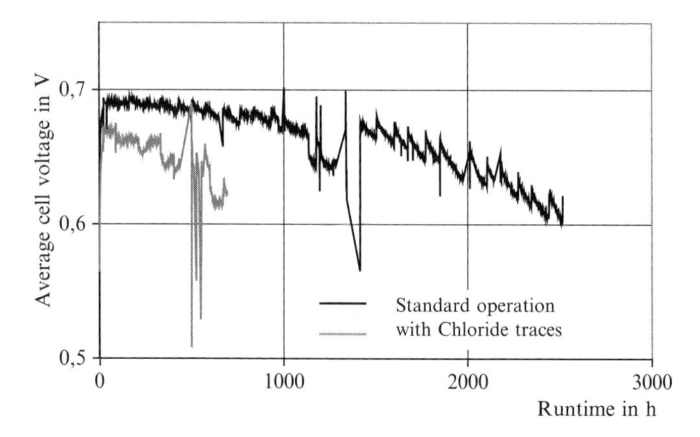

Fig. 5 Influence of chloride on uncoated 316L (DANA standard operating conditions) versus subsequent addition of 200 μmol chloride to both gas streams

are visible on the bipolar plates, predominantly located in the areas of condensing water in the flow field. Compared with this observation, the contaminated stack exhibited apparent corrosion stains, distributed all over the active area. Microscopic inspection of the bipolar plate surface unveiled severe pitting corrosion.

For the use of metallic bipolar plates in PEFC application, as a consequence, the choice of all materials used in the cell and stack setup has to be well decided. Some materials which work fine with composite bipolar plates might affect the stability of metallic bipolar plates. As an example, the influence of four different silicone gasket materials is shown in Fig. 6. In this ex situ experiment, five 316L stainless steel strips of the same size and weight were separately immersed in dilute sulfuric acid for 500 h at boiling temperature. To four of the samples, different coarsely cut silicone slabs of identical weight (RNZ 03-14, RNZ 03-10, RNZ 03-05, RNZ 03-13) were added before the test. Trace analysis of the sulfuric acid after the immersion showed a comparable number of metal ions released (per gram of steel) from the 316L for most of the gasket materials, indicating that no corrosion initiators are leached out of the silicones. Only silicone RNZ 03-14 (Fig. 6) caused significantly higher dissolution of chromium, nickel, and iron, owing to the release of corrosion catalysts from the gasket material. The rightmost column in Fig. 6, as a reference, shows the result of the steel sample without any silicone present.

6 Surface Treatment and Coating of Metallic Bipolar Plates

Corrosion-resistant stainless steels are generally able to withstand the acidic fuel cell conditions owing to their intrinsic property to from stable and well-adhering surface oxides, protecting the bulk material from further corrosive attack. On the other hand, with the growing chromium oxide layer, the contact resistance of

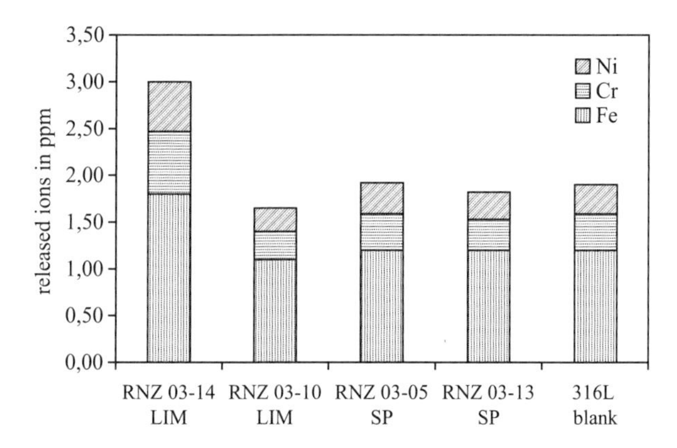

Fig. 6 Influence of contaminants released from silicone gasket materials on the corrosion behavior of 316L. Ex situ immersion test H_2SO_4, pH 2, 500 h, 100°C. *LIM* liquid injection moldable, *SP* screen printable

the stainless steel bipolar plate also grows, owing to the poor conductivity of the chromium oxide layer. To overcome this predominant factor of performance degradation, the surface of stainless steel bipolar plates can be modified or coated to minimize and stabilize its surface or contact resistance. Various techniques have been developed to reduce the formation of the chromium oxide layer (Hermann et al. 2005; Tawfik et al. 2007).

The implantation of nitrogen into the stainless steel surface, e.g., via thermal nitration (Brady et al. 2004, 2007), leads to significantly higher conductivity, which shows excellent stability. Moreover, even the corrosion resistance of the bipolar plate is noticeably increased by surface nitration. With especially high chromium content, most often nickel-based stainless alloys show very good behavior during thermal nitration by forming thin but dense chromium nitride layers. Various materials, such as Ni-50Cr or Hastelloy G35 (Ni-30Cr) have been successfully nitrided, taking advantage from the ability of Ni–Cr alloys to withstand nitrogen embrittlement owing to the good solubility of chromium in the nickel base. With presently increasing costs for alloy metals, especially for nickel, the use of nickel-based bipolar plate materials consequently leads to a increase in the overall cost of the bipolar plate. This has to be encountered as a serious argument, particularly in automotive applications. Recently, different groups have worked on nitration of less expensive iron-based alloys, which help to achieve DOE cost targets (Brady et al. 2006; Tian et al. 2007)

Various ceramic coatings which are mainly developed for tribological protection of tools have also been tested on PEFC bipolar plates (Fig. 7; RNZ 05-06). Metal nitrides such as chromium nitride, titanium nitride, or titanium aluminum nitride exhibit good electrochemical stability along with acceptable electrical conductivity (Cho et al. 2005; Li et al. 2004). Application of several nanometers to a few micrometers of such a ceramic coating leads to significantly reduced contact resistance of the steel surface. Compared with thermal nitration as described before, the discrete formation of a chromium nitride layer is viable also on more cost-competitive alloys with lower

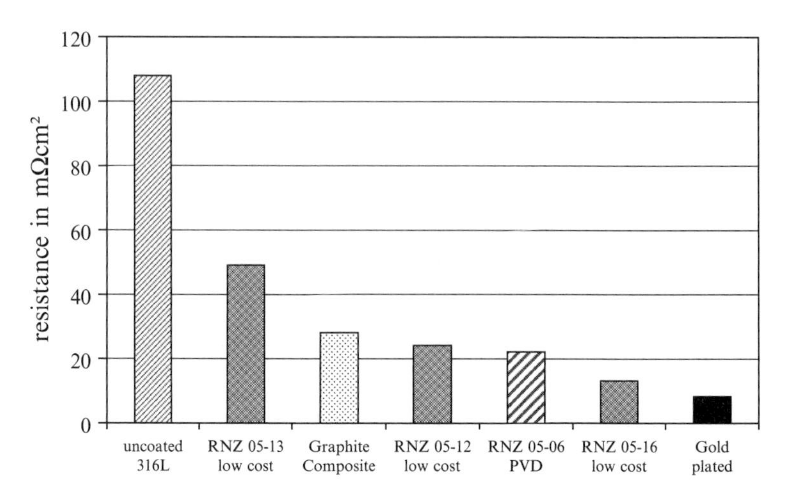

Fig. 7 Contact/through-plate resistance of identical metallic bipolar plates with different coatings between two gas-diffusion layers. The uncoated, gold-coated, and graphite composite plates are included as references

chromium content in the bulk. Therefore, the protective nature of the nitride layer helps to increase bipolar plate performance and stability. Different physical vapor deposition processes are known to apply a chromium nitride layer even on complexly shaped metallic parts. Advantageous to the different, most often variable, and hence well-controllable deposition processes is the ability to vary the constitution, microstructure, thickness, and formation rate of the nitride layer. By fine-tuning of the different parameters and thus formation of an optimized coating composition, the stability of the bipolar plate is further improved (Budinski et al. 2006; Steinwandel et al. 2005).

Most commonly for prototype-level production of metallic bipolar plates, the application of layers of noble metals such as gold is used to ensure high surface conductivity and low performance degradation (Figs. 7 and 8). With standard electroplating techniques, even thicknesses below 1 ∝m are achievable, which is necessary to keep the plate production costs low. Owing to the typical precipitation properties of gold during electroplating, the deposited gold layer starts from small, discrete clusters. Via growing to larger islands, which fill the uncoated areas in-between, finally a closed noble metal surface is formed on the base material. Because the initially "closed" gold layer is still only a few tens of nanometers thick, it is not yet thick enough to be regarded as "electrochemically dense." Hence, a typical layer thickness of several hundred nanometers of gold is required to at least ensure stable operation of coated metallic bipolar plates.

To overcome the cost issue especially for high-volume production, low-cost coatings are required which avoid the use of expensive ingredients, without jeopardizing the performance and durability of the bipolar plate. Various materials are known from the literature that implement conductive polymers or polymer-based coatings. Already used in the emerging industry of "plastic electronics," stable and intrinsically conductive polymers have been used by various groups to apply a

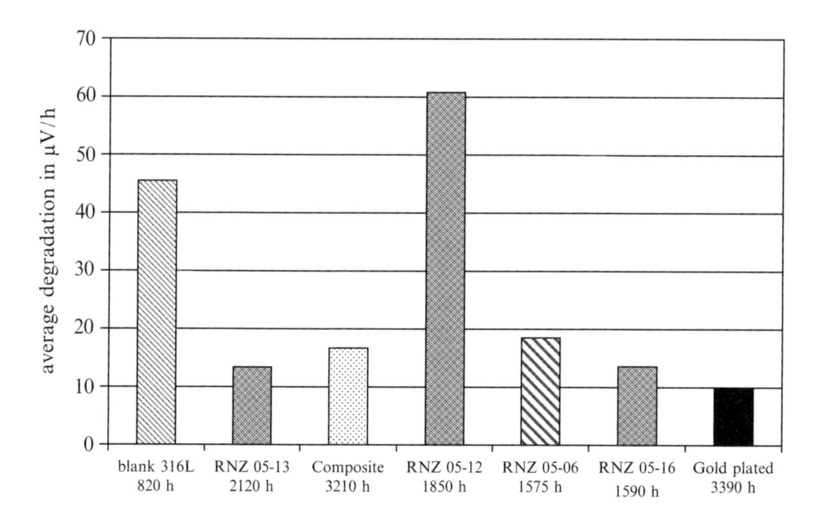

Fig. 8 Degradation of metallic bipolar plates with different coatings under constant current operation at 250m cm^{-2}, 80°C, fully humidified, anode stoichiometry 2, cathode stoichiometry 5

protective and conductive coating on metallic bipolar plates (Joseph et al. 2005). Although their conductivity and ability to protect the base material from corrosion are promising, the stability of the organic conductors under fuel cell conditions and the potential range have yet to be improved.

Figures 7 and 8 illustrate the results of a coating benchmark, which shows clear differences in performance and durability, which depend on the coating technique and composition. With different coatings the conductivity and durability of the metallic bipolar plate exceed the properties of a composite plate with comparable flow field. Diverse low-cost coatings (RNZ 05-06, RNZ 05-12, RNZ 05-13, and RNZ 05-16), which differ mainly in their composition, highlight the necessity to control the chemistry and electrochemistry of the complete system of plate and coating. It was shown that with optimized coating composition (RNZ 05-16) it is possible to approximate the performance and durability of the gold-coated reference plate.

7 Comparison with Carbon Composite Bipolar Plates

Although they lack metallic content, composite bipolar plates also undergo decomposition from corrosion-like processes. Composite bipolar plates predominantly contain carbon particles in the form of graphite or carbon black to ensure high conductivity. To achieve formability and structural stability in the molded state, the conductive fillers are embedded in thermoplastic binders (e.g., polypropylene, polyvinylidene fluoride) or thermosetting resins (e.g., phenolics, polyesters, epoxies). The composition of the composite material has to bring contrary properties into agreement: High conductivity requires an increased percentage of conductive particles, while higher binder content leads to decreasing gas permeability of the bipolar plate. Optimized composites always have to deal with some trade-offs from either of these properties (Wu and Shaw 2004). Also increasing the plate conductivity by reducing the plate thickness is limited to the minimum required material thickness, which ensures sufficiently low gas permeation through the bipolar plate. Moreover, if a thin composite plate is affected by material degradation, increasing gas breakthrough into the cooling medium could lead to dangerous operation modes of the fuel cell system.

As for metallic bipolar plates, the wet and acidic environment during fuel cell operation can also affect composite materials. Depending on the binder used, the plastic content of the composite is subject to absorption of water, which could lead to mechanical stress to the bipolar plate structure owing to minimal swelling of the material. As some of the polymer binders used contain hydrolyzable groups, such as esters, also the polymer backbone can be affected by the prevailing conditions of low pH, condensing water, and elevated temperature. Apart from this, the intermediate formation of peroxides during exceptional operating conditions could lead to slow decomposition of the composite polymer. This results in increasing porosity of the binder and ultimately the bipolar plate. Furthermore, the decomposition of the polymer binder also enhances the embrittlement of the composite material, which finally would affect the stability of the bipolar plate. Also the carbon content, which is usually considered as corrosion-stable, is accessible to oxidation by high electrode potentials and peroxides. Surface oxidation of the graphite particles results in higher contact resistance and could also change the surface properties of the bipolar plate. Introducing oxygen species into the topmost layers of the bipolar plate changes also the wettability of the plate surface.

8 Conclusions

To achieve the cost targets for high-volume production of fuel cell systems, the metallic bipolar plate will contribute a significant proportion at the stack level. In recent years, many companies and research groups made important steps forward, which verified the high performance and necessary stability of metallic bipolar plates in PEFC applications (Sommer et al. 2007; Brachmann 2004). By using corrosion-stable base materials, highly conductive but corrosion-stable coatings, as well as carefully controlling the chemical compatibility of all materials in the stack and the subsystem, one can accomplish high and stable performance of the stack.

References

Brachmann, T. (2004) Oral presentation The Honda FCX Hybrid Fuel Cell Vehicle at *Ninth Ulm Electro Chemical Talks (UECT)* May 17–18, Ulm Germany.

Brady, M.P., Weisbrod, K., Paulauskas, I., Buchanan, R.A., More, K.L., Wang, H., Wilson, M., Garzon, F., and Walker, L.R. (2004) *Scripta Materialia* 50, 1017–1022.

Brady, M.P., Yang, B., Wang, H., Turner, J.A., More, K.L., Wilson, M., and Garzon, F. (2006) *Journal of Metals* 58, 50–57.

Brady, M. P., Tortorelli, P.F., Pihl, J., More, K.L., and Meyer, H.M. (2007) *DOE Annual Progress Report*, pp. 726–727.

Budinski, M. K., Vyas, G., Kunrath, A. O., and Moore, J. J. (2006) *Patent* WO 2006/036241

Cho, E. A., Jeon, U.-S., Hong, S.-A., Oh, I.-H., and Kang, S.-G. (2005) *Journal of Power Sources* 142, 177–183.

Cleghorn, S. and Kolde, J. (2007) *Abstracts of Oral Presentations Held at the 2007 Fuel Cell Seminar*, San Antonio TX, USA, p. 179.

Endoh, E., Terazono, S., Widjaja, H., and Takimoto, Y. (2004) *Electrochemical and Solid State Letters* 7, A209–A211.

George, G. and Shaikh, H. (2002) in: Khatak, H. S. and Raj, B. (Eds.), *Corrosion of Austenitic Stainless Steels*, Narosa, New Delhi, pp. 1–36.

Hermann, A., Chaudhuri, T., and Spagnol, P. (2005) *International Journal of Hydrogen Energy* 30, 1297–1302.

Joseph, S., McClure, J.C., Chianelli, R., Pich, P., and Sebastian, P.J. (2005) *International Journal of Hydrogen Energy* 30, 1339–1344.

Kelly, M.J., Fafilek, G., Besenhard, J.O., Kronberger, H., and Nauer, G.E. (2005) *Journal Power Sources* 145, 249–252.

Li, M., Luo, S., Zeng, C., Shen, J., Lin, H., and Cao, C. (2004) *Corrosion Science* 46, 1369–1380.

Liu, W., Ruth, K., and Rusch G. (2001) *Journal of New Materials for Electrochemical Systems* 4, 227–231.

Mepsted, G. O. and Moore, J. M. (2003) in: Vielstich, W., Gasteiger, H. A., Lamm, A. (Eds.), *Handbook of Fuel Cells*, Wiley, New York, NY, pp. 385–430.

Ningshen, S., and Kamachi Mudali, U. (2002) in: Khatak, H. S. and Raj, B. (Eds.), *Corrosion of Austenitic Stainless Steels*, Narosa, New Delhi, pp. 37–73.

Pozio, A., Silva, R.F., De Francesco, M., and Giorgi (2003) *Electrochimica Acta* 48, 1543–1549.

Pupkevich, V., Glibin, V., and Karamanev, D. (2007) *Journal of Solid State Electrochemistry* 11, 1429–1434.

Shores, D.A. and Deluga, G.A. (2003) in: Vielstich, W., Gasteiger, H. A., Lamm, A. (Eds.), *Handbook of Fuel Cells*, Wiley, New York, NY, pp. 273–285.

Sommer, M., Woehr, M., and Docter, A. (2007) *Abstracts of oral presentations held at the 2007 Fuel Cell Seminar*, San Antonio TX, USA, pp. 180–183.

Steinwandel, J., Späh, R., and LaCroix, A. (2005) *Patent* DE 10139930B4.

St-Pierre, J., Wilkinson, D. P., Knights, S., and Bos, M. (2000) *Journal of New Materials for Electrochemical Systems* 3, 99–106.

Tawfik, H., Hung, Y., and Mahajan, D. (2007) *Journal of Power Sources* 163, 755–767.

Tian, R. J., Sun, J. C., and Wang, L. (2007) *Journal of Power Sources* 163, 719–724.

Wind, J., LaCroix, A., Braeuninger, S., Hedrich, P., Heller, C., and Shudy, M. (2003) in: Vielstich, W., Gasteiger, H. A., Lamm, A. (Eds.), *Handbook of Fuel Cells*, Wiley, New York, NY, pp. 385–430.

Wu, M. and Shaw, L. L. (2004) *Journal of Power Sources* 136, 37–44.

Durability of Graphite Composite Bipolar Plates

Tetsuo Mitani and Kenro Mitsuda

Abstract Highly graphite filled polymer composites were developed for use as bipolar plates in polymer electrolyte fuel cells (PEFCs). For use in PEFCs, composites should possess excellent durability in a hot and humid environment in addition to high electrical conductivity and good mechanical properties. Therefore, the stability of different composites in hot water was estimated by comparison with the initial properties and target values. On the basis of this comprehensive estimation, we obtained thermosetting composites for compression molding and thermoplastic composites for injection molding that enabled the production of precise bipolar plates. A PEFC stack assembly using the composite bipolar plates showed good, stable performance comparable to that of conventional machined graphite plates.

1 Introduction

For most polymer electrolyte fuel cell (PEFC) systems, unit cells are combined into a PEFC stack to achieve the voltage and power output level required for the application. The unit cells normally use a membrane electrode assembly, consisting of a solid polymer electrolyte with an associated gas-diffusion layer sandwiched between two bipolar plates. The bipolar plates make connections over the entire surface of one cathode and the anode of the next cell. These bipolar plates serve several functions simultaneously: (1) they create channels for fuel, air, and water, (2) they separate the unit cells in the stack, (3) they carry current away from the cell, and (4) they support the membrane electrode assembly (Lee et al. 2006; Shao et al. 2007).

Graphite is commonly used to make a bipolar plate. While graphite is electrically conductive and resistant to corrosion in the PEFC environment, the cost of both the material itself and the machining of each bipolar plate is high. To reduce the cost,

T. Mitani (✉)
Advanced Technology R&D Center, Mitsubishi Electric Corporation, 8-1-1, Tsukaguchi-Honmachi, Amagasaki, Hyogo 661-8661, Japan
e-mail: mitani.tetsuo@dr.mitsubishielectric.co.jp

F.N. Büchi et al. (eds.), *Polymer Electrolyte Fuel Cell Durability*,

graphite composite bipolar plates have been developed. These are made from a combination of graphite filler and a polymer matrix by a conventional processing method such as compression molding (Kuan et al. 2004; Radhakrishnan et al. 2006; Yin et al. 2007) or injection molding (Heinzel et al. 2004; Mighri et al. 2004; Müller et al. 2006).

2 Materials for Bipolar Plates

The composite for the bipolar plate contains a large amount of graphite powder. Figure 1 shows a cross section of the bipolar plate made from a graphite composite. The composite is packed full of graphite powder in a polypropylene matrix. Although the particle size and shape of graphite affect the rheological properties of composites, we are concerned here with moldability and durability. Polymers for composites are classified into two types: thermosetting resin and thermoplastic resin.

Thermosetting resin is cured by the crosslinking of a reactive monomer and oligomer upon heating. A low viscosity for the monomer and oligomer is advantageous to achieve high filling of graphite. For thermosetting resin, compression molding is more common than injection molding. In this case, curing can take on the order of several minutes.

A higher viscosity for thermoplastic resin increases the difficulty of kneading the graphite and molding the bipolar plates. Moldability must be improved by reducing the composite viscosity through optimization of the particle size distribution and the shape of graphite. The cost of a bipolar plate is expected to be reduced by the development of a thermoplastic composite for

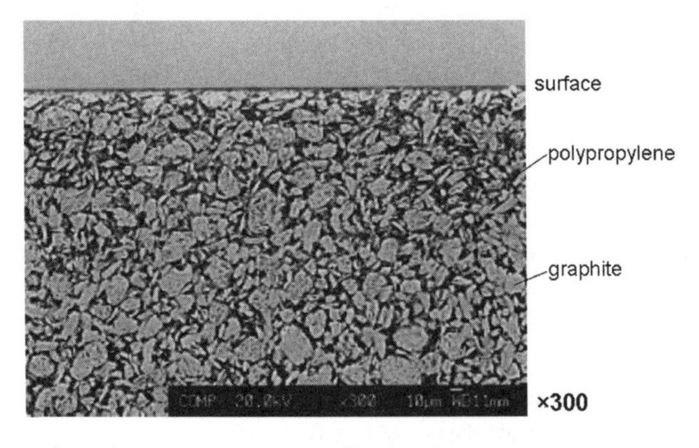

Fig. 1 Cross section of the graphite composite bipolar plate

Table 1 Target values of graphic composite bipolar plates

Properties	Measuring conditions	Target values
Electrical	Resistivity in thickness	<10 mΩ·cm
Mechanical	Flexural strength	>50 MPa
	Flexural strain at break	>1.0%
Molding precision	Thickness tolerance	±0.05 mm
Durability (150°C water immersion for 1,563 h)	Retention of flexural	>80%
	Retention of flexural weight	±1.0%
	Water conductivity	<50 μS cm^{-1}
Cost	1,000,000 Plates production	<¥200 per plate

injection molding, which would shorten the molding time to several tens of seconds. Furthermore, thermoplastic composites that are recyclable should be more environmentally benign.

As well as having high electrical conductivity and good mechanical strength, it is important that a graphite composite bipolar plate is durable for use in a PEFC. Resistance to hot water is necessary because the bipolar plates are exposed to coolant water at an operating temperature of 70–90°C. In addition to the initial properties, stability in hot water was estimated by comparison with target values (Table 1).

3 Properties Required for Bipolar Plates

It is difficult to obtain both low resistivity and good mechanical properties when using graphite composite. The intrinsically insulating polymer matrix must be filled with high concentrations of graphite powder to meet the resistivity target. Unfortunately, at this concentration, most composites become brittle. Thus, the design of graphite particles and the modification of polymers have been investigated to improve their toughness.

3.1 Electrical Resistivity

Resistivity measurements were performed on molded plates, 2 mm thick and 50 mm square. The plates were pressed at 2 MPa between two highly conductive gold-plated copper plates. The resistivity was calculated from the difference in resistance between four and two plates. The contact resistance between a sample plate and an electrode was canceled and that between two sample plates was included.

Figure 2 shows the resistivities of thermosetting and thermoplastic composite samples. Many compression-molded samples of thermosetting composites showed excellent resistivity values of less than 20 mΩ·cm. The lowest value almost reached

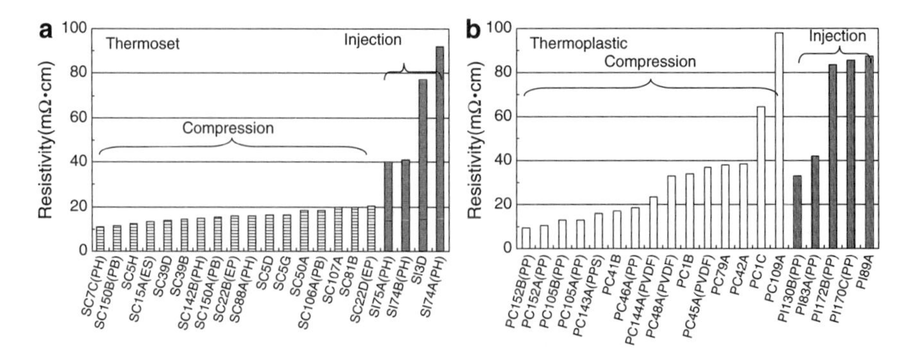

Fig. 2 Resistivities of graphite composites for a bipolar plate. In the sample names, *S*, *P*, *C*, and *I* mean "thermoset," "thermoplastic," "compression-molded," and "injection-molded," respectively

the target of 10 mΩ·cm. For thermoplastic composites, although a compression-molded sample had a value of less than 10 mΩ·cm, many samples had higher resistivity than thermosetting composites. This can be explained by the difficulty of kneading highly viscous thermoplastic resin with a large amount of graphite powder. For injection-molded samples that required more flowability, resistivity tended to increase. The most improved sample, however, showed a relatively low resistivity, 33 mΩ·cm, which seemed to be the top value for injection-molded bipolar plates.

3.2 Mechanical Properties

Flexural properties were determined using a universal testing machine. The sample dimensions were 50-mm length, 10-mm width, and 2-mm thickness.

The flexural strain of composites usually decreases with graphite content; therefore, it is difficult to satisfy both low resistivity and good mechanical properties. Figure 3 shows flexural strain at breaking plotted against flexural strength. The strains of most samples were scattered around 0.5 and more than half of the samples had strengths of less than 50 MPa. However, the values of a few samples improved, and fell in the target region.

4 Durability of Composites

Durability is very important for PEFC bipolar plates, which are used in a hot and humid environment. However, it is difficult to accurately verify the long-term stability of composites in a usual PEFC environment; thus, an accelerated test was used to estimate durability.

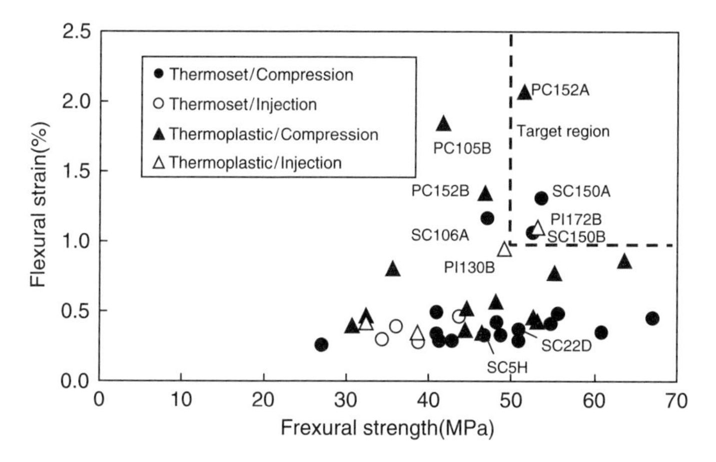

Fig. 3 Flexural strain at breaking against flexural strength

4.1 Accelerated Test

We estimated the durability of the composites by using a hot water immersion test at 150°C for more than 1,560h. This condition corresponds to 100,000h at 90°C by the empirical rule that a 10°C rise doubles the rate of reaction. Plate samples that were 2mm thick and had a total area of 36cm^2 were immersed in 60cm^3 of ion-exchanged water in polytetrafluoroethylene-lined autoclaves, as shown in Fig. 4 (Mitani et al. 2003).

4.2 Physical Stability

Physical stability was estimated by a change in resistivity and the retention of flexural strength and weight. Furthermore, dynamic moduli were measured for representative samples.

The resistivities of composites varied widely, from about 0.5 to about 3 times the initial values, since the measurement was strongly influenced by the surface condition of the samples. Generally, the change in thermosetting composites tended to be greater than that in thermoplastic composites. For a few samples, resistivity could not be measured because of degradation caused by the hydrolysis of resins.

The retention of both flexural strength and weight after the immersion test is shown in Fig. 5. Although the strength usually decreased owing to water absorption and resin degradation, several samples fell within the target region. Three samples showed a retention of weight of less than 97%, which was assumed to have been caused by hydrolysis. An increase in the strength of two samples was explained in terms of recrystallization for the thermoplastic sample and postcuring for the thermosetting sample.

Figure 6 shows the dynamic storage moduli (E') and loss moduli (E'') of four composites plotted against temperature. E' for SC5H and SC106A decreased slightly

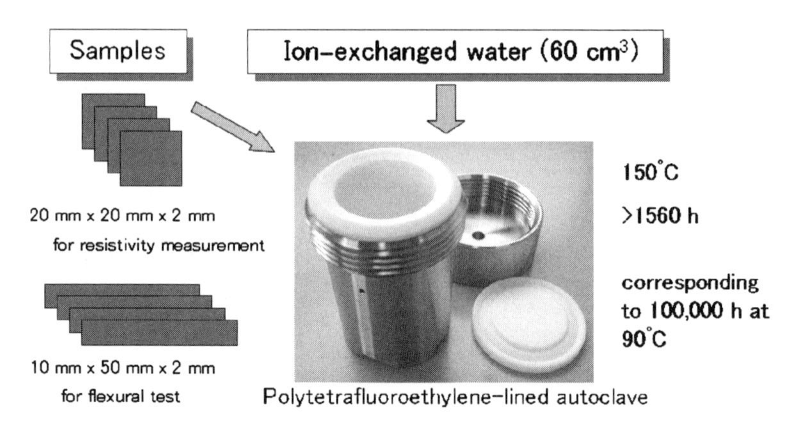

Fig. 4 Procedure for the accelerated test by hot water immersion at 150°C

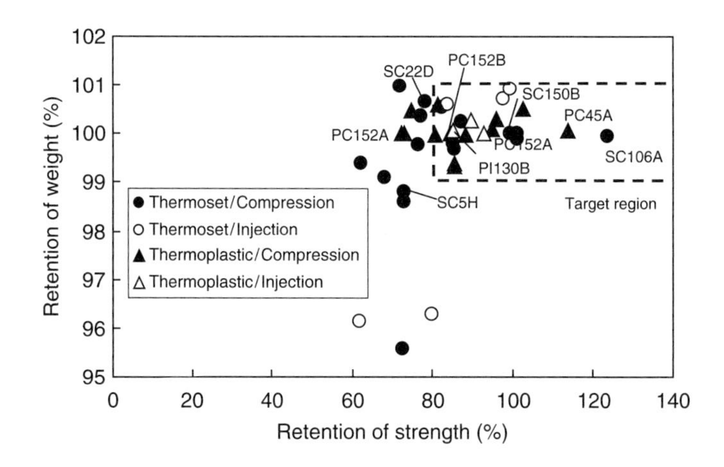

Fig. 5 Retention of weight against retention of strength

above the glass-transition temperatures of 160 and 130°C, respectively, which are indicated by peaks of E''. There were no significant differences between the initial and tested E' and E''. E' and E'' of SC22D were almost flat over the range of the measured temperature. A difference after the immersion test was also not observed with thermoplastic composite: E' of PI130B began to decrease beginning at about 160°C, which was its melting point. After the immersion test, the onset of softening shifted slightly higher, and this was thought to be caused by recrystallization.

4.3 Chemical Stability

Chemical stability was estimated by measuring water conductivity during the immersion test. Conductivity was measured using a conductivity meter and was converted to the value at 25°C. The extracted ions were analyzed by ion chromatography.

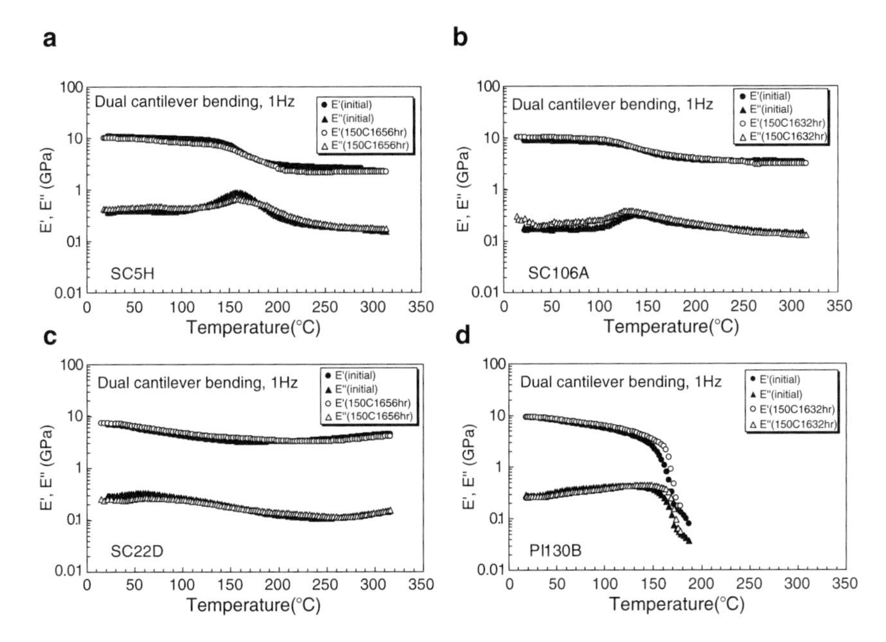

Fig. 6 Dynamic storage moduli (E') and loss moduli (E'') before and after the hot water immersion test at 150°C

Figure 7 shows the change in water conductivity during the immersion test. Many thermosetting composites that contained impurities and uncured components tended to show a rapid increase in water conductivity. For thermoplastic composites, water conductivity also increased if the graphite contained impurities or if the resin tended to hydrolyze. With the development of low-elution material, we obtained both thermosetting and thermoplastic composites that reached the target.

The extracted ions were analyzed as shown in Table 2. Conductivity increased with ion extraction. The ions NH_4^+ of PC46A, F^- of PC45A, and NH_4^+ of SC49A were caused by extraction from graphite, degradation upon molding, and extraction from resin, respectively.

4.4 *Life Estimation*

Life was predicted by the decrease in flexural strength. The times (L) for an 80% decrease at 140, 150, and 170°C were 80, 310, and 1,800 h, respectively. Figure 8 shows L plotted against reciprocal temperature ($1/T$) for SC5H. The plots were linear and the life at 90°C was predicted by the following equation

$$L = A\exp(B/T), \qquad (1)$$

where $A = 1.001 \times 10^{-16}$ h and $B = 1.802 \times 10^{4}$ K.

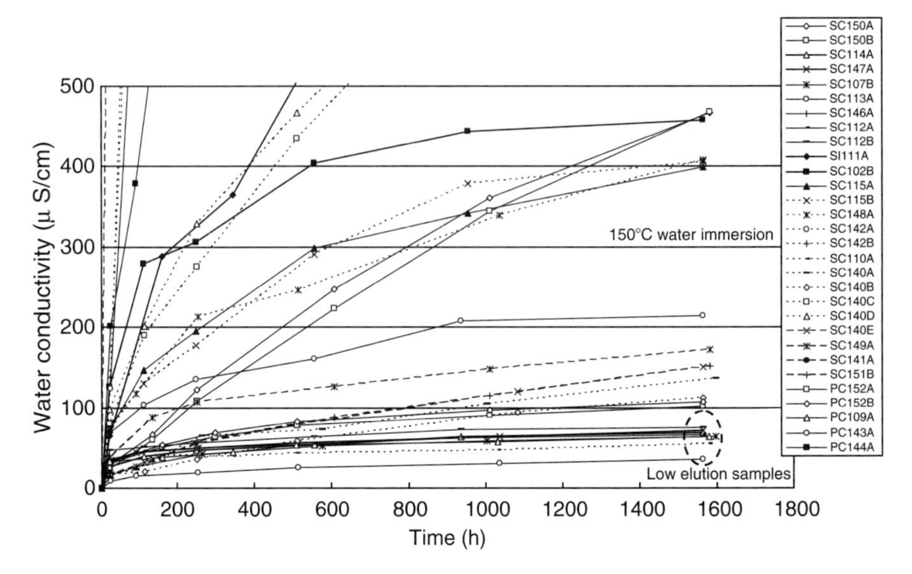

Fig. 7 Changes in water conductivity of composites during the hot water immersion test at 150°C

Table 2 Water conductivity and extracted ions detected by hot water immersion tests

Sample		C	PC46A	PC1B	PC45A	SC7C	SC49A	SI3D
Water conductivity ($\mu S\ cm^{-1}$)		14.6	427.6	40.9	954.0	134.1	>220	472.0
Anion	F^-	<0.05	3	0.5	270	4.0	<0.2	<0.1
	Cl^-	0.18	1.3	0.71	0.12	6.5	15	29
	NO_3^-	<0.01	0.075	<0.01	<0.01	<0.01	7.2	1.3
	SO_4^{2-}	0.17	9.3	0.23	<0.2	0.49	0.93	<5
Cation	Na^+	0.4	0.86	0.15	1.6	0.078	19	1.6
	NH_4^+	0.8	120	0.54	17	0.23	1,900	19
	K^+	<0.05	<0.4	<0.05	50	0.058	53	0.68
Organic acid	Formic acid	<0.1	25	<0.1	39	15	43	36
	Acetic acid	1.7	23	<0.1	42	7.6	15	34
	Tartaric acid	<0.1	<0.1	<0.1	<0.1	<0.1	<0.5	<1
	Oxalic acid	<0.1	<0.1	<0.1	<0.1	<0.1	6	<1

The predicted life was 580,000 h, which was more than 5 times the 100,000 h estimated by the empirical "ten degree rule." This suggested that the results of the hot water immersion test were valid.

4.5 Creep Estimation

To predict the deformation of bipolar plates in a PEFC stack, compression creep was measured in 90°C water. Samples were prepared by stacking five plates. A 2-mm-thick plate was equivalent to PI130B.

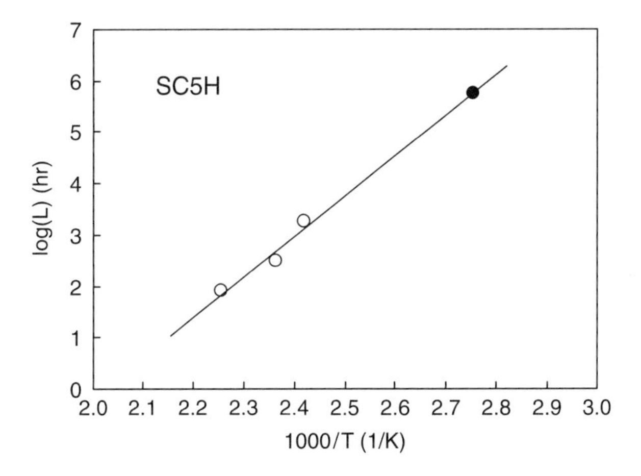

Fig. 8 Arrhenius plot of SC5H in terms of flexural strength

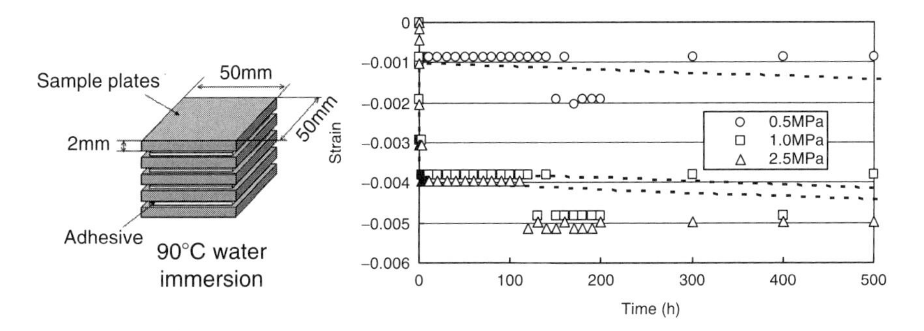

Fig. 9 Creep strain in 90°C water against time

Figure 9 shows the creep strain plotted against time. The strain increased slightly after 100 h. The creep deformation seemed to hardly progress in the stationary region because of a large amount of graphite filler. The plots of each stress were curve-fitted by the following equation

$$\varepsilon = -A[1 - \exp(-Bt)] - Ct, \tag{2}$$

where ε is strain (%) and t is time (h). A, B, and C were estimated as follows: 0.00101, 1.25, and 8.38×10^{-7} at 0.5 MPa; 0.00371, 8.43, and 8.48×10^{-7} at 1.0 MPa; and 0.00399, 9.72, and 8.76×10^{-7} at 2.5 MPa, respectively.

The bipolar plates in a PEFC stack were clamped with a pressure of about 0.5 MPa. The predicted deformation after 100,000 h using Eq. 2 was −8.5%, which was less than the allowed value of −15%. Thus, these graphite composite bipolar plates could be used to make a PEFC stack.

4.6 Comprehensive Evaluation

We estimated how well the graphite composites achieved the target values, as shown in Fig. 10. For the thermosetting composites, SC5H was estimated to be suitable because of its lower resistivity and good moldability, though its flexibility was slightly poor. The mechanical properties of SC106A increased greatly and its resistivity was improved by the development of SC150B. SC22D had the advantage of low elution and stability. Among the thermoplastic composites, which were to be used for injection molding, PI130B showed excellent low resistivity.

5 Performance of a PEFC Stack

The performance of graphite composite bipolar plates in a PEFC was validated by assembling a 1-kW-class stack that had 60 cells. The PEFC was operated at 70°C with air (70°C moisture) and a simulated reformate gas of methane (70°C moisture) that included 10 ppm CO.

As shown in Fig. 11, the current–voltage performance curve of the molded bipolar plate made from graphite composite (SC5H) was very similar to that of a machined graphite bipolar plate.

Figure 12 shows the long-term performance of the PEFC stack using the graphite composite bipolar plate. The PEFC was operated over 3,000 h at a constant current

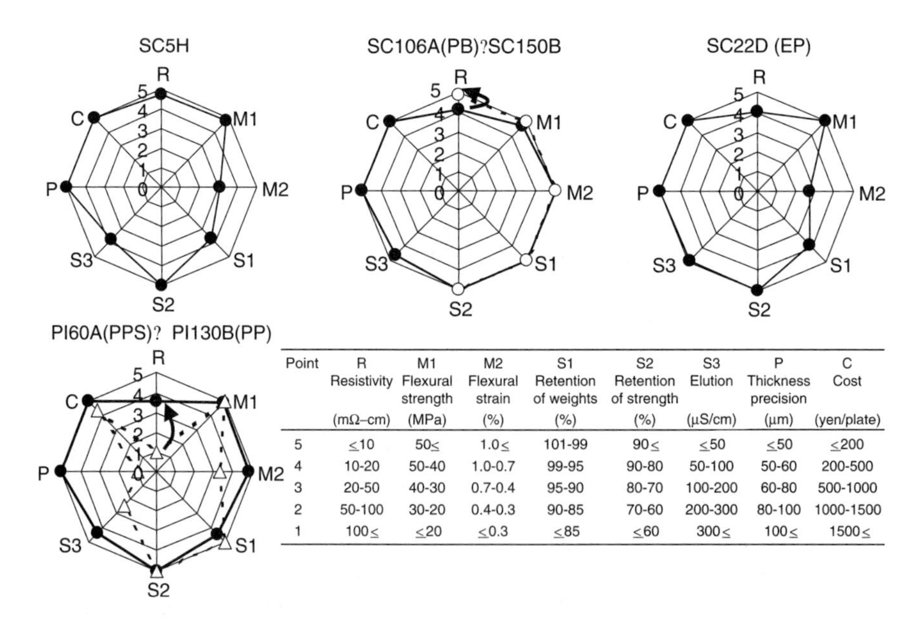

Point	R Resistivity (mΩ–cm)	M1 Flexural strength (MPa)	M2 Flexural strain (%)	S1 Retention of weights (%)	S2 Retention of strength (%)	S3 Elution (μS/cm)	P Thickness precision (μm)	C Cost (yen/plate)
5	≤10	50≤	1.0≤	101-99	90≤	≤50	≤50	≤200
4	10-20	50-40	1.0-0.7	99-95	90-80	50-100	50-60	200-500
3	20-50	40-30	0.7-0.4	95-90	80-70	100-200	60-80	500-1000
2	50-100	30-20	0.4-0.3	90-85	70-60	200-300	80-100	1000-1500
1	100≤	≤20	≤0.3	≤85	≤60	300≤	100≤	1500≤

Fig. 10 Radar charts of the compression-molded thermosetting composites and the injection-molded thermoplastic composite

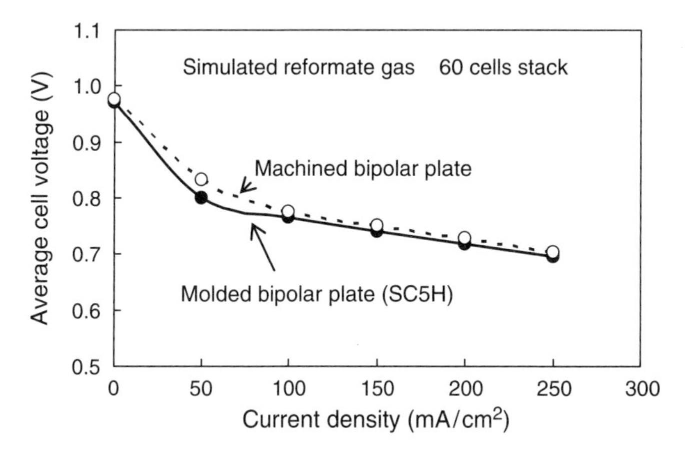

Fig. 11 Comparison of current–voltage performance of graphite composite with that of machined graphite bipolar plates

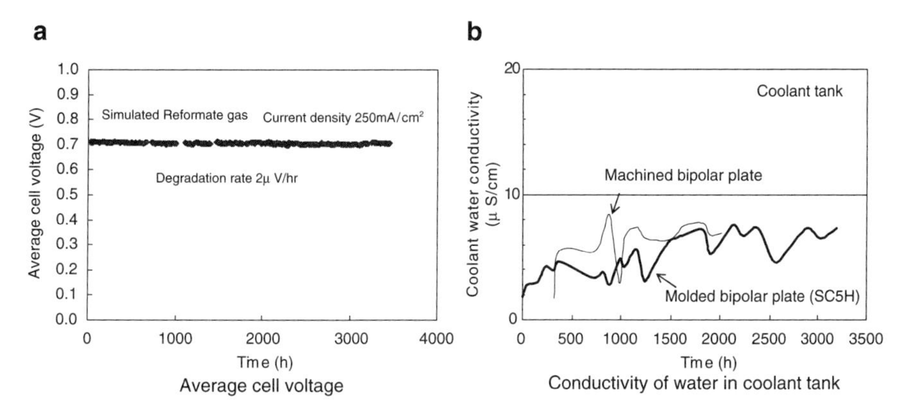

Fig. 12 Long-term performance of a polymer electrolyte fuel cell stack using the graphite composite bipolar plate

density of 250 mA cm^{-2}. The performance hardly degraded and contaminating ion elution was low, as seen with the machined graphite bipolar plate. Thus, the graphite composite bipolar plate developed appears to be stable enough for use in a PEFC.

6 Conclusions

Graphite composite materials developed for PEFC bipolar plates were investigated mainly in terms of durability. Thermosetting composites reached the target levels for both resistivity and flexural properties. Furthermore, the durability was acceptable for PEFC bipolar plates. For thermoplastic composites for injection molding, a significant reduction in resistivity was achieved. The productivity of bipolar plates is expected to greatly increase through a greater reduction in resistivity.

A bipolar plate made from thermosetting composite showed stable fuel cell performance, comparable to that of a machined graphite bipolar plate. These graphite composites are promising bipolar plate materials for use in commercial PEFCs.

Acknowledgments This work was supported by the New Energy and Industrial Technology Development Organization. The authors thank the material manufacturers and the molding-machine makers for a trial production of the molding samples.

References

Heinzel, A., Mahlendorf, F., Niemzig, O., and Kreuz, C. (2004) Injection moulded low cost bipolar plates for PEM fuel cells, J. Power Sources 131, 35–40.

Kuan, H.-C., Ma, C.-C. M., Chen, K. H., and Chen, S.-M. (2004) Preparation, electrical, mechanical and thermal properties of composite bipolar plate for a fuel cell, J. Power Sources 134, 7–17

Lee, H. S., Kim, H. J., Kim, S. D., and Ahn, S. H. (2006) Evaluation of graphite composite bipolar plate for PEM (proton exchange membrane) fuel cell: Electrical, mechanical, and molding properties. J. Mater. Process. Tech. 187–188, 425–428

Mighri, F., Huneault, M. A., and Champagne, M. F. (2004) Electrically conductive thermoplastic blends for injection and compression molding of bipolar plates in the fuel cell application, Polym. Eng. Sci. 44, 1755–1765.

Mitani, T., Hayashi, T., and Miyamoto, F. (2003) Properties of graphite–polymer composites for PEFC separators, Fuel Cell Seminar 2003, 208–211.

Müller, A., Kauranen, P., von Ganski, A., and Hell, B. (2006) Injection moulding of graphite composite bipolar plates, J. Power Sources 154, 467–471.

Radhakrishnan, S., Ramanujam, B. T. S., Adhikari, A., and Sivaram, S. (2006) High-temperature, polymer–graphite hybrid composites for bipolar plates: Effect of processing conditions on electrical properties, J. Power Sources 163, 702–707.

Shao, Y., Yin, G., Eang, Z, and Gao, Y. (2007) Proton exchange membrane cell from low temperature to high temperature: Material challenges, J. Power Sources 167, 235–242.

Yin, Q., Li, A.-J., Wang, W.-Q., Xia, L.-G., and Wang, Y.-M. (2007) Study on the electrical and mechanical properties of phenol formaldehyde resin/graphite composite for bipolar plate, J. Power Sources 165, 717–721.

7
Sealings

Gaskets: Important Durability Issues

Ruth Bieringer, Matthias Adler, Stefan Geiss, and Michael Viol

Abstract In the past, the construction and optimization of single fuel cell components was often considered most important and relatively little attention has been paid to the sealing of the cells. Enduring sealing solutions though are a prerequisite for functionality, continuous operation and achievement of high efficiencies. The requirements for the applied sealing materials are multifarious; some of them are common to all types of polymer electrolyte fuel cells (PEFCs), while others depend on the type of fuel cell in question. Besides the usual requirements for sealing materials such as optimized relaxation behavior and a good processability allowing for inexpensive mass production, all suitable sealing materials must have a general fuel cell compatibility. First, the materials must not contain potential catalyst poisons which might migrate and deactivate the catalyst layer of the PEFC; second the materials must not contain any substances which might reduce the performance of the PEFC; and finally the materials must not contain any components which might be eluted and thus have the potential to block pores of the gas-diffusion layer, coat other active surfaces, or interfere in whatever way with the electrochemistry of the cell. The differences among the three main types of polymer-electrolyte-based fuel cells (PEFC, direct methanol fuel cell, high-temperature PEFC) for the sealing material are, on the one hand, the different temperatures at which the cells are operated and on the other hand, the different media against which the materials need to be resistant (water, fuel: H_2, O_2, reformate, methanol, formic acid, phosphoric acid, coolants). The resulting catalogue of requirements necessitates an in-depth understanding of the material behavior within the cell; therefore fundamental investigations need to emphasize a profound understanding of the deterioration mechanisms (e.g., oxidative and thermal processes, hydrolysis, chemical nature of the neighboring parts, influence of surrounding media, etc.). Many times the existing and commonly employed methods for evaluating the sealing performance of a gasket are found not to be sufficient, so either known methods have to be adapted

---

R. Bieringer (✉)
Freudenberg Forschungsdienste KG, Höhnerweg 2-4,
Weinheim 69469, Germany
e-mail: ruth.bieringer.adler@freudenberg.de

F.N. Büchi et al. (eds.), *Polymer Electrolyte Fuel Cell Durability*,

or completely new methods have to be set up. With the resulting knowledge base optimized sealing solutions can be developed, including new materials and composites as well as innovative gasket designs.

1 Introduction

Polymer electrolyte fuel cell (PEFC) components generally have to fulfill the following four technical requirements to allow for an effective long-term fuel cell operation: functionality, which is mostly determined by the product design; service life which is influenced by the material as well as the design; reliability, essentially determined by the manageability and process stability; and sustainability, which is predominantly influenced by function integration and series production.

When focusing in particular on the sealing, three prime functions can be distinguished: sealing of anode and cathode areas, sealing of the cooling plates and the prevention of gas crossover.

The operational reliability of a fuel cell sealing is only given when all prime functions are met under all circumstances; especially after assembly, for all conceivable tolerances as well as for all possible service conditions and when secondly those prime functions can be upheld over the entire service life of the cell.

Despite the ongoing and still intensifying industrial and academic research and development activities in the field of PEFC components, the sealing of the cell often is a neglected area of concern. Owing to the cost and significance of components such as membrane, catalyst, and bipolar plates, extensive research has been devoted to improve the quality and performance and to reduce the cost of these components. Although the importance of a robust and durable fuel cell seal is out of question, far less attention has been paid to the durability and performance of the fuel cell sealing and often only minor efforts seem to be undertaken to come up with adequate sealing solutions. The importance of and the background to fuel cell sealing was reviewed in 2004 (Dillard et al. 2004)

Often long-term or even lifetime cell performance tests – many times of entire stacks – are conducted with only sparse attention being paid to cell and stack sealing. The risk that these expensive and time-consuming experiments fail early because of inappropriate sealing is becoming more pronounced with the other fuel cells components gaining quality and longevity.

A failure of the fuel cell sealing eventually will lead to leakage of the reactant gases, at best resulting in fuel losses and at worst resulting in reactant gas crossover generating hot spots and finally leading to a deterioration of the membrane. This then further accelerates gas crossover, so in the end the overall performance of the cell will be considerably reduced or ended by a catastrophic failure.

Besides a gasket failure resulting in leakage, inadequate sealing material can lower the fuel cell performance in various other ways. To understand the impact of the sealing to the cell/stack performance the required profile of a polymer electrolyte membrane fuel cell (PEMFC) sealing has to be depicted in more detail. Therefore, the mechanical, chemical, and thermal environments to which fuel cell seals are subjected have to be investigated.

2 Requirements for PEFC Sealing

When considering the mechanical requirements, the compensation of all tolerances within the stack – besides the general sealing function – is the most important task of the gasket. The different parts of a fuel cell, particularly bipolar plates, gas-diffusion layers and the sealing itself, show dimensional tolerances owing to processes involved in their manufacture. Furthermore an assembled stack may bow because of the applied compacting forces, single components can show a distinct permanent set, and all components expand/contract when exposed to temperature changes. The resulting overall tolerances which a single gasket has to compensate can sum up to 300 μm.

To avoid a mechanical overload, the appropriate gasket design as well as a suitable gasket material have to be chosen. The required strains of the gasket materials in question as well as the expected mechanical deformations have to be determined in advance.

To reduce the internal electric resistance of the stack and to allow compression of the stack to promote intimate contact of all components while at the same time bridging tolerances brought in by the different fuel cell components and with only limited space available, flexible sealing systems show distinct advantages over rigid solutions such as duromeric glues or soft metal (a thin metal layer partially or fully coated with an elastomeric layer typically of less than 50 μm). However, they must also be designed to help prevent overcompression of stack components, above all of the gas-diffusion layers. These requirements often necessitate materials with low hardness and stiffness (low compression force) which at the same time have an optimized relaxation behavior. Profiled gaskets are favorable design solutions and are generally superior to gasket solutions which completely fill out the available space, such as flat gaskets or formed-in-place gaskets. A classification of the above-mentioned sealing design concepts is shown in Fig. 1.

In Fig. 2 the line load of different sealing design concepts is compared. It is clearly seen that the force–displacement characteristics of a profiled elastomeric polymer gasket are most favorable. An elastic metallic design (e.g., bead) is always more rigid and therefore much less advantageous.

The advantages of a profiled gasket design as compared with a design of a flat gasket can be seen from Fig. 3. The range between insecure sealing on the one hand, and fuel cell part damage caused by excessive assembly forces or line loads on the other hand, as indicated by the minimum and maximum acceptable contact strain, is only small for a flat gasket design owing to the unfavorable maxima always occurring at the edges of the gasket.

Depending on the type of PEFC, the requirements for the sealing as far as chemical and thermal resistances are concerned are different:

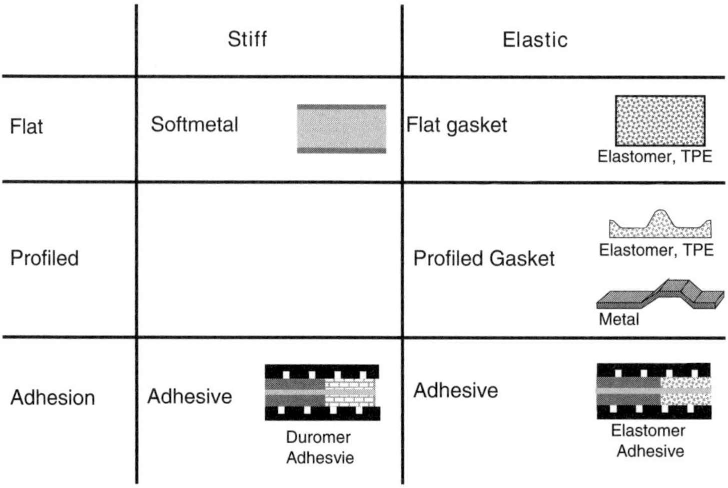

	Stiff	Elastic
Flat	Softmetal	Flat gasket Elastomer, TPE
Profiled		Profiled Gasket Elastomer, TPE Metal
Adhesion	Adhesive Duromer Adhesvie	Adhesive Elastomer Adhesive

Fig. 1 Classification of different sealing design concepts

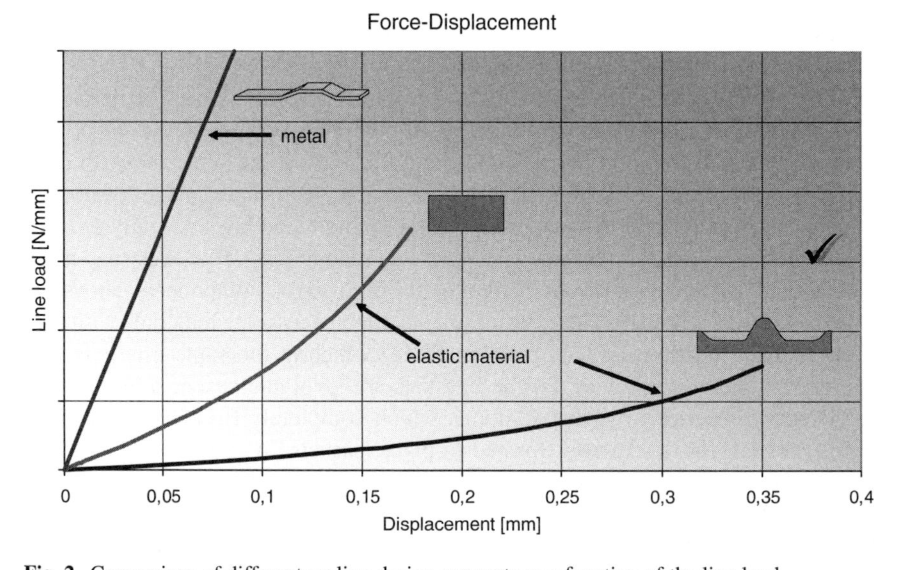

Fig. 2 Comparison of different sealing design concepts as a function of the line load

For a standard PEMFC the temperature range at which the cells are typically operated lies between –40 and 100°C; currently the most common membrane materials are perfluorinated sulfonic acids (e.g., Nafion® from DuPont) and the media against which the sealing needs to be resistant are air, hydrogen gas, water and traces of hydrogen fluoride (HF).

The temperature range of the direct methanol fuel cell is from –40 to 80°C; the membrane material is the same as for the standard PEFC and the media in question are air, hydrogen gas, water, methanol, formic acid and traces of HF.

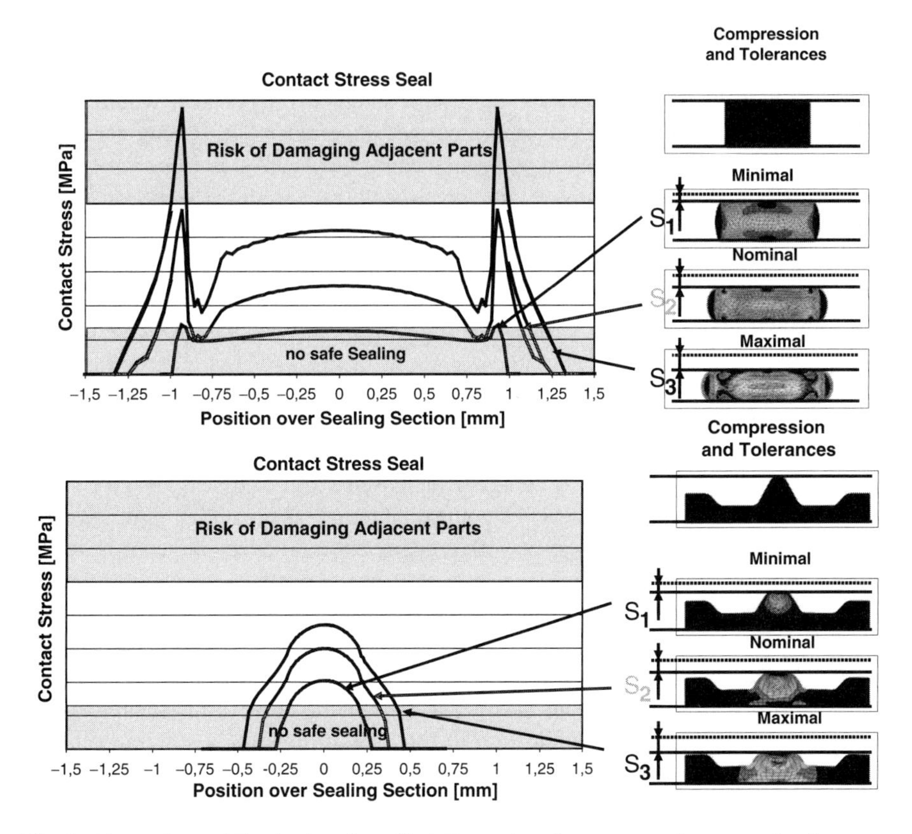

Fig. 3 Comparison of flat (*top*) and profiled (*bottom*) gaskets: range of compression between secure sealing and damage of adjacent parts

For the high-temperature PEFC the temperature limits range from –40 to 200°C, the membrane needs to be of a temperature-resistant type (e.g., polybenzimidazole doped with phosphoric acid as developed, e.g., by Celanese) and the media in contact with the sealing are air, hydrogen gas, water, water vapor, phosphoric acid and polyphosphoric acid.

What has to be kept in mind is that quick temperature cycles to which a sealing material might be exposed are generally more demanding and more difficult to cope with than a maximum long-term service temperature.

Depending on the design, the sealing can also come into contact with media used for cooling of the cell (e.g., water–glycol mixtures, hydrocarbon-based coolants) so sufficient resistance against these cooling media also needs to be taken care of.

The temperature and media resistance represent prerequisites which sealing materials for a given application generally have to fulfill. Appropriate sealing materials for a PEFC must furthermore exhibit a general "fuel cell compatibility." This means that the sealing material must be compatible with all other materials used in the cell (sufficient chemical resistance) and must not interfere with the electrochemistry.

Depending on cell design and construction, the sealing material can come into contact with the catalyst layer of the membrane electrode assembly; in these cases care must be taken that the material does not contain any substances which in the presence of the catalyst lead to unwanted side reactions. The sealing materials must neither contain any components which can act as catalyst poisons which will contaminate and thus deactivate the electrocatalyst, leading to reduced cell efficiency.

Furthermore the sealing materials may not contain any components which might interfere with the function of the membrane; prominent examples here are bivalent cations which can irreversibly complex the ionic sites, resulting in reduced membrane conductivity.

Also no elutable components can be tolerated in the sealing material as they might diffuse through the cell and clog pores in the gas-diffusion layer or deposit in other areas of the cell. This will result in cell contamination and an increase in contact resistance, ultimately leading to increased Ohmic losses. Examples for these substances are plasticizers, oligomers, internal and external process aids employed during manufacturing, as well as degradation products of inappropriately chosen sealing materials.

As the sealing can contact potential-bearing parts of the cell, the materials in question must be highly electrically insulating; in order not to lose fuel and because of safety concerns, high gas tightness is often desirable.

The above-listed requirements all refer to the functionality of the sealing material in an operating PEFC. In addition to these requirements, a suitable sealing material also has to meet requirements related to the process of gasket manufacturing as well as to the gasket design.

Loose gaskets are convenient for the assembly of single cells or prototype stacks and provided that a simple gasket design is chosen, the shape can be altered quickly to change the sealing conditions or optimize the geometry. For cost-effective mass production, however, the sealing function has to be integrated on other fuel cell components, such as on bipolar plates, soft goods (gas-diffusion layer or membrane electrode assembly) and suitable subfilms. Often the substrates for the integration of the sealing function are thermally and mechanically not very withstanding, thus necessitating production processes and material compositions which can compensate for these shortcomings.

3 Material Selection

As explained in Sect. 2, the material class which meets best the aforementioned requirements is the class of elastomeric materials.

Elastomers are a versatile class of polymer materials with distinct advantages in sealing applications over other polymer classes such as thermoplastics, duromers, and thermoplastic elastomers. In particular, the superior relaxation behavior, together with the availability of low-hardness types, makes elastomers the most

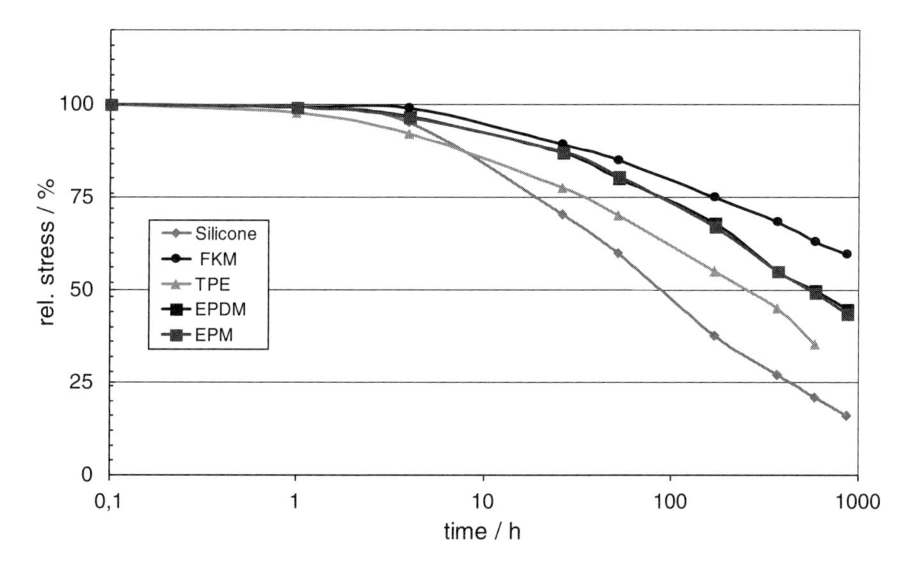

Fig. 4 Compression stress relaxation characteristics of different polymeric materials at 100°C and 100% relative humidity

interesting material class for PEFC sealing. Figure 4 shows a comparison of different polymeric materials with respect to their long-term compression stress relaxation behavior.

To better understand the advantageous properties of elastomers but also to point out potential pitfalls, deeper insight into this unique class of materials is necessary.

Elastomers are multi-ingredient compounds typically consisting of 6–20 different raw materials in various amounts; each component is employed to either contribute to the desired material properties or facilitate the manufacturing process utilized.

Through the choice of a particular base rubber, the thermal properties and the media resistances are largely determined. Fillers (carbon black, mineral fillers, silicas) are added to achieve sufficient material strength, while plasticizers are added to reduce the compound viscosity and influence low-temperature flexibility. The aging behavior can be influenced by addition of a variety of antidegradants, which help protect the part from influences such as heat, UV light and ozone. Moreover processing aids are generally employed to enable easy handling of the compounds as well as a reliable and reproducible manufacturing process. For an elastomer to behave truly elastically, vulcanizing agents must be added, which in the vulcanization process under the influence of pressure and heat induce chemical cross-linking of the material, resulting in the entropy elastic behavior unique for this material class.

As nearly each of the raw material classes in a rubber compound can have an adverse effect on the PEFC performance if wrongly chosen, great care must be taken when formulating an elastomeric sealing material and broad rubber expertise is necessary in order not to risk premature losses of fuel cell efficiency owing to unwanted side effects contributing to fuel cell degradation.

Not only the composition of the sealing material itself needs to be considered, but careful consideration of all associated manufacturing steps is required as well: mixing of the rubber compound, manufacturing of the gasket, its after-treatment, and packing and assembly.

Though it is clearly beyond the scope of this chapter to explain the advantages and risks of the single raw material classes, the choice of an appropriate base rubber class is discussed in some more detail.

A first indication as to which types of base rubbers are particularly suitable for PEFC sealing is given by the temperature and media requirements. In Fig. 5 the main elastomer classes are shown according to their low-temperature flexibility and maximum long-term service temperature.

Whereas most of the R-class rubbers (BR, NR, IR, SBR, CR, NBR) do not possess sufficient high-temperature resistance, many of the polar rubber classes (CM, EVM, ACM, HNBR) show unsuitable low-temperature properties.

When, furthermore, the resistance against the aforementioned media is taken into account only a few elastomer classes remain. Most suitable are the following: IIR (butyl rubber) and EPDM (ethylene propylene diene rubber) or more generally hydrocarbon-based rubbers with highly saturated polymer backbones for sufficient temperature resistance and chemical inertness as well as FKM (fluoro rubber), which is especially suited for high-temperature applications.

Today frequently VMQ (silicone rubber) is employed for PEFC sealing. The reason seems to be its softness, its simple processability and widespread availability more than its unique material properties. For less demanding, mainly short-term

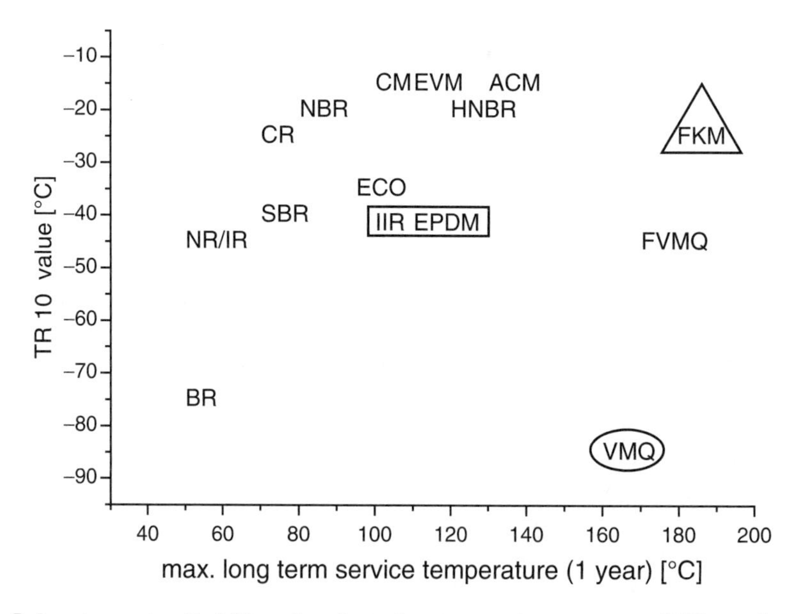

Fig. 5 Low-temperature flexibility and maximum long-term service temperature of different elastomer classes

applications silicone generally seems to be sufficient but for long-term applications silicone-based materials are clearly inferior to hydrocarbon-based rubbers or FKM (see also the findings of Tan et al. 2007).

From our understanding today this is mainly caused by the insufficient stability of silicones towards the humid and warm and somewhat acidic environment found in all types of PEFCs. Silicone-based materials under these conditions are subjected to degradation processes, resulting in material deterioration. As a consequence this material class has two main disadvantages. First, the degradation processes lead to the generation of depositions of silicone fragments on inner surfaces of the cell as seen in Fig. 6, disturbing the wetting character of the electrodes and the water balance of the cell as well as reacting with the electrocatalyst at the cathode (Schulze et al. 2004).

Second, the material deterioration causes a distinct permanent set, largely increasing the risk of premature failure; compare with the data given in Fig. 4.

Although there exists a substantial amount of literature dealing with the degradation of elastomeric gasket materials, only a few results on degradation and the mechanisms behind have been reported in a PEFC environment. A recent fuel cell review does not even mention seals (Steele and Heinzel 2001).

Silicone deterioration in a PEMFC has been described by various groups (Ahn et al. 2002, Schulze et al. 2004, Hinds 2004, Cleghorn et al. 2006, Husar et al. 2007 and Tan et al. 2007). On the other hand, there is also literature suggesting that there are suitable silicone materials for PEFC sealing, but in these cases only ex situ laboratory tests were performed, indicating that the test conditions chosen might not depict reality in an operating cell (Frisch 2003).

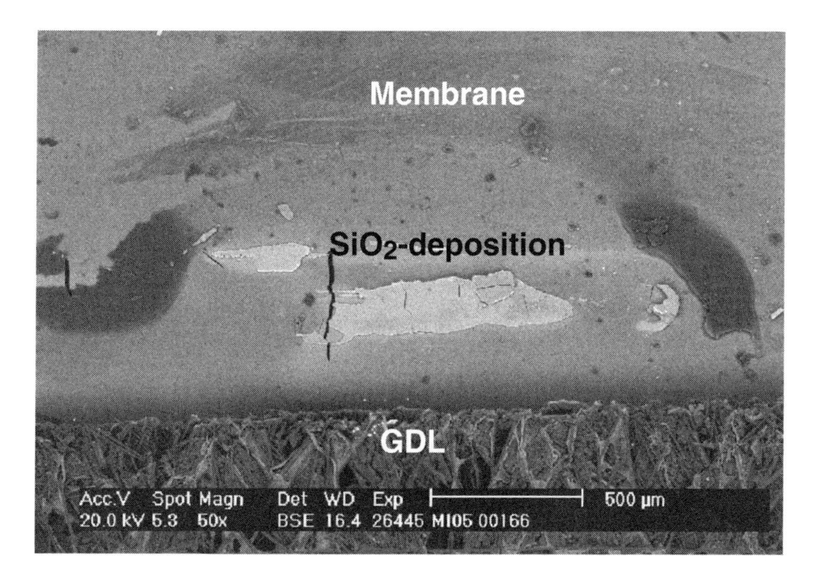

Fig. 6 Silicone fragments (SiO_2) of a silicone gasket on the membrane observed with scanning electron microscopy; after 2,000 h of cell operation. *GDL* gas-diffusion layer

4 Test Methods

To be able to choose the best-suited sealing material or to reliably judge whether a given material is suited for a PEFC sealing application, much attention has to be devoted to the selection of appropriate material test methods.

The long-term change of mechanical properties, especially the long-term relaxation behavior of sealing materials under investigation has to be determined under close-to-reality conditions. Not only has the temperature a major influence, but also all the influencing media as described in Sect. 2 need to be taken into consideration.

To ensure sufficient fuel cell compatibility of a potential elastomeric sealing material, each raw material used as well as the manufacturing processes applied have to be examined critically.

When conducting one's own material compounding all raw materials employed have to be selected with great care and their "cleanliness" has to be checked in advance.

If gaskets are bought from suppliers or the elastomer composition is not known for other reasons, it is highly advisory to invest in a range of analytical tests to prove the fuel cell compatibility. Important from our point of view are solvent extractions to get information on type and quantity of elutable components as well as thermodesorption analyses to obtain information about volatile ingredients.

Whenever dealing with unknown elastomeric sealing materials, a thorough analytical investigation is the only reliable way to distinguish between fuel-cell-compatible and fuel-cell-incompatible material.

Furthermore the suitability of a fuel cell sealing material should be checked in cell tests where the cell performance can be recorded over time, where all conceivable operation modes such as temperature cycles can be employed, and where afterwards the gaskets as well as all other parts of the cell can be investigated carefully.

5 Conclusion

During fuel cell operation the electrochemical performance decreases and the operation behavior changes. These effects are induced by a degradation of fuel cell components, which makes the study of the structural and chemical changes of single components induced by electrochemical stress necessary to understand the underlying degradation mechanisms. Today there are no standardized ex situ laboratory tests available to investigate these open questions with respect to fuel cell sealing.

Many times degradation effects or a general fuel cell incompatibility are observed, but the basic mechanisms causing these findings remain largely unexplained.

Another currently unresolved topic is the incomplete understanding of the influence of the individual raw materials of a sealing material on the functionality of the entire cell or stack. A definition of thresholds regarding the nature and quantity of contaminants which can be tolerated in the fuel cell environment is missing; also there is no general agreement as to which laboratory tests should be employed to reliably evaluate the capability of a potential material for PEFC sealing.

The spectrum of service conditions of a PEFC is quite large and – if at all – is only known by the customer. To fully understand the customer's requirements and therefore to be able to come up with the optimal sealing solution, intimate cooperation is indispensable.

Above all, to find the often cited balance of cost, performance and durability for the particular PEFC sealing application in mind, much more fundamental work has to be carried out.

References

Ahn, S.-Y., Shin, S.-J., Ha, H.Y., Hong, S.-A., Lee, Y.-C., Lim, T. W. and Oh, I.-H. (2002) Performance and lifetime analysis of the kW-class PEMFC stack. *J. Power Sources*, 106, 295

Cleghorn, S.J.C., Mayfield, D.K., Moore, D.A., Moore, J.C., Rusch, G., Sherman, T.W., Sisofo, N.T. and Beuscher, U., (2006) A polymer electrolyte fuel cell life test: 3 years of continuous operation. *J. Power Sources*, 158, 446–454

Dillard, D.A., Guo, S., Ellis, M.W., Lesko, J.J., Dillard, J.G., Sayre, J. and Vijayendran, B. (2004) *Seals and Sealants in PEM Fuel Cell Environments: Material, Design and Durability Challenges*, Presented at the Second International Conference on Fuel Cell Science, Engineering and Technology, June 14–16, Rochester, New York, USA

Frisch, L. (2003) *PEMFC Stack Sealing Using Silicone Elastomers. Fuel Cell Power for transportation 2003 (SP-1741)*, Society of Automotive Engineers, Warrendale, PA

Hinds, G. (2004) *Performance and Durability of PEM Fuel Cells: A Review*. NPL Report DEPC-MPE 002

Husar, A., Serra, M. and Kunusch, C. (2007) Description of gasket failure in a 7 cell PEMFC stack. *J. Power Sources*, 169, 85–91

Schulze, M., Knöri, T., Schneider, A. and Gülzow, E.(2004) Degradation of sealings for PEFC test cells during fuel cell operation. *J. Power Sources*, 127, 222–229

Steele, B.C.H., Heinzel, A. (2001) Materials for fuel cell technologies. *Nature*, 414, 345–352

Tan, J., Chao, Y.J., Van Zee, J.W. and Lee, W.K. (2007) Degradation of elastomeric gasket materials in PEM fuel cells. *Materials Science and Engineering A*, 445–446, 669–675

Part II
Cells and Stack Operation

1. Introduction

Editors

The cell and stack level is an intermediate stage between the component-based analysis and the system level. On the way to the durability targets of 5,000 h for automotive applications, as formulated by the US DOE for 2010/2015, or the Japanese NEDO's lifetime targets for stationary applications of 40,000 and 90,000 h in 2010 and 2015, respectively, investigations on a technical cell level are required.

On the cell and stack level, therefore, more complex boundary and operating conditions need to be investigated than on the component level. Combinations and coupling of the degradation of different components complicate the analysis. New boundary conditions on the cell and stack level include the effects of contamination from different sources or the effect of freezing. Effects of contamination from the ambient air at the cathode or from a fuel reforming process on the anode side are investigated. Also the effect of ambient temperature, here mainly the exposure to subfreezing conditions, is a challenge for reaching the durability targets. On the level of technical cells and stacks, the complexity and interaction of the processes is also higher than in small single cells used for component analysis owing to lateral gradients in the cells, i.e., in water vapor pressure. These effects are important and may be lifetime-limiting. On the technical level, the influence of parameters such as flow-field design, cell or stack compression, and interaction between cells in the stack can be of decisive importance for the rate of degradation and, therefore, for the lifetime of the stack.

In Part II, the most important degradation phenomena on the cell and stack level are analyzed and discussed by highly rated experts in the field from industry and academia.

F.N. Büchi et al. (eds.), *Polymer Electrolyte Fuel Cell Durability*,

2
Impact of Contaminants

Air Impurities

Jean St-Pierre

Abstract Commercialization of the proton exchange membrane fuel cell, an efficient energy-conversion device, requires additional gains in system lifetime. Contamination represents a key degradation mode. Its status is summarized and analyzed to identify research needs. Contaminant sources include ambient air, system components located upstream of the fuel cell stack, and fuel and coolant loops. The number of reported contaminants was conservatively estimated at 97, but many contaminant compositions are still unclear and many gaps remain to be explored, including airstream system components and coolant and fuel streams. For the latter cases, contaminants may reach the cathode compartment by diffusion through the membrane or as a result of seal or bipolar plate failure, thus representing potential interaction sources. In view of this large potential inventory of contaminants, recommendations were made to accelerate studies, including the addition of identification tests performed by material developers, development of standard tests, and definition of an exposure scale for ranking purposes. Because anions are excluded from the membrane in contact with weak solutions (Donnan exclusion), mechanisms involving anions need to be reevaluated. Contaminant mechanisms were synthesized, resulting in only eight separate cases. This situation favors the development of two key simple mathematical models addressing kinetic and ohmic performance losses that are expected to positively impact the development of test plans, data analysis, model parameter extraction, contaminant classification (use of apparent rate constants), and hypothetical scenario evaluation. Many mitigation strategies were recorded (41) and were downselected by elimination of untimely material-based solutions. The remaining strategies were grouped into three generic approaches requiring further quantitative evaluation and optimization: cathode compartment wash, cathode potential variations, and manufacturing material and processing specifications.

J. St-Pierre
Department of Chemical Engineering, University of South Carolina, SC, USA
e-mail:jeanst@cec.sc.edu

F.N. Büchi et al. (eds.), *Polymer Electrolyte Fuel Cell Durability*,

1 Introduction

Proton exchange membrane fuel cells are expected to play a key role in the energy-generation sector for portable, stationary, and motive applications owing to significant advantages that include better energy efficiency, potential for fuel independence, and reduced environmental concerns (US Department of Energy, http://www1.eere.energy.gov/hydrogenandfuelcells). Portable applications are on the verge of commercialization, but increased system lifetime remains an essential common development effort across different designs. Many degradation mechanisms have already been identified, such as catalyst agglomeration, ionomer decomposition, and carbon support corrosion. Contamination has also drawn significant attention and is notably a difficult problem to tackle considering the almost innumerable contaminant compositions that can be drawn from the atmosphere and entrained into the open cathode compartment.

A recent review summarizes contamination sources, impacts, mechanisms, and mitigations (Cheng et al. 2007); however, the review is largely incomplete and partially misleading in relation to air contaminants. The contaminants identified are limited and were directly copied from the US Department of Energy documentation without any attempt at improvement. Mechanisms to isolate the most likely contamination path were not synthesized or criticized even if in many cases information is available from multiple research groups. Mathematical models were not reviewed for the air contaminants. Only one mitigation strategy was cited and future trends were briefly discussed with a few recommendations that are partially misleading. A more focused review dedicated to ionic contaminants was also published (Okada 2003).

The objectives of the present review include an improved synthesis and evaluation of the current air contamination state of the art, including their sources, their effects and associated predictive mathematical models, and mitigation strategies. The information collected is also discussed at a high level to identify the most relevant future research activities, with special attention given to the fact that many contaminants still need to be evaluated in a timely fashion to meet an aggressive automotive fuel cell system development schedule.

2 Literature Review

Contaminant ingress within the cathode compartment of proton exchange membrane fuel cells fueled by hydrogen, the preferred system for automotive applications and the most technically challenging, represented the focus of this analysis. Therefore, proton exchange membrane fuel cells based on other fuels, such as methanol and reformate, were excluded from the present analysis to focus the discussion. For these other fuels, additional contamination routes (contaminant leaching in the liquid methanol solution fuel followed by dissolution in the ionomer and transport to the cathode, etc.) and contaminants (CO, CO_2, CH_4, etc.) exist.

Contaminants are defined here as any substances which do not contribute to the basic fuel cell reactions, converting oxygen and hydrogen into water, heat, and electricity:

$$O_2 + 4H^+ + 4e^- \rightarrow 2H_2O, \tag{1}$$

$$H_2 \rightarrow 2H^+ + 2e^- \tag{2}$$

Therefore, O_2, H^+, e^-, and H_2O within the cathode compartment and H_2, H^+, and e^- within the anode compartment are the only species that are not considered contaminants. The assumption is therefore made that other substances, including fuel cell and system materials and incoming air, are potential sources of harmful species to cell performance. Consideration of all contaminant sources rather than just incoming air has the advantage of highlighting potential interactions between compartments. As an example, nitrogen present in air contributes to mass-transport losses at high current densities and its diffusion to the anode compartment through the ionomer dilutes the fuel, especially for recirculation-based designs, thus creating additional mass-transport losses. From this standpoint, research efforts were devoted to its reduction by pressure swing adsorption (St-Pierre and Wilkinson 2001). Even platinum, the prevalent catalyst, dissolves at high potentials, penetrates the ionomer by ion exchange, and increases the ohmic drop (Yu et al. 2005; Iojoiu et al. 2007). The broad contaminant definition is used to capture as many potential contaminants as possible from the literature search. This strategy was successful because the number of potential contaminants was significantly increased in comparison with the US Department of Energy's list as previously reported (Cheng et al. 2007).

The information collected to assess the present status of the contaminant sources, contaminant effects, and mitigation strategies is succinctly summarized and collected in three tables representing a dense information synthesis (119 papers/881 pages), which are separately analyzed in the following sections. Table content was minimized to concentrate the analysis on trends and defining future directions rather than specifically analyzing the technical merits of specific experimental procedures, mechanisms, mathematical model assumptions, or efficiency of mitigation strategies. The reader is referred to the cited literature for more details on these topics.

3 Contaminant Propagation Routes

Typical fuel cell system and fuel cell stack layouts are first introduced to facilitate the identification of contaminant sources. Figure 1 illustrates the main fuel cell system elements. Ambient air is admitted in the oxidant loop, compressed, cooled, and humidified before being directed to the fuel cell stack inlet port. Atmospheric

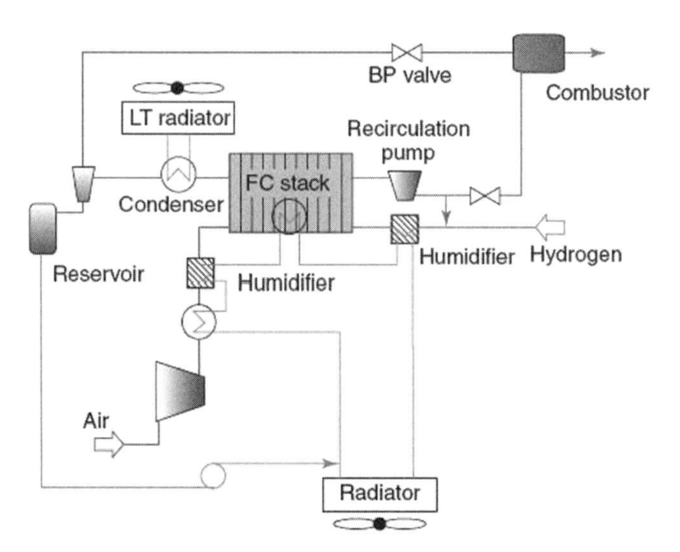

Fig. 1 Proton exchange membrane fuel cell air/hydrogen system process flow diagram. *BP* bypass, *FC* fuel cell, *LT* low temperature. (From Masten and Bosco 2003. Copyright John Wiley & Sons Ltd. Reproduced with permission)

air can carry contaminants, but can also entrain additional ones from any of the subsystems located upstream of the fuel cell stack, including piping and seals connecting all of these elements. A significant fraction of the contaminants is removed from the incoming airstream by a filter (not indicated in Fig. 1). The presence of an air filter does not invalidate the need to research the effect of contaminants on fuel cell performance because the air filter can fail and this reduces energy efficiency (pressure drop increase), contaminant transfer can still occur from subsystem components located upstream of the fuel cell stack, fundamental understanding of contamination mechanisms can lead to improved or new mitigation strategies, and validated mathematical models derived from known mechanisms can be used for predictive purposes, including establishment of tolerance limits. Figure 1 also illustrates that the fuel and coolant streams can interact with the airstream at the fuel cell stack level.

Figure 2a illustrates a representative bipolar plate with its associated inlet and outlet ports, and serpentine flow field channel. Figure 2b shows a generalized cross section of a complete unit cell taken along the different lines indicated in Fig. 2a. The essential difference between these cross sections is the composition of the environment contacting the external part of the seal leading to either a port or the atmosphere (Fig. 2a). As a result of seal degradation or breakage, fuel or coolant can penetrate into the porous gas-diffusion electrode and subsequently the airstream. A similar situation can arise as a result of a bipolar plate or membrane breakage. Component failure is not a necessary requirement for the fuel stream to interact with the airstream since the membrane is permeable to gases, liquids, and cations. The airstream can also be contaminated by any fuel cell stack components

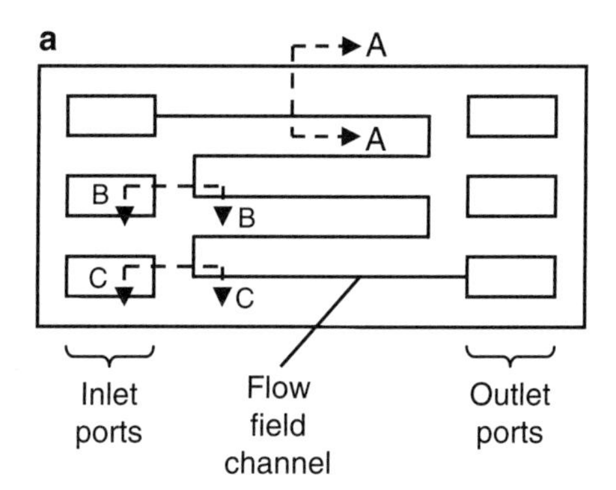

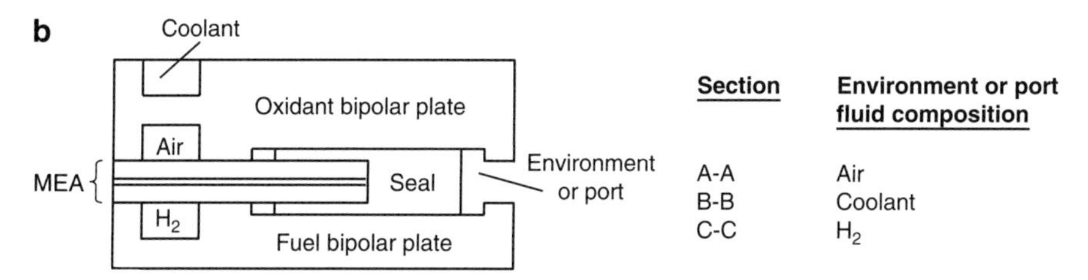

Fig. 2 **(a)** Face view of a representative bipolar plate (elements are not necessarily shown to actual scale). **(b)** Cross-section view of a representative unit cell design (elements are not necessarily shown to actual scale). *MEA* membrane/electrode assembly

especially downstream of the filter. Therefore, the fuel cell system and fuel cell stack layouts indicate that the airstream can include contaminants from three main sources: ambient air, oxidant stream loop subcomponents, and transfer from the fuel and coolant loops. A robust fuel cell system design tolerant to contaminants requires proper consideration of all these sources.

Table 1 summarizes the results of the literature survey in terms of detected contaminants and their respective source. A significant number of contaminants were added to the US Department of Energy's list as reported (Cheng et al. 2007), from 23 to a conservative estimate of 97 (three radicals, six anions, 14 cations, 14 elements, 49 molecules, and 11 not clearly identified classes such as diesel emissions and seal decomposition products). Despite this increase, many potential contaminant sources have not yet been investigated, including most of the fuel cell system airstream components (air filters, compressor oils, heat exchanger and humidifier materials, pipes, and seals), coolant (ethylene glycol, perfluorocarbons; St-Pierre et al. 2005) and coolant loop materials, and fuel stream components. The present large knowledge gap is further compounded by current fuel cell stack material development activities aimed at reducing cost and extending lifetime, which can alter both material degradation mechanisms (Sect. 4) and resulting products/contaminants. For example, these numerous alternative materials, easily reaching several tens in each case, include platinum alloys and non-noble-metal-based catalysts (Gasteiger et al. 2005), metal bipolar plates (Tawfik et al. 2007), and ionomers (Hickner et al. 2004). As a result, efficient means to rapidly

Table 1 Summary of contaminants and their sources

Source			
Component	Subcomponent	Contaminant	References
Ambient air		NH_3, NH_4^+ (created from NH_3 in contact with the membrane), CO, CO_2, H_2S, SO_2, C_6H_6, C_3H_8, NO, NO_2, CNCl, HCN, $CH_3POFOCH(CH_3)_2$, $ClCH_2CH_2SCH_2CH_2Cl$, Na^+, diesel emissions and rock-derived particles (containing Na, Mg, Al, Si, P, S, Cl, K, Ca, Ti, Cr, Fe, Ni, Cu, Zn, Pt, Pb), Cl^-	Bétournay et al. (2004), Garsany et al. (2007), Halseid et al. (2004), Halseid et al. (2006a, b), Jing et al. (2007), Mohtadi et al. (2004), Moore et al. (2000), Okada et al. (1998c), Yadav et al. (2007), Yang et al. (2006)
Fuel cell	Ni and Fe alloy bipolar plate thermal conversion coatings (nitrides). Al and Ni alloys, Cu, Cu coated (Ni, Cr, Sn, TiW, Ni/Sn, Ni/Au), Ti, stainless steel, and coated stainless steel (TiN, nitride, F doped SnO_2, proprietary) bipolar plates	Ni, Ni^{2+}, Cr, Cr^{3+}, Cu, Al^{3+}, TiO_2, Ti^{+3}, TiOO· (?), Fe, Fe^{2+}, Fe^{3+}, Mo, oxides, passivation film/corrosion products, Sn	Bosnjakovic and Schlick (2004), Brady et al. (2004, 2006), Davies et al. (2000a), Ma et al. (2000), Makkus et al. (2000), Pozio et al. (2003), Schmitz et al. (2004), Silva and Pozio (2007), Wang et al. (2003, 2007a), Wang and Northwood (2006), Wind et al. (2002)
	Nafion solution	Organic decomposition products	Laporta et al. (2000)
	Ionomer	HF, F^-, SO_2, SO_4^{2-}, short chain fragments or fluorinated alkyl radicals, perfluoro(3-oxapentane)-1-sulfonic-4-carboxylic diacid, CO_2, SO, H_2SO_2, H_2SO_3, anionic SO_x- or F-containing species	Aoki et al. (2006), Bosnjakovic and Schlick (2004), Curtin et al. (2004), Guilminot et al. (2007b), Healy et al. (2005), Inaba et al. (2006), Kadirov et al. (2005), Kinumoto et al. (2006), Liu et al. (2001), Liu and Case (2006), Mittal et al. (2006, 2007), Ohma et al. (2007), Panchenko et al. (2004), Qiao et al. (2006), Teranishi et al. (2006), Trogadas and Ramani (2007), Xie et al. (2005), Yadav et al. (2007), Zhang et al. (2007)
	Pt catalyst	Cl^-, SO_4^{2-} (?), Pt^{2+}, Pt^{4+}, other Pt cation (?), H_2O_2, HO•, HO_2•, O_2^-, $PtCl_4^{2-}$, $PtCl_6^{2-}$	Aoki et al. (2006), Bi et al. (2007), Bosnjakovic and Schlick (2004), Dam and de Bruijn (2007), Darling and Meyers (2003, 2005), Ferreira et al. (2005), Guilminot et al. (2007a), Inaba et al. (2006), Liu

			and Zuckerbrod (2005), Mitsushima et al. (2007), Mittal et al. (2006), Ohma et al. (2007), Panchenko et al. (2004), Schmidt et al. (2001), Swider and Rolison (1996), Teranishi et al. (2006), Wang et al. (2006), Yadav et al. (2007), Yasuda et al. (2006)
	Pt alloy catalysts	Fe, Co, Cr, Ni, Pt	Antolini et al. (2006), Colón-Mercado and Popov (2006), Koh et al. (2007), Seo et al. (2006), Wan et al. (2002), Xie et al. (2005), Yu et al. (2005)
	PtMo alloy anode catalyst	Mo	Lebedeva and Janssen (2005)
	C catalyst support	S, SO_4^{2-} (?), CO_2	Gülzow et al. (2000), Maass et al. (2008), Reiser et al. (2005), Swider and Rolison (1996, 1999, 2000)
	C catalyst support, sublayer and porous substrate	CO, CO_2, HCOOH, H_2O_2	Cai et al. (2006), Chaparro et al. (2006), Roen et al. (2004)
	Gas-diffusion electrode	Polytetrafluoroethylene degradation products	Gülzow et al. (2002), Schulze et al. (2001, 2007), Schulze and Christenn (2005)
	Gas-diffusion layer	Cr	Xie et al. (2005)
	Silicone, ethylene propylene diene monomer and fluoroelastomer seals	Ca, Mg, Si, Si oxide and decomposition products, $[SiO(C_6H_5)_2]_4$	Ahn et al. (2002), Cleghorn et al. (2006), Davies et al. (2000b), Schulze et al. (2004), Sethuraman et al. (2007), Tan et al. (2007)
Fuel cell	Anode compartment	NH_3, mercaptans, sulfides, unsaturated hydrocarbons, amines, isocyanides, pyrazines, ketones, esters, aldehydes, carboxylic acids, furanones, lactones, ionones, Ru^{n+}	Gancs et al. (2007), Imamura et al. (2005), Uribe et al. (2002)
	Deionized water coolant	Si, Al, S, K, Fe, Cu	Ahn et al. (2002)
Ambient air, fuel cell	System pipes, C-based components	NaCl, Na^+, K^+, Ca^{2+}, Fe^{3+}, Ni^{2+}, Cu^{+2}	Mikkola et al. (2007), Okada et al. (1999a, b, 2000)
Ambient air, fuel cell, coolant	Fuel cell stack materials	Na^+, Ni^{2+}, Cu^{2+}, Fe^{3+}	Kelly et al. (2005a)
System, fuel cell	Piping, humidification bubblers, cell components	Fe^{2+}, Cu^{2+}, Cl^-	Kinumoto et al. (2006), Yadav et al. (2007)

identify contaminants, evaluate their effects, and classify contaminants are urgently needed to meet fuel cell development timelines. Such a daunting task can be partly addressed by requesting from component and material developers the nature of the degradation products to diffuse the work to many fuel cell industry participants. Clearly, this solution does not address the potential coordination issue but could provide more information if some researchers and developers are willing to add composition identification tests during component or material evaluations. This could be facilitated by developing standard contamination identification and testing protocols. Fuel standards development is an ongoing activity within the US Fuel Cell Council (http://www.usfcc.com) that can provide a template to create similar air activities. Establishment of an exposure scale (contamination event probability and severity) to rank contaminant evaluation priorities represents another avenue to reduce immediate research program needs. However, the severity of a contaminant is currently difficult to assess owing to different operating conditions/fuel cell designs used and poorly documented performance effects. Only 30 of the 88 papers cited in Table 1 reported operating cell voltages. Also, few of those 30 papers established a direct link between the performance loss and the contaminant concentration, mostly because techniques able to identify or quantify a contaminant cannot always be implemented within an operating fuel cell. Accelerated tests using an identified and concentrated contaminant level represent an alternative. Other contaminant identification, evaluation, or classification suggestions are provided in the next section.

The need for increased contaminant identification activities is also highlighted in Table 1 by the significant presence of unknown or imprecise compositions. The exact form and composition of a contaminant is necessary to test its effect on fuel cell performance in accelerated tests (e.g., by the use of higher concentrations than expected during normal operation). Several anions were also identified, raising questions with respect to published contamination mechanisms. Donnan exclusion, the reduction in concentration of mobile ions within an ion-exchange membrane due to the presence of fixed ions of the same sign as the mobile ions, dictates at the concentration expected for these contaminants (much less than 0.1N), trace amounts diluted by product water for most of the reported anions except potentially Cl^- (seawater ingress?), that they should be virtually excluded from the ionomer (Helfferich 1962). More recently, it was observed that Nafion exhibits an even greater permselectivity than predicted by Donnan's theory (Lehmani et al. 1997), effectively raising the 0.1N limit. This observation requires confirmation as activity coefficient corrections for high concentrations represent a potential explanation. Donnan exclusion represents an additional clue or an opportunity to further unravel contamination mechanisms. For example, does the sulfur contained in the catalyst carbon support lead to SO_4^{2-} only at the exposed platinum surface? In this case, the loss in performance is expected to be minimal since this catalyst is not electrochemically active. Is F^- present within the ionomer during its degradation or is it generated somewhere outside the ionomer after transport by a precursor species such as HF? What is the effective Donnan exclusion limit for anionic contaminants in Nafion and alternative ionomers (anion concentration resulting in a significant decrease of the membrane proton permselectivity from its ideal value of 1)?

Diffusion of hydrogen to the cathode and its reaction with oxygen on platinum, leading to an ionomer degrading contaminant species (H_2O_2?), represents a well-documented interaction between compartments. Many others have not yet been investigated in detail. There are two major cathode compartment contamination cases worthy of consideration: coolant rated for freezing temperatures (ethylene glycol?) and hydrogen odorants to facilitate leak detection.

4 Contaminant Mechanisms Including Mathematical Models

Table 2 summarizes contaminant performance losses types, mechanisms, and mathematical models obtained from the literature survey. Table 2 is further graphically synthesized to contamination pathways (Fig. 3). At the unit cell level, contaminants originate from the anode compartment (H_2 and cation such as NH_4^+), the seal (seal fillers and decomposition products), the bipolar plate (metal cations and corrosion products), and the air flow field channel (particles, pollution gases, cations from marine environments, etc.). Most of these contaminants reach the catalyst surface via the ionomer or the membrane, but some accumulate at the interface between the bipolar plate and the gas-diffusion layer (metallic bipolar plate corrosion products) or within the gas-diffusion electrode (seal fillers and decomposition products). At the catalyst layer level, contaminants also originate from the catalyst support (sulfur in carbon), but all penetrate the ionomer. Some species may also accumulate on the ionomer surface (hydrophilic seal decomposition product). At the catalyst level, contaminants originate from the catalyst itself (alloying elements, leftover synthesis precursors), are deposited on the catalyst or its substrate (hydrophilic seal decomposition product), or are produced by the action of the catalyst on already present contaminants (NH_3 oxidation, ruthenium deposition, generation of peroxide from O_2 and H_2, creation of carbon surface groups from radicals subsequently decomposed to CO or CO_2, synthesis of radicals from an ionomer sulfonate group by the catalytic action of Fe^{3+} or Cu^{2+} subsequently decomposed to SO_2, etc.). Figure 4 illustrates the different types of performance loss mechanisms associated with the contamination pathways identified in Fig. 3 (Table 2). The different elements of Fig. 4 follow a systematic sequence with fuel cell stack performance loss mechanisms illustrated first (kinetic, ohmic, mass transport, and mixed mechanism sequence) and followed by fuel cell system performance loss mechanisms. Figure 4a shows the effect of a contaminant on oxygen reduction. The contaminant occupies space either by adsorption or by deposition, reducing the effective catalyst surface area (surface coverage $\theta \neq 0$). The contaminant reduces faradaic efficiency by using some of the electrons available for an electrochemical reaction (faradaic efficiency $\phi_f < 100\%$). Finally, the presence of the contaminant affects the elementary oxygen reduction steps (steric hindrance, electric double layer structure change, etc.), thus modifying the proportion of the water and contaminant peroxide products (selectivity $\phi_2 \neq$ selectivity ϕ_1). Figure 4b shows the long-term effect of sulfur present either in the catalyst support or in the

Table 2 Summary of contaminant performance losses, mechanisms, and mathematical models

Performance loss				
Main cause	Fuel cell performance impact	Mechanism	Mathematical model	References
Catalyst surface poisoning	Kinetic	Contaminant adsorption (competitive in mixtures with preferential adsorption of the largest-affinity contaminant), contaminant decomposition/electrochemical reaction intermediates production, O_2 reduction reaction pathway modification (atop O_2 adsorption favored rather than bridged O_2, electric double layer structure change induced by cation insertion in ionomer, Pt oxide modification including kinetics, changes in proton activity) or contaminant deposition reduces the catalyst area, increases the O_2 reduction reaction overpotential, decreases faradaic efficiency, and increases product selectivity (increased H_2O_2 contaminant production)		Cai et al. (2006), Chaparro et al. (2006), Cleghorn et al. (2006), Gancs et al. (2007), Garsany et al. (2007), Halseid et al. (2006a, b, 2007), Jing et al. (2007), Lebedeva and Janssen (2005), Mohtadi et al. (2004), Moore et al. (2000), Okada et al. (1999a, b, 2000, 2001, 2003), Schmidt et al. (2001), Schulze et al. (2004), Swider and Rolison (1996, 2000), Tan et al. (2007), Yang et al. (2006)
Catalyst loss	Ohmic	Pt particle dissolution acceleration by adsorbed S on Pt from SO_2 or other sources decreasing ionomer ionic conductivity		Garsany et al. (2007), Sung et al. (1997), Swider and Rolison (1996, 2000), Yang et al. (2006)
			Zero-dimensional transient kinetic rate model for Pt dissolution including electrochemical Pt dissolution, electrochemical PtO formation, and chemical PtO dissolution (not linked to cell performance)	Darling and Meyers (2003)
Catalyst loss	Ohmic		Transient one-dimensional model for the Pt^{2+} concentration and	Darling and Meyers (2005)

			potential through the membrane/electrode assembly based on electrochemical/chemical reactions, mass and charge balances. Model is not linked to ohmic or flooding loss	
			Transient two-dimensional model for the Pt^{2+} concentration and potential through the membrane/electrode assembly based on electrochemical/chemical reactions, mass and charge balances. Model is not linked to ohmic or flooding loss	Franco and Tembely (2007)
Metallic bipolar plate degradation	Ohmic	Passive film formation and growth on Ni alloys, Al alloy, Ti, nitrided Ni and Fe alloys, coated stainless steel (nitride, proprietary), and stainless steel bipolar plate surfaces, decreasing electronic conductivity		Brady et al. (2006), Cunningham et al. (2002), Davies et al. (2000a), Ma et al. (2000), Makkus et al. (2000), Silva and Pozio (2007), Wang et al. (2003, 2006b), Wind et al. (2002)
Ionomer charge carrier replacement	Kinetic, ohmic, mass transport	Ion-exchange replacing H^+ in ionomers (higher affinity for most other cations) affecting both thermodynamic and transport properties. Membrane water content is reduced by contaminant cations (lower cation hydrophilicity, potential for cross linking and ion pairing with sulfonate sites, formation of salt-like structures favored by drying/wetting cycles). Transport of cations (through hydrophilic domains) does not proceed by hopping unlike that of protons and results in a larger water transference number. Larger cations also have the ability to pump water in the constrained channels. Ion and water transport are coupled since ions and water move in the same hydrophilic channels; therefore, the conductivity		Antolini et al. (2006), Colón-Mercado and Popov (2006), Cunningham et al. (2002), Halseid et al. (2004, 2006b), Iojoiu et al. (2007), Iyer et al. (1996), Kelly et al. (2005a, b), Ma et al. (2000), Makkus et al. (2000), Mikkola et al. (2007), Mohtadi et al. (2004), Okada et al. (1997, 1998a, b, 1999a, b, 2000, 2001, 2003), Pourcelly et al. (1990, 1996), St-Pierre et al. (2000), Shi et al. (1997), Wan et al. (2002), Wang et al. (2006b), Xie and Okada (1995, 1996), Yu et al. (2005), Zaluski and Xu (1994)

(continued)

Table 2 (continued)

Performance loss				
Main cause	Fuel cell performance impact	Mechanism	Mathematical model	References
		and the water diffusion coefficient correlate with channel cross section. By extension, O_2 diffusivity is affected in the same manner; therefore, cations penetrating the membrane reduce the ionic conductivity and the water and O_2 diffusivities and increase water transference (no hopping mechanism, less water creating more confinement, larger impurity cations), creating dehydration near the anode (ohmic loss), potential flooding near the cathode (mass-transport loss), and reduced O_2 permeability (another mass-transport loss). A kinetic effect although observed has not been explained		
			Concentration-dependent diffusion control exchange between foreign cation and proton (not linked to cell performance)	Samec et al. (1997)
			Equilibrium ion membrane absorption model from binary cation solutions	Tandon and Pintauro (1997)
			Steady-state one-dimensional model for membrane water transport (electroosmotic drag, diffusion) and associated ohmic drop based on a mass balance. Selected prop	Okada et al. (1998c), Okada (1999)

			erty dependencies on membrane water content lead to some analytical solutions	
Ionomer charge carrier replacement	Kinetic, ohmic, mass transport		Transient one-dimensional model of water transport in membranes in the presence of localized contamination by cations (diffusion and electroosmotic drag). Model is not linked to ohmic loss or flooding loss	Chen et al. (2004)
Seal degradation	Mass transport (?)	Leaching of seal material fillers, and, seal backbone de-cross-linking and chain scission catalyzed by high ionomer acidity affecting gas-diffusion electrode and catalyst layer hydrophilicity (?) and leading to reduced O_2 transport rates		Cleghorn et al. (2006), Schulze et al. (2004), Sethuraman et al. (2007), Tan et al. (2007)
Ionomer degradation	Kinetic, ohmic, mass transport	Impurity cation (Fe^{2+}, Cu^{2+}, Ti^{3+}) or Pt^{2+} from decomposition of Pt at high potentials ion exchange with the membrane catalyzing the release of radicals from the O_2 reduction reaction side reaction peroxide product (either at the catalyst layer or redeposited Pt within the membrane) degrading the membrane structure side and main chains by depolymerization (e.g., by formation of cross-linking S–O–S groups, C–O–C and C–S bonds breaking) (Cr^{2+}, Co^{2+}, Li^{+}, Na^{+}, K^{+}, Ca^{2+}, do not catalyze radicals release). In the absence of peroxide, Fe^{3+} and Cu^{2+} can also produce radicals (cation directly reacts with the sulfonate group, creating a radical which by recombination leads to SO_2 and O_2 in addition to another side chain radical). Loss of sulfonic groups negatively affects conduc		Aoki et al. (2006), Bosnjakovic and Schlick (2004), Curtin et al. (2004), Inaba et al. (2006), Kadirov et al. (2005), Kinumoto et al. (2006), Liu et al. (2001), Liu and Zuckerbrod (2005), Ohma et al. (2007), Pozio et al. (2003), Qiao et al. (2006), Teranishi et al. (2006)

(continued)

Table 2 (continued)

Performance loss				
Main cause	Fuel cell performance impact	Mechanism	Mathematical model	References
		tivity, water diffusivity, and membrane water content, whereas C–F bond decomposition leads to thinning, pinholes, and loss of catalyst active area. Ionomer decomposition can also initiate at residual H-containing terminal bonds (a result of polymer processing). Peroxide is also formed by reactant crossover recombination followed by subsequent diffusion in ionomers and decomposition by specific impurity cations		
Compartment containment loss	Kinetic (?), ohmic (?), mass transport (?)	Leaching of seal material fillers, and seal backbone de-cross-linking and chain scission catalyzed by high ionomer acidity affecting both chemical and mechanical seal properties (reactant dilution and additional path for fuel and coolant loops contaminants?)		Cleghorn et al. (2006), Schulze et al. (2004), Tan et al. (2007)
Particulate accumulation in filtering system	None	Oxidant pressure drop increase and system efficiency decrease		Bétournay et al. (2004)

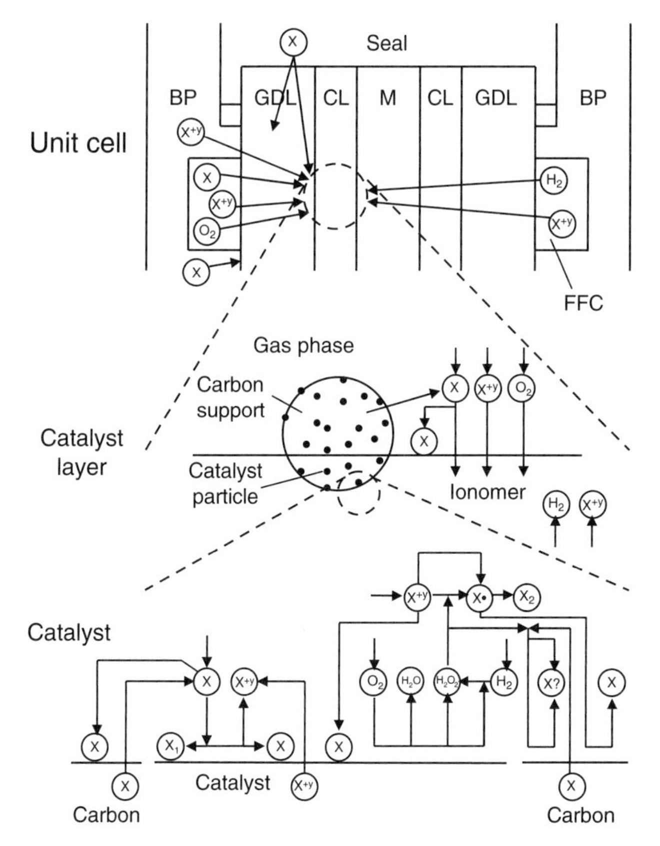

Fig. 3 Contaminant pathways summary. Contaminants are generically identified as X, X_1, X_2, and X? (a molecule), X^{y+} (a cation), and X• (a radical). The *arrows* for both unit cell and catalyst layer levels refer to species movements. At the catalyst level, *arrows* indicate either species adsorption/desorption or reactions between species. *BP* bipolar plate, *CL* catalyst layer, *FFC* flow-field channel, *GDL* gas-diffusion layer, *M* membrane

reactant stream. The larger platinum dissolution rate favors a decrease in membrane ionic conductivity. Figure 4c depicts the long-term negative effect of metallic-based bipolar plate corrosion or passivation product accumulation on interfacial resistance R and the development of an ohmic drop ($\Delta V = IR$). Figure 4d shows the effect of the inclusion of foreign cations within the ionomer owing to their greater affinity for the ion-exchange sites. Both transport and thermodynamic properties are affected. The ionomer water content decreases as a result of a reduced cation hydrophilicity in comparison with H^+ (hydration enthalpy), formation of ionic pairs with the sulfonate site (electrostatic shielding reducing interactions with water dipoles), or sulfonate sites cross-linking (electrostriction). As a result, the ionomer hydrophilic channel size is significantly reduced (steric hindrance, interactions with channel walls) and negatively impacts O_2 and H_2O diffusivities, and ionic conductivity. Additionally, foreign cations do not possess the proton capability to move by hopping (hydrogen bond), which further increases the negative

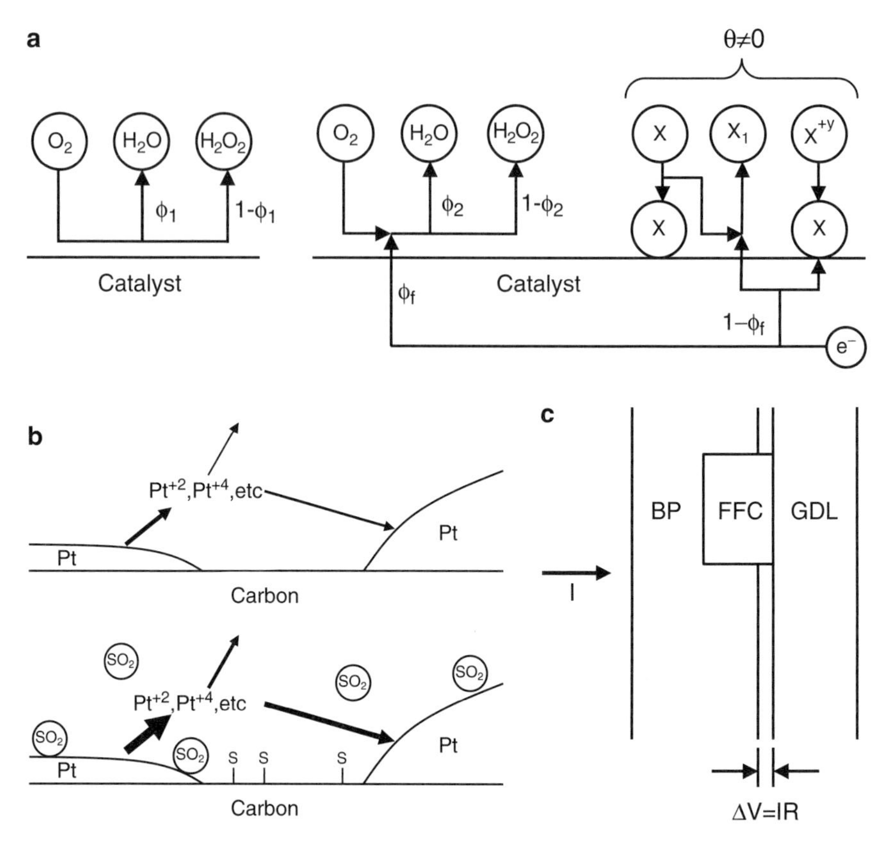

Fig. 4 (**a**) Oxygen reduction reaction pathways in the absence (*left*) and in the presence (*right*) of a contaminant. The presence of a contaminant affects the surface coverage θ, oxygen reduction reaction product selectivity ϕ, and faradaic efficiency ϕ_f. (**b**) Platinum dissolution rate in the absence (*top*) and in the presence (*bottom*) of sulfur contamination either from the air intake or from the carbon support. The increased platinum dissolution rate favors a decrease in ionomer ionic conductivity. (**c**) Metallic bipolar plate corrosion products and/or passivation layer contribution to ohmic losses. (**d**) Representative composition and species flux directions of Nafion hydrophilic channels in the absence (*top*) and in the presence (*bottom*) of a cationic contaminant X^{y+}. The *dotted lines* in the *top diagram* indicate a probable proton path by the hopping mechanism (hydrogen bond). Insertion of foreign cations with lower hydrophilicity than H^+ leads to a reduced water content and steric hindrance negatively affecting transport properties such as ionic conductivity, O_2 diffusivity, and H_2O diffusivity. Electroosmotic drag is increased owing to the pumping action of the larger and solvated foreign cations. (**e**) Oxygen flow towards the catalyst in the absence (*left*) and in the presence (*right*) of hydrophilic contaminants in the gas-diffusion electrode (*GDE*). (**f**) Ionomer degradation effects by C–F bond scission leading to thinning, pinholes, and loss of effective catalyst area, and by sulfonate site loss leading to a water content decrease (impact similar to foreign cation insertion, (**d**). Elements are not necessarily to actual scale (e.g., charge neutrality is not respected in the schematic ionomer channels). (**g**) Cathode compartment containment loss as a result of seal degradation. Fluid ingress causes either dilution (kinetic and mass-transport losses because the compartment pressure $p_2 < p_1$) or creates additional paths for contamination. (**h**) Air intake filter clogging by solid particles leads to an increased pressure drop and decreased system efficiency. *BP* bipolar plate, *FFC* flow-field channel, *GDL* gas-diffusion layer, *M* membrane

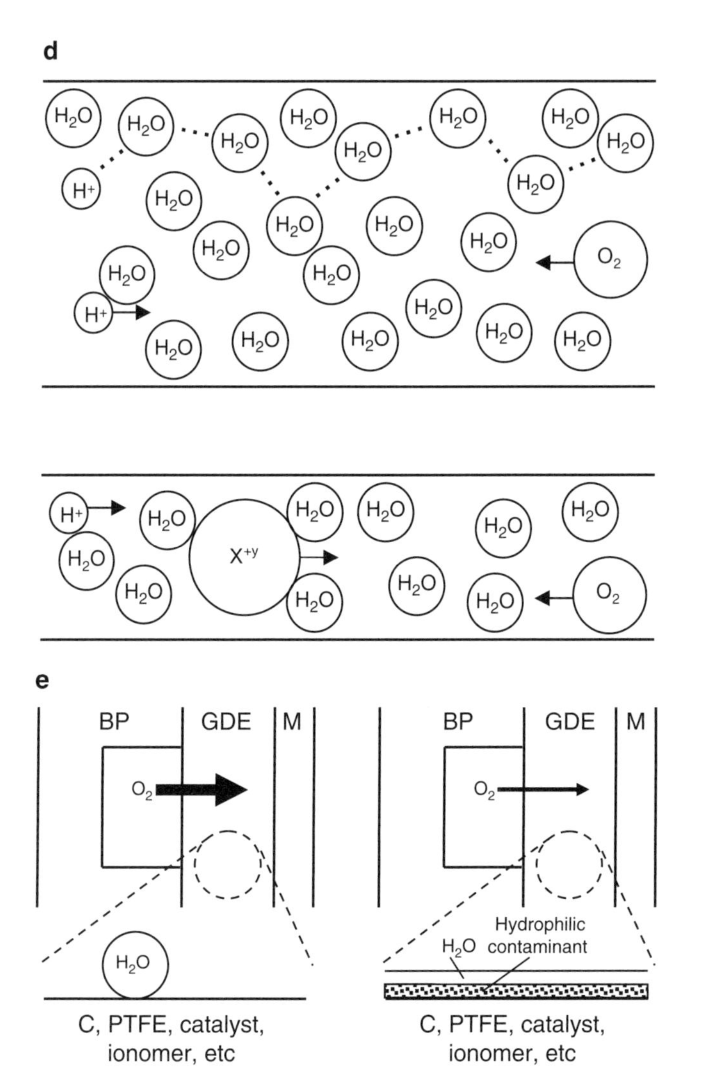

Fig. 4 (continued)

impact on ionic conductivity. The increased confinement also accounts for a significantly larger electroosmotic drag by foreign cations (pumping effect in addition to attached solvated water molecules). The net effect is an increased ohmic drop, including ionomer dehydration at the anode and mass-transport losses by cathode electrode flooding and reduced O_2 permeability in the ionomer. Figure 4e demonstrates the effect of a hydrophilic contaminant deposited within the gas-diffusion electrode which affects the material wettability (contact-angle change). The increased liquid water presence in the electrode limits the O_2 access to the catalyst (mass-transport loss). Figure 4f shows the effect of the ionomer disintegration by peroxide and radical attack. Scission of the C–F bond leads to ionomer loss (mem-

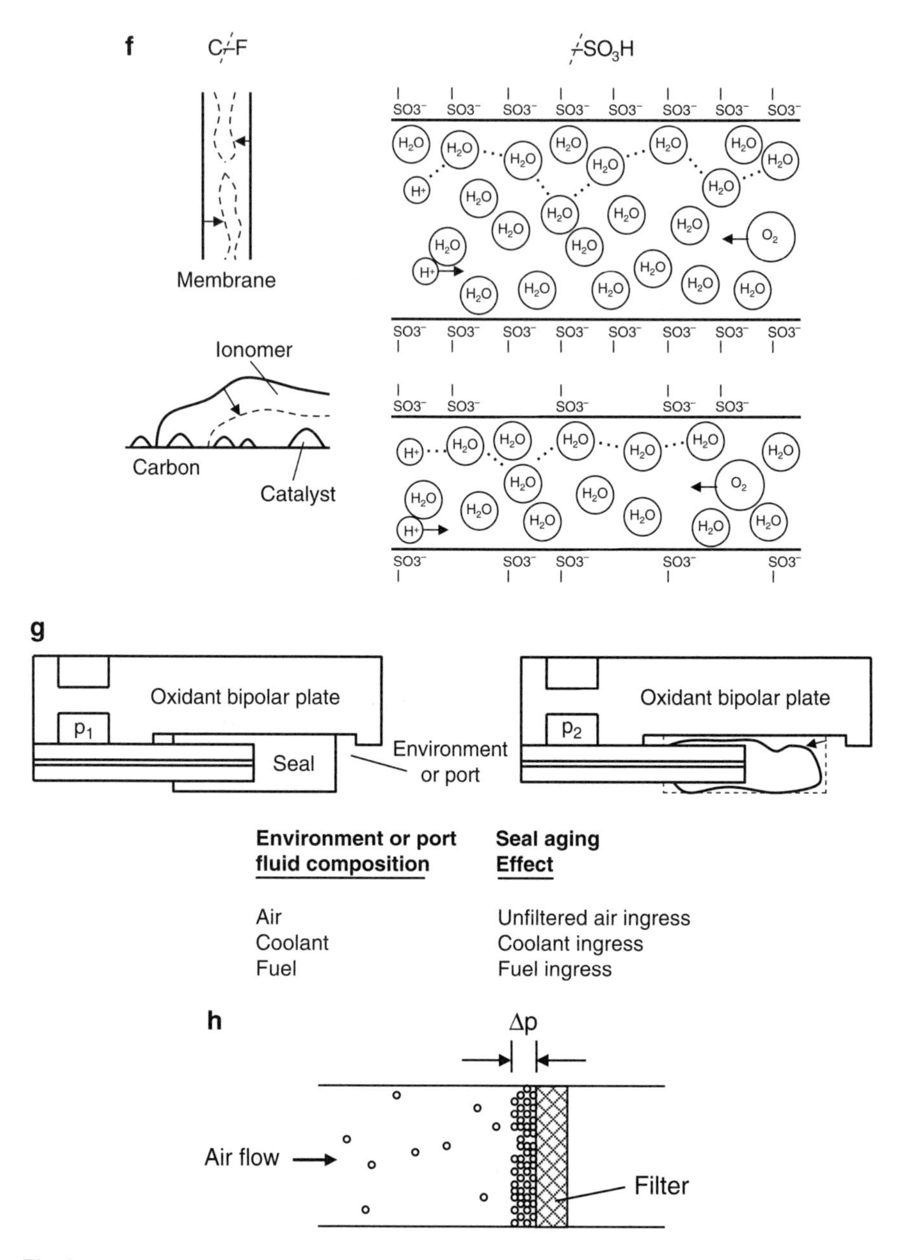

Fig. 4 (continued)

brane thinning and holes), which leads to a deficiency in membrane integrity (loss of reactant separation function creating an additional path for contamination) and an effective catalyst area decrease (kinetic loss). Loss of the sulfonate group leads to a reduced water content, with an associated decrease in membrane conductivity (ohmic loss) and H_2O diffusivity (mass-transport loss?). Presumably, other transport properties are also affected (electroosmotic drag, O_2 diffusivity) but additional work is required to establish the net mass-transport effect. Figure 4g depicts the effect of a loss in reactant containment as a result of a seal failure. In general terms, all eventualities lead to a potential oxidant stream dilution (kinetic and mass-transport losses) and increase in contamination levels (some mechanisms have already been mentioned). For the air ingress case, the probability that the contaminant level increases is significantly reduced during operation (the oxidant pressure is higher than the atmospheric pressure) as opposed to a shutdown condition (equalized pressures). As already indicated, the coolant and fuel contaminant effects on fuel cell cathode performance have apparently not been reported (Fig. 3). As for the fuel ingress case, H_2 recombines with O_2 forming peroxide (Fig. 3) and fuel contaminants have another route in addition to the membrane to contaminate the cathode compartment. Figure 4h represents the clogging of an air intake filter by solid particles (dust, diesel emissions, etc.) leading to a pressure drop Δp and a reduction in fuel cell efficiency.

Contaminant pathways and their associated performance loss mechanisms are currently relatively few (eight) considering the number of contaminants presently identified (97). Although there is no guarantee that future research activities will not reveal additional degradation mechanisms (several mechanisms are unresolved; Table 2) and many species have not been identified (Table 1), it is fortunate that this situation exists because only a few mathematical models are needed for representation. An additional decrease is possible by prioritization. If long-term degradation modes (Fig. 4b, f, g), modes that are not sufficiently general (metallic bipolar plates may not necessarily be integrated in future products; Fig. 4c), modes related to undesirable current regimes (mass-transport loss at high current densities; Fig. 4e), and modes easily preventable (Fig. 4h) are temporarily eliminated from the priority list to address more immediate concerns, only two general modes subsist (catalyst and membrane contamination; Fig. 4a, d). Development of simple, relatively accurate (best accuracy achievable in the least amount of development time) models with preferably analytical solutions is perceived as a potential solution to support contamination test plan elaboration, and to accelerate data analysis, extraction of model parameters with physical meaning (including apparent rate constants), contaminant classification (observable fingerprint behavior, use of apparent rate constants), and analysis of scenarios (predictions including definition of tolerance levels). Efforts in this direction are already in progress (St-Pierre et al. 2008) and were based on the extension of a Langmuirian-based model for reversible adsorption by adding competition with adsorbed O_2 undergoing reduction (Bhugun and Anson 1997). A simpler approach without competitive adsorption was considered for fuel contaminants (Narusawa et al. 2003), but is likely only applicable at low contaminant coverage (small change in oxygen reduction reaction overpotential) leading to negligible interactions with adsorbed oxygen and related intermediates. These approaches are in stark contrast with some existing fuel contamination models (Zhang et al. 2005), which, given the current context, do not positively contribute to the current need for fast contaminant behavior processing because they are too detailed and specific, thus significantly complicating analysis and increasing the risk that the model will no longer be relevant owing to fuel cell material evolution.

Table 2 reveals that few mathematical models were formulated, addressing only two degradation mechanisms, with even fewer able to predict an actual performance loss (some initial elements are provided, but most models are incomplete). This situation supports the proposal presented in the previous paragraph. More collaboration is also apparently needed between experimentalists and mathematical modelers considering that there is little overlap between these two groups. The US Fuel Cell Council can be cited again as an example fostering interaction between experimentalists and modelers (http://www.usfcc.com).

5 Mitigation Strategies

Table 3 shows that a much larger number of mitigation strategies (41) are available than reported (one; Cheng et al. 2007). However, significant additional work needs to be completed as many effects and mechanisms are still unclear or unknown. This information is required for optimization. Because many mitigation strategies are specific, it is also important to establish priorities. Many material approaches are listed in Table 3, ranging from new catalysts to catalyst supports, binders, bipolar plate compositions, coatings, material pretreatments, additives, and synthesis routes. These approaches are notoriously more time intensive to develop and implement. Additionally, there are already significant activities worldwide to develop new materials to decrease fuel cell cost and increase functionality (high-temperature operation, reduced humidification, etc.). In this context, selecting a material approach for study may not be the best solution as its relevance may not necessarily last. Therefore, material approaches are ranked at a low priority. There are fewer fuel cell designs and operation-related mitigation approaches listed in Table 3, which are easier to implement either in the fuel cell system itself or as part of a potential maintenance schedule or repair tool box (the assumption is made that these will eventually be available at car dealers or other car repair shops). These are further grouped into three separate generic approaches consisting of cathode compartment wash, cathode potential changes, and manufacturing material and processing specifications. Three are disregarded here since implementation is already being considered (oxidant and coolant filters, cell designs avoiding contact between metallic bipolar plate and membrane such as the example depicted in Fig. 2b).

Several approaches are related to the removal of contaminants by circulating a fluid within the cathode compartment with the option of imposing a potential at the same time (Fig. 5a). The exact nature of the fluid (gas, liquid, or two-phase mixture, composition) needs to be optimized because both the effect and the mechanism are currently unclear. Contaminant desorption from the catalyst and carbon components, and removal of contaminant cation from the membrane were reported but quantitative studies have apparently not been completed. Several other approaches are based on the application of a potential at the cathode either directly or by redirecting the reactant streams with the objective to either remove (desorption) or

Table 3 Summary of contamination mitigation types, approaches, effects, and mechanisms

Type		Approach	Effect	Mechanism	References
Material strategies	Bipolar plate	Thermal conversion coatings (nitride, carbide, boride including ternary nitride)	Unknown		Brady et al. (2004)
		Materials containing less Fe	Less membrane degradation	Reduction in Fe centers in the membrane catalyzing radical production	Pozio et al. (2003)
		Stainless steel with higher Cr and Ni content	Decreased contact resistance		Davies et al. (2000b), Wang et al. (2003)
		Stainless steel with lower Cr+Fe content	Decreased contact resistance	Passivation/corrosion film comprises more conductive oxides	Silva and Pozio (2007)
		Stainless steel with a nitride coating	Decreased contact resistance and corrosion rate	Absence of corrosion products on treated plate surface	Silva and Pozio (2007), Wang and Northwood (2007)
		Stainless steel pretreatment	Surface structure modification reducing contact resistance		Makkus et al. (2000)
		Stainless steel with a C coating (pyrolysis of high C content polymers)	Reduced corrosion and contaminant level	Isolation of bipolar plate surface from the corrosive environment	Cunningham et al. (2002)
		Stainless steel with an oxide coating	Decreased contact resistance or corrosion rate		Wang et al. (2007a)
	End plate	Anodization of Al alloy followed by sealing with epoxy resin	Significantly reduced corrosion rate	Segregation of plate from corrosive environment	Fu et al. (2007)
	Membrane	Addition of an additive (cis-C_2H_2–$(COOH)_2$, trans-$C_2H_2(COOH)_2$, $C_6H_4(COOH)_2$, H_2NCH_2COOH, $CH_3CH(NH_2)COOH$)	Contaminant effect reduced	Pt/Nafion interface changes by substance with opposite charge than contaminant (ion pairing, intermediate layer with liquidlike structure)	Okada et al. (2003)

(continued)

Table 3 (continued)

Type		Approach	Effect	Mechanism	References
		Reduction in ionomer end groups and vulnerability (e.g.,, treatment with elemental fluorine)	Reduction in ionomer degradation rate	Reduction in sites vulnerable to peroxide and radical attack	Curtin et al. (2004)
		Ionomer coating with an anion exchange layer (cross-linked polyethylene-like polymer with NH_2 and $CONH_2$ groups)	Unknown	Enhanced ionomer composite H^+ selectivity	Zeng et al. (2000)
		Adding Pt or Ag/SiO_2	Reduced ionomer degradation	Catalyzed decomposition of H_2O_2 and, HO• and HO_2• radicals scavenging	Aoki et al. (2006), Xing et al. (2007)
		Select membrane with faster rate of water absorption	Reduced membrane ohmic drop	Reduce membrane dehydration near the anode by the larger cation contaminant induced electroosmotic drag	Chen et al. (2004), Okada et al. (1998c)
		Select a thinner membrane (<50–60 μm)	Reduced membrane dehydration	Water drag from anode to cathode compensated by diffusion from cathode to anode	Okada et al. (1998c), Okada (1999)
		Adapted laminated or bipolar designs	Reduced membrane dehydration	Water drag from anode to cathode compensated by diffusion from cathode to anode at most sensitive locations (near the cathode)	Okada (1999)
		Less Ru permeable membranes	Reduced contaminant level		Gancs et al. (2007)
Material strategies	Catalyst	Other catalyst (favors either NH_4^+ oxidation, noble metals less affected by NH_3)	Unknown		Halseid et al. (2006b)
		Improved mixing homogeneity of alloy components	Proportion of segregated alloying element phase reduced	Reduced alloying element dissolution rate	Antolini et al. (2006), Lebedeva and Janssen (2005)

		Use of ordered alloy phases	Removal of less stable disordered alloy phases	Reduced alloying element dissolution rate	Koh et al. (2007)
		Alloying Pt catalyst	Reduced membrane ohmic drop, C support corrosion and contaminant level	Reduced Pt cation insertion into ionomer and, decrease in C defects (surface heterogeneity) and corrosion initiation sites	Ball et al. (2007), Yu et al. (2005)
		Ru tolerant catalyst	Unknown		Gancs et al. (2007)
		Adding MnO_2	Reduced ionomer degradation	Increased H_2O_2 decomposition rate reducing its effective concentration	Trogadas and Ramani (2007)
	Catalyst support	Acetylene black	Increase in O_2 reduction reaction activity for Pt	Reduction in S levels	Swider and Rolison (2000)
		Low surface area carbons	Reduced corrosion and contaminant level		Ball et al. (2007)
		High temperature treatment (graphitization)	Reduced corrosion and contaminant level	Decrease in C defects (surface heterogeneity) and corrosion initiation sites	Ball et al. (2007), Maass et al. (2008)
	Gas diffusion electrode binder	More chemically stable hydrophobic agent than polytetrafluoroethylene	Unknown		Schulze et al. (2005, 2007)
Design and operation strategies	Cathode potential change	Potential scanning, open circuit potential	Unclear after NH_3 exposure, partial recovery after C_6H_6, H_2S, or mixture (NO, NO_2, SO_2) exposure, full recovery after NO_2 and SO_2 exposure	H_2S, NO_2, SO_2, C_6H_6 oxidation at high cathode potentials. Oxidation of S at 0.7 V vs Ag/AgCl (1 M)	Garsany et al. (2007), Halseid et al. (2006b), Jing et al. (2007), Mohtadi et al. (2004), Moore et al. (2000), Okada et al. (2001), Sung et al. (1997)
		Temporarily switching reactants from one compartment to the other	Breakdown of passive layer on stainless steel in anode environment reducing contact resistance		Makkus et al. (2000)

(continued)

Table 3 (continued)

Type		Approach	Effect	Mechanism	References
	Cathode potential change, cathode compartment wash	Exposure to aqueous acid solution with or without electrode potential cycling	Removal of contaminant cation from ionomer		Okada et al. (1999b, 2000), Shi and Anson (1997)
	Cathode compartment wash	Use higher anode reactant flow relative humidity	Reduced membrane ohmic drop	Reduce membrane dehydration near the anode by the larger cation contaminant induced electroosmotic drag	Chen et al. (2004)
		Circulate liquid water within the cathode compartment	Unclear	Removal of adsorbed hydrophilic NaCl from fuel cell carbon components? Removal or redistribution of NaCl in the ionomer?	Mikkola et al. (2007)
		Wet fuel and oxidant circulation within respective reactant compartments	Partial to complete recovery after exposure		St-Pierre et al. (2000)
	Membrane wash	Immersion in aqueous acid solution	Removal of contaminant cation from ionomer	Ion exchange between protons and contaminant cations	Iojoiu et al. (2007)
	System component, cathode compartment wash	Contaminant source removal including the addition of an acid trap, a filter system or the use of a N_2 purge	Partial to complete recovery after exposure	Oxidation of NH_4^+ to N_2 and/or NO at >0.7 V vs SHE, CO oxidation at operating cathode potentials in the presence of O_2, contaminant desorption from the catalyst, other contaminant mechanisms unclear	Bétournay et al. (2004), Halseid et al. (2006b), Mohtadi et al. (2004), Moore et al. (2000), Yang et al. (2006)

Design and operation strategies	Manufacturing materials and processing specifications	Cleanliness of membrane/electrode assembly preparation and reactant stream	Reduced O_2 reduction reaction overpotential and H_2O_2 production	Reduction in Cl^- levels	Schmidt et al. (2001)
		Low-temperature oxidation of carbon black on Pt oxide in oxygenated aqueous acidic media	Increase in O_2 reduction reaction activity for Pt	Reduction in S levels	Swider and Rolison (1999)
		Use of water rather than organic solvents to solubilize ionomers	Unknown	Reduction in potential contaminant sources	Laporta et al. (2000)
	Unit cell design	Avoid contact between metallic bipolar plate and ionomer (gasket)	Reduced corrosion and contaminant level	Reduction in acidity of the medium in contact with the bipolar plate	Davies et al. (2000b), Makkus et al. (2000)
	System component	Adding a filtering unit to the coolant loop comprised of activated carbon and, cation and anion exchange resins	Unclear	Reduced contaminant concentration in the coolant in contact with the membrane leading to reduced penetration	St-Pierre and Jia (2002)
		Adding a filter at the cathode loop inlet	Reduced contaminant level	Contaminants reaching the fuel cell reduced by adsorption on suitable materials	Kennedy et al. (2007), Ma et al. (2008)

SHE Standard hydrogen electrode

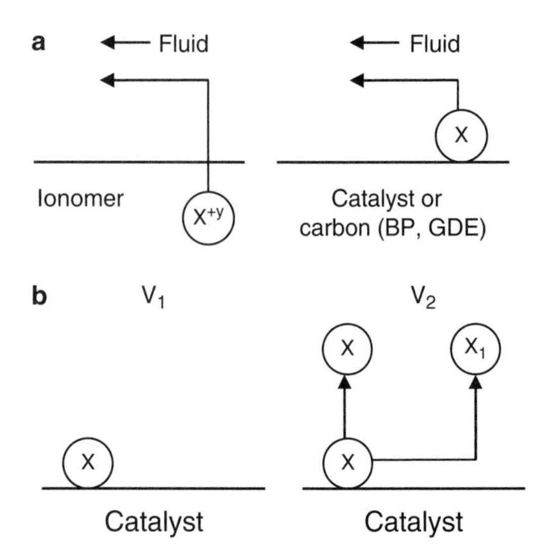

Fig. 5 (**a**) Mitigation by fluid circulation in the cathode compartment leading to ion exchange of foreign cations X^{y+} from the ionomer and contaminant X desorption from the catalyst and carbon components. (**b**) Mitigation by imposed potential changes at the cathode leads to contaminant X desorption or the creation of a product X_1 removed from the catalyst surface. *BP* bipolar plate, *GDE* gas-diffusion electrode

oxidize/reduce contaminants (Fig. 5b). For this latter approach, both effects and mechanisms are unclear in some cases. Therefore, additional work is required for efficiency optimization, including detection of products to determine the likelihood of additional contamination (product X_1 in Fig. 5b). Both of the preceding generic mitigation approaches are consistent with the suggested mechanism modeling strategy exposed in the previous section, which also targets initially kinetic and ohmic effects. A few mitigation approaches are concerned with the incoming material purity prior to cell assembly or to the cleanliness of the manufacturing process itself. A detailed analysis and testing plan represents a beneficial development activity to reduce the impact of contaminants and an opportunity to refine incoming material and processing specifications.

6 Conclusion

A high-level summary and analysis of contamination sources, mechanisms, and mitigations led to several specific research directions to address the daunting task of a timely evaluation of an unknown and largely unidentified pool of fuel cell cathode contaminants. From this standpoint, the additional problem of a stream containing multiple contaminants requires careful consideration because the possibilities are even more staggering than for single contaminants (the number of combinations is conservatively estimated at $97 != 9.6 \times 10^{151}$, a large number) creating a difficulty to determine the mixtures that have the most likeliness or interest. Because there is already significant work to be completed for single contaminants, the recommendation to study air contaminant mixtures (Cheng et al. 2007) does not make sense unless it is supplemented by an efficient procedure to reduce the number of candidates. Such procedures are urgently needed and, for example, are beyond the capabilities of combinatorial approaches (Smotkin and Díaz-Morales 2003). Other recommendations by this research group (Cheng et al. 2007) are also misleading. Transient operating condition tests were suggested to bring understanding closer to applications. Contamination time scales vary widely but are usually relatively long, reaching tens or hundreds of hours partly owing to the dilute concentrations. Under these conditions, a transient regime will very likely not reveal any additional information since the contamination time scale is longer than the time scale of any other process taking place within the fuel cell. The longest fuel cell time scales are related to liquid water movement and heat transfer, reaching up to 1 h in length. During a change in operating condition, the contamination process is virtually at a steady state (rate-determining step); therefore, the fuel cell response to contamination is in this regard a summation of responses obtained under steady-state operating conditions but with different operating parameter values. Development of efficient filters was also advocated to levels lower than 5 ppb (Cheng et al. 2007); however, a justification for such a low contamination level was not provided and is not expected to be uniformly applicable to all contaminants.

References

Ahn, S.-Y., Shin, S.-J., Ha, H. Y., Hong, S.-A., Lee, Y.-C., Lim, T. W. and Oh. I.-H. (2002) Performance and lifetime analysis of the kW-class PEMFC stack. J. Power Sources 106, 295–303

Antolini, E., Salgado, J. R. C. and Gonzalez, E. R. (2006) The stability of Pt-M (M = first row transition metal) alloy catalysts and its effect on the activity in low temperature fuel cells – A literature review and tests on a Pt-Co catalyst. J. Power Sources 160, 957–968

Aoki, M., Uchida, H. and Watanabe, M. (2006) Decomposition mechanism of perfluorosulfonic acid electrolyte in polymer electrolyte fuel cells. Electrochem. Commun. 8, 1509–1513

Ball, S. C., Hudson, S. L., Thompsett, D. and Theobald, B. (2007) An investigation into factors affecting the stability of carbons and carbon supported platinum and platinum/cobalt alloy catalysts during 1.2 V potentiostatic hold regimes at a range of temperatures. J. Power Sources 171, 18–25

Bétournay, M. C., Bonnell, G., Edwardson, E., Paktunc, D., Kaufman, A. and Lomma, A. T. (2004) The effects of mine conditions on the performance of a PEM fuel cell. J. Power Sources 134, 80–87

Bhugun, I. and Anson, F. C. (1997) A generalized treatment of the dynamics of the adsorption of Langmuirian systems at stationary or rotating disk electrodes. J. Electroanal. Chem. 439, 1–6

Bi, W., Gray, G. E. and Fuller, T. F. (2007) PEM fuel cell Pt/C dissolution and deposition in Nafion electrolyte. Electrochem. Solid-State Lett. 10, B101–B104

Bosnjakovic, A. and Schlick, S. (2004) Nafion perfluorinated membranes treated in Fenton media: Radical species detected by ESR spectroscopy. J. Phys. Chem. B. 108, 4332–4337

Brady, M. P., Weisbrod, K., Paulauskas, I., Buchanan, R. A., More, K. L., Wang, H., Wilson, M., Garzon, F. and Walker L. R. (2004) Preferential thermal nitridation to form pin-hole free Cr-nitrides to protect proton exchange membrane fuel cell metallic bipolar plates. Scripta Materialia 50, 1017–1022

Brady, M. P., Yang, B., Wang, H., Turner, J. A., More, K. L., Wilson, M. and Garzon, F. (2006) The formation of protective nitride surfaces for PEM fuel cell metallic bipolar plates. JOM 58, 50–57

Cai, M., Ruthkosky, M. S., Merzougui, B., Swathirajan, S., Balogh, M. P. and Oh, S. H. (2006) Investigation of thermal and electrochemical degradation of fuel cell catalysts. J. Power Sources 160, 977–986

Chaparro, A. M., Mueller, N., Atienza, C. and Daza, L. (2006) Study of electrochemical instabilities of PEMFC electrodes in aqueous solution by means of membrane inlet mass spectrometry. J. Electroanal. Chem. 591, 69–73

Chen, F., Su, Y.-G., Soong, C.-Y., Yan, W.-M. and Chu, H.-S. (2004) Transient behavior of water transport in the membrane of a PEM fuel cell. J. Electroanal. Chem. 566, 85–93

Cheng, X., Shi, Z., Glass, Z., Zhang, L., Zhang, J., Song, D., Liu, Z.-S., Wang, H. and Shen, J. (2007) A review of PEM hydrogen fuel cell contamination: Impacts, mechanisms, and mitigation. J. Power Sources 165, 739–756

Cleghorn, S. J. C., Mayfield, D. K., Moore, D. A., Moore, J. C., Rusch, G., Sherman, T. W., Sisofo, N. T. and Beuscher, U. (2006) A polymer electrolyte fuel cell life test: 3 years of continuous operation. J. Power Sources 158, 446–454

Colón-Mercado, H. R. and Popov, B. N. (2006) Stability of platinum based alloy cathode catalysts in PEM fuel cells. J. Power Sources 155, 253–263

Cunningham, N., Guay, D., Dodelet, J. P., Meng, Y., Hlil, A. R. and Hay, A. S. (2002) New materials and procedures to protect metallic PEM fuel cell bipolar plates. J. Electrochem. Soc. 149, A905–A911

Curtin, D. E., Lousenberg, R. D., Henry, T. J., Tangeman, P. C. and Tisack, M. E. (2004) Advanced materials for improved PEMFC performance and life. J. Power Sources 131, 41–48

Dam, V. A. T. and de Bruijn, F. A. (2007) The stability of PEMFC electrodes – Platinum dissolution vs potential and temperature investigated by quartz crystal microbalance. J. Electrochem. Soc. 154, B494–B499

Darling, R. M. and Meyers, J. P. (2003) Kinetic model of platinum dissolution in PEMFCs. J. Electrochem. Soc. 150, A1523–A1527

Darling R. M. and Meyers, J. P. (2005) Mathematical model of platinum movement in PEM fuel cells. J. Electrochem. Soc. 152, A242–A247

Davies, D. P., Adcock, P. L., Turpin, M. and Rowen, S. J. (2000a) Bipolar plate materials for solid polymer fuel cells. J. Appl. Electrochem. 30, 101–105

Davies, D. P., Adcock, P. L., Turpin, M. and Rowen, S. J. (2000b) Stainless steel as a bipolar plate material for solid polymer fuel cells. J. Power Sources 86, 237–242

Ferreira, P. J., la O', G. J., Shao-Horn, Y., Morgan, D., Makharia, R., Kocha, S. and Gasteiger, H. A. (2005) Instability of Pt/C electrocatalysts in proton exchange membrane fuel cells – A mechanistic investigation. J. Electrochem. Soc. 152, A2256–A2271

Franco, A. A. and Tembely, M. (2007) Transient multiscale modeling of aging mechanisms in a PEFC cathode. J. Electrochem. Soc. 154, B712–B723

Fu, Y., Hou, M., Yan, X., Hou, J., Luo, X., Shao, Z. and Yi, B. (2007) Research progress of aluminum alloy endplates for PEMFCs. J. Power Sources 166, 435–440

Gancs, L., Hult, B. N., Hakim, N. and Mukerjee, S. (2007) The impact of Ru contamination of a Pt/C electrocatalyst on its oxygen-reducing activity. Electrochem. Solid-State Lett. 10, B150–B154

Garsany, Y., Baturina, O. A. and Swider-Lyons, K. E. (2007) Impact of sulfur dioxide on the oxygen reduction reaction at Pt/Vulcan carbon electrocatalysts. J. Electrochem. Soc. 154, B670–B675

Gasteiger, H. A., Kocha, S. S., Sompalli, B. and Wagner, F. T. (2005) Activity benchmarks and requirements for Pt, Pt-alloy, and non-Pt oxygen reduction catalysts for PEMFCs. Appl. Catal. B 56, 9–35

Guilminot, E., Corcella, A., Charlot, F., Maillard, F. and Chatenet, M. (2007a) Detection of Pt^{z+} ions and Pt nanoparticles inside the membrane of a used PEMFC. J. Electrochem. Soc. 154, B96–B105

Guilminot, E., Corcella, A., Chatenet, M., Maillard, F., Charlot, F., Berthomé, G., Iojoiu, C., Sanchez, J.-Y., Rossinot, E. and Claude, E. (2007b) Membrane and active layer degradation

upon PEMFC steady-state operation I. Platinum dissolution and redistribution within the MEA. J. Electrochem. Soc. 154, B1106–B1114
Gülzow, E., Schulze, M. and Steinhilber, G. (2002) Investigation of the degradation of different nickel anode types for alkaline fuel cells (AFCs). J. Power Sources 106, 126–135
Gülzow, E., Schulze, M., Wagner, N., Kaz, T., Reissner, R., Steinhilber, G. and Schneider, A. (2000) Dry layer preparation and characterization of polymer electrolyte fuel cell components. J. Power Sources 86, 352–362
Halseid, R., Vie, P. J. S. and Tunold, R. (2004) Influence of ammonium on conductivity and water content of Nafion 117 membranes. J. Electrochem. Soc. 151, A381–A388
Halseid, R., Bystroň, T. and Tunold, R. (2006a) Oxygen reduction on platinum in aqueous sulphuric acid in the presence of ammonium. Electrochim. Acta 51, 2737–2742
Halseid, R., Vie, P. J. S. and Tunold, R. (2006b) Effect of ammonia on the performance of polymer electrolyte membrane fuel cells. J. Power Sources 154, 343–350
Halseid, R., Wainright, J. S., Savinell, R. F. and Tunold, R. (2007) Oxidation of ammonium on platinum in acidic solutions. J. Electrochem. Soc. 154, B263–B270
Healy, J., Hayden, C., Xie, T., Olson, K., Waldo, R., Brundage, M., Gasteiger, H. and Abbott, J. (2005) Aspects of the chemical degradation of PFSA ionomers used in PEM fuel cells. Fuel Cells 5, 302–308
Helfferich, F. (1962) *Ion exchange*. McGraw-Hill, New York
Hickner, M. A., Ghassemi, H., Kim, Y. S., Einsla, B. R. and McGrath, J. E. (2004) Alternative polymer systems for proton exchange membranes (PEMs). Chem. Rev. 104, 4587–4612
Imamura, D., Akai, M. and Watanabe, S. (2005) Exploration of hydrogen odorants for fuel cell vehicles. J. Power Sources 152, 226–232
Inaba, M., Kinumoto, T., Kiriake, M., Umebayashi, R., Tasaka, A. and Ogumi, Z. (2006) Gas crossover and membrane degradation in polymer electrolyte fuel cells. Electrochim. Acta 51, 5746–5753
Iojoiu, C., Guilminot, E., Maillard, F., Chatenet, M., Sanchez, J.-Y., Claude, E. and Rossinot, E. (2007) Membrane and active layer degradation following PEMFC steady-state operation II. Influence of Pt^{z+} on membrane properties. J. Electrochem. Soc. 154, B1115–B1120
Iyer, S. T., Nandan, D. and Venkataramani, B. (1996) Alkaline earth metal ion-proton-exchange equilibria on Nafion-117 and Dowex 50W X8 in aqueous solutions at 298 ± 1 K. React. Funct. Polym. 29, 51–57
Jing, F., Hou, M., Shi, W., Fu, J., Yu, H., Ming, P. and Yi, B. (2007) The effect of ambient contamination on PEMFC performance. J. Power Sources 166, 172–176
Kadirov, M. K., Bosnjakovic, A. and Schlick, S. (2005) Membrane-derived fluorinated radicals detected by electron spin resonance in UV-irradiated Nafion and Dow ionomers: Effect of counterions and H_2O_2. J. Phys. Chem. B 109, 7664–7670
Kelly, M. J., Egger, B., Fafilek, G., Besenhard, J. O., Kronberger, H. and Nauer, G. E. (2005a) Conductivity of polymer electrolyte membranes by impedance spectroscopy with microelectrodes. Solid State Ionics 176, 2111–2114
Kelly, M. J., Fafilek, G., Besenhard, J. O., Kronberger, H. and Nauer, G. E. (2005b) Contaminant absorption and conductivity in polymer electrolyte membranes. J. Power Sources 145, 249–252
Kennedy, D. M., Cahela, D. R., Zhu, W. H., Westrom, K. C., Nelms, R. M. and Tatarchuk, B. J. (2007) Fuel cell cathode air filters: Methodologies for design and optimization. J. Power Sources 168, 391–399
Kinumoto, T., Inaba, M., Nakayamaa, Y., Ogata, K., Umebayashi, R., Tasaka, A., Iriyama, Y., Abe, T. and Ogumi, Z. (2006) Durability of perfluorinated ionomer membrane against hydrogen peroxide. J. Power Sources 158, 1222–1228
Koh, S., Leisch, J., Toney, M. F. and Strasser, P. (2007) Structure-activity-stability relationships of Pt–Co alloy electrocatalysts in gas-diffusion electrode layers. J. Phys. Chem. C 111, 3744–3752
Laporta, M., Pegoraro, M. and Zanderighi, L. (2000) Recast Nafion-117 thin film from water solution. Macromol. Mater. Eng. 282, 22–29
Lebedeva, N. P. and Janssen, G. J. M. (2005) On the preparation and stability of bimetallic PtMo/C anodes for proton-exchange membrane fuel cells. Electrochim. Acta 51, 29–40

Lehmani, A., Turq, P., Périé, M., Périé, J. and Simonin, J.-P. (1997) Ion transport in Nafion 117 membrane. J. Electroanal. Chem. 428, 81–89

Liu, D. and Case, S. (2006) Durability study of proton exchange membrane fuel cells under dynamic testing conditions with cyclic current profile. J. Power Sources 162, 521–531

Liu, W. and Zuckerbrod, D. (2005) In situ detection of hydrogen peroxide in PEM fuel cells. J. Electrochem. Soc. 152, A1165–A1170

Liu, W., Ruth, K. and Rusch, G. (2001) Membrane durability in PEM fuel cells. J. New Mater. Electrochem. Syst. 4, 227–232

Ma, L., Warthesen, S. and Shores, D. A. (2000) Evaluation of materials for bipolar plates in PEMFCs. J. New Mater. Electrochem. Syst. 3, 221–228

Ma, X., Yang, D., Zhou, W., Zhang, C., Pan, X., Xu, L., Wu, M. and Ma, J. (2008) Evaluation of activated carbon adsorbent for fuel cell cathode air filtration. J. Power Sources 175, 383–389

Maass, S., Finsterwalder, F., Frank, G., Hartmann, R. and Merten, C. (2008) Carbon support oxidation in PEM fuel cell cathodes. J. Power Sources, 176, 444–451

Makkus, R. C., Janssen, A. H. H., de Bruijn, F. A. and Mallant, R. K. A. M. (2000) Use of stainless steel for cost competitive bipolar plates in the SPFC. J. Power Sources 86, 274–282

Masten, D. A. and Bosco, A. D. (2003) System design for vehicle applications: GM/Opel. In: Vielstich, W., Gasteiger, H. and Lamm, A. (Ed.), *Handbook of Fuel Cells – Fundamentals, Technology and Applications, Vol. 4 – Fuel Cell Technology and Applications, Part 3*. Wiley, Chichester, West Sussex, England, pp. 714–724

Mikkola, M. S., Rockward, T., Uribe, F. A. and Pivovar, B. S. (2007) The effect of NaCl in the cathode air stream on PEMFC performance. Fuel Cells 7, 153–158

Mitsushima, S., Kahawara, S., Ota, K.-i. and Kamiya, N. (2007) Consumption rate of Pt under potential cycling. J. Electrochem. Soc. 154, B153-B158

Mittal, V. O., Kunz, H. R. and Fenton, J. M. (2006) Is H_2O_2 involved in the membrane degradation mechanism in PEMFC?. Electrochem. Solid-State Lett. 9, A299–A302

Mittal, V. O., Kunz, H. R. and Fenton, J. M. (2007) Membrane degradation mechanisms in PEMFCs. J. Electrochem. Soc. 154, B652–B656

Mohtadi, R., Lee, W.-K. and Van Zee, J. W. (2004) Assessing durability of cathodes exposed to common air impurities. J. Power Sources 138, 216–225

Moore, J. M., Adcock, P. L., Lakeman, J. B. and Mepsted, G. O. (2000) The effects of battlefield contaminants on PEMFC performance. J. Power Sources 85, 254–260

Narusawa, K., Hayashida, M., Kamiya, Y., Roppongi, H., Kurashima, D. and Wakabayashi, K. (2003) Deterioration in fuel cell performance resulting from hydrogen fuel containing impurities: poisoning effect of CO, CH_4, HCHO and HCOOH. JSAE Rev. 24, 41–46

Ohma, A., Suga, S., Yamamoto, S. and Shinohara, K. (2007) Membrane degradation behavior during open-circuit voltage hold test. J. Electrochem. Soc. 154, B757–B760

Okada, T. (1999) Theory for water management in membranes for polymer electrolyte fuel cells – Part 2. The effect of impurity ions at the cathode side on the membrane performances. J. Electroanal. Chem. 465, 18–29

Okada, T. (2003) Effect of ionic contaminants. In: Vielstich, W., Gasteiger, H. and Lamm, A. (Ed.), *Handbook of Fuel Cells – Fundamentals, Technology and Applications, Vol. 3 – Fuel Cell Technology and Applications, Part 1*. Wiley, Chichester, West Sussex, England, pp. 627–646

Okada, T., Nakamura, N., Yuasa, M. and Sekine, I. (1997) Ion and water transport characteristics in membranes for polymer electrolyte fuel cells containing H^+ and Ca^{+2} cations. J. Electrochem. Soc. 144, 2744–2750

Okada, T., Møller-Holst, S., Gorseth, O. and Kjelstrup, S. (1998a) Transport and equilibrium properties of Nafion membranes with H^+ and Na^+ ions. J. Electroanal. Chem. 442, 137–145

Okada, T., Xie, G., Gorseth, O., Kjelstrup, S., Nakamura, N. and Arimura, T. (1998b) Ion and water transport characteristics of Nafion membranes as electrolytes. Electrochim. Acta 43, 3741–3747

Okada, T., Xie, G. and Meeg, M. (1998c) Simulation for water management in membranes for polymer electrolyte fuel cells. Electrochim. Acta 43, 2141–2155

Okada, T., Ayato, Y., Yuasa, M. and Sekine, I. (1999a) The effect of impurity cations on the transport characteristics of perfluorosulfonated ionomer membranes. J. Phys. Chem. B 103, 3315–3322
Okada, T., Dale, J., Ayato, Y., Asbjørnsen, O. A., Yuasa, M. and Sekine, I. (1999b) Unprecedented affect of impurity cations on the oxygen reduction kinetics at platinum electrodes covered with perfluorinated ionomer. Langmuir 15, 8490–8496
Okada, T., Ayato, Y., Dale, J., Yuasa, M., Sekine, I. and Asbjørnsen, O. A. (2000) Oxygen reduction kinetics at platinum electrodes covered with perfluorinated ionomer in the presence of impurity cations Fe^{3+}, Ni^{2+} and Cu^{2+}. Phys. Chem. Chem. Phys. 2, 3255–3261
Okada, T., Ayato, Y., Satou, H. Yuasa, M. and Sekine, I. (2001) The effect of impurity cations on the oxygen reduction kinetics at platinum electrodes covered with perfluorinated ionomer. J. Phys. Chem. B 105, 6980–6986
Okada, T., Satou, H. and Yuasa, M. (2003) Effects of additives on oxygen reduction kinetics at the interface between platinum and perfluorinated ionomer. Langmuir 19, 2325–2332
Panchenko, A., Dilger, H., Kerres, J., Hein, M., Ullrich, A., Kaz, T. and Roduner, E. (2004) In-situ spin trap electron paramagnetic resonance study of fuel cell processes. Phys. Chem. Chem. Phys. 6, 2891–2894
Pourcelly, G., Oikonomou, A., Gavach, C. and Hurwitz, H. D. (1990) Influence of the water content on the kinetics of counter-ion transport in perfluorosulphonic membranes. J. Electroanal. Chem. 287, 43–59
Pourcelly, G., Sistat, P., Chapotot, A., Gavach, C. and Nikonenko, V. (1996) Self diffusion and conductivity in Nafion membranes in contact with NaCl + $CaCl_2$ solutions. J. Membr. Sci. 110, 69–78
Pozio, A., Silva, R. F., De Francesco, M. and Giorgi, L. (2003) Nafion degradation in PEFCs from end plate iron contamination. Electrochim. Acta 48, 1543–1549
Qiao, J., Saito, M., Hayamizu, K. and Okada, T. (2006) Degradation of perfluorinated ionomer membranes for PEM fuel cells during processing with H_2O_2. J. Electrochem. Soc. 153, A967–A974
Reiser, C. A., Bregoli, L., Patterson, T. W., Yi, J. S., Yang, J. D., Perry, M. L. and Jarvi, T. D. (2005) A reverse-current decay mechanism for fuel cells. Electrochem. Solid-State Lett. 8, A273–A276
Roen, L. M., Paik, C. H. and Jarvi, T. D. (2004) Electrocatalytic corrosion of carbon support in PEMFC cathodes. Electrochem. Solid-State Lett. 7, A19–A22
Samec, Z., Trojánek, A., Langmaier, J. and Samcová, E. (1997) Diffusion coefficients of alkali metal cations in Nafion from ion-exchange measurements – An advanced kinetic model. J. Electrochem. Soc. 144, 4236–4242
Schmidt, T. J., Paulus, U. A., Gasteiger, H. A. and Behm, R. J. (2001) The oxygen reduction reaction on a Pt/carbon fuel cell catalyst in the presence of chloride anions. J. Electroanal. Chem. 508, 41–47
Schmitz, A., Wagner, S., Hahn, R., Uzun, H. and Hebling, C. (2004) Stability of planar PEMFC in printed circuit board technology. J. Power Sources 127, 197–205
Schulze, M. and Christenn, C. (2005) XPS investigation of the PTFE induced hydrophobic properties of electrodes for low temperature fuel cells. Appl. Surf. Sci. 252, 148–153
Schulze, M., Gülzow, E. and Steinhilber, G. (2001) Activation of nickel-anodes for alkaline fuel cells. Appl. Surf. Sci. 179, 251–256
Schulze, M., Knöri, T., Schneider, A. and Gülzow, E. (2004) Degradation of sealings for PEFC test cells during fuel cell operation. J. Power Sources 127, 222–229
Schulze, M., Wagner, N., Kaz, T. and Friedrich, K. A. (2007) Combined electrochemical and surface analysis investigation of degradation processes in polymer electrolyte membrane fuel cells. Electrochim. Acta 52, 2328–2336
Seo, A., Lee, J., Han, K. and Kim, H. (2006) Performance and stability of Pt-based ternary alloy catalysts for PEMFC. Electrochim. Acta 52, 1603–1611
Sethuraman, V. A., Weidner, J. W. and Protsailo, L. V. (2007) Effect of diphenyl siloxane on the catalytic activity of Pt on carbon. Electrochem. Solid-State Lett. 10, B207–B209

Shi, M. and Anson, F. C. (1997) Dehydration of protonated Nafion coatings induced by cation exchange and monitored by quartz crystal microgravimetry. J. Electroanal. Chem. 425, 117–123
Silva, R. F. and Pozio, A. (2007) Corrosion study on different types of metallic bipolar plates for polymer electrolyte membrane fuel cells. J. Fuel Cell Sci. Technol. 4, 116–122
Smotkin, E. S. and Díaz-Morales, R. R. (2003) New electrocatalysts by combinatorial methods. Annu. Rev. Mater. Res. 33, 557–579
St-Pierre, J. and Wilkinson, D. P. (2001) Fuel cells: A new, efficient and cleaner power source. AIChE. J. 47, 1482–1486
St-Pierre, J. and Jia, N. (2002) Successful demonstration of Ballard PEMFCs for space shuttle applications. J. New Mater. Electrochem. Syst. 5, 263–271
St-Pierre, J., Jia, N. and Rahmani, R. (2008) Proton exchange membrane fuel cell contamination model – Competitive adsorption demonstrated with NO_2. J. Electrochem. Soc. 155, B315-B320
St-Pierre, J., Wilkinson, D. P., Knights, S. and Bos, M. (2000) Relationships between water management, contamination and lifetime degradation in PEFC. J. New Mater. Electrochem. Syst. 3, 99–106
St-Pierre, J., Roberts, J., Colbow, K., Campbell, S. and Nelson, A. (2005) PEMFC operational and design strategies for sub zero environments. J. New Mater. Electrochem. Syst. 8, 163–176
Sung, Y.-E., Chrzanowski, W., Zolfaghari, A., Jerkiewicz, G. and Wieckowski, A. (1997) Structure of chemisorbed sulfur on a Pt(111) electrode. J. Am. Chem. Soc. 119, 194–200
Swider, K. E. and Rolison, D. R. (1996) The chemical state of sulfur in carbon-supported fuel-cell electrodes. J. Electrochem. Soc. 143, 813–819
Swider, K. E. and Rolison, D. R. (1999) Catalytic desulfurization of carbon black on a platinum oxide electrode. Langmuir 15, 3302–3306
Swider, K. E. and Rolison, D. R. (2000) Reduced poisoning of platinum fuel-cell electrocatalysts supported on desulfurized carbon. Electrochem. Solid-State Lett. 3, 4–6
Tandon, R. and Pintauro, P. N. (1997) Divalent/monovalent cation uptake selectivity in a Nafion cation-exchange membrane: experimental and modeling studies. J. Membr. Sci. 136, 207–219
Tan, J., Chao, Y. J., Van Zee, J. W. and Lee, W.-K. (2007) Degradation of elastomeric gasket materials in PEM fuel cells. Mater. Sci. Eng. A, 445–446, 669–675
Tawfik, H., Hung, Y. and Mahajan, D. (2007) Metal bipolar plates for PEM fuel cell – A review. J. Power Sources 163, 755–767
Teranishi, K., Kawata, K., Tsushima, S. and Hirai, S. (2006) Degradation mechanism of PEMFC under open circuit operation. Electrochem. Solid-State Lett. 9, A475–A477
Trogadas, P. and Ramani, V. (2007) Pt/C/MnO_2 hybrid electrocatalysts for degradation mitigation in polymer electrolyte fuel cells. J. Power Sources 174, 159–163
Uribe, F. A. Gottesfeld, S. and Zawodzinski, T. A. (2002) Effect of ammonia as potential fuel impurity on proton exchange membrane fuel cell performance. J. Electrochem. Soc. 149, A293–A296
Wan, L.-J., Moriyama, T., Ito, M., Uchida, H., and Watanabe, M. (2002) In situ STM imaging of surface dissolution and rearrangement of a Pt-Fe alloy electrocatalyst in electrolyte solution. Chem. Commun. 58–59
Wang, Y. and Northwood, D. O. (2006) An investigation on metallic bipolar plate corrosion in simulated anode and cathode environments of PEM fuel cells using potential-pH diagrams. Int. J. Electrochem. Sci. 1, 447–455
Wang, Y. and Northwood, D. O. (2007) An investigation of the electrochemical properties of PVD TiN-coated SS410 in simulated PEM fuel cell environments. Int. J. Hydrogen Energy 32, 895–902
Wang, H., Sweikart, M. A. and Turner J. A. (2003) Stainless steel as bipolar plate material for polymer electrolyte membrane fuel cells. J. Power Sources 115, 243–251
Wang, X., Kumar, R. and Myers, D. J. (2006) Effect of voltage on platinum dissolution – Relevance to polymer electrolyte fuel cells. Electrochem. Solid-State Lett. 9, A225–A227
Wang, H., Turner, J. A., Li, X. and Bhattacharya, R. (2007) SnO_2:F coated austenite stainless steels for PEM fuel cell bipolar plates. J. Power Sources 171, 567–574
Wind, J., Späh, R., Kaiser, W. and Böhm, G. (2002) Metallic bipolar plates for PEM fuel cells. J. Power Sources 105, 256–260

Xie, G. and Okada, T. (1995) Water transport behavior in Nafion 117 membranes. J. Electrochem. Soc. 142, 3057–3062

Xie, G. and Okada, T. (1996) Pumping effects in water movement accompanying cation transport across Nafion 117 membranes. Electrochim. Acta 41, 1569–1571

Xie, J., Wood III, D. L., Wayne, D. M., Zawodzinski, T. A., Atanassov, P. and Borup, R. L. (2005) Durability of PEFCs at high humidity conditions. J. Electrochem. Soc. 152, A104–A113

Xing, D., Zhang, H., Wang, L., Zhai, Y. and Yi, B. (2007) Investigation of the Ag-SiO_2/sulfonated poly(biphenyl ether sulfone) composite membranes for fuel cell. J. Membr. Sci. 296, 9–14

Yadav, A. P., Nishikata, A. and Tsuru, T. (2007) Effect of halogen ions on platinum dissolution under potential cycling in 0.5 M H_2SO_4 solution. Electrochim. Acta 52, 7444–7452

Yang, D., Ma, J., Xu, L., Wu, M. and Wang, H. (2006) The effect of nitrogen oxides in air on the performance of proton exchange membrane fuel cell. Electrochim. Acta 51, 4039–4044

Yasuda, K., Taniguchi, A., Akita, T., Ioroi, T. and Siroma, Z. (2006) Platinum dissolution and deposition in the polymer electrolyte membrane of a PEM fuel cell as studied by potential cycling. Phys. Chem. Chem. Phys. 8, 746–752

Yu, P., Pemberton, M. and Plasse, P. (2005) PtCo/C cathode catalyst for improved durability in PEMFCs. J. Power Sources 144, 11–20

Zaluski, C. S. and Xu, G. (1994) AC impedance and conductivity study of alkali salt form perfluorosulfonate ionomer membranes. J. Electrochem. Soc. 141, 448–451

Zeng, R., Pang, Z. and Zhu, H. (2000) Modification of a Nafion ion exchange membrane by a plasma polymerization process. J. Electroanal. Chem. 490, 102–106

Zhang, J., Litteer, B. A., Gu, W., Liu, H. and Gasteiger, H. A. (2007) Effect of hydrogen and oxygen partial pressure on Pt precipitation within the membrane of PEMFCs. J. Electrochem. Soc. 154, B1006–B1011

Zhang, J., Wang, H., Wilkinson, D. P., Song, D., Shen, J. and Liu, Z.-S. (2005) Model for the contamination of fuel cell anode catalyst in the presence of fuel stream impurities. J. Power Sources 147, 58–71

Impurity Effects on Electrode Reactions in Fuel Cells

Tatsuhiro Okada

Abstract The oxygen reduction reaction (ORR) on platinum-based catalysts in the cathode catalyst layer is affected by several kinds of impurities, such as impurity cations or organic impurities in membrane electrode assemblies and in polymer electrolyte membranes. These impurities may come from outside the cathode chamber, or may be generated inside as decomposition products on the catalyst surface or as crossed-over fuels from the anode. In this chapter the effect of inorganic and organic impurities of 0.1–10 mmol dm^{-3} on the kinetics of the ORR investigated by electrochemical measurements is discussed. Cationic species, aldehydes, and alcohols are found to degrade strongly the ORR current. A method to cope with such impurity problems is proposed where small amounts of additives in the membrane electrode assembly or in the membrane suppress the degradation and affect positively the ORR performances at the catalyst surface.

1 Introduction

The membrane electrode assembly (MEA) is a delicate component in low-temperature fuel cells based on polymer electrolyte membranes. Its condition is affected by many factors (1) selection and preparation of MEA materials (catalysts, supporting carbon powder, membrane materials, binder for MEA hot pressing, etc.), (2) history of MEA usage, (3) fuel cell operation parameters, and so on. The resulting MEA condition exerts a strong influence on the fuel cell performance, which is also a function of running time.

Elucidation of the degradation mechanism of anode or cathode reactions by impurity materials in polymer electrolyte fuel cells (PEFCs) is a crucial topic, to

T. Okada
National Institute of Advanced Industrial Science and Technology, Higashi 1-1-1, Central 5, Tsukuba, Ibaraki 305-8565, Japan,
email: okada.t@aist.go.jp

F.N. Büchi et al. (eds.), *Polymer Electrolyte Fuel Cell Durability,*

attain its longevity. Especially, since its kinetics is very slow, the cathode reaction (oxygen reduction reaction, ORR; $O_2 + 4H^+ + 4e^- \rightarrow 2H_2O$) will be affected much by environmental factors. In this work, the cathode catalyst layer in fuel cells is modeled by electrochemical measuring systems with a three-electrode cell, and the degradation mechanism of the oxygen electrode by impurity materials is studied for several cases.

First, we discuss the ORR on a platinum plate rotating disk electrode (RDE) covered with Nafion® film in the presence of various kinds of impurity cations (Okada 2003), such as alkali and alkaline earth metal cations, transition metal cations, or alkylammonium cations, in H_2SO_4 aqueous solutions.

Second, materials that may affect the ORR on platinum, organic impurities that come from the environment, MEA components, or fuels, are discussed. Alcohols or aldehydes are dissolved in H_2SO_4 solutions, and the platinum RDE is investigated electrochemically under the ORR. The blockage of the platinum catalyst surface by these impurity molecules turns out to be a major cause of the degradation.

A new method is proposed to cope with the impurity problems, where small amounts of additives are added to the electrolyte that contacts the catalyst layer. The mechanism of inhibition of the impurity effects is discussed on the basis of the picture of geometric effects or electric double layer modifications by adsorbed molecules.

2 Experimental Procedures

2.1 *Platinum RDE with Nafion® Film Exchanged with Cation Species*

A platinum RDE covered with Nafion® polymer was used as the model of the catalyst surface (Zecevic et al. 1997). After it had been cleaned and dried, the platinum disk electrode was coated with Nafion® polymer by putting a 10:1 mixture of 5 wt% Nafion® solution (Aldrich) and dimethylformamide on a spin-coater (Moore and Martin 1986). The film was dried in air at 80 °C for 30 min, and then annealed in a vacuum chamber at 130 °C for 12 h (film thickness ca. 10 μm).

A three-electrode glass cell was used with the temperature maintained at 25 °C. Current–potential curves of the ORR, cyclic voltammograms (CVs), and linear-sweep voltammograms, first for the noncontaminated condition, were obtained for the Nafion® film covered platinum RDE.

For contaminated film, various amounts of alkali and alkaline earth metal sulfates (Li_2SO_4, Na_2SO_4, $CaSO_4$), transition metal sulfates [$Fe_2(SO_4)_3$, $NiSO_4$, $CuSO_4$] or alkylammonium sulfates [$(NR_4)_2SO_4$, (R is H, CH_3), NR_4HSO_4 (R is CH_3, C_2H_5, C_3H_7, C_4H_9)], at 0.1–10% compared with H^+ in the solution, were dissolved in the solution, and the ORR kinetics were measured likewise. The kinetic measurements were done for different soaking times: 2 h, 1 day, and 3 days.

2.2 *Analysis of Pt/Nafion® Film RDE*

The ORR on a Nafion® film covered RDE is analyzed as reported elsewhere (Okada et al. 1999). The ORR on a Nafion® film covered RDE is governed by three steps (1) diffusion of O_2 in the solution, (2) diffusion of O_2 in Nafion® film, and (3) charge transfer at the platinum surface. Then the kinetic current $j_{k,c}/j_{k,r}$ corresponding to the maximum O_2 concentration in the film (as the ratio corresponding to the contaminated and pure conditions) is obtained from the Koutecky–Levich plots in noncontaminated and contaminated conditions. Transport parameters, C_f^* and D_f, the maximum concentration, and the diffusion coefficient of oxygen in the Nafion® film, respectively, are obtained in the forms $C^*_{f,c}D_{f,c}/C^*_fD_f$ and $C^*_{f,c}D_{f,c}^{1/2}/C^*_fD_f^{1/2}$ from the Koutecky–Levich plots and from linear-sweep voltammetry at various scan rates. Also the parameter αn_a, where α is the transfer coefficient and n_a is the number of electrons involved in the rate-determining step, is obtained from linear-sweep voltammetry.

2.3 *Platinum RDE with Organic Additives*

The ORR electrochemistry in O_2 gas saturated 0.05 mol dm^{-3} H_2SO_4 was investigated at 25 °C using a platinum RDE, with or without organic impurities in the solution. These organic impurities are supposed to come into the cathode catalyst layer through the catalyst ink or MEA binders (2-propanol, Triton-X 100), from decomposition products from MEA binders and membranes (acetone, 1-hexanal, and 1-octanal) or from crossed-over anode fuel (methanol) through the polymer electrolyte in the case of direct methanol fuel cells.

Organic compounds were added to 0.05 mol dm^{-3} H_2SO_4 at 1, 10, and 100 mmol dm^{-3}, and the change of the ORR current was evaluated as the measure of the ORR degradation in the presence of organic impurities. Cyclic voltammetry was performed to obtain the platinum active surface from the hydrogen adsorption/desorption peaks.

2.4 *Effect of Pyridyl Compounds on the ORR on Platinum in the Presence of Methanol*

A rotating ring-disk electrode (RRDE) with a platinum ring and platinum-disk was used in this experiment in a three-compartment electrochemical cell. Pyridyl compounds were added at 0.1 mmol dm^{-3} to the solution, x mol dm^{-3} CH_3OH + 0.05 mol dm^{-3} H_2SO_4 (x = 0, 0.1, 1) saturated with O_2 gas, and polarization measurements were performed (Shiroishi et al. 2004). The open circuit potential, limiting current density, yield of H_2O formation from O_2 (percentage of H_2O), and the current of the methanol oxidation reaction (MOR; $CH_3OH + H_2O^+ \rightarrow CO_2 + 6H^+ + 6e^-$) were evaluated with and without the additives.

A new in situ Fourier transform IR (FTIR) method, surface-enhanced IR absorption spectroscopy (SEIRAS; Osawa 1997), was employed that could observe the surface of the platinum electrode on a silicon prism while electrochemically polarizing it in 0.1 M $HClO_4$. A thin platinum film was chemically deposited on the base plane of a hemicylindrical silicon prism.

After SEIRAS had been performed at each potential in 0.1 M $HClO_4$, 2,2'-bipyridine (bpy) was added to the spectroelectrochemical cell. The potential of the working electrode was maintained at 0.1 V for 30 min and then the FTIR measurements were conducted. Finally, methanol was added to the cell, and the potential was held at 0.1 V for 30 min before the FTIR measurements.

3 The Effect of Impurities on the ORR Kinetics

3.1 Degradation Phenomena of the ORR in the Presence of Metal Cation Impurities

The kinetic current $j_{k,c}$ for oxygen reduction is calculated for Pt–Nafion® electrodes in the presence of impurity cations. Figure 1 shows the time course of the ratios of the ORR kinetic current $j_{k,c}/j_{k,r}$ for the contaminated condition with reference to the pure condition, with several levels of Na^+, Ca^{2+}, Fe^{3+}, and Ni^{2+} impurity ions.

Decay of the kinetics by 40–50% occurred as a very slow process. The decay of the transport parameters, $C^*_{f,c}D_{f,c}$, and $C^*_{f,c}D_{f,c}^{1/2}$, occurred in the same way, but was less than that of the charge-transfer parameters, showing a decay of about 30–40%. Some of the parameters measured after 3–5 days (nearly steady state) are listed in Table 1(Okada et al. 2001).

It is seen that impurity cations, even at a level as low as 0.1% H^+ ions, affect negatively the cathode performance. Some trends in the degradation are observed:

1. Compared with the penetration rate of cations through the Nafion® film (within 1 h), the decay rate is very slow, in the time span of 3–5 days.
2. Concerning the amount of impurities, the decay becomes gradual over 0.1%, and does not increase further beyond the level of 1–10%.
3. Cations of higher charge are exchanged more into the Nafion® film, but the amount of penetration is not directly related to the degradation.
4. Seeing that degradation is low for Li^+ cation (Table 1), the decay is connected to the decrease in the water content in the Nafion® film.

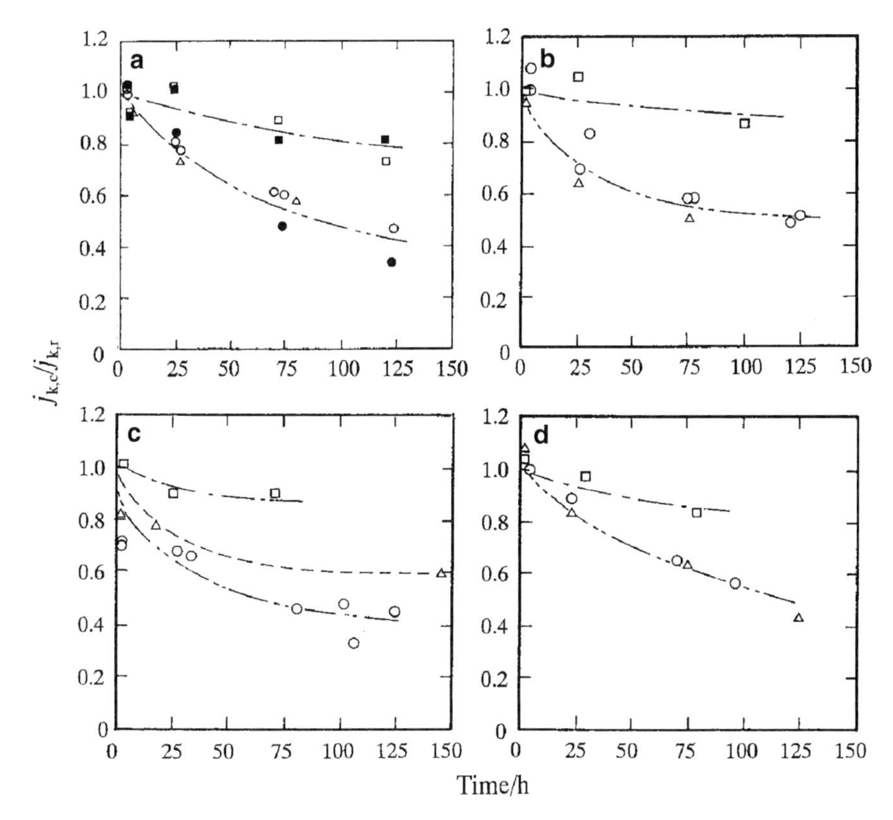

Fig. 1 Kinetic current of oxygen reduction on platinum covered with Nafion® film in 0.1N H_2SO_4 solution containing various kinds of impurity cations. The kinetic current $j_{k,c}$ for contaminated film normalized by the kinetic current $j_{k,r}$ for noncontaminated film is plotted as a function of soaking time. Impurity levels of 0.1% (*open square*), 1% (*open triangle*), and 10% (*open circle*) as compared with H^+. (**a**) Na^+, (**b**) Ca^{2+}, (**c**) Fe^{3+}, (**d**) Ni^{2+}. (Reproduced from Okada et al. 2001, copyright the American Chemical Society)

These facts strongly indicate that the degradation is linked with the change in polymer structures as the time elapses, because the process of water removal from the Nafion® film, which accompanies the relaxation of polymer chain, would occur much more slowly compared with the ion-exchange processes.

It should be noted that degradation did not occur on the bare platinum immersed in liquid electrolyte, indicating that this was a specific phenomenon on the catalyst surface–polymer electrolyte interface (Okada et al. 2000). Unlike liquid electrolytes, the polymer electrolyte has immobile cation-exchange sites (sulfonic acid groups on side chains), so when they bind impurity metal cations, the "constrained electric double layer" lowers the electric field in the electrochemical double layer (Okada et al. 2001).

Table 1 The effect of the impurity level in solution on the impurity level in Nafion® film (% of SO_3), water content (λ)(n_{H2O} / $n_{SO_3^-}$), density in the wet state [d(wet)]/g cm^{-3}, and kinetic current ratio {$j_{k,c}/j_{k,r}$} between contaminated and pure conditions

Impurity level in solution (%)	Impurity level in Nafion membrane (% of SO_3^-), λ, d(wet), $j_{k,c}/j_{k,r}$						
	Li^+	Na^+	K^+	Ca^{2+}	Cu^{2+}	Ni^{2+}	Fe^{3+}
0.1	0.1 (22.1) [1.56]	0.1 (20.8) [1.64] {0.8}	0.2 (22.1) [1.56]	1 (20.9) [1.64] {0.9}	1.1 (22.1) [1.61]	1.5 (22.1) [1.61] {0.8}	1.6 (22.1) [1.61] {0.9}
1.0	0.8 (22.1) [1.57] {0.9}	2 (20.6) [1.64] {0.5}	3 (21.3) [1.59] {0.5}	10 (20.9) [1.66] {0.5}	17 (22.0) [1.63] {0.6}	16 (22.0) [1.61] {0.4}	17 (22.0) [1.62] {0.6}
10	6 (22.2) [1.57] {0.8}	14 (20.3) [1.64] {0.5}	31 (19.8) [1.62]	66 (18.8) [1.70] {0.5}	65 (21.2) [1.65]	64 (21.0) [1.64] {0.4}	77 (19.8) [1.68] {0.4}

Reproduced from Okada et al. 2001, copyright the American Chemical Society

This unique effect in the metal–polymer electrolyte interface must be considered in the design of fuel cells, because metal cations can easily penetrate into the cathode chamber, either from the airstream or through corrosion of stack or tubing materials.

3.2 *Degradation Phenomena of the ORR in the Presence of Alkylammonium Ion Impurities*

It is reported that even 10 ppm NH_3 produced during the reforming process in the H_2 fuel streams irreversibly deteriorates the fuel cell performance (Uribe et al. 2002). Whether this is due to the catalytic effect of NH_3 on the ORR at the cathode or is due to the conductivity loss of the ionomer by exchanged ammonium ion NH_4^+ is an unresolved current debate (Soto et al. 2003; Halseid et al. 2006), and the effect of NH_4^+ or derivatives on the ORR may relate to this subject. The ORR was studied on Pt–Nafion® electrodes in 0.05 mol dm^{-3} H_2SO_4 in the presence of alkylammonium cations R_4N^+ (R is H, CH_3, C_2H_5, C_3H_7, and C_4H_9, abbreviated as NH_4^+, MeN^+, EtN^+, PrN^+, and BuN^+, respectively) (Okada et al. 2003).

CVs in O_2-saturated H_2SO_4 solution containing NH_4^+, MeN^+, and EtN^+ ions did not show significant changes from those in pure H_2SO_4, as long as the amount of impurity was not large (1% of H^+). In the case of PrN^+ or BuN^+, hydrogen adsorption/desorption peaks as well as platinum oxide formation/reduction peaks were distorted even for the lower amount of 0.1%.

In Fig. 2, the decay of the kinetic current in comparison with that for the noncontaminant condition, $j_{k,c}/j_{k,r}$, is plotted as a function of time. Different trends

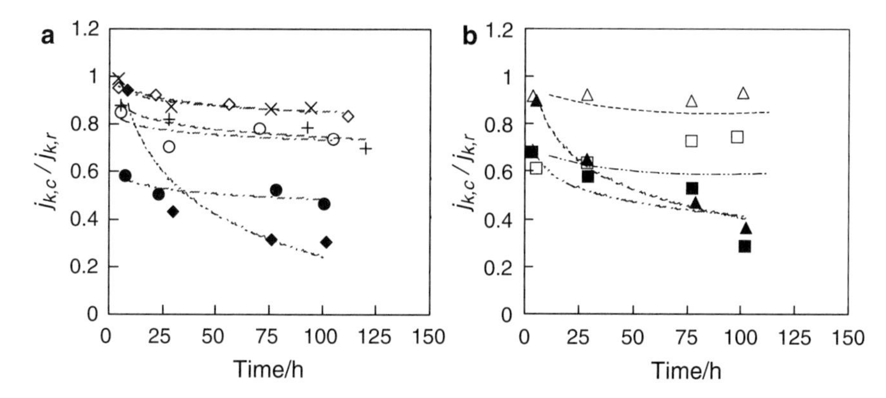

Fig. 2 Kinetic current of oxygen reduction on platinum covered with Nafion® film in 0.05 mol dm^{-3} H_2SO_4 solution containing (**a**) NH_4^+ (*cross, plus*), $(CH_3)_4N^+$ (*open circle, filled circle*), or $(C_2H_5)_4N^+$ (*open diamond, filled diamond*) and (**b**) $(C_3H_7)_4N^+$ (*open triangle, filled triangle*) or $(C_4H_9)_4N^+$ (*open square, filled square*) ions. Impurity amount *cross, open circle, open diamond* 1%, *plus, filled circle, filled diamond* 10%, *open triangle, open square* 0.1%, and *filled triangle, filled square* 1% as compared with H^+ in the solution. (Reproduced from Okada et al. 2003, copyright the American Chemical Society)

are observed in the decay curves of $j_{k,c}$ between two groups of ions. For NH_4^+, MeN^+, and EtN^+ ions, slow decay of $j_{k,c}$ occurred with the soaking time, and this trend was larger for a larger amount of alkylammonium ions. For an amount of 10% as compared with H^+ in the solution, the decay was 60–70% after about 3–5 days. On the other hand, for PrN^+ or BuN^+, the decay occurred rapidly in about several hours, and even in the case of 1% contamination, it reached about 70%. This indicates that the impurity ion BuN^+ adsorbs on the platinum surface from an early stage, and blocks the passage of current on platinum.

The remarkable contrast between the group NH_4^+, MeN^+, and EtN^+ and the group PrN^+ and BuN^+ is also seen in parameters $C^*_{f,c}D_{f,c}$ and $C^*_{f,c}D_{f,c}^{1/2}$ (Fig. 3). In the former group, both parameters decreased gradually over the time course, while in the latter group an abrupt change occurred. For the former group, the decrease of C^*_f prevailed over the time course, whereas for the latter group, a decrease of D_f occurred. The ions in the former group occupy the ionic channel and drive out water molecules, and C^*_f will decrease there. On the other hand, the latter group brings about the irreversible change of the polymer network not only in the ionic channel but also in the polymer backbone. Once this happens, the path for oxygen diffusion will be blocked and D_f will decrease from its value in the initial stage.

For NH_4^+, MeN^+, EtN^+, and PrN^+ ions, αn_a stayed at about 0.5±0.05, with no significant changes with time, while for BuN^+, a negative shift was observed. The last case infers that the reaction route of the ORR is affected by the specific adsorption of large alkylammonium ions.

Concerning the degradation of the ORR, the electric double layer effect and the platinum oxide effect are thought to be most plausible reasons. The fact that

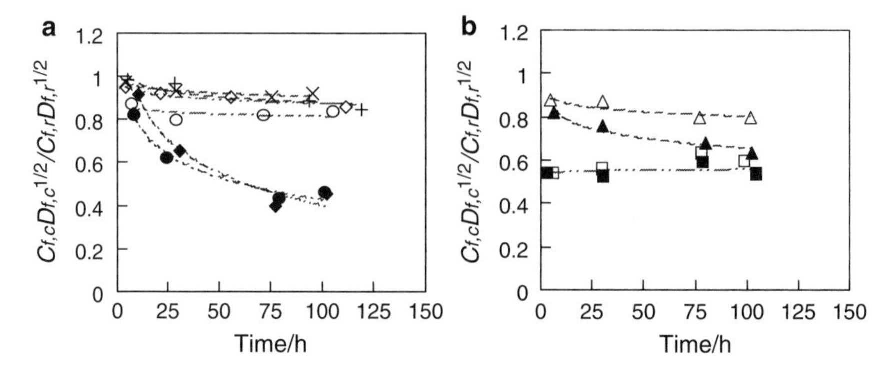

Fig. 3 The parameter $C^*_{f,c}D_{f,c}^{1/2}$ for oxygen transport in Nafion® film in contact with 0.05 mol dm^{-3} H_2SO_4 containing (**a**) NH_4^+, $(CH_3)_4N^+$, or $(C_2H_5)_4N^+$ ions and (**b**) $(C_3H_7)_4N^+$ or $(C_4H_9)_4N^+$ ions. The ratios of these values for the contaminated film against those for the noncontaminated film are shown over the time course. The symbols have the same meaning as in Fig. 2. (Reproduced from Okada et al. 2003, copyright the American Chemical Society)

decay of the charge-transfer kinetics was greater (about 60–70% decrease) than that of the oxygen transport parameters in the Nafion® film (about 20–30%) indicated that the effect was connected rather to the platinum–Nafion® interface than to the polymer bulk. The hysteresis in the oxygen reduction current between anodic and cathodic scans is related to the state of the platinum surface. The "blockage" of active platinum surfaces by alkylammonium ions appeared as the third mechanism. CVs and polarization curves revealed specific changes caused by PrN^+ or BuN^+ ions, probably due to the blockage of the platinum surface.

Table 2 summarizes changes in the ORR and the characteristics of Nafion® films in the presence of impurity ions. It is seen that the amount of water in the membrane differs very much for different ammonium ion species. For higher molecular weight alkylammonium ions, the membrane drying and conductivity decrease are much greater than those for lower molecular weight ones. Especially the membrane drying might bring about almost irreversible changes in the polymer structure, and this would cause a high barrier for electron transfer from platinum to the oxygen molecule at the platinum–polymer interface.

3.3 Degradation Phenomena of the ORR in the Presence of Organic Impurities

Polarization curves of the ORR in the presence of several kinds of organic impurities in 0.05 mol dm^{-3} H_2SO_4 are shown in Fig. 4. With acetone, methanol, and 2-propanol, the ORR started at slightly more negative potentials than in pure

Table 2 Characteristics of the oxygen reduction reaction (ORR) and film properties on Nafion®-filmed platinum electrodes, in the presence of various kinds of impurity ions

Parameters[a]	Impurity level in solution (%)	Impurity ions				
		NH_4^+	$(CH_3)_4N^+$	$(C_2H_5)_4N^+$	$(C_3H_7)_4N^+$	$(C_4H_9)_4N^+$
λ		–	–	–	$18._5$	$19._0$
d(wet)		–	–	–	1.60	1.62
κ/κ_H	0.1	–	–	–	0.92	0.86
$[j_k]$		–	–	–	0.9_3	0.7_5
$[CD]$		–	–	–	0.2_7	0.09
$[CD^{1/2}]$		–	–	–	0.8_0	0.6_0
λ		20_5	16_1	14_7	13_8	1.4
d(wet)		1.57	1.69	1.64	1.59	1.70
κ/κ_H	1.0	0.96	0.65	0.52	0.33	0.0
$[j_k]$		0.8_7	0.7_4	0.8_4	0.3_6	0.2_9
$[CD]$		0.9_7	0.8_5	0.8_0	0.2_3	0.04
$[CD^{1/2}]$		0.9_2	0.8_4	0.8_6	0.6_4	0.5_4
λ		19_8	13_0	13_4	–	–
d(wet)		1.64	1.66	1.62	–	–
κ/κ_H	10	0.80	0.49	0.38	–	–
$[j_k]$		0.7_8	0.4_7	0.3_0	–	–
$[CD]$		0.9_1	0.6_8	0.6_5	–	–
$[CD^{1/2}]$		0.8_9	0.4_6	0.4_6	–	–

Reproduced from Okada et al. 2003, copyright the American Chemical Society

[a] $\lambda(=n_{H2}O / n_{SO3}^-)$, water content; d(wet), density in the wet state; κ/κ_H, conductivity ratio with reference to H-form membrane, for Nafion® membranes in equilibrium with 0.05 mol dm^{-3} H_2SO_4 containing various kinds of alkylammonium ions. For H-form Nafion® membrane, l = 20.5 and d(wet) = 1.61. $[j_k]$, $[CD]$, and $[CD^{1/2}]$ designate $j_{k,c}$, $C^*_{f,c}D_{f,c}$, and $C^*_{f,c}D_{f,c}^{1/2}$ parameters for Nafion®-filmed platinum electrodes after 4 days in the presence of impurity ions as compared with no-impurity conditions

H_2SO_4, but the magnitude of the current density was not affected largely up to a concentration of 10 mmol dm^{-3}. With 2-propanol, the current due to the alcohol oxidation was superimposed on the ORR current, and the potential shifted by approximately 0.5 V at 0.1 mol dm^{-3}. Both 2-propanol and acetone are MEA components because 2-propanol is added to the catalyst ink, and acetone is produced by oxidation of 2-propanol on the platinum cathode. The effects are, however, not serious because their concentration is less than 10 mmol dm^{-3}. Methanol is used as a fuel in direct methanol fuel cells, and crossover through the polymer electrolyte membrane is known to cause a degradation of the ORR at the cathode, when the concentration of the fuel is as high as 5 mol dm^{-3}.

Aldehydes and Triton-X had very serious effects on the ORR. Even at a low level of 1 mmol dm^{-3}, they reduced the ORR current to one-half to one-seventh. As the hydrogen oxidation/reduction peaks in the CVs indicate, this is caused by strong adsorption of these organic materials on platinum. These organic materials are either decomposition products from the membrane materials or additives in the catalyst ink, and their effect should not be ignored when these compounds exist at 1 mmol dm^{-3}.

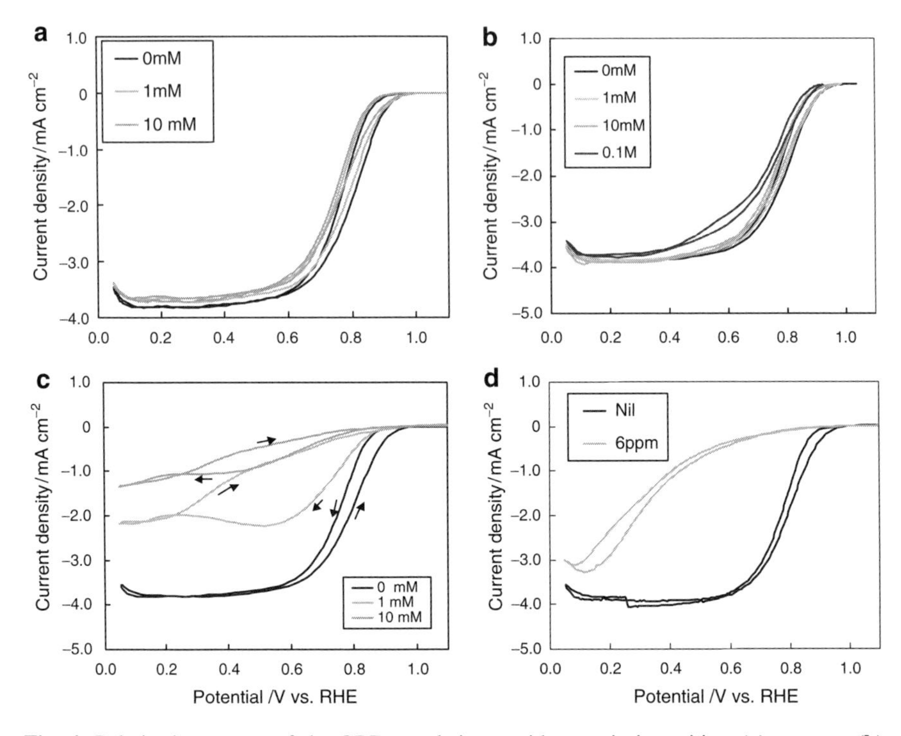

Fig. 4 Polarization curves of the ORR on platinum with organic impurities; (**a**) acetone, (**b**) 2-propanol, (**c**) hexanal, and (**d**) Triton-X100. *RHE* reversible hydrogen electrode

4 Inhibitor Additives and the ORR Kinetics

4.1 Inhibitor Additives to Prevent ORR Degradation by Impurity Cations

A new method to suppress the ORR degradation at the platinum–ionomer interface is proposed on the basis of its sensitivity to the interface structure and the polymer orientation. Some additives that may counteract this structural change at the interface, owing to the opposite charge from impurity cations, are tested as additives in the ionomer film. ORR kinetics is investigated in the presence of impurity cations, and it is evaluated whether or not these additives can inhibit the degradation of the charge-transfer step effectively (Okada et al. 2003).

Two to four percent of additives, with reference to the polymer mass, were incorporated into the Nafion® film, and electrochemical measurements were performed on the Pt–Nafion® disk electrode in 0.05 mol dm^{-3} H_2SO_4 containing 10% of Na^+ or Ca^{2+} ions. Additives tested were carboxylic acids or amino acids, i.e., maleic acid, fumaric acid, phthalic acid, glycine, and D-α-alanine.

In all cases, very small changes in the CVs for the Pt–Nafion® disk electrode were observed in comparison with bare platinum, and with or without impurity ions, indicating that these additives did not block the platinum surface. These results indicated a promising effect of additives, because with additives the CVs did not deteriorate owing to the presence of Na^{+} or Ca^{2+} ions.

Polarization curves of the ORR on platinum disk electrodes covered with Nafion® containing fumaric acid are shown in Fig. 5. Without the additive, the ORR current was degraded greatly by the presence of 10% Ca^{2+} ion, but incorporation of fumaric acid in the film drastically reduced this degradation. In some cases the additives in the film caused a decrease in the limiting current (3–10% decrease as compared with Nafion®-filmed platinum with no additive), but even in such cases the decrease in the current after adding the impurity ions was drastically reduced. Phthalic acid, glycine, and D-α-alanine were also good examples of inhibiting the degradation behavior of the ORR on Nafion®-filmed platinum.

The improvement was apparent when $j_{k,c}/j_{k,r}$ was plotted as a function of time. Without the additives, $j_{k,c}/j_{k,r}$ dropped about 40–50% from the initial value, after 4–5 days. With additives, the charge-transfer kinetics was protected from degradations, except for maleic acid.

Transport parameters were plotted against time course in the presence of 10% Ca^{2+}. Without the additives, both $C^{*}_{f,c}D_{f,c}$ and $C^{*}_{f,c}D_{f,c}^{1/2}$ parameters decreased about 20–30% from the initial value. With additives, the decrease of $C^{*}_{f,c}D_{f,c}^{1/2}$ was not large (about 10%), but the decrease of $C^{*}_{f,c}D_{f,c}$ was enhanced for the Nafion® film. This means that the diffusion coefficient of oxygen in the film dropped owing to the additives.

The αn_a parameter remained almost constant with time (αn_a =0.5), and no substantial change in the rate-determining step was anticipated with additives.

There may be several reasons for this improvement caused by additives. One is the anchoring effect of additives against reorientation of polymer network at the Pt–ionomer interface. Second, the carboxylic acid group formed an ion pair with the impurity cations, and decreased the effective concentration of the latter ions in the polymer. The fact that CVs obtained with additives did not alter significantly

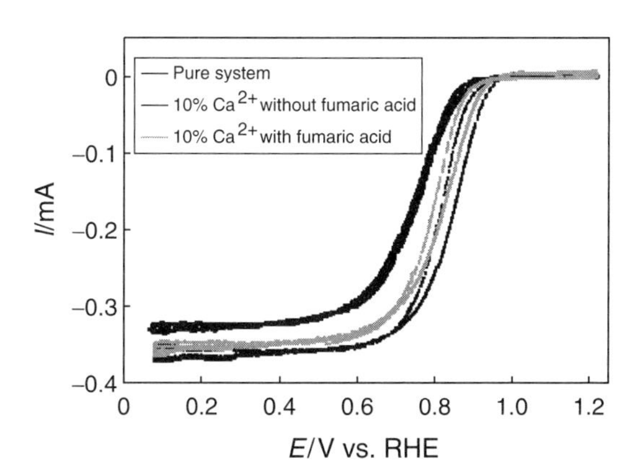

Fig. 5 ORR polarization curves of a Pt–Nafion® electrode in the presence of 10% Ca^{2+} without and with fumaric acid. Scan rate 0.01 V s^{-1}

from those obtained with no additives supports this idea. The third reason is that these additives formed an intermediate layer between platinum and ionomer, and created a liquid-electrolyte-like structure where no degradation by impurity ions occurred.

The unfortunate point was that the oxygen transport in the polymer was degraded, albeit no degradation occurred in the charge-transfer processes. Probably this is due to the interaction of the additives with the bulk polymer structure. In practical applications, the molecular structure and the amount of additives should be carefully optimized so that this adverse effect is minimized, without losing the inhibitor function.

4.2 Effect of Additives on the ORR in the Presence of Methanol

As a measure to the ORR degradation by organic impurities, an idea emerges that additives that are insensitive to the ORR but can block the organic impurities from adsorption on the platinum surface may be effective for this degradation. For this purpose, 2,2'-bipyridine (bpy), 1,10-phenanthroline (phen), 2,2':6',2'-terpyridine (terpy), 4,6-diphenyl-1,10-phenanthroline (dpphen), and the metal complexes were tested as additives (Shiroishi et al. 2004).

Polarization curves for oxygen reduction on the platinum disk electrode are shown in curves d–f in Fig. 6, at various methanol concentrations without bpy. The

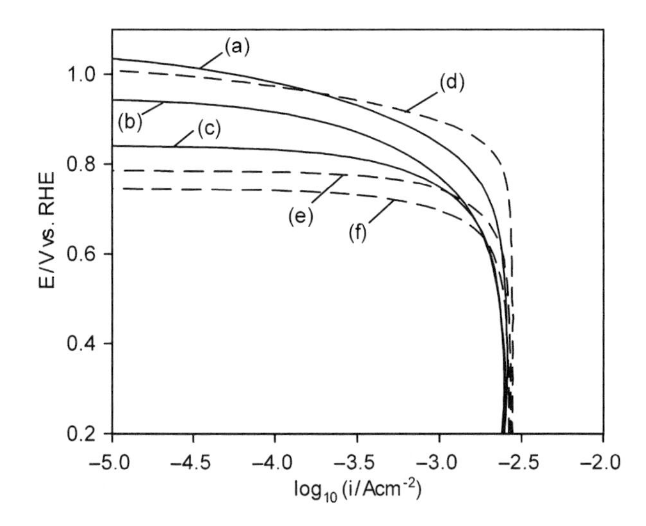

Fig. 6 ORR polarization curves in 0.1 mol dm^{-3} $HClO_4$. Scan rate 0.05 V s^{-1}, rotation speed 300 rpm. (**a**) 0 mol dm^{-3} CH_3OH + 0.1 mmol dm^{-3} 2,2'-bipyridine (bpy); (**b**) 0.1 mol dm^{-3} CH_3OH + 0.1 mmol dm^{-3} bpy; (**c**) 1 mol dm^{-3} CH_3OH + 0.1 mmol dm^{-3} bpy; (**d**) 0 mol dm^{-3} CH_3OH + 0 mol dm^{-3} bpy; (**e**) 0.1 mol dm^{-3} CH_3OH + 0 mol dm^{-3} bpy; (**f**) 1 mol dm^{-3} CH_3OH + 0 mol dm^{-3} bpy. (Reproduced from Shiroishi et al. 2003, copyright the American Chemical Society)

onset potential of cathodic current E_c dropped by 0.207 V for 0.1 mol dm^{-3} CH_3OH and by 0.295 V for 1 mol dm^{-3} CH_3OH compared with that without methanol. Curves a–c in Fig. 6 show ORR polarization curves at various methanol concentrations with 0.1 mmol dm^{-3} bpy. E_c shifted by −0.06 V for 0.1 mol dm^{-3} CH_3OH and by −0.177 V for 1 mol dm^{-3} CH_3OH. Thus, E_c improved by more than +0.13 V compared with the case without bpy, and it was demonstrated that bpy adsorbed on platinum suppresses methanol oxidation selectively.

Table 3 summarizes polarization data for oxygen reduction with various additives. In all cases the additives shift E_c in positive directions. The limiting current densities for oxygen reduction (j_L) decreased slightly with additives, except for $Cu_2(phen)_4OH$, where it decreased by 75%. The peak currents for methanol oxidation with additives significantly reduced with additives. In the absence of methanol, the order of E_c is as bpy > $Ni_2(bpy)_4OH$ > phen > none > $Cu_2(phen)_4OH$ > terpy > dpphen > $Fe(terpy)Cl_x$. This order does not correspond to that of the methanol oxidation peak. This is because E_c is affected by two opposing phenomena arising from the adsorption of additives. The additives adsorbed onto platinum suppress the formation of platinum oxide species and shift E_c positively, while the additives occupy active sites for oxygen reduction on platinum to shift E_c negatively. The extent of these counteracting effects may depend on the structure of the additives and interaction between the additives and platinum surfaces. This might be attributed to the inhibition of oxygen diffusion to the surface of platinum probably owing to barrier formation by additives.

Table 3 Effect of additives on polarization curves for oxygen reduction and methanol oxidation[a]

	0 M CH_3OH		0.1 M CH_3OH			1 M CH_3OH		
Additive[b]	$E_c/(V)$[c]	j_L (mA cm^{-2})[d]	$E_c/(V)$[c]	j_L/(mA cm^{-2})[d]	j_m/(mA cm^{-2})[f]	$E_c/(V)$[c]	j_L/(mA cm^{-2})[d]	j_m/(mA cm^{-2})[f]
–	0.968	2.58	0.775	2.63	0.50	0.692	2.53	1.90
bpy	1.002	2.62	0.942	2.54	0.084	0.825	2.37	0.49
bpy (0.4 mM)	0.987	2.54	0.943	2.50	0.040	0.837	2.49	0.26
phen	0.991	2.58	0.932	2.52	0.054	0.826	2.43	0.25
terpy	0.958	2.47	0.881	2.43	0.049	0.789	2.39	0.34
dpphen	0.952	2.27	0.798	1.99	0.077	0.729	1.67	0.20
$Cr_2(phen)_4OH$	0.976	2.47	0.887	2.47	0.097	0.792	2.31	0.49
$Fe(terpy)Cl_x$[e]	0.851	2.41	0.817	2.35	0.003	0.784	2.30	0.04
$Ni_2(bpy)_4OH$	0.991	2.55	0.918	2.53	0.078	0.810	2.45	0.47
$Cu_2(phen)_4OH$	0.962	0.65	0.886	0.60	0.044	0.783	0.55	0.19

Reproduced from Shiroishi et al. 2004, copyright the Chemical Society of Japan

[a]Measurements were performed in 0.05 mol dm^{-3} H_2SO_4 at room temperature under 1 atm. oxygen or nitrogen. Scan rate 0.005 V s^{-1}, rotation speed 300 rpm

[b]The concentrations of the additives are 0.1 mmol dm^{-3}

[c]Potential at cathodic current of 1×10^{-5} A cm$^{-2}_{real}$.

[d]Diffusion-limiting current of oxygen reduction at 0.3 V vs. RHE normalized by the geometrical area of a platinum disk electrode

[e]$x = 2 - 3$

[f]A peak current of the methanol oxidation reaction under 1 atm nitrogen

4.3 SEIRAS Study of Adsorbed Pyridyl Additives on a Platinum surface

Figure 7 shows SEIRAS spectra at various potentials that are referenced to those without methanol and additives in 0.1 mol dm^{-3} $HClO_4$. The bands derived from bpy molecules were observed at 3,100 cm^{-1} (C–H stretching), 1,440 cm^{-1} (symmetric ring stretching), and 1,535 cm^{-1} (asymmetric ring stretching). In the absence of methanol, the addition of bpy caused a decrease in the intensity of OH stretching peaks around 3,520 cm^{-1}, which consisted of OH stretching modes for hydrogen-bonded water and oxonium ion. Also a decrease of HOH bending oscillators around 1,630 cm^{-1} occurred in the region from 0.1 to 1.1 V, suggesting that adsorbed water molecules and those existing near the surface of platinum were released from the surface by bpy.

After addition of methanol, the intensity of OH stretching increased a little over the whole potential range. The band around 3,670 cm^{-1} observed in the absence of bpy, attributable to monomeric H_2O and non-hydrogen-bonded OH, disappeared in the presence of bpy. The band assigned to linear CO stretching consisted of two

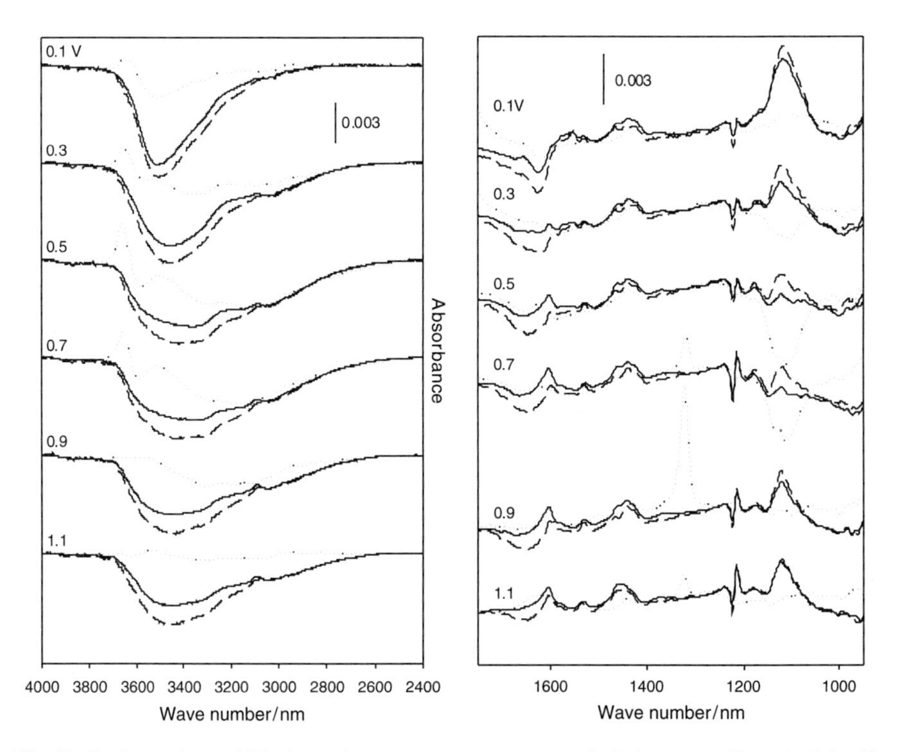

Fig. 7 Surface-enhanced IR absorption spectroscopy spectra of platinum on silicon in 0.1 mol dm^{-3} $HClO_4$ at various potentials under N_2.(*dotted line*), in 0.1 mol dm^{-3} CH_3OH + 0 mmol dm^{-3} bpy; (*dashed line*), in 0 mol dm^{-3} CH_3OH + 0.1 mmol dm^{-3} bpy; (*line*), in 0.1 mol dm^{-3} CH_3OH + 0.1 mmol dm^{-3} bpy. *Left* OH stretching band, *right* HOH bending band, formate and perchlorate bands. (Reproduced from Shiroishi et al. 2003, copyright the American Chemical Society)

components, and its intensity was decreased to 1.5% by bpy as compared with that without bpy at 0.5 V. A band around 1,320 cm^{-1} assigned as the symmetric OCO stretching mode of the formate species, the intermediate of the non-CO path for methanol oxidation, diminished in the presence of bpy. It is thus concluded that bpy interferes with the MOR intermediates, and eventually suppresses methanol oxidation on platinum, where bpy hinders both non-CO and CO paths on platinum (Shiroishi et al. 2005).

The fraction of platinum sites occupied by bpy, x_{bpy}, increased with increasing concentration in the solution, and reached a plateau at 1 mmol dm^{-3}, where 36% of platinum sites can adsorb a hydrogen atom. The fraction of "free sites" is defined as the fraction of platinum sites that are not occupied by bpy:

$$x_{\text{free}} = 1 - x_{\text{bpy}}.$$

The MOR currents in the absence of oxygen gas were divided by x_{free} to estimate the methanol oxidation activity per "free site." In contrast to the expectation that methanol oxidation activity per platinum site does not change, the normalized current decreased with increasing x_{bpy}, indicating that methanol oxidation activity per "free site" decreased with increasing fraction of bpy sites.

There are two possible reasons why bpy reduces methanol oxidation on platinum more than oxygen reduction. One is the geometrical effect of bpy adsorbed on platinum, since a methanol molecule is larger than a dioxygen molecule, and the other reason is an electronic effect by bpy on a platinum orbital. A simple Monte Carlo simulation study indicates that selective oxygen reduction is caused by the difference in the number of required adsorption sites between methanol (four) and dioxygen molecules (two) (Shiroishi et al. 2005). The suppression of platinum oxide species by bpy is another factor that enhances the oxygen reduction.

5 Concluding Remarks

Many impurities other than those discussed here may also affect both the anode and the cathode reactions in fuel cells. Gaseous impurities are known to be the most serious factors (see also chapters "Air Impurities" and "Performance and Durability of PEM Fuel Cells Operating with Reformate") at the fuel cell anode and the cathode. In this chapter, those impurities were excluded from discussion, and only the impacts of cationic and organic substances (that may occur as water-soluble species) on the ORR were considered, but the results indicated that these contaminants were equally serious problems for fuel cell degradation.

The problem is that the impurities discussed here can penetrate into the catalyst layer by way of either the gas phase or the liquid phase, and once they have entered into the polymer electrolyte in contact with the catalyst surface they are difficult to remove. Some are also intentionally added to the catalyst ink or to the binder solution in the MEA fabrication processes. The countermeasure to cope with this kind of impurity, therefore, is to control the catalyst–ionomer interface by use of the third component, which would be added to the MEA in its preparation stage.

Addition of the third component that can adsorb on and modify the catalyst–ionomer interface or control the access of harmful compounds to the catalyst surface would be a good solution to this problem. Some carboxylic acid or amino acid compounds, or pyridyl compounds are proposed for this purpose, and have turned out to be good candidates to minimize these impurity problems. To find the best solution, further selection and optimization of additives are needed, but it is expected that the method proposed in this work would be applicable in many cases.

Acknowledgments The author wishes to acknowledge all the people who worked on this project. Especially, useful discussions with the late Prof. Odd Andreas Asbϕrnsen, Reidar Tunold, Signe Kjelstrup, and Bjϕrn Hafskjord of the Norwegian University of Science and Technology (NTNU), Isao Sekine and Makoto Yuasa of Tokyo University of Science, and Masatoshi Osawa of Hokkaido University are greatly acknowledged. Thanks are due to Jϕrgen Dale, Yusuke Ayato, Hiroki Satou, Hidenobu Shiroishi, and Keiji Kunimatsu, who have devoted very much effort to the exciting project of impurity effects on electrode reactions in PEFCs. The author must add the names of Steffen Mϕller-Holst, Rune Halseid, and Preven J.S. Vie, who offered him helpful discussions and encouragements on this topic when he visited NTNU several times.

References

Halseid, R., Bystroň, T. and Tunold. R. (2006) Oxygen Reduction on Platinum in Aqueous Sulphuric Acid in the Presence of Ammonium. Electrochim. Acta 51, 2737–2742

Moore, R. B. and Martin, C. R. (1986) Procedure for Preparing Solution-cast Perfluorosulfonate Ionomer Films and Membranes. Anal. Chem. 58, 2569–2570

Okada, T. (2003) Effect of Ionic Contaminants. In: Vielstich, V., Lamm, A. and Gasteiger, H. (Eds.) Handbook of Fuel Cells – Fundamentals, Technology and Applications. Wiley, England, Vol. III, Chap. 48, pp. 627–646

Okada, T., Dale, J., Ayato, Y., Asbjørnsen, O. A., Yuasa, M. and Sekine, I. (1999) Unprecedented Effect of Impurity Cations on the Oxygen Reduction Kinetics at Platinum Electrodes Covered with Perfluorinated Ionomer. Langmuir 15, 8490–8496

Okada, T., Ayato, Y., Dale, J., Yuasa, M., Sekine, I. and Asbjørnsen, O. A. (2000) Oxygen Reduction Kinetics at Platinum Electrodes Covered with Perfluorinated Ionomer in the Presence of Impurity Cations Fe^{3+}, Ni^{2+} and Cu^{2+}. Phys. Chem. Chem. Phys. 2, 3255–3261

Okada, T., Ayato, Y., Satou, H., Yuasa, M. and Sekine, I. (2001) The Effect of Impurity Cations on the Oxygen Reduction Kinetics at Platinum Electrodes Covered with Perfluorinated Ionomer. J. Phys. Chem. B 105, 6980–6986

Okada, T., Satou, H. and Yuasa, (2003) M. Effects of Additives on Oxygen Reduction Kinetics at the Interface between Platinum and Perfluorinated Ionomer. Langmuir 19, 2325–2332

Osawa, M. (1997) Dynamic Processes in Electrochemical Reactions Studied by Surface-Enhanced Infrared Absorption Spectroscopy (SEIRAS). Bull. Chem. Soc. Jpn. 70, 2681–2880

Shiroishi, H., Ayato, Y., Kunimatsu, K. and Okada, T. (2004) Effect of Additives on Electrochemical Reduction of Oxygen in the Presence of Methanol. Chemistry Lett. 33, 792–793

Shiroishi, H., Ayato, Y., Okada, T. and Kunimatsu, K. (2005) Mechanism of Selective Oxygen Reduction on Platinum by 2–2'-Bipyridine in the Presence of Methanol. Langmuir 21, 3037–3043

Soto, H. J., Lee, W., Van Zee, J. W. and Murthy, M. (2003) Effect of Transient Ammonia Concentrations on PEMFC Performance. Electrochem. Solid State Lett. 6, A133–A135

Uribe, F. A., Gottesfeld, S. and Zawodzinski, Jr., T. (2002) Effect of Ammonia as Potential Impurity on Proton Exchange Membrane Fuel Cell Performance. J. Electrochem. Soc. 149, A293–A296

Zecevic, S. K., Wainright, J. S., Litt, M. H., Gojkovic, S. Lj. and Savinell, R. F. (1997) Kinetics of O_2 Reduction on a Pt Electrode Covered with a Thin Film of Solid Polymer Electrolyte. J. Electrochem. Soc. 144, 2973–2982

Performance and Durability of a Polymer Electrolyte Fuel Cell Operating with Reformate: Effects of CO, CO_2, and Other Trace Impurities

Bin Du, Richard Pollard, John F. Elter, and Manikandan Ramani

Abstract The performance and durability of a polymer electrolyte fuel cell (PEFC) operating with reformate is discussed. Brief overviews are given on how dilution affects the thermodynamic driving force and how diffusion of N_2 and CO_2, two major components in a typical reformate mix, affects the overall voltage. The primary focus is on the impact of CO on the voltage performance of the PEFC, i.e., the anode overpotential at different CO levels. Specifically, the effects of CO concentration and the impact of various CO mitigation methods on durability and degradation are presented. CO/air bleed interactions are discussed in connection with peroxide-induced membrane/ionomer degradation rates. Furthermore, the possibility of in situ anode CO formation from CO_2 via the reverse water-gas-shift reaction is assessed for realistic PEFC operating conditions. The discussion includes results obtained at high CO levels and the stability of Pt–Ru catalysts. The impact of trace impurities such as NH_3, H_2S, and small organic molecules is also described.

1 Introduction

In the absence of a national hydrogen infrastructure, on-site hydrogen generation is expected to be the choice for most polymer electrolyte fuel cell (PEFC) applications in the foreseeable future (Vielstich et al. 2003; DOE 2005a). The most viable technology for on-site H_2 generation today is reforming technology utilizing natural gas or other hydrocarbon fuels that are readily available through existing distribution channels. Currently, there is no nationwide supply infrastructure for methanol, considered by some as the ultimate liquid fuel for transportation applications because of its high energy density and a low conversion temperature for steam reforming (250–400 °C). On the other hand, the US military has an extensive network of fuel supply depots (most of them for diesel and kerosene, a jet fuel). It is

B. Du (✉)
Plug Power Inc., 968 Albany Shaker Road, Latham, NY 12110, USA
e-mail: bin_du@plugpower.com

F.N. Büchi et al. (eds.), *Polymer Electrolyte Fuel Cell Durability*,

unlikely that the military will change its existing fuel supply infrastructure to accommodate fuel cell applications. Therefore, for military applications, the reforming subsystem of a fuel cell system must be able to handle hydrocarbons with high sulfur content (catalyst deactivation) and high molecular weight (conversion temperature above 1,200 °C). Diesel/jet fuel reforming technology is still in the development stage.

A fuel generated by hydrocarbon reforming is called "reformate." The most commonly used technologies for natural gas reforming are steam reforming, partial oxidation reforming, or autothermal reforming (a combination of steam and partial oxidation reforming). The use of reformate as a fuel presents special challenges for a PEFC system in terms of system efficiency, reliability, and durability (Ralph and Hogarth 2002; Vielstich et al. 2003; DOE 2005a; Du et al. 2006a–c, 2008). Many of these issues can be attributed to the detrimental effects of impurities (such as CO, CO_2, and H_2S) in the hydrocarbons or as the by-products of the reforming process. The presence of the CO_2 and N_2 also reduces the driving force for the reactions and lowers the overall system efficiency. The state-of-the-art PEFC stack life exceeds 12,000 h using simulated reformate gas (Higashiguchi et al. 2003; Knights et al. 2004). Plug Power has demonstrated a stack life of over 13,000 h in its latest field systems operating with an integrated natural gas fuel processor (Du et al. 2006a).

System durability is often defined as the maximum lifetime of a system with no more than 10% loss in efficiency at the end of life (de Bruijn 2005). The US Department of Energy's Multi-Year Hydrogen Program stipulates that a stationary PEFC system must last longer than 40,000 h to compete with other distributed power generation systems (DOE 2005b). It is estimated that, when operating with neat H_2, the maximum voltage degradation rate (VDR) needs to be less than 2 μV/h to achieve 40,000 h of stack life (de Bruijn 2005). An even lower VDR, approximately 1.6 μV/h, is required when operating on a reformate with 50% H_2 (Du et al. 2006a). This is because a fuel cell operating on reformate has a lower beginning-of-life cell voltage than one operating with neat H_2 under otherwise the same operating conditions. Part of this voltage loss stems from the reduced partial pressure of hydrogen at the reaction interface, caused by the presence of N_2 and CO_2 in the bulk reformate combined with an increase in the resistance for diffusion of hydrogen from the channels to the electrode surface. The remainder of the voltage loss stems from a higher anode overpotential η_a caused by the reduced hydrogen concentration and by trace amounts of impurities such as CO, NH_3, and H_2S. The effects of non-hydrogen components in a reformate mixture are discussed in the following in the context of PEFC performance and durability.

2 Measurement of Anode Overpotential

Two approaches are commonly employed to separate the contributions to η_a and, hence, assess the effects of individual reformate components on the cell voltage of a PEFC (Fig. 1). The simple approach uses cell polarization data for different anode

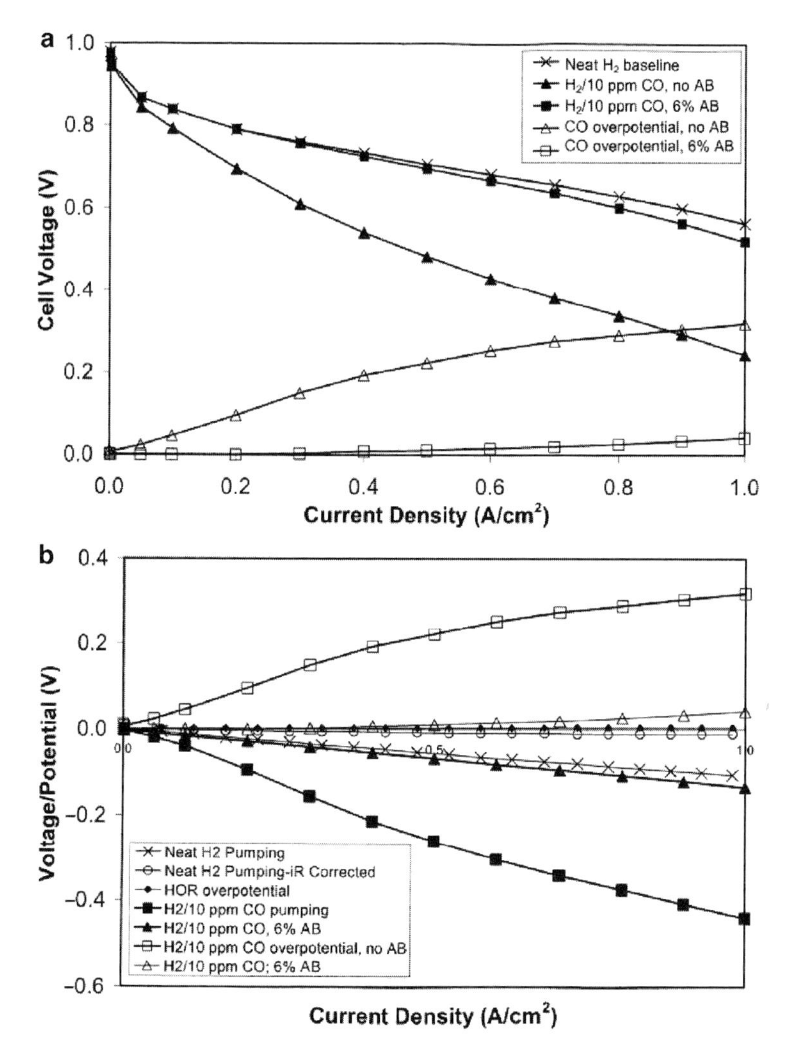

Fig. 1 Estimation of anode overpotential from (**a**) fuel cell polarization data and (**b**) H_2 and reformate pumping polarization data. *AB* air bleed

gas compositions as well as baseline polarization data for neat hydrogen under otherwise the same operating conditions. For example, to estimate the impact of H_2 with 10 ppm CO on the value of η_a, the polarization of H_2 + 10 ppm CO is subtracted from the baseline neat H_2 polarization (Fig. 1a). The difference is the net overpotential loss as a result of exposing a Pt/C electrode to 10 ppm CO. Similarly, the effect of air bleed (AB) on the value of η_a can be deduced from the difference between the polarization data of H_2/10 ppm CO with and without AB. Here, one of the assumptions is that the contribution of the hydrogen oxidation reaction (HOR) to η_a is independent of the H_2 concentration because of the extremely fast HOR kinetics.

Another assumption is that the ohmic loss remains constant for all measurements using different reformate mixtures. Cathode overpotential is assumed to be fuel-independent.

The approach described above is acceptable for obtaining rough estimates of η_a values. However, it is not reliable for accurate measurements and may introduce a significant standard deviation if η_a accounts for less than 5% of the overall cell voltage. This is especially true for measurements taken in the low current density (CD) region. For example, the η_a values of H_2/10 ppm CO with AB are 3 and 16 mV at 0.3 and 0.6 A/cm^2, respectively, but the corresponding cell voltages are 0.755 and 0.671 V (Fig. 1a). The η_a value only accounts for 2.4% of the cell voltage at 0.6 A/cm^2 and is less than 0.4% of the cell voltage at 0.3 A/cm^2. The latter clearly falls within the standard deviation of cell voltage measurements. As discussed in the following section, this is also true for the contributions of "inert" components (such as N_2) at low CD and for poisoning species at very low concentrations. Furthermore, the method described above cannot be used to extract the contribution of the HOR to η_a values.

A more accurate but slightly more complicated approach is to obtain hydrogen pumping polarization data (Fig. 1b). In this case, hydrogen/reformate is fed to the anode side of a PEFC but no oxidant gas is provided at the cathode side. A power supply is used to provide an external current to drive the HOR at the anode and the hydrogen evolution reaction at the cathode. After correction for ohmic losses, the pumping polarization for neat hydrogen is found to be less than 10 mV for CDs below 1 A/cm^2. The net contribution from the HOR can be easily derived from the *iR*-corrected hydrogen pumping polarization data assuming that the absolute values of η_c (hydrogen evolution reaction) and η_a (HOR) are approximately equal (Fig. 1b). The contributions of various reformate components to the overall η_a value can be taken as the difference between the pumping polarizations on reformate and hydrogen.

3 CO Poisoning Effects and Mitigation Methods

3.1 The CO Poisoning Effects

The first stage of fuel processing, by either steam reforming or by autothermal reforming, produces a gas mixture with considerably more than 1% CO (Vielstich et al. 2003). Subsequently, the CO content is reduced through a series of high- and low-temperature water-gas-shift (WGS) reactors and a preferential oxidation (PROX) reactor to bring the CO level to less than 10 ppm. Even this low CO concentration, however, can still result in a significant increase in η_a (see Fig. 1), especially when a platinum-only electrode is used (Gasteiger et al. 1995a, b; Schmidt et al. 1997, 1998; Schmidt 1999; Ralph et al. 2002; Du et al. 2008). The ubiquitous CO in a reformate is a major concern for PEFCs because even a few parts per million of CO can induce a considerable cell voltage loss (Oetjen et al. 1996). The Pt–CO adlayer coverage can reach over 98% even when just a few parts per million of CO are present in a reformate (Ralph and Hogarth 2002). This is because the Pt–CO adlayer formation is much more exothermic than the energetically neutral Pt–H adatom formation (Springer et al. 2001; Baschuk and Li 2003). In this situation, the HOR can occur at only a few bare platinum sites in a compact CO monolayer (Gasteiger et al. 1995b). Trace amounts of CO in a gas mixture gradually accumulate on the platinum surface and compete with H_2 molecules for available platinum sites, resulting in an elevated value of η_a (Davies et al. 2004). This quantity can continue to increase until the CO coverage reaches a threshold (approximately 0.79 V for a typical platinum electrode) that triggers electrochemical CO oxidation (Gasteiger et al. 1995a):

$$CO + H_2O \rightarrow CO_2 + 2H^+ + 2e^- \tag{1}$$

3.2 *CO Mitigation Methods*

3.2.1 CO-Tolerant Electrode

When operating with a CO-containing fuel, a PEFC often employs a platinum alloy electrode to reduce the anode overpotential loss from CO poisoning. To date, the most effective anode material is a Pt–Ru alloy with a 1:1 atomic ratio (Ianniello et al. 1994; Ralph and Hogarth 2002; Du et al. 2008). The Pt–Ru alloy electrode catalyzes the electrochemical CO oxidation reaction (1) through a bifunctional mechanism (Wang et al. 1996; Lin et al. 1999; Camara et al. 2002). It is generally accepted that this process involves the formation of either a ruthenium-activated H_2O molecule or a Ru(OH) surface complex adjacent to Pt–CO sites:

$$\mathrm{Ru} + \mathrm{H_2O} \rightarrow \mathrm{Ru} - \mathrm{H_2O}, \tag{2}$$

$$\mathrm{Ru} - \mathrm{H_2O} + \mathrm{Pt} - \mathrm{CO} \rightarrow \mathrm{CO_2} + 2\mathrm{H^+} + 2e^-, \tag{3}$$

or

$$\mathrm{Ru} + \mathrm{H_2O} \rightarrow \mathrm{Ru(OH)} + \mathrm{H^+} + e^-, \tag{4}$$

$$\mathrm{Ru(OH)} + \mathrm{Pt} - \mathrm{CO} \rightarrow \mathrm{CO_2} + \mathrm{H^+} + e^-. \tag{5}$$

There appears to be a close link between the actual CO oxidation mechanism and the Pt–CO bonding configuration: the route involving Ru–H_2O is associated with a CO molecule bridge-bonded to two adjacent platinum sites (Camara et al. 2002) whereas the route involving Ru(OH) is linked to a CO molecule linearly bonded to a single platinum site (Lin et al. 1999). Density functional calculations on cluster models were carried out to compare the characteristics of the CO oxidation reactions on a PtRu alloy and pure platinum (Saravanan et al. 2003). The results indicate that, while pure platinum forms a more stable complex with OH, this Pt–OH complex is less efficient in catalyzing the CO oxidation reaction than the Ru–OH complex formed on the surface of PtRu. This is consistent with the experimental observations on a platinum-modified Ru(0001) electrode (Lei et al. 2005).

Even a PtRu electrode can suffer from a significant CO overpotential loss, especially when a fuel cell operates at high CD. This is because the peak CO oxidation current occurs near 0.39 V for a state-of-the-art PtRu electrode. At its onset potential (below 0.1 V), the CO oxidation current of a PtRu anode is capable of oxidizing only a few parts per million of CO. The ignition potential, defined as the potential at which the CD increases by approximately 2 orders of magnitude within a narrow potential range, was found to be as high as 0.45 V in some cases (Schmidt et al. 1997). This anodic potential could lead to ruthenium oxidization (see Sect. 3.5) and, subsequently, loss of CO tolerance.

3.2.2 AB Method

It has been reported that CO can be chemically oxidized to CO_2 (6) by feeding in, or "bleeding," a small amount of air into the anode (Gottesfeld and Pafford 1988; Gottesfeld 1990):

$$CO + \frac{1}{2} O_2 \rightarrow CO_2 \tag{6}$$

This reaction is catalyzed by platinum and promoted electrochemically (Adcock et al. 2005; Tsiplakides et al. 2005). Unlike the electrochemical CO oxidation catalyzed by PtRu (1), which prefers a hydrophilic environment to facilitate proton and water transport, this AB process prefers a hydrophobic environment for facile O_2 diffusion (Du et al. 2008). The amount of air required for AB to be effective depends on both the CO and the H_2 concentrations. For two reformates with the same CO concentration, the one with a lower H_2 concentration brings more CO into the anode if the fuel cell is operated at a specific anode reactant stoichiometry relative to the minimum required stoichiometry. Excess O_2, usually at a O_2 to CO ratio greater than 100 (stoichiometry for (6) is greater than 200), is required for effective CO removal. The required O_2 to CO ratio increases as the CO concentration decreases because of the dilution effect. The excess O_2 consumes valuable H_2 and results in lowered system efficiency. Furthermore, it was found that AB leads to the formation of a significant amount of H_2O_2 at a Pt/C anode (Jusys et al. 2003; Jusys and Behm 2004; Stamenkovic et al. 2005). The yield of H_2O_2, the product of the two-electron oxygen reduction reaction (ORR), is substantially higher (58% on carbon and 15% on platinum) at the working potential of a PEFC anode (below 0.1 V) than at the working potential of a cathode (above 0.4 V) (Antoine and Durand 2000). At the cathode, the ORR is dominated by the four-electrode process with H_2O as the main product (more than 99.5%). Free radicals (HO•, HOO•) generated from H_2O_2 have been identified as the primary cause for ionomer/membrane chemical degradation (Baldwin et al. 1990; Pianca et al. 1999; Curtin et al. 2003; Schiraldi 2006). The presence of O_2 at the anode (through AB or O_2 crossover) represents a major source of H_2O_2 generation. The increased rate of membrane/ionomer degradation that results from use of AB as a CO mitigation method is a primary cause for the increased VDR and shortened lifetime for a PEFC operating on a CO-containing reformate (Du et al. 2006a–c).

It should be noted that, even though the yield of H_2O_2 is less than 0.5% at the cathode, the amount of H_2O_2 generated at the cathode is higher than that at the anode because the amount of O_2 available there is much greater. For example, an anode with 2% AB has approximately 800–1,000 ppm O_2 available (including crossover O_2) and the corresponding H_2O_2 level is expected to be less than 150 ppm. At an air cathode with a stoichiometry of 2, the O_2 available is more than 10% which, at a yield of 0.5%, would give more than 500 ppm H_2O_2. Therefore, for a PEFC operating with a CO-containing reformate, cathodic peroxide production is still the primary concern. This is consistent with the fluoride release rate (FRR) data presented in Sect. 3.3.

3.2.3 Pulsed AB Method

We have developed a pulsed AB (PAB) technology to reduce the amount of air introduced to the anode. This minimizes H_2O_2 formation and, hence, reduces the rate of membrane/ionomer degradation (Du et al. 2006b). This technique is based on the observation that voltage recovery in the presence of AB is much faster than CO poisoning in the absence of AB. For a fuel cell operating with PAB, AB is shut off periodically to reduce the amount of air introduced into the anode, thereby reducing H_2O_2 generation and preserving valuable H_2. For a reformate with 10 ppm CO, PAB reduced the amount of air needed for AB by more than 80% relative to a continuous AB under otherwise the same operating conditions. This leads to over 70% reduction in FRR, which is a good indicator of the degradation rate of the fluorinated membrane/ionomer (Baldwin et al. 1990). Both single cell and short stack endurance tests demonstrated improved cell performance when PAB was used instead of continuous AB (Du et al. 2006b, 2008).

PAB has also been shown (Fig. 2) to be a simple yet effective CO mitigation strategy for systems with variable CO concentrations or transient high CO concentrations because it automatically adjusts the pulsing frequency in response to changes in CO concentration (Du et al. 2006b). This is especially valuable for a system running on a low CO reformate that has constantly changing CO concentrations. The system automatically adjusts to the CO concentration changes by changing the time intervals between each AB period. This can be achieved readily by setting two control parameters: a maximum cell voltage (AB off) and a minimum cell voltage (AB on). When properly implemented, the PAB method allows a

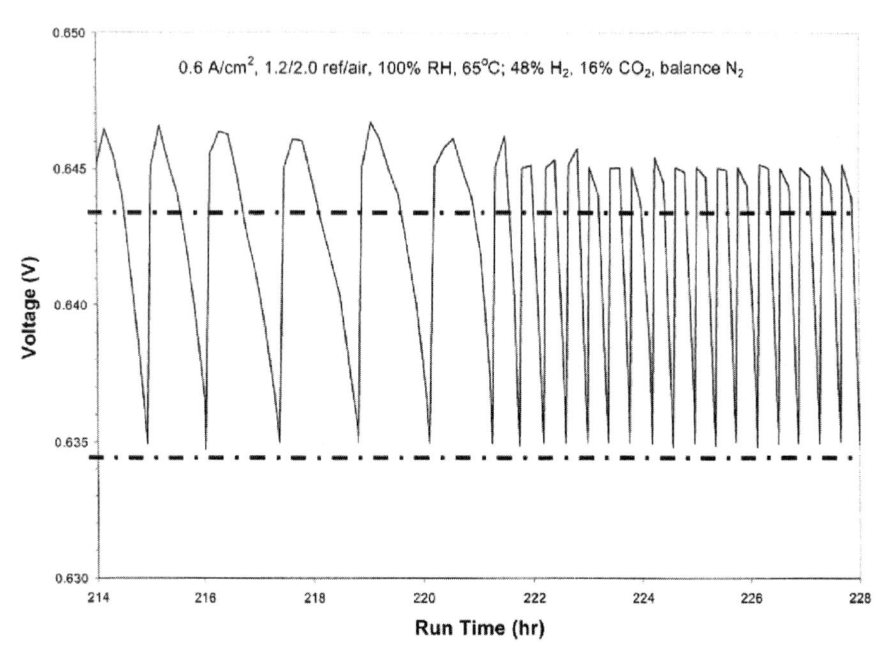

Fig. 2 Example of air bleed pulsing frequency in response to changing CO level during pulsed air bleed. (From Du et al. 2006, with permission)

system to sustain stable operation even during a period of high transient CO concentration.

3.2.4 Current Pulsing Method

The anode CO oxidation overpotential varies depending on the Pt–Ru composition and catalyst fabrication process. The CO coverage of platinum sites increases as the percentage of AB decreases when there is not enough O_2 to fully oxidize CO to CO_2. This is manifested by a steady decline in the cell voltage as AB is reduced (Fig. 3). When the anode overpotential approaches the CO oxidation potential, the cell undergoes a "self-cleansing" process which results in a quick recovery of the voltage (Zhang and Datta 2002). This is evident when a cell is operating on 100 ppm CO reformate, with insufficient AB (Du et al. 2006b). The voltage oscillation frequency depends on the AB level: the lower the AB, the greater the frequency of the cycles. A similar oscillation pattern was reported on PtRu electrodes exposed to H_2 gases containing 100–1,000 ppm CO (Zhang et al. 2002, 2004; Zhang and Datta 2005). The oscillation occurs at a fuel cell temperature below 70 °C and its onset is dependent on CO concentration, anode flow rate, CD, and temperature. With a simple model based on the temperature dependence of the oscillation frequency, the apparent activation energy for the "self-cleansing" process was estimated to be around 60 kJ/mol (Zhang and Datta 2002). Several groups have

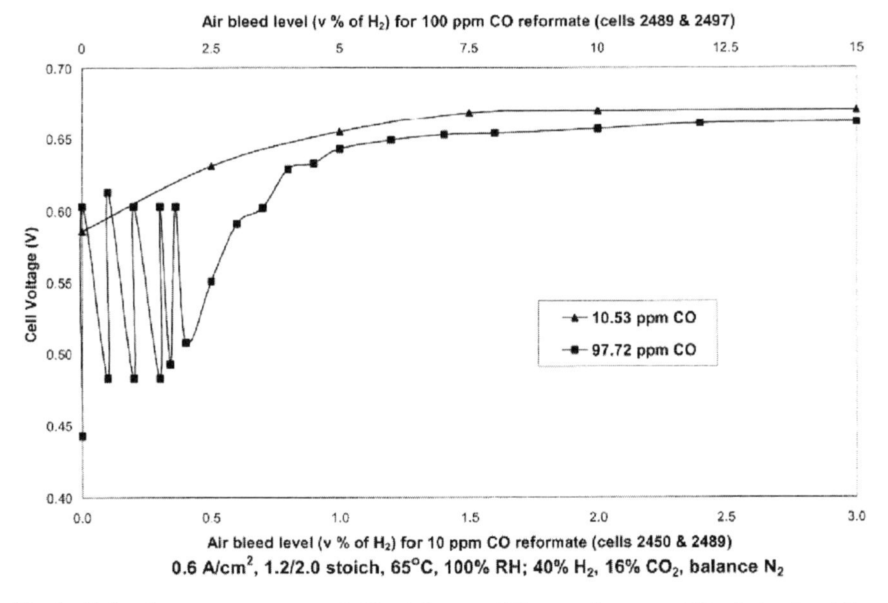

Fig. 3 Cell voltage vs. percentage air bleed for two different reformates. *RH* relative humidity, *stoich* stoichiometry. (From Du et al. 2006, with permission)

proposed the use of a current pulsing technique (electrochemical PROX) to induce voltage oscillation as a CO mitigation strategy (Carrette et al. 2001; Thomason et al. 2004; Zhang and Datta 2004; Adams et al. 2005; Jimenez et al. 2005). It was suggested that voltage oscillation might be a desirable mode of operation for a PEFC using a CO-containing reformate (Zhang et al. 2004). It was found that the time-averaged cell voltage, cell efficiency, and power density were higher in such a mode than in stable steady-state operation because of a decrease in the time-averaged anode overpotential on oscillation. However, this mode of operation presents a special challenge for the power conditioning subsystem.

3.2.5 Reconfigured Anode

An anode design that has a basis closely related to the AB approach is the so-called reconfigured anode in which a thin layer of metal (such as Pt/C or Ru/C) and/or metal oxide (such as FeO_x or RuO_xH_y) is added to the outside of the anode gas diffusion layer facing the flow field (Eberle et al. 2002; Uribe et al. 2004; Adcock et al. 2005; Santiago et al. 2005). Unlike a normal anode electrode layer that is impregnated with ionomers for facile proton transport, this ionomer-free CO-oxidation layer is made to be hydrophobic to facilitate interaction of CO and O_2 through improved gas diffusion. With AB, this configuration is capable of operating with a reformate containing up to 500 ppm CO. The extra catalyst layer serves as a "CO filter" (in situ PROX) so that the anode electrode is not exposed to high concentrations of CO. A Pt–Ru alloy is still recommended as the anode electrode material to deal with the trace amounts of CO that can penetrate this extra layer. The amount of air needed for the AB is expected to be considerably less for the reconfigured anode than for a standard PtRu anode.

3.2.6 Ex Situ CO Filter

Significant progress has been made in the area of CO-tolerant electrodes, but the anode electrode cannot be expected to be robust against all possible impurities in the window of its operating potential. It is preferred in some system architectures to place a filter upstream to clean the fuel prior to reaching the stack. In addition to CO, such a filter system can also remove other organic and inorganic contaminants. For example, Donaldson's point-of-use chemical filtration systems are effective in the removal of SO_2 and NH_3, with close to 100% removal under dry conditions (Donaldson Company, Inc. 2003). However, filters working on the principle of chemical adsorption have to be regenerated off-site and/or replaced periodically. This adds additional service cost to the system. On-site electrochemical regeneration is possible for the CO filters with an external power supply (Karuppaiah and Lakshmanan 2003). CO is selectively adsorbed onto platinum-based filters and a predetermined duty cycle is applied at a certain potential. The control scheme allows manipulation of the duty cycle to enable selective CO adsorption and oxidation using an external power supply. This type of CO filter is practically an electrochemical PROX (Lakshmanan and Weidner 2002). Such a filter can handle a reformate fuel with high CO concentration (much more than 100 ppm), hence significantly shortening the start-up time of current reformer-based systems. There is added power consumption during the cleansing cycles. It should be noted that such an approach is similar in principle to commercial CO detectors that utilize amperometric detection technique with a three-electrode design (Stetter and Pan 1994).

3.2.7 High-Temperature PEFC

Another approach to mitigate CO effects is to operate a fuel cell at an elevated temperature. The equilibrium CO coverage on platinum decreases as the temperature increases because CO adsorption on platinum is an exothermic process (Davies et al. 2004). At temperatures above 180 °C, one can operate with a reformate containing 1% CO or even higher. Nafion®-based membrane electrode assemblies (MEAs) require liquid water for proton transport and can only operate at temperatures below 80 °C at ambient pressure. On the other hand, polybenzimidazole-based systems, which use impregnated phosphoric acid as the electrolyte, have an operating temperature range of 160–180 °C (Staudt et al. 2005). Schmidt et al. demonstrated over 18,000 h of stable fuel cell operation using PEMEAS Celtec-P® series MEAs (Schmidt and Baurmeister 2006). With the enhanced CO tolerance, platinum electrodes can be used instead of PtRu alloy electrodes. This is fortuitous because Pt–Ru has poor chemical stability in an acidic medium at the elevated temperatures that are used.

3.3 AB Optimization and PEFC Durability

As mentioned already, AB improves cell voltage performance in the presence of CO but also increases the formation of H_2O_2 which is detrimental to the chemical and mechanical integrity of PEFC MEAs (Baldwin et al. 1990; Pianca et al. 1999; Curtin et al. 2003; Schiraldi 2006). Therefore, it is imperative to optimize the AB level to achieve maximum cell performance with minimum H_2O_2 production. We define AB level as a volumetric percentage of dry air based on dry H_2 content of a given fuel. Knowledge of the CO–AB sensitivity, as illustrated in Fig. 3, is useful for selecting the most desirable AB ranges for an optimization process (Du et al. 2006b). When combined with polarization curves acquired at different AB levels, it provides a quick screening method to study the AB sensitivity over the full CD range. It should be pointed out that a typical polarization measurement does not allow enough time for a system to reach steady state when reformate of low CO concentration is used or when the AB level is low. To compensate for this, the selected AB range should always be confirmed by steady-state measurements at specific CDs.

The CO–AB optimization process involves three steps (1) determine an AB range for a reformate of certain CO and H_2 concentrations; (2) measure VDR and FRR at selected AB levels; (3) select the optimum AB set point on the basis of VDR/FRR data and cell lifetime expectations. The optimization should consider two additional factors: hydrogen safety and anode stoichiometry. The hydrogen safety requirement (Air Products and Chemicals 2004) limits the use of AB to less than 15% as an effective mitigation method for no higher than 200 ppm CO at a Pt–Ru anode. When high AB is used, the anode stoichiometry should be adjusted to compensate for the H_2 consumed by the excess O_2 to avoid fuel starvation.

Both VDR and FRR are used to evaluate the effects of different AB levels on the long-term cell performance and cell life. VDR should be calculated over a period of at least several hundred hours. Water samples can be collected two or three times a week for fluoride analysis. The resulting FRR data are tabulated for performance evaluation (Table 1). To facilitate the selection process, two charts are also generated for each CO concentration: AB level versus FRR (Fig. 4a) and AB level versus VDR (Fig. 4b).

Figure 4a shows the AB-FRR chart for a 10 ppm CO reformate. A possible "sweet spot," a minimum FRR at 1.5% AB, is picked as the optimum AB level. The lowest FRR indicates a minimum in H_2O_2 production at 1.5%, hence the lowest membrane/ionomer degradation rate. The amount of H_2O_2 formation increases with an increase in the amount of O_2 and/or the percentage yield of H_2O_2 from the ORR. For example, the H_2O_2 formation is augmented by an increase in O_2 when

Table 1 Fluoride release rate data From reformate CO–air bleed endurance tests

Air bleed (% of H_2)	Anode F^- ($\times10^{-8}$ g/h cm^2)	Cathode F^- ($\times10^{-8}$ g/h cm^2)	Total F^- ($\times10^{-8}$ g/h cm^2)
6* (397 h)	0.17(4)	1.19(14)	1.36(17), 6.81*
	0.13(3)[a]	0.63(14)[a]	0.76(14), 4.98[a]
8 (674 h)	0.47(10)	1.29(18)	1.77(28), 2.73
	0.33(6)[b]	1.15(20)[b]	1.48(24), 3.44[b]
	0.22(5)[a]	0.79(21)[a]	1.01(25), 3.51[a]
10 (678 h)	1.42(11)	2.88(31)	4.30(38), 2.04
	0.59(12)[a]	1.16(19)[a]	1.75(29), 1.96[a]
12* (166 h)	0.29(5)	1.35(22)	1.64(24), 472
	0.17(4)[b]	1.14(21)[b]	1.32(24), 6.62[b]
2% (10 ppm CO)	0.39(4)	2.51(40)	2.90(41)643
8% (50 ppm CO)	0.86(19)	2.82(39)	3.35(50), 3.27
12% (200 ppm CO) Baseline	1.36(33)	13.0(3.8)	14.4(4.1), 9.56

*Numbers in *parentheses* are standard deviations
[a]After CO reformate test
[b]Before CO reformate test

the AB is increased from 1.5 to 2%. This is implied by the increased FRR for 2% AB over that for 1.5% AB. On the other hand, there is not enough O_2 to prompt complete CO removal at 1% AB, as evidenced by an increase in anode overpotential (see Fig. 3). The result of an insufficient O_2 supply is an increase in CO coverage at 1% AB which, in turn, leads to an increase in the yield of H_2O_2 production.

It has been demonstrated that the yield of H_2O_2 production is directly linked to the CO coverage on platinum: the higher the CO coverage, the higher the yield of H_2O_2 (Jusys et al. 2003; Jusys and Behm 2004). The reason is that an increase in CO coverage reduces the availability of vacant platinum sites adjacent to each other. The four-electron ORR pathway requires two adjacent platinum sites, whereas the two-electron ORR pathway requires only a single platinum site. Therefore, compared with 1.5% AB, there is also an increase in FRR at 1% AB because an increase in the yield of H_2O_2 leads to an increase in H_2O_2 production even with a decrease in the total amount of O_2.

Figure 4b shows an AB-VDR chart from endurance tests using a 100 ppm CO reformate. It shows a distinct minimum near 10% AB. The minimum FRR is between 9 and 10% (see Table 1). At 12% AB, the drastic increase in VDR is attributed to the stoichiometry effect: the actual anode stoichiometry is approximately 1.09, significantly lower than the 1.2 set point because of significant H_2 consumption by the excess O_2 from the AB. At an anode stoichiometry of 1.09, the cell is at the edge of fuel starvation. Indeed, the cell tripped multiple times because of low cell voltage as a result of partial fuel starvation.

At CO concentrations above 100 ppm, the hydrogen flammability limit and fuel starvation (stoichiometry) limit the AB selection. Under these circumstances, the maximum AB allowed can only partially mitigate the poisoning effect of CO. This leads to unstable cell performance, elevated FRR, and, in turn, shortened cell life expectancy (Baldwin et al. 1990).

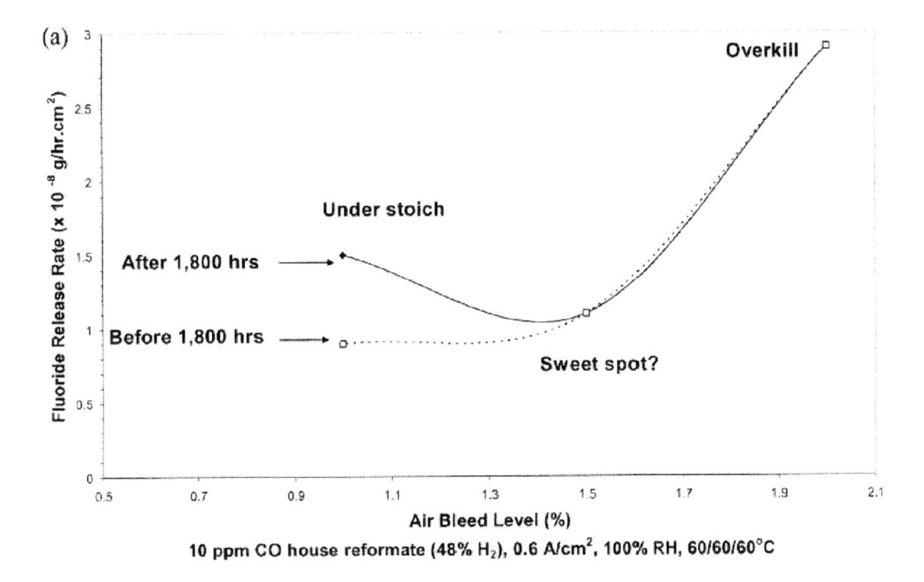

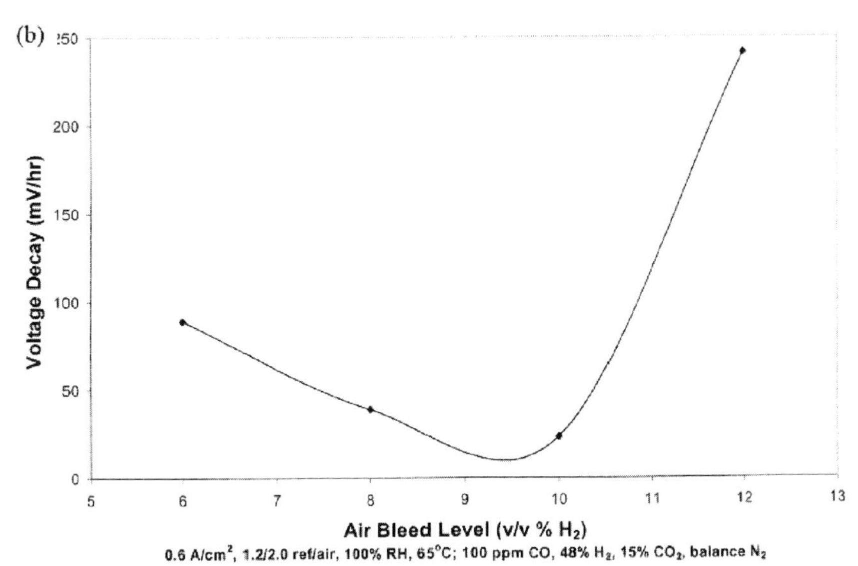

Fig. 4 (**a**) Fluoride release rate chart from the 10 ppm CO reformate endurance tests. (**b**) Voltage degradation rate chart from the 100 ppm CO reformate endurance tests. (From Du et al. 2006, with permission)

3.4 *Effect of Crossover O_2*

Although the amount of crossover O_2 from the cathode side is, on paper, sufficient for a reformate with 10 ppm CO, we found that such O_2 is not equivalent to the O_2 from AB in terms of CO removal (Du et al. 2006b). One possible explanation is that the crossover O_2 does not provide a sufficiently high O_2 stoichiometry locally, where it is required for effective CO removal. For example, at low CO concentrations, CO will begin by first adsorbing onto the platinum sites near the anode inlet and AB injection results in at least 200:1 O_2 to CO stoichiometry at the inlet. On the other hand, the crossover O_2 arrives at the anode electrode uniformly over the entire active area. Furthermore, these O_2 molecules must pass through the anode electrode where they can be partially consumed by H_2 to form H_2O and/or H_2O_2 (Antoine and Durand 2000). Therefore, the effective O_2 stoichiometry at the anode inlet is low as a result of O_2 crossover.

3.5 *Effect of Transient High CO Concentration*

A CO transient is defined as a finite period (a few minutes to a few hours) in which the CO concentration from the reformer deviates significantly from the steady-state specification. We used simulated reformate compositions to study transient CO effects by switching between a baseline operation (10 ppm CO reformate) and "high CO transients" at otherwise the same conditions (Du et al. 2006b). We found that high CO transients often lead to a recoverable voltage loss. The time required for recovery depends on the actual CO concentration and the AB level at which a system is operated prior to and during the transient. The voltage recovery starts as soon as the "high transient CO" is over. Full recovery can be achieved within 15 min with a sufficiently high AB level. The temporary voltage "degradation" is caused by insufficient AB that results in slow catalyst poisoning. This voltage loss can be recovered fully by a temporary increase in the AB level. The recovery time depends on the AB level but it is largely independent of the transient CO level, which primarily affects VDR. However, fuel cells suffer permanent loss in their CO mitigation capability when they are exposed to very high (1–5%) CO levels for a long time (2–3 h). This is attributed to the formation of RuO_xH_y as a result of high anode overpotential (more than 270 mV) at open circuit potential (Du et al. 2006b). While it is possible to reduce the freshly formed RuO_xH_y back to ruthenium, aged RuO_xH_y is very stable electrochemically and inactive for catalysis of CO oxidation (Ramani et al. 2001; Ramani et al. 2001; Kim and Popov 2002; Jang et al. 2003; Halseid et al. 2007).

The transient CO effect can be described using a lumped model of H_2 and CO adsorption, desorption, and electro-oxidation coupled with an *I–V* relationship for fuel cell performance (Bhatia and Wang 2004). It was found that, for reformate with low H_2 concentration, the hydrogen dilution amplifies the CO poisoning problem even though the dilution alone does not have an appreciable impact on the cell polarization. Transient CO effect vs. different recovery procedures were compared experimentally by alternating the anode feed between neat hydrogen and CO-containing hydrogen (Jimenez et al. 2005). For CO concentrations below 100 ppm, current pulsing is effective with no lasting impact on cell performance. At higher CO concentrations (e.g., above 1,000 ppm), neat H_2 cycling combined with current pulsing is recommended to prevent gradual performance degradation.

4 CO_2 and Reverse WGS Reaction

CO_2 is a major component in a reformate mixture, typically in the range 15–25%. CO_2 is not completely "inert" because it can bind loosely with the platinum surface and it is the product of CO oxidation. Therefore, it could conceivably be reduced back to CO either chemically or electrochemically. As discussed already, the latter is a known electrode poison even at trace (parts per million) levels. Therefore, the negative effect of CO_2 on the performance and durability of a PEFC might be closely related to the CO poisoning effect.

The extent of the CO_2 reduction at a PEFC anode is highly dependent on the nature of the electrode and the fuel cell operating conditions (de Bruijn et al. 2002; Papageorgopoulos and de Bruijn 2002; Urian et al. 2003; Gu et al. 2004, 2005; Smolinka et al. 2005; Du et al. 2006b). The chemical reduction process can be written as

$$CO_2 + H_2 \rightarrow CO + H_2O. \tag{7}$$

This is known as the reverse WGS (RWGS) reaction. Platinum is a known catalyst for the WGS reaction; hence, it also catalyzes the reverse reaction in (7) and brings it closer to equilibrium.

CO_2 can also be reduced to CO electrochemically, through the following cathodic process:

$$CO_2 + 2H^+ + 2e^- \rightarrow CO + H_2O. \tag{8}$$

The reaction in (8) does not occur at the anode under normal operating conditions but it might take place at the cathode if CO_2 is transported across the membrane or if CO_2 reaches the cathode through other pathways.

The equilibrium CO concentration is estimated to be between 60 and 170 ppm for typical PEFC operating temperatures (Gu et al. 2004, 2005). The CO coverage of both a platinum and a PtRu electrode at open circuit was measured after exposure to a H_2/CO_2 (50:50, v/v) gas mixture at a flow rate of 100 sccm for 10 h. The CO stripping cyclic voltammetry (CV) data indicated a substantial CO buildup on the anode surface (Fig. 6). The CO coverage was higher on platinum than on PtRu because of several competing reactions (e.g., the reactions in (1) and (6)). As expected from the equilibrium constant for the reaction in (7), the CO coverage was found to be a function of temperature, pressure, and the CO concentration (de Bruijn et al. 2002; Gu et al. 2004, 2005).

The RWGS reaction was also studied on a platinum model electrode using a combination of in situ surface-enhanced infrared spectroscopy, mass spectrometry, and fuel cell performance data (Smolinka et al. 2005). The results indicate that CO_2 reduction proceeds through coadsorbed hydrogen adatoms and is favored in the hydrogen underpotential deposition region. Furthermore, the reaction is kinetically hindered by self-poisoning (product CO) and blocking of platinum sites. The resulting CO coverage on a platinum electrode is found to be close to 0.45 of a monolayer. This CO adlayer has little effect on the rate of the HOR, which proceeds

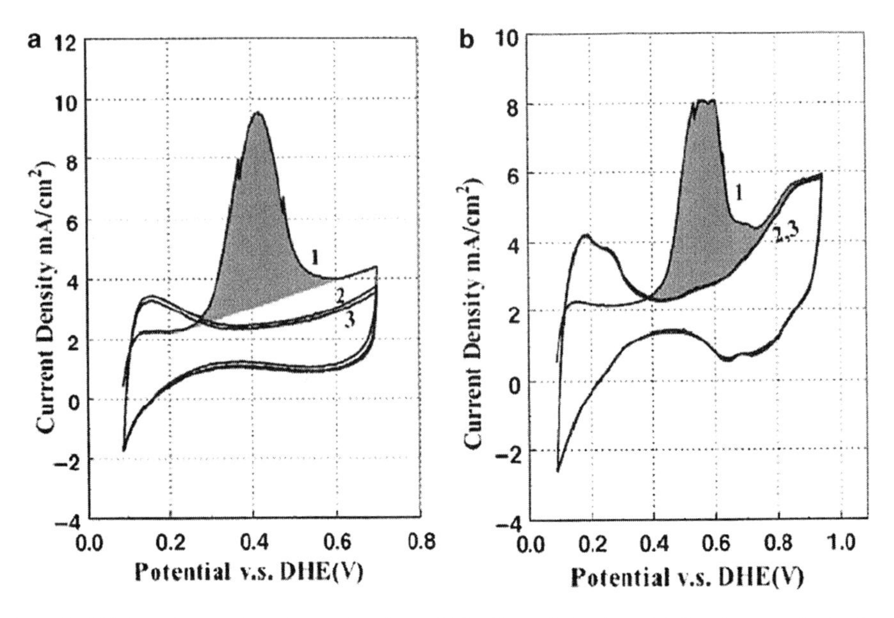

Fig. 5 CO stripping cyclic voltammetry data of (**a**) a PtRu electrode and (**b**) a platinum electrode. The electrodes were exposed to 100 sccm CO_2/H_2 (50:50 v/v) for 10 h before being flushed with N_2 for 5 min. Scan rate 10 mV/s; 100% RH at T_{cell} = 70 °C. *DHE* dynamic hydrogen electrode. (Adapted from: Gu et al. 2005, with permission)

through vacant platinum sites, i.e., holes in the CO adlayer. However, these holes are too small for CO_2 reduction to take place. This explains the relatively small performance loss (less than 5 mV at 0.6 A/cm^2 compared with the H_2/N_2 feed) for a PEFC operating with a H_2/CO_2 feed (Fig. 5), implying that there was no substantial formation of CO as the RWGS product. We found that, even feeding a hydrogen fuel containing 10 ppm CO, the overpotential of a platinum-only anode (0.4 mg Pt/cm^2) reached as high as 250 mV at 0.6 A/cm^2 in the absence of AB after it was exposed to CO for more than 1 h (Du 2004). It was found that, when a platinum electrode was exposed to CO_2 in the double-layer region, it did not lead to significant poisoning of the platinum electrode (de Bruijn et al. 2002). From this observation, it was inferred that adsorption of CO_2 itself is not important in the RWGS reaction since adsorption of neutral molecules (such as CO_2) is generally greatest in the double-layer region where competition with charged species and water molecules is minimal. It was reported that the reaction of CO_2 with hydrogen adatoms was suppressed on a PtRu electrode (Papageorgopoulos and de Bruijn 2002). Thin-film electrodes were found to be more tolerant to CO_2 than the standard porous electrodes, particularly when coated with Nafion® ionomers (Papageorgopoulos et al. 2002). A study on the effect of CO_2 on platinum alloy electrodes containing either ruthenium or molybdenum found that PtRu has the smallest overpotential loss, whereas PtMo behaves in a manner similar to that of Pt/C (Urian et al. 2003).

In an operating fuel cell the gas streams have a finite residence time. The RWGS reaction does not attain thermodynamic equilibrium because of slow reaction kinetics, the existence of competing reactions, and other factors (de Bruijn et al. 2002; Papageorgopoulos et al. 2002; Urian et al. 2003; Smolinka et al. 2005; Du et al. 2006a–c). As a result, the increase in η_a at a CD below 0.8 A/cm^2 is very small (less than 10 ± 5 mV) for a platinum electrode exposed to a H_2/CO_2 mixture over one exposed to a H_2/N_2 mixture under otherwise the same conditions (see Fig. 6). For example, the increase in η_a is less than 10 mV at 200 mA/cm^2, which is less than 5% of the overpotential loss for a platinum anode exposed to H_2 gases containing 40 and 100 ppm CO (200 and 350 mV, respectively) (Ralph and Hogarth 2002). However, even with 60% CO_2, the contribution of CO_2 to η_a is less than 40 mV (Fig. 6) at a CD of 1.6 A/cm^2 (Gu et al. 2004).

We found that with a PtRu alloy electrode the effect of the RWGS reaction is negligible in an operating fuel cell (Du et al. 2006b). As mentioned earlier, the PtRu alloy catalyzes electrochemical CO oxidation (1). Another competing reaction is chemical oxidation of CO in the presence of O_2 (6), which may be present at the anode as a result of deliberate addition of air (e.g., AB) or by permeation through the membrane. This reaction is catalyzed by platinum. Commercial MEAs with a PtRu anode showed no detectable RWGS reaction using simulated reformate with 10 ppm CO and 15–17% CO_2, at either fuel cell conditions or open circuit (Du et al. 2006a–c). Furthermore, when a CO-free reformate with 16% CO_2 was used, no CO was detected over the entire test range with or without AB. Finally, there was no detectable CO at the anode exhaust under either fuel cell conditions or open circuit when simulated reformate containing either 0 or 2.4 ppm CO was tested for over 13 h.

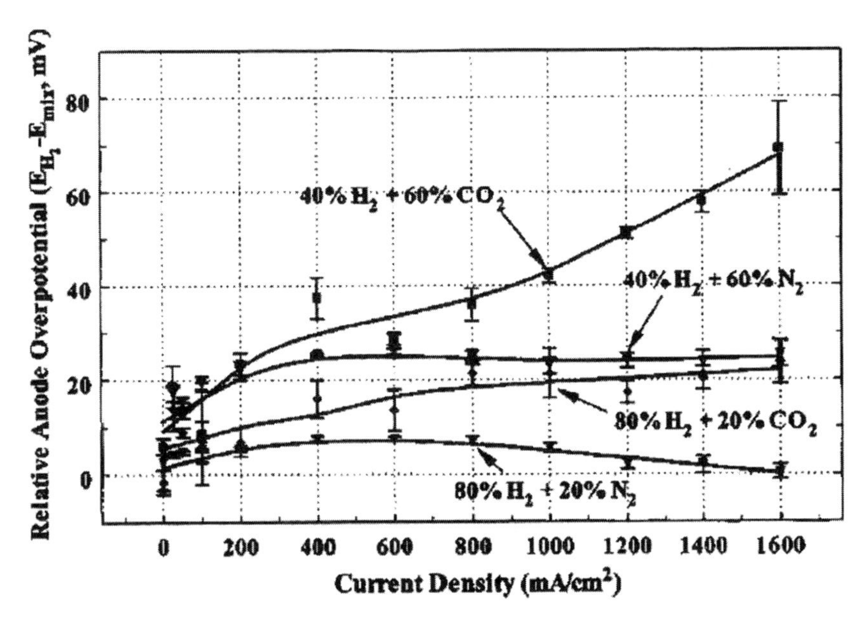

Fig. 6 Relative anode overpotential due to CO_2 and N_2 dilution for a platinum anode. (From Gu et al. 2004, with permission)

The absence of a detectable RWGS reaction at the Pt–Ru anode electrode has a profound implication for the fuel reforming process. Should there have been significant CO production from the RWGS, it would have rendered futile any efforts to produce reformate with a CO concentration lower than that which would be generated in situ by the RWGS reaction. Fortunately, this is not the case. The slow RWGS kinetics justifies the effort to generate reformate with a low CO concentration.

5 N_2 Dilution and the Inertial Effect

Depending on the specific reforming conditions, a reformate feed may contain 25–75% nonreactive components such as N_2 and water vapor. This results in a small but noticeable overpotential loss that can be precisely assigned to the concentration (dilution) and inertial (diffusion) effects. In addition, CO_2, a major component of the reformate mixture, has the same dilution effect as N_2 and H_2O, but has a greater impact on the rate of H_2 diffusion because M_{CO2} (44) is higher than that of both N_2 (M_{N2} =28) and H_2O ($M_{H2}O$ = 18) (Bird et al. 1960; Gu et al. 2004). Figure 6 compares the anode polarizations of H_2/CO_2 mixtures with those of H_2/N_2 mixtures under the same operating conditions.

The thermodynamic value of η_a that results from N_2 dilution can be readily calculated (Gu et al. 2004). For a simulated reformate with 60:40 (v/v) H_2/N_2, the calculated concentration overpotential is 7.9 mV at 70 °C and ambient pressure. This is consistent with η_a at open circuit (Gu et al. 2004). When a load is applied, however, η_a (more than 20 mV) is more than twice the calculated value and it is nearly a constant in the low-CD (below 0.8 A/cm^2) region. With a pure H_2 feed, the only gas-phase diffusion resistance arises from water vapor that has either been added to the inlet stream (to mitigate membrane degradation) or been transported through the membrane from the cathode. With deliberate additions of an inert gas such as nitrogen, the diffusion resistance becomes more substantial because the effective diffusion coefficient D_{1m} of H_2 in the gas mixture m is reduced. This is known as the Maxwell–Stefan effect and can be quantified using the following approximate relationship (Bird et al. 1960):

$$(1-x_1) / D_{1m} = \Sigma_{2,n} (x_j / D_{ij}),$$

where x_j is the mole fraction of species j and D_{ij} is the binary diffusion coefficient for transport of species i through species j. This coefficient is proportional to $(1/M_i + 1/M_j)^{1/2}$, where M denotes the molecular weight. Hence, diffusion of hydrogen is more difficult in the presence of high molar fractions of heavy molecules, an observation that is sometimes referred to as the inertial effect. This diffusion resistance lowers the hydrogen partial pressure at the reaction sites. At low CDs, the main effect is to increase the local overpotential, in accordance with the Butler–Volmer equation. As the CD is increased, the mass-transport resistance becomes more significant. The effect is a function of H_2 concentration: the lower the H_2 concentration, the greater the mass-transport loss and the smaller the limiting CD. A reformate may contain 20–50% H_2, depending on the reforming process and the system design. It should be mentioned that relative humidity affects the anode potential the same way N_2 does but that the net effect of N_2 dilution on the thermodynamic, kinetic, and mass-transport driving forces is approximately independent of relative humidity (Gu et al. 2004).

The discussion above explains how N_2, an inert gas, lowers the beginning-of-life cell voltage relative to neat hydrogen under the same operating conditions. This change, in turn, necessitates the use of additional cells and/or lowers the maximum acceptable VDR since a stack must meet the specified power requirements throughout its lifetime. Also, the presence of N_2 limits the choices of anode flow-field design, precluding the use of dead-header designs and impacting fuel recirculation systems. Finally, it has been observed that the presence of N_2 can make adsorbed CO more difficult to remove from the surface of the anode catalyst. Specifically, the extent of poisoning became more pronounced when H_2 was diluted with N_2 and the ratio of CO to H_2 was held constant (Wang et al. 2006). This, combined with the performance degradation associated with CO and other impurities in a reformate, presents a special challenge to the design and operation of a PEFC system operating on reformate.

On the other hand, a diluted anode feed can improve the flow dynamics because of the increased flow rate at the same hydrogen stoichiometry, albeit at a cost of reduced system efficiency because of higher auxiliary power consumption. In addition, use of N_2 reduces the dew-point difference between the anode inlet and outlet. These factors can have a positive impact on water management, e.g., anode flooding and water back-diffusion (Du et al. 2005, 2006a–c).

6 Other Potential Electrode-Poisoning Species

There are other potential impurities in reformate that could have a profound impact on fuel cell performance and durability. These include inorganic species, such as NH_3 and H_2S, and small organic molecules such as CH_4, HCHO, and HCOOH (Uribe et al. 2002; Mohtadi et al.2003; Narusawa et al. 2003; Soto et al. 2003; Garzon et al. 2006).

6.1 *Effect of NH_3*

Trace NH_3 is formed during the reforming stage in which the coexistence of H_2 and N_2 at high temperature and in the presence of catalysts can produce as much as 30–90 ppm NH_3 (Little 1994; Uribe et al. 2002). The actual amount of NH_3 varies depending on the type of the reforming process and whether or not the fuel processor includes shift reactors, PROX reactors, and other fuel-cleaning steps. The effect of NH_3 on PEFC performance depends on both NH_3 concentration and exposure time. Tests with hydrogen gases containing 30 and 130 ppm NH_3 demonstrated that the performance loss can be fully recovered if a fuel cell is exposed to NH_3 for 1–3 h (Uribe et al. 2002). However, it was found that long-term exposure (15 h) resulted in unacceptable cell voltage loss which could not be recovered even after several days of operating on neat H_2. Uribe et al. examined several possible mechanisms through which NH_3 may affect the cell performance. Experimental results indicated that the primary cause is the replacement of H^+ by NH_4^+ within the ionomer present in the anode catalyst layer. This leads to a decrease in proton conductivity.

The recovery process is thought to involve proton generation by an anodic HOR current. Long-term exposure to NH_3 causes a further drop in membrane conductivity through the same replacement mechanism. CV data showed no changes at both the anode and the cathode, indicating that there is no direct poisoning of either electrode by NH_3. Soto et al. examined the effect of high transient NH_3 (200–1,000 ppm) (Soto et al. 2003). They found that a fuel cell exposed to 200-ppm transient NH_3 for 10 h can be completely recovered using neat H_2. At higher NH_3 concentrations (500 ppm for 4 h or 1,000 ppm for 2 h), the recovery is incomplete even though the net NH_3 dose is the same for all three NH_3 concentrations. The performance loss is also higher for 1,000 ppm NH_3 than for 500 ppm NH_3 at the same dose. A two-step recovery process is proposed to account for the two slopes observed during the recovery process (Fig. 7). Soto et al. suggest that the first step involves the relatively

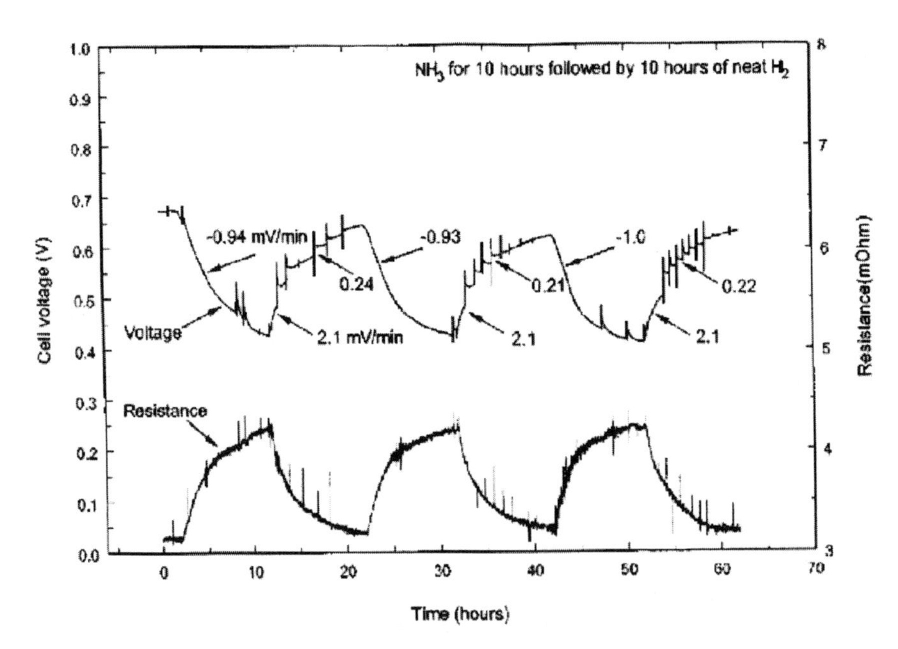

Fig. 7 Transient performance with neat H_2 and 200 ppm NH_3 in H_2 with 0.6 A/cm^2 at 101 kPa and 70 °C. (From Soto et al. 2003, with permission)

fast neutralization of NH_4^+ at the electrode/ionomer interface inside the electrode layer. The second step involves the slow propagation of this neutralization front towards the membrane region. This is consistent with the two slopes in voltage recovery and a first-order recovery in cell resistance.

6.2 *Effect of H_2S*

Many sulfur compounds are notorious poisons for metal catalysts because of their high affinity for metals. For safety purposes, organic sulfur-containing compounds are often added as odorants to natural gas, liquefied propane gas, and other fuels. When these fuels are used for hydrogen production, they must undergo desulfurization to remove the sulfur-containing odorants, which are poisonous to the reforming catalysts; however, it is difficult to completely remove these materials. Some trace sulfur compounds may be converted to H_2S during the reforming process. However, H_2S is considered harmful to PEFC electrodes even at parts per billion levels because it adsorbs readily on platinum and ruthenium (Paal et al. 1997; Uribe et al. 2001; Wang et al. 2001; Mohtadi et al. 2003). Detection of H_2S at parts per billion levels is very challenging because H_2S contaminates any metallic surfaces it comes in contact with, and these surfaces are difficult to clean afterwards. Consequently, measurements can become unreliable even after one experiment. Because of this, no reliable data are available for the effect of H_2S on fuel cell performance at low parts per billion levels. A permanent voltage loss was observed when a PEFC was exposed to H_2S (Uribe et al. 2001). When a hydrogen gas containing 50 ppm H_2S was used to study the effects of H_2S on platinum and PtRu electrodes, higher partial recovery of the cell performance was observed after multiple CV scans as compared with recovery in neat H_2 (Mohtadi et al. 2003). The degree of recovery is determined by the electrochemical oxidation of the surface species Pt–S to SO_3 or SO_4^{2-}, corresponding to two distinct oxidation peaks at 0.89 and 1.09 V, respectively. Ruthenium does not increase the H_2S tolerance and plays no role in the recovery process.

6.3 *Effect of Small Organic Molecules*

CH_4, HCHO and HCOOH were found in reformate generated from methanol (Narusawa et al. 2003). All three small organic molecules are present in minute quantities from either steam reforming or autothermal reforming. No adverse effect was observed from CH_4 adsorption on platinum electrodes. On the other hand, both HCHO and HCOOH had a negative impact on fuel cell performance. It is estimated that the poisoning coefficient for HCHO is about 0.1 times that for CO, whereas the poisoning coefficient for HCOOH is only 0.004 times that for CO.

7 Summary

The durability and degradation of PEFCs operating with reformate depend on the exact composition of a given reformate. N_2 and CO_2, two major components in a typical reformate mixture, affect the overall system performance and durability through the dilution effect (reduced hydrogen partial pressure) and the inertial effect (reduced rate of hydrogen diffusion). The RWGS reaction is not significant for a PtRu anode under operating conditions that are typical for a PEFC system. The presence of CO in the reformate, even at trace (parts per million) levels, has a significant impact on PEFC performance and durability because the yield of H_2O_2 is increased. The presence of CO can also adversely affect cell voltage if the availability of vacancies is attenuated significantly, and this occurs more readily for some catalysts than others. The most common CO mitigation method, namely, AB, promotes a further increase in H_2O_2 by introducing excess amounts of O_2 into the anode, which has a working potential conducive to H_2O_2 formation. PAB improves cell performance and durability and enhances system reliability by minimizing AB time and being responsive to high CO transients. Reconfigured anodes also improve performance with reduced air requirement. Trace impurities, such as NH_3, H_2S, and small organic molecules, can have a profound and often irreversible impact on a PEFC.

Much remains to be studied on the effect of trace amounts of impurities on PEFC performance. This represents an excellent area for precompetitive research at universities and national laboratories. For example, such research is currently being conducted by the NSF Fuel Cell Consortium at the University of South Carolina. Also, the US Department of Energy's Multi-Year Hydrogen Program recently awarded three multimillion-dollar projects on this topic to the University of Connecticut, Clemson University, and Los Alamos National Laboratory, respectively.

Acknowledgement Part of this work was supported by a grant from the Department of Energy (DE-FC36-03GO13098 as a subcontractor to the 3M Corporation).

References

Adams, W. A., Blair, J., Bullock, K. R., and Gardner, C. L. (2005). Enhancement of the performance and reliability of CO poisoned PEM fuel cells. J. Power Sources 145(1): 55–61

Adcock, P. A., Pacheco, S. V., Norman, K. M., and Uribe, F. A. (2005). Transition metal oxides as reconfigured fuel cell anode catalysts for improved CO tolerance: Polarization data. J. Electrochem. Soc. 152(2): A459–A466

Air Products and Chemicals, Inc. (2004). Gaseous Hydrogen SafetyGram Sheet, Air Products and Chemicals, Inc

Antoine, O. and Durand, R. (2000). RRDE study of oxygen reduction on Pt nanoparticles inside Nafion®: H_2O_2 production in PEMFC cathode conditions. J. Appl. Electrochem. 30(7): 839–844

Baldwin, R., Pham, M., Leonida, A., McElroy, J., and Nalette, T. (1990). Hydrogen – oxygen proton-exchange membrane fuel cells and electrolyzers. J. Power Sources 29(3–4): 399–412

Baschuk, J. J. and Li, X. G. (2003). Modelling CO poisoning and O_2 bleeding in a PEM fuel cell anode. Intl. J. Energy Res. 27(12): 1095–1116

Bhatia, K. K. and Wang, C.-Y. (2004). Transient carbon monoxide poisoning of a polymer electrolyte fuel cell operating on diluted hydrogen feed. Electrochim. Acta 49(14): 2333–2341

Bird, R. B., Stewart, W. E., and Lightfoot, E. N. (1960). *Transport Phenomena*. New York, Wiley

Camara, G. A., Ticianelli, E. A., Mukerjee, S., Lee, S. J., and McBreen, J. (2002). The CO poisoning mechanism of the hydrogen oxidation reaction in proton exchange membrane fuel cells. J. Electrochem. Soc. 149(6): A748–A753

Carrette, L. P. L., Friedrich, K. A., Huber, M., and Stimming, U. (2001). Improvement of CO tolerance of proton exchange membrane (PEM) fuel cells by a pulsing technique. Phys. Chem. Chem. Phys. 3(3): 320–324

Curtin, D. E., Lousenberg, R. D., Henry, T. J., Tangeman, P. C., and Tisack, M. E. (2003). *Advanced Materials for Improved PEMFC Performance and Life*. The 8th Grove Fuel Cell Symposium, London, UK

Davies, J. C., Nielsen, R. M., Thomsen, L. B., Chorkendorff, I., Logadottir, A., Lodziana, Z., Norskov, J. K., Li, W. X., Hammer, B., Longwitz, S. R., Schnadt, J., Vestergaard, E. K., Vang, R. T., and Besenbacher, F. (2004). CO Desorption rate dependence on CO partial pressure over platinum fuel cell catalysts. Fuel Cells 4(4): 309–319

de Bruijn, F. (2005). Current status of fuel cell technology for mobile and stationary applications. Green Chem. 7(3): 132–150

de Bruijn, F. A., Papageorgopoulos, D. C., Sitters, E. F., and Janssen, G. J. M. (2002). The influence of carbon dioxide on PEM fuel cell anodes. J. Power Sources 110(1): 117–124

DOE (2005a). Fuel Cell Handbook, 7th Ed. US Department of Energy, National Energy Technology Laboratory. US Department of Energy, National Energy Technology Laboratory, Morgantown, WV, EG&G Technical Services, Inc

DOE (2005b). Multi-Year Research, Development and Demonstration Programs for Hydrogen, Fuel Cells & Infrastructure Technologies Program, US Department of Energy

Donaldson Company, Inc. (2003). Point of Use (POU) Filtration for Optical Elements in Semiconductor Lithography Tools

Du, B. (2004). unpublished results

Du, B., Jacobson, D. L., Wang, G. Q., Eldrid, S., Elter, J. F., Eisman, G. A., and Arif, M. (2005). *Tuning Hydrogen Content for Improved PEMFC Water Management: A Neutron Radiography Study*. The 2nd International. Conference Green & Sustainable Chem., Washington, DC

Du, B., Guo, Q., Pollard, R., Rodriguez, D., Smith, C., and Elter, J. F. (2006a). PEM fuel cells: Status and challenges for commercial stationary power applications. JOM 58(8): 44–48

Du, B., Pollard, R., and Elter, J. F. (2006b). CO–air bleed interaction and performance degradation study in proton exchange membrane fuel cells. ECS Trans. 3(1): 705–713

Du, B., Wang, G., Elter, J. F., Pollard, R., Jacobson, D. L., Hussey, D., Arif, M., and Eisman, G. (2006c). *Applications of Neutron Radiography in PEM fuel cell Research and Development.* The 8th World Conference on Neutron Radiography, Gaithersburg, MD

Du, B., Guo, Q., Qi, Z., Mao, L., Pollard, R., and Elter, J. F. (2008). Chapter 12. Materials for Proton Exchange Membrane Fuel Cells. in R. H. Jones and G. J. Thomas. *Materials for Hydrogen Economy*. New York, Taylor & Francis: 251–309

Eberle, K., Bernd, R., Joachim, S., and Raimund, S. (2002). Device and Method for Combined Purification and Compression of Hydrogen Containing CO and the Use Thereof in Fuel Cell Assemblies. US Patent No. 6,361,896

Garzon, F., Uribe, F. A., Rockward, T., Urdampilleta, I. G., and Brosha, E. L. (2006). The impact of hydrogen fuel contaminates on long-term PMFC performance. ECS Trans. 3(1): 695–703

Gasteiger, H. A., Markovic, N. M., and Ross, P. N. (1995a). H_2 and CO electrooxidation on well-characterized Pt, Ru, and Pt–Ru. 1. Rotating disk electrode studies of the pure gases including temperature effects. J. Phys. Chem. 99(20): 8290–8301

Gasteiger, H. A., Markovic, N. M., and Ross, P. N. (1995b). H_2 and CO electrooxidation on well-characterized Pt, Ru, and Pt–Ru. 2. Rotating disk electrode studies of CO/H_2 mixtures at 62 °C. J. Phys. Chem. 99(45): 16757–16767

Gottesfeld, S. D. (1990). Preventing CO poisoning in fuel cells. US Patent No. 4,910,099

Gottesfeld, S. and Pafford, J. (1988). A new approach to the problem of carbon monoxide poisoning in fuel cells operating at low temperatures. J. Electrochem. Soc. 135(10): 2651–2652

Gu, T., Lee, W. K., Van Zee, J. W., and Murthy, M. (2004). Effect of reformate components on PEMFC performance. J. Electrochem. Soc. 151(12): A2100–A2105

Gu, T., Lee, W. K., and Van Zee, J. W. (2005). Quantifying the 'reverse water gas shift' reaction inside a PEM fuel cell. Appl. Catal. B 56(1–2): 43–50

Halseid, R., Wainright, J. S., Savinell, R. F., and Tunold, R. (2007). Oxidation of ammonium on platinum in acidic solutions. J. Electrochem. Soc. 154(2): B263–B270

Higashiguchi, S., Hirai, K., Shinke, N., Ibe, S., Yamazaki, O., Yasuhara, K., Hamabashiri, M., Koyama, Y., and Tabata, T. (2003). *Development of Residential PEFC Cogeneration Systems at Osaka Gas*. 2003 Fuel Cell Seminar, Miami Beach, FL, Mira Digital Publishing, St. Louis, MO

Ianniello, R., Schmidt, V. M., Stimming, U., Stumper, J., and Wallau, A. (1994). CO adsorption and oxidation on Pt and Pt–Ru alloys: dependence on substrate composition. Electrochim. Acta 39(11–12): 1863–1869

Jang, J. H., Han, S., Hyeon, T., and Oh, S. M. (2003). Electrochemical capacitor performance of hydrous ruthenium oxide/mesoporous carbon composite electrodes. J. Power Sources 123(1): 79–85

Jimenez, S., Soler, J., Valenzuela, R. X., and Daza, L. (2005). Assessment of the performance of a PEMFC in the presence of CO. J. Power Sources 151: 69–73

Jusys, Z. and Behm, R. J. (2004). Simulated 'air bleed': Oxidation of adsorbed CO on carbon supported Pt. Part 2. Electrochemical measurements of hydrogen peroxide formation during O_2 reduction in a double-disk electrode dual thin-layer flow cell. J. Phys. Chem. B. 108(23): 7893–7901

Jusys, Z., Kaiser, J., and Behm, R. J. (2003). Simulated 'air bleed': oxidation of adsorbed CO on carbon supported Pt. J. Electroanal. Chem. 554–555: 427–437

Karuppaiah, C. and Lakshmanan, B. (2003). Carbon Monoxide Filter. US Patent No. 6517963

Kim, H. and Popov, B. N. (2002). Characterization of hydrous ruthenium oxide/carbon nanocomposite supercapacitors prepared by a colloidal method. J. Power Sources 104(1): 52–61

Knights, S. D., Colbow, K. M., St-Pierre, J., and Wilkinson, D. P. (2004). Aging mechanisms and lifetime of PEFC and DMFC. J. Power Sources 127(1–2): 127–134

Lakshmanan, B. and Weidner, J. W. (2002). Electrochemical CO filtering of fuel-cell reformate. Electrochem. Solid-State Letts. 5: A267–A270

Lei, T., Zei, M. S., and Ertl, G. (2005). Electrocatalytic oxidation of CO on Pt-modified Ru(0001) electrodes. Surf. Sci. 581(2–3): 142–154

Lin, W. F., Iwasita, T., and Vielstich, W. (1999). Catalysis of CO electrooxidation at Pt, Ru, and PtRu alloy: An in situ FTIR study. J. Phys. Chem. B 103(16): 3250–3257

Little, A. D. (1994). Multi-Fuel Reformers for Fuel Cells Used in Transportation, Phase I, Final Report, Arthur D. Little, Inc

Mohtadi, R., Lee, W. K., Cowan, S., Van Zee, J. W., and Murthy, M. (2003). Effects of hydrogen sulfide on the performance of a PEMFC. Electrochem. Solid-State Letts. 6(12): A272–A274

Narusawa, K., Hayashida, M., Kamiya, Y., Roppongi, H., Kurashima, D., and Wakabayashi, K. (2003). Deterioration in fuel cell performance resulting from hydrogen fuel containing impurities: poisoning effects by CO, CH_4, HCHO and HCOOH. JSAE Review 24(1): 41–46

Oetjen, H. F., Schmidt, V. M., Stimming, U., and Trila, F. (1996). Performance data of a proton exchange membrane fuel cell using H_2/CO as fuel gas. J. Electrochem. Soc. 143(12): 3838–3842

Paal, Z., Matusek, K., and Muhler, M. (1997). Sulfur adsorbed on Pt catalyst: Its chemical state and effect on catalytic properties as studied by electron spectroscopy and n-hexane test reactions. Appl. Catal. A 149(1): 113–132

Papageorgopoulos, D. C. and de Bruijn, F. A. (2002). Examining a Potential Fuel Cell Poison. J. Electrochem. Soc. 149(2): A140–A145

Pianca, M., Barchiesi, E., Esposto, G., and Radice, S. (1999). End groups in fluoropolymers. J. Fluorine Chem. 95(1–2): 71–84

Ralph, T. R. and Hogarth, M. P. (2002). Catalysis for low temperature fuel cell: Part II. The anode challenges. Platinum Metal Rev. 46(3): 117–135

Ramani, M., Haran, B. S., White, R. E., and Popov, B. N. (2001a). Synthesis and characterization of hydrous ruthenium oxide-carbon supercapacitors. J. Electrochem. Soc. 148(4): A374–A380

Ramani, M., Haran, B. S., White, R. E., Popov, B. N., and Arsov, L. (2001b). Studies on activated carbon capacitor materials loaded with different amounts of ruthenium oxide. J. Power Sources 93(1–2): 209–214

Santiago, E. I., Paganin, V. A., do Carmo, M., Gonzalez, E. R., and Ticianelli, E. A. (2005). Studies of CO tolerance on modified gas diffusion electrodes containing ruthenium dispersed on carbon. J. Electroanal. Chem. 575(1): 53–60

Saravanan, C., Dunietz, B. D., Markovic, N. M., Somorjai, G. A., Ross, P. N., and Head-Gordon, M. (2003). Electro-oxidation of CO on Pt-based electrodes simulated by electronic structure calculations. J. Electroanal. Chem. 554–555: 459–465

Schiraldi, D. A. (2006). Perfluorinated polymer electrolyte membrane durability. Polym. Rev. 46(3): 315

Schmidt, T. J. and Baurmeister, J. (2006). Durability and reliability in high-temperature reformed hydrogen PEFCs. ECS Trans. 3(1): 861–869

Schmidt, T. J., Noeske, M., Gasteiger, H. A., Behm, R. J., Britz, P., Brijoux, W., and Bonnemann, H. (1997). Electrocatalytic activity of PtRu alloy colloids for CO and CO/H_2 electrooxidation: Stripping voltammetry and rotating disk measurements. Langmuir 13(10): 2591–2595

Schmidt, T. J., Noeske, M., Gasteiger, H. A., Behm, R. J., Britz, P., and Bonnemann, H. (1998). PtRu alloy colloids as precursors for fuel cell catalysts. J. Electrochem. Soc. 145(3): 925–931

Schmidt, T. J., Gasteiger, H. A., and Behm, R. J. (1999). Rotating disk electrode measurements on the CO tolerance of a high-surface Area Pt/vulcan carbon fuel cell catalyst. J. Electrochem. Soc. 146(4): 1296–1304

Smolinka, T., Heinen, M., Chen, Y. X., Jusys, Z., Lehnert, W., and Behm, R. J. (2005). CO_2 reduction on Pt electrocatalysts and its impact on H_2 oxidation in CO_2 containing fuel cell feed gas-A combined in situ infrared spectroscopy, mass spectrometry and fuel cell performance study. Electrochim. Acta 50(25–26): 5189–5199

Soto, H. J., Lee, W.-K., Van Zee, J. W., and Murthy, M. (2003). Effect of transient ammonia concentrations on PEMFC performance. Electrochem. Solid-State Letts. 6(7): A133–A135

Springer, T. E., Rockward, T., Zawodzinski, T. A., and Gottesfeld, S. (2001). Model for polymer electrolyte fuel cell operation on reformate feed: effects of CO, H_2 dilution, and high fuel utilization. J. Electrochem. Soc. 148(1): A11–A23

Stamenkovic, V., Grgur, B. N., Ross, P. N., and Markovic, N. M. (2005). Oxygen reduction reaction on Pt and Pt-bimetallic electrodes covered by CO. J. Electrochem. Soc. 152(2): A277–A282

Staudt, R., Boyer, J., and Elter, J. F. (2005). *Development, design, and performance of high temperature fuel cell technology*. Extended Abstracts for 2005 Fuel Cell Seminar, Palm Springs, CA, Courtesy Associates, Washington, DC

Stetter, J. R. and Pan, L. (1994). Amperometric carbon monoxide sensor module for residential alarms US Patent No. 5331310

Thomason, A. H., Lalk, T. R., and Appleby, A. J. (2004). Effect of current pulsing and 'self-oxidation' on the CO tolerance of a PEM fuel cell. J. Power Sources 135(1–2): 204–211

Tsiplakides, D., Balomenou, S., Katsaounis, A., Archonta, D., Koutsodontis, C., and Vayenas, C. G. (2005). Electrochemical promotion of catalysis: mechanistic investigations and monolithic electropromoted reactors. Catal. Today 100(1–2): 133–144

Uribe, F. A., Zawodzinski, J. T. A., and Gottesfeld, S. (2001). *Abstract 339. The Impact of Hydrogen Fuel Contaminants on Long-Term PEMFC Performance*. The 200th Electrochemical Society Meeting, San Francisco, CA

Uribe, F. A., Gottesfeld, S., and Zawodzinski, J. T. A. (2002). Effect of ammonia as potential fuel impurity on proton exchange membrane fuel cell performance. J. Electrochem. Soc. 149(3): A293–A296

Urian, R. C., Gulla, A. F., and Mukerjee, S. (2003). Electrocatalysis of reformate tolerance in proton exchange membranes fuel cells: Part I. J. Electroanal. Chem. 554–555: 307–324

Uribe, F. A., Valerio, J. A., Garzon, F. H., and Zawodzinski, T. A. (2004). PEMFC reconfigured anodes for enhancing CO tolerance with air bleed. Electrochem. Solid-State Letts. 7(10): A376–A379

Vielstich, W., Lamm, A., and Gasteiger, H. A., Eds. (2003). Handbook of Fuel Cells: Fundamentals, Technology, and Applications. West Sussex, England, Wiley

Wang, K., Gasteiger, H. A., Markovic, N. M., and Ross, P. N. (1996). On the reaction pathway for methanol and carbon monoxide electrooxidation on Pt–Sn alloy versus Pt–Ru alloy surfaces Electrochim. Acta 41(16): 2587–2593

Wang, Y., Yan, H., and Wang, E. F. (2001). The electrochemical oxidation and the quantitative determination of hydrogen sulfide on a solid polymer electrolyte-based system. J. Electroanal. Chem. 497(1–2): 163–167

Wang, W., Van Zee, J. W., and Lee, W. K. (2006). The effect of N_2 dilution on CO poisoning in a proton exchange membrane fuel cell. ECS Trans. 1(6): 541–547

Zhang, J. and Datta, R. (2002). Sustained potential oscillations in proton exchange membrane fuel cells with PtRu as anode catalyst. J. Electrochem. Soc. 149(11): A1423–A1431

Zhang, J. and Datta, R. (2004). Higher power output in a PEMFC operating under autonomous oscillatory conditions in the presence of CO. Electrochem. Solid-State Letts. 7(3): A37–A40

Zhang, J. and Datta, R. (2005). Electrochemical preferential oxidation of CO in reformate. J. Electrochem. Soc. 152(6): A1180–A1187

Zhang, J., Fehribach, J. D., and Datta, R. (2004). Mechanistic and bifurcation analysis of anode potential oscillations in PEMFCs with CO in anode feed. J. Electrochem. Soc. 151(5): A689–A697

3
Freezing

Subfreezing Phenomena in Polymer Electrolyte Fuel Cells

Jeremy P. Meyers

Abstract One of the most critical aspects of proper fuel cell design is water management: too little water, and the membrane will dry out; too much water, and the catalyst layer will flood and block access of the reactant gases to the electrocatalyst surface. Developers have put considerable effort into the optimization of three-dimensional structures that can accommodate the simultaneous demands of membrane hydration and gas access to the catalyst layer under normal operating conditions, as well as of the power-plant-level designs that can ensure water is properly distributed and managed. In fuel cell power plants that operate intermittently and are exposed to atmospheric conditions, such as in automotive applications, the challenges of water management are complicated by the fact that the system will frequently have to be started from subfreezing conditions. Given the volume change associated with freezing water, one expects that any water retained in the pores of the catalyst layer or at the catalyst layer interface with the gas-diffusion layers will expand and can therefore create considerable stresses on the porous materials, possibly deforming them from their initial state. To design a cell that can accommodate these changes, a thorough understanding of the physics of water movement and freezing is necessary. Researchers have seen degradation associated with freeze/thaw cycles and, more specifically, with drawing current from the fuel cell when the temperature of the cell itself is below freezing, a procedure that will be frequently experienced for fuel cell power plants deployed in automotive applications. Experiments demonstrate a marked increase in the mean pore size and width of the pore size distribution of the catalyst layer after thermal cycling. Modeling of the phase transition associated with thawing and freezing, however, as well as the coupled phenomena of water management and thermal management under partially frozen conditions is rather limited in the open literature. In this chapter, we examine the limits of current understanding, as well as the data that suggest freezing-point

J.P. Meyers
Department of Mechanical Engineering, College of Engineering,
The University of Texas at Austin, 1 University Station C2200, Austin, TX 78712, USA
e-mail: jeremypmeyers@mail.utexas.edu

F.N. Büchi et al. (eds.), *Polymer Electrolyte Fuel Cell Durability*,

depression in polymer electrolyte fuel cell materials and the implications of lowered temperatures on fundamental kinetic processes.

1 Introduction

As developers attempt to perfect fuel cell components and systems for transportation applications, they must ensure that these power plants perform to specification in the wide range of atmospheric conditions in which the vehicles will operate. Some products in which fuel cells might some day act as the primary power source might legitimately be specified to operate under a more restricted range of atmospheric conditions than an automobile is likely to experience, but the very fact that conventional light-duty vehicles are used to transport people and material over long distances implies that automakers cannot be sure that these power plants will not experience the extremes of temperatures to which conventional internal-combustion-engine vehicles are routinely subjected.

As such, it is necessary not only for fuel cell power plants to survive and operate in subfreezing conditions, but also for them to be able to start up from conditions in which the water retained in the stack has undergone the phase transition from vapor or liquid to solid ice. Different automotive stack developers and government agencies have different requirements for an unassisted start, but all share a requirement that the stack and system must be able to start up and operate in subzero temperature conditions (DOE 2007; Mathias et al. 2005; NEDO).

As polymer electrolyte fuel cells (PEFCs) generate water as a product of the electrochemical energy conversion process and require humidification to impart adequate membrane conductivity, we must anticipate that liquid water will be present in these systems, and that, furthermore, water can then undergo a phase transition to form ice when the stack components drop below freezing on shutdown (Fuller, and Newman 1993; Ishikawa et al. 2007). Because the saturation vapor pressure of water is a strong function of temperature, the water-carrying capacity of the exit gas stream is a strong function of stack temperature. Therefore, even without upstream humidification, the liquid-water content of the cell will increase when the stack temperature is low (Ahmed et al. 2002; Mao and Wang 2007).

Even though most configurations will almost certainly generate sufficient waste heat to overcome overboard heat losses and raise internal stack temperatures above 0 °C while operating, the dynamics of heating the stack from a subfreezing idle state must be considered when determining how to design a stack and system that can operate dynamically. In this chapter, we consider the requirements of automotive fuel cells, the physical state of water in the various components and materials within the stack, and the coupled heat- and mass-transport phenomena that dictate performance under subfreezing conditions. We then examine the approaches that researchers and developers have taken to improve the robustness of fuel cell performance under subfreezing conditions.

2 Requirements

One of the most basic requirements related to fuel cell usage in subfreezing conditions is simply for the fuel cell to be able to survive at and start up from subfreezing temperatures. The Department of Energy's PEFC stack technical targets for 2010 include survivability down to –40 °C (DOE 2007). Customers must be confident that their automobiles will survive and start up under all the conditions they are likely to experience. This requirement ensures that the stack and system will be robust under these conditions, but developers have further specified the startup procedure by restricting both the time for startup and the amount of energy that is required to start up the system.

The Department of Energy has recently added specific freeze-related requirements to its goals and objectives for automotive fuel cell development, and has stated that systems must be able to start up to 50% rated power from –20 °C in as little as 30 s with consumption of less than 5 MJ of energy, including system shutdown and startup. The energy requirement does warrant brief discussion: a maximum energy requirement essentially precludes heating the stack with energy from an external combustion chamber or some other form of stored energy. The restrictions on startup/shutdown energy also preclude an extended shutdown procedure to thoroughly dry out the cell components and prepare the cell for startup in a cold or frozen state. It has been shown that drying out the stack can lead to more reproducible startup times, and might allow a system to meet the startup time requirement more easily (Pesaran et al. 2005), but an extended blow-down procedure could violate the overall energy consumption requirement.

The startup energy requirement springs both from the need to ensure that the driver will not be stranded in a startup scenario in which the energy required to heat the stack substantially depletes the stored-onboard hydrogen, and from the need to retain the fuel cell's key advantage of tank-to-wheels fuel efficiency over the course of an entire drive cycle. The vehicle-level requirement on startup time and energy drives system- and component-level requirements to ensure that the stack is able to start up on its own. The vehicle startup requirements dictate that the stack components have sufficiently low thermal masses that internal cell temperatures can be raised quickly, that waste heat generated in the electrochemical processes associated with fuel cell operation can effectively be retained within the stack, and that the porous bodies within the stack can accommodate the quantities of liquid water that are generated at the lower temperatures (Mao and Wang 2007; Sundaresan and Moore 2005a, b).

While there has been significant activity in this area in the patent literature, relatively little effort has been reported in peer-reviewed publications until quite recently. Several factors complicate an investigation into subfreezing phenomena. The first of these is that, other than mapping the nature and extent of phase changes as a function of temperature, freeze-related phenomena are inherently transient problems, as waste heat will raise the internal cell temperatures and stack exit temperatures well above ambient conditions. While there are many papers describing thermal and water balance in PEFC systems, many of these early papers describe steady-state phenomena, and so cannot be directly adapted to examine the dynamics of cooling or starting up from a cold state (Bernardi and Verbrugge 1992; Fuller and Newman 1993). Other factors relevant to low-temperature operation that were neglected in previous models are the kinetics of phase change and modes of water transport that are only significant at or near freezing conditions. As these complications can be safely neglected except when temperatures approach the freezing point of water, the earlier studies appropriately excluded these phenomena. Further complicating the questions surrounding freezing is the simple fact that the details of freeze-related phenomena are highly dependent upon the cell and system design, and, as a consequence, experimental results have differed greatly among different research groups. It will take considerable effort to develop models that are sufficiently descriptive to aid in the design of freeze-tolerant components across the range of cell and system designs that are currently under development.

3 Performance Degradation

Early literature suggested that there was little degradation from freezing fuel cells to subfreezing temperatures (Simpson et al. 1995; Wilson et al. 1995), though admittedly for a limited number of cycles. Wilson et al. demonstrated that there was no significant change in performance of fuel cells subjected to three cycles to −10 °C; Simpson et al. demonstrated stable performance over ten cycles. However, a more recent paper suggested that there could be significant degradation in the performance of PEFCs subjected to cycling from −10 °C (Cho et al. 2003). An 11% drop in the current at 0.6 V was observed after four thermal cycles of the cell from −10 °C to a typical steady-state operating temperature of 80 °C. It was also reported that the electrochemical surface area of the cathode, as measured by the cyclic voltammogram, dropped by 25% during these cycles, while significant increases in ohmic resistance were noted. These changes were attributed to ice formation that resulted in the eventual delamination of the catalyst layer from the membrane and a loss of electrical contact with the external circuit. Similarly large losses in performance have been demonstrated by another study that also observed significant degradation after ten subfreezing starts from −10 °C (Oszcipok et al. 2005); specifically, degradation associated with each cold start cycle from −10 °C was observed. These results were attributed to the formation of ice resulting in structural changes to the catalyst and gas-diffusion layer (GDL).

It is worth noting, however, that other developers have shown that there is negligible performance loss for a cell started up and shut down multiple times, though their experimental protocol included a purging step to prepare the cell components for startup from freezing (Knights et al. 2004). As discussed before, extended dry-out procedures can require considerable amounts of energy and some developers have been reticent to rely on these extended shutdown procedures or to bring a system online during a shutdown to dry out the stack as ambient temperatures drop below freezing.

These apparent discrepancies (from no degradation to measurable and significant performance loss with each cycle) in the various literature results can perhaps be attributed to differences in the experimental protocol and in the details of the cell assembly. In particular, one expects that differences in membrane electrode assembly (MEA) fabrication technique can have considerable impact on the location of where water collects in the cold state, and on the relative strength of adhesion between adjacent layers.

It is important to note that while there are differences among experiments relative to freeze/thaw cycling, additional complications are associated when the fuel cell is started from a subfreezing condition. In this case, not only will the water in the cell be redistributed and change phases, but water will actually be generated in a supercooled state as it is generated at the cathode catalyst layer in the time before the cell core temperature is raised above freezing by waste heat generation. As such, the startup and shutdown procedure will have a large effect upon the quantity and location of water (and consequently of ice) in the cell components.

Moreover, the degradation is also a function of the rate of heating/cooling perhaps to an even greater degree than the temperature to which the cell equilibrates. The nature of water movement under these transient conditions will be addressed later. That rates of cooling and heating might have an effect on degradation is demonstrated by the fact that rapid cycling to −80 °C leads to delamination of the electrode, while normal cycling to −40 °C (even for 100 cycles) shows no such delamination in an identically prepared MEA, even though one would expect only modest differences in the equilibrium state of water for those two temperatures (Mukundan et al. 2005). Furthermore, whether ice formation can lead to structural changes of the GDL will likely depend upon the nature of the GDL and whether it is a carbon cloth of carbon paper; the component materials used in the fuel cell assembly will also play a vital role in determining the durability of fuel cells subjected to multiple freeze/thaw cycling.

4 Changes to Material Properties and Morphology

There is limited, and sometimes contradictory, data in the literature available on the durability of the catalyst layer and the GDL under freeze/thaw cycling. Studies have revealed that even a free-standing hydrated catalyst layer can be subjected to cracking and peeling while cycling (six cycles) from −30 °C (Guo and Qi 2006). This damage was associated with a loss in the electrochemical surface area of the catalyst that can be avoided by drying the catalyst.

While there are some data characterizing membrane properties at low temperature, to be discussed in greater detail in the following section, there is little information on the long-term effect of freeze/thaw cycling on the membrane properties. A recent study has revealed that extensive (385 cycles) cycling between −80 and +40 °C can result in a change in the mechanical and chemical properties of the dry membrane, including lower oxygen permeability, higher through-plane conductivity, and decreased strength (McDonald et al. 2004). This result is a little bit surprising, as calorimetry data suggest that most of the water in the membrane remains in an unfrozen state at these temperatures, as will be examined in a subsequent section. The detailed characterization of the durability of Nafion and other ionomeric polymers under fuel cell conditions, namely, in a high relative humidity atmosphere and under compressive load, during freeze/thaw cycling is still lacking.

Some recent results have also revealed that the GDL properties can change when the fuel cell subjected to freeze/thaw cycling (Lee and Merida 2007). It was observed that 50 freeze/thaw cycles from −35 °C resulted in a change in the air permeability of the GDL material. This was attributed to a degradation of the attachment of the microporous layer during the freeze cycling. This study also found that there were no changes in the resistivity, contact angle, and porosity of the material during the freeze cycling. Another study of the GDL material under in situ conditions revealed that the external contact angle can change from 131° to 112° when the fuel cell is subjected to ten cold starts from −10 °C (Oszcipok et al. 2005). This loss in hydrophobicity of the material resulted in a loss in fuel cell performance.

These studies indicate that while there might not be catastrophic failure associated with freeze/thaw cycling of fuel cells, there may be slow changes in component properties and MEA integrity that may lead to degradation in fuel cell performance when the fuel cell is subjected to extensive freeze/thaw cycling. These changes could include delamination, loss in catalyst electrochemical surface area, and changes in GDL and membrane physical/chemical properties. The effects of undergoing a phase transition on the various component properties still need to be evaluated in greater detail if we are to understand the degradation that can result from freeze/thaw cycling. It is therefore important to understand the thermodynamic states that will be experienced, as well as the transient phenomena associated with water movement and the thermal profiles associated with changing temperatures.

5 State of Water

Within the range of temperatures that a hydrogen PEFC will experience between a subzero temperature shutdown and operation at rated power, water is the only substance that is likely to undergo significant phase changes. The remainder of the reactant and product streams will remain in the gaseous state; the fuel cell components themselves will remain in a solid state. As such, we focus our attention on the behavior of water in the fuel cell. To understand and, eventually, to predict the nature of performance degradation and of the dynamics of startup from a cold state, it is critical to understand the nature and phase behavior of water in PEFC materials, and to understand how the water is distributed among the components of the cell. We therefore turn our attention first to the thermodynamic equilibrium state of water in the fuel cell components.

Water is expected to behave essentially like bulk water in the larger pores of the GDL and the adjacent flow fields, and it is therefore anticipated that this water will undergo the liquid–solid phase transition at precisely 0 °C. The dynamics of this process, however, will depend upon the pore-size distribution, the degree of connectedness, and the hydrophilicity of the surfaces. Capillary forces and the specific chemical interactions between water and the ionomer will result in freezing-point depression, and will lead to a distribution of freezing points throughout the components of the cell. Most of the available data on the equilibrium condition are on Nafion and other perfluorinated sulfonic acids, which, until recently, had been the only polymer materials widely used in PEFCs.

The state of water in Nafion has been characterized and described as three distinct types of water: nonfreezing water with very strong chemical interactions with the ionic groups in the polymer, bound freezing water, and free water (Yoshida and Miura 1992). The free water behaves like bulk water and freezes at 0 °C, while the bound freezing water is water that is trapped in the channels of the ionomer. The freezing point of this type of water depends upon the size and nature of the internal pores of the polymer, which in turn depend on the degree of hydration of the polymer (Cappadonia et al. 1994). The chemically bound water does not freeze all the way down to −120 °C, and is thought to be the source of the relatively high conductivity of Nafion, even in the frozen state. A recent study has reported the detailed characterization of water in Nafion at subfreezing temperatures as a function of the initial water content per sulfonic acid group (Thompson et al. 2006). This study revealed that the water in the center of the Nafion clusters tends to freeze first as it is shielded from the acid groups; after this water freezes, it leaves a more concentrated acid adjacent to the acid groups.

The conductivity of Nafion is highly dependent upon the state of water in the polymer, and has been shown to have an increased activation energy at lower temperatures, where the bulk-like water in the membrane is likely in the frozen state (Mukundan et al. 2005). These data are reproduced in Fig. 1. Even though a considerable fraction of water in the polymer is frozen, the fuel cell can still carry current through the polymer, even at subzero temperatures. While the ionic conductivity does not differ sharply at the normal freezing point, as one would expect if the entire polymer transitioned from a liquid to a solid state, the increased ionic resistance does limit the maximum current density at which the cell can operate under its own power. Consequently, startup procedures must consider control algorithms that gradually increase power as the cell components heat up and the maximum self-sustaining current increases (Blair 2005; Meyers 2005). This limitation on deliverable current adds complexity to the challenge of meeting fuel cell startup requirements and might require complex control strategies to determine how much current to draw from the cell during the transient.

6 Water Movement in Porous Media

While the ionic conductivity depends upon the water content and temperature, the mass-transport characteristics of the fuel cell can be considerably more complicated. In general, a fuel cell is optimized to deliver optimized performance at its preferred

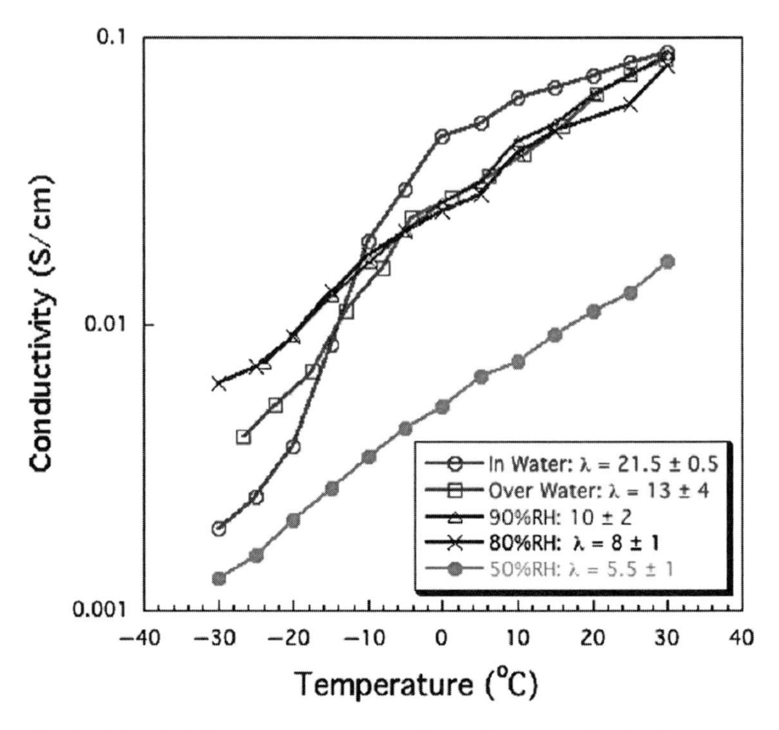

Fig. 1 Conductivity of Nafion membranes exposed to various saturation levels versus temperature near the freezing point. (Reproduced with permission from Mukundan et al. (2005). Copyright 2005, The Electrochemical Society)

operating temperature. The combination of macroscopic and microscopic porous layers in the GDL are selected to ensure that the membrane retains water and ionic conductivity, while providing contiguous paths for gas transport and liquid product removal (Park et al. 2006; Weber and Newman 2005; Ye and Wang 2007). With the current class of membrane materials, it is necessary to operate at very close to fully humidified conditions at rated power, to prevent the membrane from becoming dehydrated and suffering from unacceptably high ionic resistance. Under colder conditions, the cell will be considerably wetter, as the vapor-carrying capacity of the reactant gases will be lowered as the saturation pressure of water drops. As such, properties that are optimized for performance at rated power will likely compromise performance at colder temperatures. A thorough understanding of water movement is therefore necessary to inform designs that will perform well both at steady-state operating temperature and during the transitions between rated power and a subfrozen shutdown.

The gas-diffusion medium is the component most responsible for optimal water management in the PEFC. Without good water management, the fuel cell performance is decreased owing to transport losses of the reactants. Transport processes in a fuel cell involve the movement of various species through and between the cell's adjacent layers, each of which consists of different combinations of phases. The low temperatures (below 100 °C) and high relative humidities at which PEFCs operate imply that the product water is generally formed as a liquid. This liquid water can collect in the pores of the catalyst layer. The liquid water must be removed to maintain clear gas pathways to the active sites of the catalysts. As a result, the gas phase and liquid phase often compete to occupy the same pores of the catalyst layers and the gas-diffusion medium of typical fuel cells (Weber and Newman 2004). Liquid water in the pores of the gas-diffusion medium can therefore affect the fuel cell's mass transport in the gas phase by blocking access to that phase and reducing gas-phase connectivity and effective permeabilities.

The first fuel cell models to employ porous-electrode theory assumed a constant level of liquid-water content throughout the thickness of the GDL (Bernardi and Verbrugge 1992) or they assumed the liquid-water content was zero (Fuller and Newman 1993). Subsequent models added complexity to the model: Springer et al. tried to determine the effective path lengths of diffusing gas molecules in the cathode backing as functions of water content current density, but they assumed that the degree of saturation was constant through the thickness of the layers. (Springer et al. 1996) These early models showed that the water content in the GDL strongly affects the rates of gas-phase mass transport to the catalyst surface and that water content is a function not only of operating conditions but also of position within the catalyst layer.

Weber et al. (2004) described four separate phases in the catalyst layer: the solid phase, the ionomer phase, the gas phase, and the liquid phase. This last phase is the liquid water in the pores of the catalyst layer. Their model allows for water content that can vary with the capillary pressure in the pores of the catalyst layer and the adjacent GDL; liquid water's permeability and the effective gas diffusivities are strong functions of the water content in the pores. Low water content suggests only limited liquid-phase transport occurs through the GDL; high water content implies that liquid can move relatively easily but gases cannot. The relationship between water content and capillary pressure depends on the physical properties of the porous media, such as the composite contact angle of the pores and the pore size distribution. Models have also been developed to describe the dependence of water content on capillary pressure (Kumbur et al. 2006). These models suggest that changes to surface properties have profound effects on mass transport in fuel cells.

At present, no models precisely describe how the properties of porous-media surfaces in a fuel cell change as its operating conditions change. However, existing models that predict fuel cell performance for a given set of porous-media surface properties suggest that changing these properties dramatically affects fuel cell performance, especially in ranges of operating conditions for which mass-transport rates are important. Changing the pore size or the surface properties can change the water content in the pores, making the fuel cell prone to either dry-out or flooding, and presumably complicating further the challenge of ensuring high performance at rated temperature and power while maintaining a structure that is robust under subfreezing soak and startup conditions.

7 Ice Formation and Frost Heave

As mentioned previously, the water content in the catalyst layer will be determined by the extent of drying of the fuel cell before freezing, so the various literature studies may have widely different starting water contents in the catalyst layer. Furthermore, because the saturated vapor pressure curve for water is a strong function of temperature, even cells that operate at undersaturated conditions can exhibit condensation and freezing as the temperature drops after shutdown. It is expected that liquid-water and ice contents will be highest at the periphery of the cell, as the cell should cool first at the outer perimeter, and water will be transported to the coldest regions, condense, and subsequently freeze.

Another factor that can affect the state of hydration of the catalyst layer in the fuel cell is the movement of liquid water towards a freezing front, commonly referred to as "frost heave." In this phenomenon, water is drawn to a freezing front by interfacial forces. This is shown schematically in Fig. 2. Primary frost heave occurs when there is a sharp interface between frozen and unfrozen regions; secondary frost heave occurs when there is a fringe between frozen and unfrozen regions in a porous material. Just as models of water movement in fuel cells suggest that there can be varying levels of liquid-water saturation in the pores of diffusion media as water and vapor compete for the same pores, there can also be varying degrees of ice content within the pores of the media, even at temperatures removed from the normal freezing point of water.

Excess energy from capillary forces implies that smaller pores will have lower freezing points than larger pores or planar interfaces, thereby ensuring a range of freezing points for porous bodies with a range of pore sizes. Pores that are mostly filled with ice have a smaller effective pore diameter than regions that do not have ice, which means that water can be drawn into these smaller pores by the lowering of the liquid-water pressure in those regions. A full description of the coupled heat transfer, pressure-driven flow, and phase transition is quite complex.

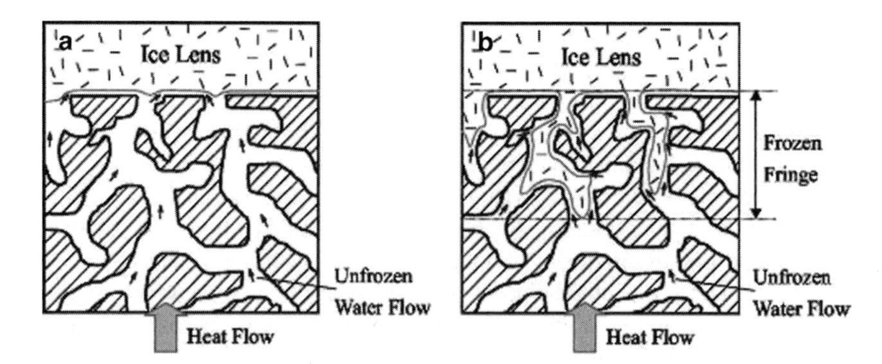

Fig. 2 Frost heave. (**a**) Primary heave. (**b**) Secondary heave. (Reproduced with permission from He and Mench (2006). Copyright 2006, The Electrochemical Society)

This movement has been studied primarily in the geology and soil literature in attempts to describe the water being drawn towards freezing surfaces in soils, which can lead to large displacements and considerable forces acting on the pore structure of the soils (Rempel et al. 2004). He and Mench (2006) presented a one-dimensional model of this effect and their results suggest that liquid water can be drawn from adjacent layers of the cell to form ice lenses adjacent to the catalyst layers that can block gas access to portions of the catalyst and induce delamination. Their paper also does a good job of summarizing the competing models of the phenomenon, and some of the debate about what forces need to be considered to quantify the movement: specifically, whether capillary forces, disjoining pressures, or refreezing phenomena need to be considered. Given the relatively sparse coverage of frost heave in the fuel cell literature and the continued uncertainty about the forces at play, it seems unlikely that all of the implications of this phenomenon have been considered, and further study will be required before an optimized cell structure can be recommended.

8 Dynamics of Startup

Most literature studies have indicated that PEFCs are capable of self-starting from subfreezing temperatures as low as −20 °C without any external heating (Yan and Wu 2005). There are a few reports of the inability of fuel cells to self-start from −5 °C (Hishinuma et al. 2004) or −15 °C (Qiangu et al. 2006), which may be related to the authors' failure to dry out the fuel cells to the desirable extent or, more likely, the high thermal mass of their cells. In practical fuel cell stacks, only the end cells will be in contact with the plates that provide contact to the external circuit and distribute the axial load; these plates tend to have high thermal masses without anywhere close to the rates of self-heating that the cells themselves provide. As such, end cells tend to lag center cells in performance, and single cell test results can skew interpretation of what will happen at the stack level (Meyers 2005).

When the fuel cell is operated at subfreezing temperatures, the water generated at the cathode will form ice that can result in a loss in performance of the fuel cell. One study has revealed that this water is initially present in a supercooled state and then its temperature rises to 0 °C at the time of freezing (Ishikawa et al. 2007). The charge-transfer resistance during a cold start from −10 °C, as measured by AC impedance, has been shown to increase with time, providing evidence for ice buildup in the catalyst layer (Oszcipok et al. 2005). Water formation in the membrane during the startup of a dried cell can result in membrane hydration, however, resulting in a lowering of the high-frequency resistance of the cell (Oszcipok et al. 2005). These competing aspects have been modeled using either empirical models (Oszcipok et al. 2006) or detailed models of the various parameters affecting the cell potential (Ahluwalia and Wang 2006). These models reveal that the fuel cell should be operated at a sufficiently high load to generate enough heat for an unassisted start. Most, if not all, of the fuel cell performance models that have been developed to date, however, have been constructed to describe steady-state performance at temperature, and the effects of higher water contents and lowered temperatures have not been incorporated into a transient model to describe a startup.

Furthermore, the cell has to heat up to above freezing temperature fast enough to avoid the ice formation from completely shutting down the electrochemical reaction at the cathode catalyst layer. To assist in this, the cathode gas flow rate can be increased (to blow ice away and to carry any excess water) and the inlet gases can be heated (to limit time below freezing temperatures). Because of the low vapor pressures of water at low temperatures, however, and the low heat capacities of the reactant gases, these approaches have limited utility in affecting the startup profiles of full-sized stacks.

9 Mitigation Strategies

The patent literature has over 100 patents of various mitigation strategies to use while operating/storing fuel cells at subfreezing temperatures. A detailed analysis of this patent literature has been conducted by the National Renewable Energy Laboratory and is available online (Pesaran et al. 2005). In essence, these mitigation strategies fall into three categories: (1) those that keep the fuel cell warm, thus preventing ice formation; (2) those that prevent ice formation by either drying out the fuel cell or by replacing the water with a nonfreezing liquid; and (3) those that prevent ice formation during startup by providing heat. Fuel cell stacks can be kept warm by providing insulation, or by providing heat either through a battery or by operating the cell intermittently in a low power mode, and while this approach might be used in conjunction with a robust design to improve the customer's experience, the so-called keep it warm approaches are incompatible with automotive requirements.

The freezing water can be avoided by eliminating carrying a water tank on board and by running the fuel cell at lower inlet relative humidities and reclaiming the exhaust water, though it is not clear if the current class of membrane materials will operate stably under such conditions. The water inside the stack can be minimized by running the cell under dry reactant gases (H_2, air) or dry nitrogen before shutdown or by vacuum-drying the fuel cell. Moreover, during startup, extra heat can be provided from a battery, by catalytically combusting hydrogen, or by preheating the reactant gases, and the heat-carrying capacity of the reactant gases is rather low. All these strategies aim to avoid/minimize the ice formation that can result in fuel cell performance loss, though they add complexity, and are likely incompatible with the Department of Energy's 5-MJ energy target.

While efforts to minimize freezing and maximize the rate at which stacks can self-heat during startup can improve performance of startup after short shutdowns, they are insufficiently robust to meet automotive requirements. A stack that operates at lower relative humidity, if such operation were enabled by novel membranes currently under development, would also simplify the problem by lowering the amount of water retained in the cell during shutdown, but it is clear that water redistribution and freezing are almost certain to occur in the wide range of environmental conditions that passenger vehicles are likely to experience. It is clear that detailed understanding of water distribution and phase behavior under subfreezing conditions will be necessary for stack developers to meet automotive requirements.

References

Ahluwalia, R.K., and Wang, X. (2006) Rapid self-start of polymer electrolyte fuel cell stacks from subfreezing temperatures, *J. Power Sources*, 162, 502–512.

Ahmed, S., Kopasz, J., Kumar, R., and Krumpelt, M. (2002) Water balance in a polymer electrolyte fuel cell system, *J. Power Sources*, 112, 519–530.

Bernardi, D.M., and Verbrugge, M.W. (1992) A mathematical model for the solid–polymer–electrode fuel cell, *J. Electrochem. Soc.*, 139, 2477–2491.

Blair, L. (2005) Fundamental Issues in Subzero PEMFC Startup and Operation, Presentation at DOE Workshop on Fuel Cell Operations at Sub-Freezing Temperatures, Phoenix, AZ, February 1–2, 2005, http://www1.eere.energy.gov/hydrogenandfuelcells/fc_freeze_workshop.html

Cappadonia, M., Erning, J.W., and Stimming, U. (1994) Proton conduction of Nafion-117 membrane between 140 K and room temperature, *J. Electroanal. Chem.*, 376, 189–193.

Cho, E., Ko, J.-J., Ha, H.Y., Hong, S.-A., Lee, K.-Y., Lim, T.-W., and Oh, I.-H. (2003) Characteristics of the PEMFC repetitively brought to temperatures below 0DGC, *J. Electrochem. Soc.*, 150, A1667–A1670.

DOE, U.S. (2007) Multi-Year RD&D Plan http://www1.eere.energy.gov/hydrogenandfuelcells/mypp/.

Fuller, T.F., and Newman, J. (1993) Water and thermal management in solid-polymer electrolyte fuel cells, *J. Electrochem. Soc.*, 140, 1218–1225.

Guo, Q.H., and Qi, Z.G. (2006) Effect of freeze-thaw cycles on the properties and performance of membrane-electrode assemblies, *J. Power Sources*, 160, 1269–1274.

He, S., and Mench, M. (2006) One-dimensional transient model for frost heave in polymer electrolyte fuel cells. *J. Electrochem. Soc.*, 153, A1724–1731.

Hishinuma, Y., Chikahisa, T., Kagami, F., and Ogawa, T. (2004) The design and performance of a PEFC at a temperature below freezing, *JSME Int. J. Ser. B, Fluids Therm. Eng.*, 47, 235–241.

Ishikawa, Y., Morita, T., Nakata, K., Yoshida, K., and Shiozawa, M. (2007) Behavior of water below the freezing point in PEFCs, *J. Power Sources*, 163, 708–712.

Knights, S.D., Colbow, K.M., St-Pierre, J., and Wilkinson, D.P. (2004) Aging mechanisms and lifetime of PEFC and DMFC, *J. Power Sources*, 127, 127–134.

Kumbur, E.C., Sharp, K.V., and Mench, M.M. (2006) Liquid droplet behavior and instability in a polymer electrolyte fuel cell flow channel, *J. Power Sources*, 161, 333–345.

Lee, C., and Merida, W. (2007) Gas diffusion layer durability under steady-state and freezing conditions, *J. Power Sources*, 164, 141–153.

Mao, L., and Wang, C.Y. (2007) Analysis of cold start in polymer electrolyte fuel cells, *J. Electrochem. Soc.*, 154, B139–B146.

Mathias, M.F., Makharia, R., Gasteiger, H.A., Conley, J.J., Fuller, T.J., Gittleman, C.J., Kocha, S.S., Miller, D.P., Mittelsteadt, C.K., Xie, T., Yan, S.G., and Yu, P.T. (2005) Two fuel cell cars in every garage?, *Interface*, 14, 24.

Meyers, J.P. (2005) Fundamental Issues in Subzero PEMFC Startup and Operation, Presentation at DOE Workshop on Fuel Cell Operations at Sub-Freezing Temperatures, Phoenix, AZ, February 1–2, 2005, http://www1.eere.energy.gov/hydrogenandfuelcells/fc_freeze_workshop.html

McDonald, R.C., Mittelsteadt, C.K., and Thompson, E.L. (2004) Effects of deep temperature cycling on Nafion 112 membranes and membrane electrode assemblies, *Fuel cells*, 4, 208–213.

Mukundan, R., Kim, Y.S., Garzon, F., and Pivovar, B. (2005) Freeze/thaw effects in PEM fuel cells, *ECS Trans.*, 1, 403–413.

NEDO, homepage: http://www.nedo.go.jp/nenryo/gijutsu/index.html.

Oszcipok, M., Riemann, D., Kronenwett, U., Kreideweis, M., and Zedda, A. (2005) Statistic analysis of operational influences on the cold start behaviour of PEM fuel cells, *J. Power Sources*, 145, 407–415.

Oszcipok, M., Zedda, M., Riemann, D., and Geckeler, D. (2006) Low temperature operation and influence parameters on the cold start ability of portable PEMFCs, *J. Power Sources s*, 154, 404–411.

Park, S., Lee, J.W., and Popov, B.N. (2006) Effect of carbon loading in microporous layer on PEM fuel cell performance, *J. Power Sources*, 163, 357–363.

Pesaran, A.A., Kim, G.-H., and Gonder, J.D., (2005) PEM fuel cell freeze and rapid startup investigation, in *Milestone Report*, http://www.nrel.gov/hydrogen/pdfs/pem_fc_freeze_milestone.pdf, NREL Editor, NREL.

Qiangu, Y., Toghiani, H., Young-Whan, L., Kaiwen, L., and Causey, H. (2006) Effect of sub-freezing temperatures on a PEM fuel cell performance, startup and fuel cell components, *J. Power Sources*, 160, 1242–1250.

Rempel, A.W., Wettlaufer, J.S., and Worster, M.G. (2004) Premelting dynamics in a continuum model of frost heave, *J. Fluid Mech.*, 498, 227–244.

Simpson, S.F., Salinas, C.E., Cisar, A.J., and Murphy, O.J. (1995) Factors affecting the performance of proton exchange membrane fuel cells. In A. Landgrebe, S. Gottesfeld, and G. Halpert (Eds.), *Proceedings of the First Symposium of the Electrochemical Society, First International Symposium on Proton Conducting Membrane Fuel Cells*, The Electrochemical Society, Pennington, NJ, pp. 182–192.

Springer, T.E., Zawodzinski, T.A., Wilson, M.S., and Gottesfeld, S. (1996) Characterization of polymer electrolyte fuel cells using AC impedance spectroscopy, *J. Electrochem. Soc.*, 143, 587–599.

Sundaresan, M., and Moore, R.M. (2005a) Polymer electrolyte fuel cell stack thermal model to evaluate sub-freezing startup, *J. Power Sources*, 145, 534–545.

Sundaresan, M., and Moore, R.M. (2005b) PEM fuel cell stack cold start thermal model, *Fuel Cells*, 5, 476–485.

Thompson, E.L., Capehart, T.W., Fuller, T.J., and Jorne, J. (2006) Investigation of low-temperature proton transport in Nafion using direct current conductivity and differential scanning calorimetry, *J. Electrochem. Soc.*, 153, A2351–A2362.

Weber, A.Z., and Newman, J. (2004) Modeling transport in polymer-electrolyte fuel cells, *Chem. Rev.*, 104, 4679–4726.

Weber, A.Z., and Newman, J. (2005) Effects of microporous layers in polymer electrolyte fuel cells, *J. Electrochem. Soc.*, 152, A677–A688.

Weber, A.Z., Darling, R.M., and Newman, J. (2004) Modeling two-phase behavior in PEFCs, *J. Electrochem. Soc.*, 151, A1715.

Wilson, M.S., Valerio, J.A., and Gottesfeld, S. (1995) Low platinum loading electrodes for polymer electrolyte fuel cells fabricated using thermoplastic ionomers, *Electrochim. Acta*, 40, 355–363.

Yan, Q., and Wu, J. (2005), *207th Meeting Electrochemical Society*, MA2005-01, Abstract # 1514.

Ye, X.H., and Wang, C.Y. (2007) Measurement of water transport properties through membrane electrode assemblies – Part II. Cathode diffusion media, *J. Electrochem. Soc.*, 154, B683–B686.

Yoshida, H., and Miura, Y. (1992) Behavior of water in perfluorinated ionomer membranes containing various monovalent cations, *J. Membrane Sci.*, 68, 1–10.

4
Reliability Testing

Application of Accelerated Testing and Statistical Lifetime Modeling to Membrane Electrode Assembly Development

Michael Hicks and Daniel Pierpont

Abstract Accelerated testing and statistical lifetime modeling are important tools in the development of durable membrane electrode assemblies (MEAs). There are several reasons for using accelerated tests, such as demonstrated durability improvement, marketing a product's competitive advantage, and reduced product development time. Three types of accelerated testing are often used; screening tests, mechanistic tests, and lifetime tests. Accelerated lifetime tests are particularly useful when combined with statistical analysis to provide predictive capability for MEA lifetimes in "real life" conditions. This contribution outlines the main techniques for accelerated testing and important rules to follow for accurate results, such as observing the MEA failure modes for consistency.

1 Introduction

Accelerated testing and statistical lifetime modeling are important tools in the development of durable membrane electrode assemblies (MEAs). There are several reasons for using accelerated tests and three of the main reasons are listed here:

1. *Marketing*. "Our product lasts X times longer than the competition" is a powerful statement; it provides a competitive advantage over your competition.
2. *Customers*. Reliability and durability are key requirements that need to be met to supplement an existing technology with an emerging technology.
3. *Product development*. Time is of the essence and you cannot wait X years to find out if your latest product advancement lasts X years; you need an accurate estimate of lifetime within a short period of time.

The field of accelerated testing can be loosely divided into three types of tests: screening tests, mechanistic tests, and lifetime tests (Meeker and Escobar 1998). Screening

D. Pierpont (✉)
3M Company, Fuel Cell Components Program, 3M Center, Building 0201-BN-33, St. Paul, MN 55144-1000, USA
e-mail: dmpierpont@mmm.com

F.N. Büchi et al. (eds.), *Polymer Electrolyte Fuel Cell Durability*,

tests are typically ex situ component (a component is any item that is assembled into the final product, e.g., a membrane is a component of an MEA) tests designed to quantify incremental improvements during component development. Screening tests are very useful in determining the relative durability of one component design compared with another. Mechanistic tests are designed such that the degradation pathway and kinetic parameters can be determined. Mechanistic tests are useful when it is important to know the exact degradation pathway in order to mitigate it. Lifetime tests are typically performed on the final product to determine its durability and reliability under a given set of operating conditions. When accelerated testing is combined with statistical lifetime analysis, it may be possible to predict the lifetime of the product under various operating conditions. Of the three types of accelerated tests, screening tests are the easiest to set up and are the most widely used, whereas lifetime tests often take long periods of time and are difficult to administer properly and consistently. There is one cardinal rule that needs to be followed in all accelerated tests – the failure mode in the accelerated test *must be* the same as the failure mode in "normal" operating conditions; otherwise the test results will be misleading and possibly useless.

2 Accelerated Testing

Before developing accelerated tests, the performance of the product over time must be determined. An example of MEA performance over time is shown in Fig. 1. Once the time-dependant performance has been established, the next step is to postulate failure modes followed by appropriate accelerated tests designed to mimic the failure mode. For example, the loss in cell voltage at 0.05 A cm^{-2} may be due to increasing H_2 crossover, the loss in current at 0.8 V may be due to loss of catalyst activity or surface area, and the loss in current at 0.4 V may be due to increasing mass-transport resistance from gas-diffusion-layer (GDL) flooding. Once the failure mode is understood, the next step is to establish an accelerated test to accelerate that particular failure mode.

2.1 Screening Tests

Screening tests are exactly what the name implies – a "brute-force" method to "screen" numerous components quickly and efficiently to determine relative component durability. Screening tests do not provide insight into the failure kinetics or mechanism. The following will discuss examples of various screening tests for membranes, catalysts, and GDLs.

2.1.1 Membrane

The primary failure mode for membranes is hole formation, which can be caused by chemical degradation, material fatigue due to mechanical stresses, or a combination

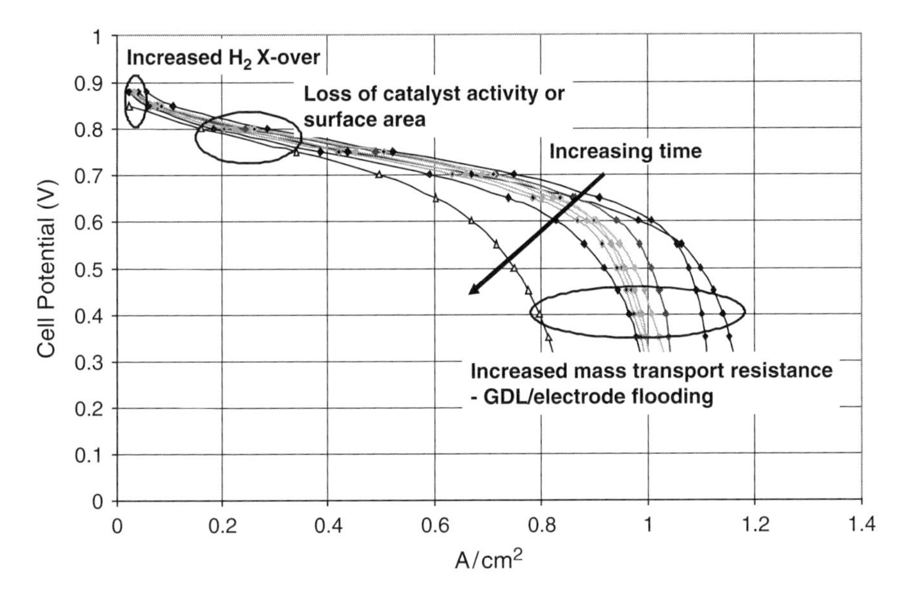

Fig. 1 Typical membrane electrode assembly (MEA) performance over time. *GDL* gas-diffusion layer

of chemical and mechanical stresses. Fenton's test is a common ex situ test to measure membrane chemical stability and involves soaking the membrane in a peroxide solution and measuring either the membrane weight loss or fluoride release over time. Another common test is the MEA open circuit voltage (OCV) test (Escobedo 2006) to measure membrane chemical stability. Unlike Fenton's test, the MEA OCV test is an in situ test that uses an MEA instead of only a membrane. In the OCV test, an MEA is placed in a cell and exposed to H_2 on one electrode and O_2 (or air) on the other electrode. By holding the MEA at OCV, gas crossover can occur, which results in peroxide formation, which in turn degrades the membrane. Just like Fenton's test, fluoride release over time is the important metric to track (Fig. 2a). The main difference between Fenton's test and the OCV test is that Fenton's test only measures chemical stability of the membrane, while the OCV test measures both the membrane's ability to inhibit the production of peroxide (by affecting the gas crossover rates) and the chemical stability of the material. As a result, while the OCV test may be more realistic in terms of fuel cell operation, it is also more difficult to interpret the results owing to the two mechanisms in comparison with Fenton's test.

A humidity cycle (Gittleman et al. 2005) is an example of a simple test to mechanically stress the membrane. In this test, the membrane or MEA is placed in a cell and both electrodes are exposed to humidified N_2. The humidity level of the N_2 gas varies over time from 0 to 100% relative humidity, or even supersaturated conditions (Fig. 2b). The rapid change in humidification causes the membrane to swell or shrink, which introduces mechanical stresses in the membrane. Periodically,

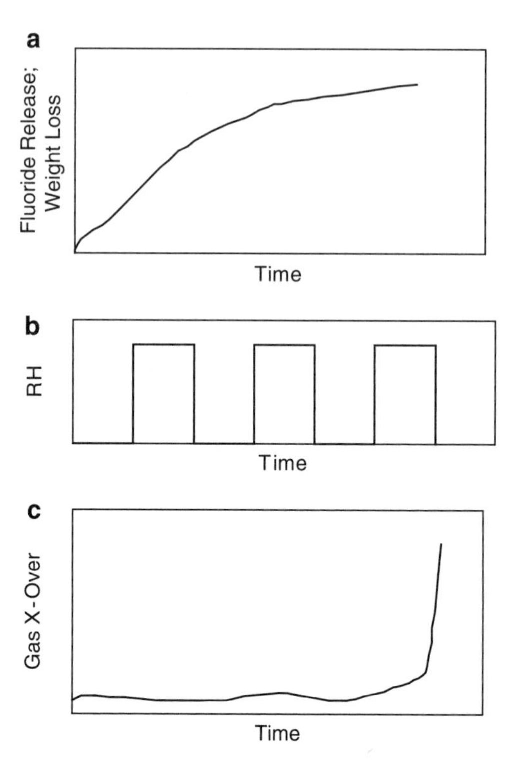

Fig. 2 Examples of membrane screening tests. (**a**) Typical Fenton's test results. (**b**) Relative humidity cycle profile. (**c**) Typical gas crossover data resulting from the relative humidity (*RH*) test

gas crossover from one electrode to another is measured as a function of time to determine when membrane breech occurs (Fig. 2c). Since the feed gas is N_2, no chemical degradation occurs during this test. By changing the gas feed to H_2 on one electrode and O_2 on the other electrode and introducing and load profile cycle on top of the humidity cycle, one can expose the membrane to chemical and mechanical stresses simultaneously. In this coupled test, fluoride release, gas crossover, and OCV decay are typically monitored over time.

2.1.2 Catalyst

There are three primary methods to age catalysts and each addresses different failure modes. The first method is an ex situ thermal aging technique that measures support stability (Stevens et al. 2005). In this method, catalyst powder is weighed and placed in an oven. The weight loss of the powder is monitored over time (Fig. 3a). This technique studies the stability of the support to platinum-catalyzed chemical combustion and is very useful for determining the relative stability of different carbon supports. The second method is an in situ technique to measure the electrochemical

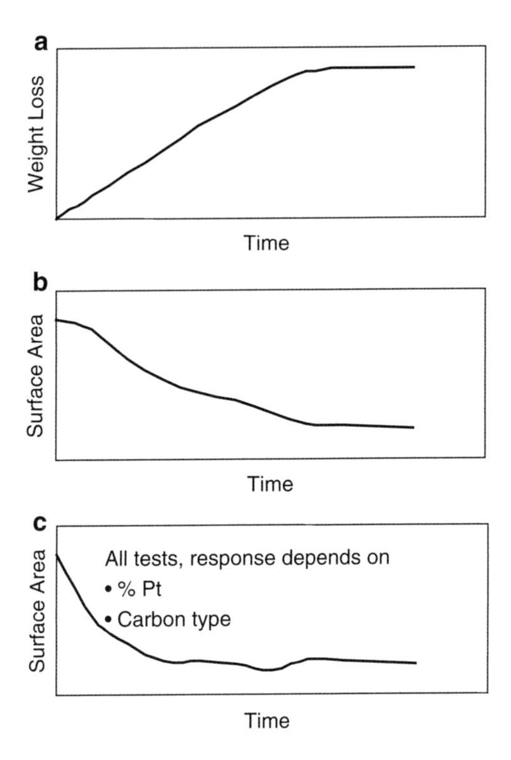

Fig. 3 Examples of catalyst screening tests. Typical results from (**a**) thermal aging, (**b**) constant-potential aging, and (**c**) potential cycling

stability of the support at a given potential. In this test, an MEA is typically held at approximately 1.0 V under H_2/N_2. Periodically, the surface area is measured via cyclic voltammetry and the change in surface area over time is recorded as a measure of the catalyst stability (Fig. 3b). The time frame to observe a loss in surface area greatly depends on the carbon type and cell voltage. Typically, the higher the carbon support surface area or the higher the cell voltage, the easier it is to age the support and observe a change in surface area. The last method, potential cycling, is the most complex to analyze as it accelerates electrochemical degradation of the support, platinum agglomeration, and platinum dissolution. The potential cycle is typically from 0.05 to 1.0 or 1.2 V with the cell under H_2/N_2 gas feeds. The potential cycle causes the most damage to the catalyst because it accelerates three degradation pathways. Typically, a rapid decrease in catalyst surface area is observed (Fig. 3c).

2.1.3 Gas-Diffusion Layer

The primary failure of the GDL is loss of hydrophobicity, which results in increased mass-transport resistance, typically referred to as MEA flooding. Three different screening tests have been used to accelerate this failure mode (Frisk et al. 2004).

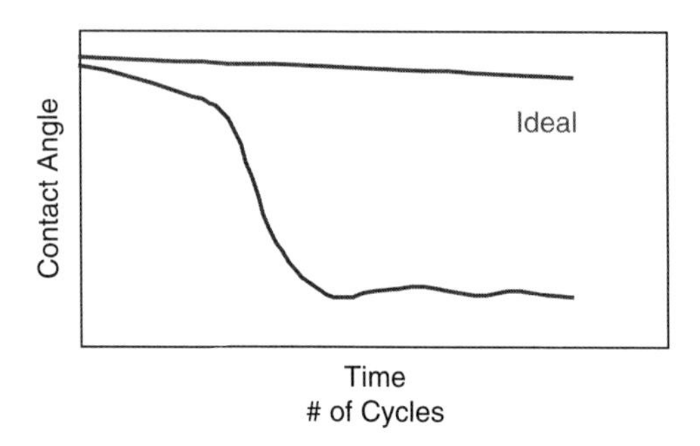

Fig. 4 Typical results from GDL screening tests

The first test studies carbon chemical stability by soaking the GDL in a hydrogen peroxide solution at elevated temperature. The other two tests study carbon electrochemical stability via constant-potential or potential cycling tests. In either electrochemical test, the GDL is placed in a sulfuric acid solution. For the constant-potential tests, a voltage outside that of normal fuel cell operation is applied to the GDL; generally the voltage is greater than 1 V. For the potential-cycling test, the GDL voltage is cycled between 0.3 and 0.8 V (a voltage range within typical fuel cell operation). For either electrochemical test, the current or charge is monitored and the higher the value, the more unstable the GDL. Throughout the chemical and electrochemical accelerated tests, the contact angle of the GDL is measured (Fig. 4). The material is considered to fail when there is a sudden drop in the contact angle or when the contact angle drops below a critical threshold. To quantify the effect of change in contact angle, the aged GDL can be assembled into an MEA and evaluated. In this manner, a relationship between contact angle and fuel cell performance may be established.

2.2 *Mechanistic Tests*

Mechanistic test are very similar to screening tests with one important exception – mechanistic tests are interested in determining the failure kinetics and pathway, whereas screening tests are only interested in the relative durability of component A compared with component B. In this regard, mechanistic tests are designed to answer the question "How does component A fail?" while screening tests answer the question "Is component A more durable than component B?" Understanding the reason behind the difference in durability between component A and component B is at the heart of any mechanistic test. Furthermore, mechanistic tests may even incorporate screening tests combined with detailed quantitative analysis of the component to determine failure kinetics and pathways.

For a PFSA membrane, it is generally accepted that the primary degradation pathway is –COOH end group unzipping (Curtin et al. 2004) due to attack from

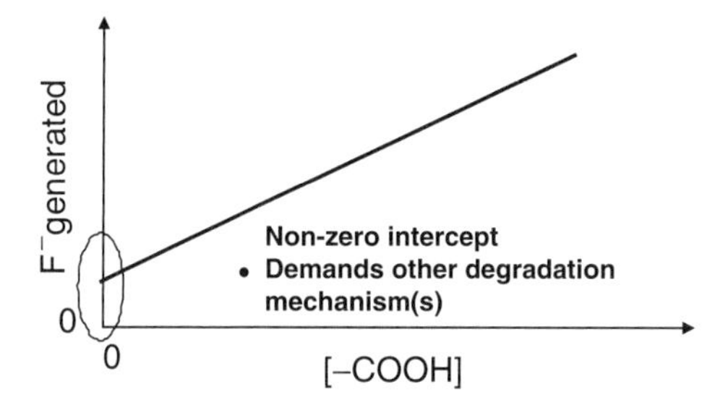

Fig. 5 Impact of –COOH end group concentration on fluoride ion generation (Schwiebert et al. 2005)

OH• or other radicals. However, this might not be the only degradation pathway. A plot of fluoride ion generation versus –COOH end group concentration (Fig. 5) (Schwiebert et al. 2005) results in a nonzero fluoride ion generation intercept, which strongly suggests that there is a secondary degradation mechanism. To investigate a secondary membrane failure mode, a mechanistic test was implemented using small model compounds with representative functional groups found in the polymer electrolyte membrane (Schiraldi 2006; Schiraldi et al. 2006). Studying the small model compounds offer two main advantages over the studying the polymer: (1) it is relatively easy to track chemical/structural changes in small molecules in comparison with a polymer and (2) by utilizing small molecules, one can isolate and better study the reactivity of various functional groups. Through the use of the model compound mechanistic tests, a better understanding of membrane degradation resulted; namely, that there is a secondary failure mechanism centering on the ether linkage in the side chain (Schiraldi 2006; Schiraldi et al. 2006).

2.3 Lifetime Tests

As the name implies, lifetime tests are designed to determine the lifetime of the final product (not of the individual components) under a given set of operating conditions. The operating conditions can be "end-use" conditions or "accelerated" conditions. "End-use" refers to the desired operating conditions of the product: for example, 70 °C cell temperature and 100% inlet gas relative humidity are often used for stationary power use conditions. "Accelerated" refers to any condition different from the "end use" that will accelerate the product failure modes and thereby shorten the lifetime of the product during the test: for example, 90 °C cell temperature and 50% inlet gas relative humidity. The first step in establishing lifetime tests is to define the variables involved. A partial list of fuel cell system variables is as follows:

- Cell temperature
- Gas dew point (inlet and outlet)
- Load setting

- Gas pressure
- Flow-field design
- Contaminants
- Gas composition
- Cell compression

The list is by no means all inclusive and it will vary depending upon fuel composition and system design. After the variable list has been generated, the next step is to identify the one to three most important variables to investigate. Then a series of experiments can be designed around the selected variables before testing is started. Since lifetime tests can have high sample-to-sample variability, multiple replicas (more than five) of each sample at each condition must be run. A postmortem analysis can help identify or confirm failure modes. For example, if there is a particular MEA that died earlier than expected it would be important to look for an edge failure as opposed to a hole in the active area.

One of the important variables affecting MEA/system lifetime is load setting. To determine the effect of load on lifetime, three different load profiles were investigated (Fig. 6). In load profile A, the current is cycled from 0.01 to 0.51 A cm^{-2} in 0.25 A cm^{-2} increments. In load profile B, the load is held constant at 0.35 A cm^{-2}. In load profile C, the load is cycled from 0.26 to 0.51 A cm^{-2} in 0.25 A cm^{-2} increments. By comparing lifetimes from load profiles A and C, one can determine the impact of near-OCV operation on lifetime. By comparing the lifetimes from load profiles B and C, one can determine the impact of load cycling on lifetime. The results of the load profile lifetime test example are shown in Fig. 7. Clearly, the load profile has a significant impact on MEA/system lifetime.

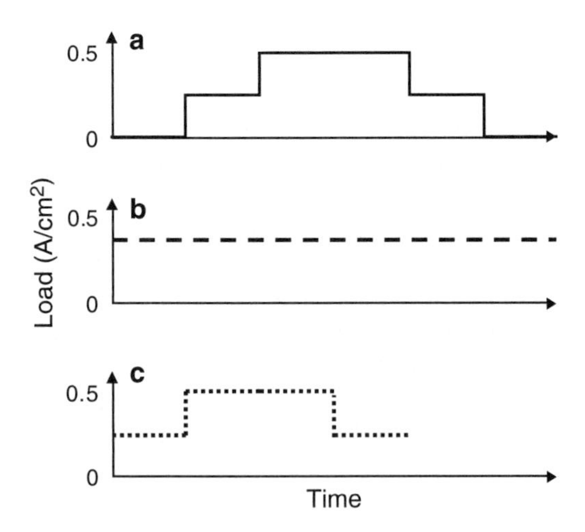

Fig. 6 MEA accelerated lifetime tests load profile

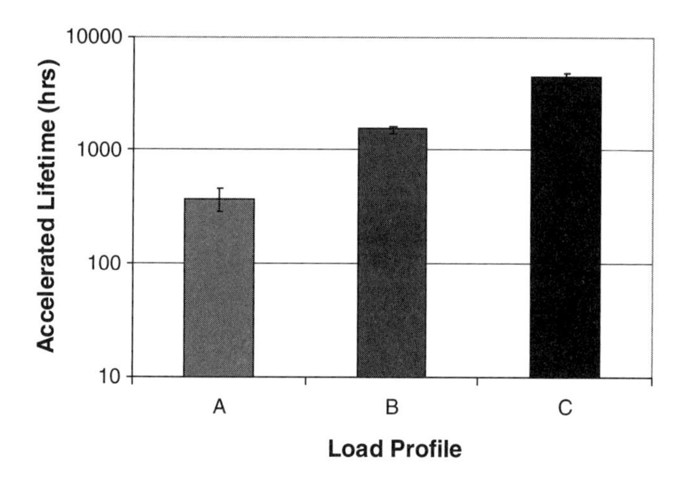

Fig. 7 Effect of load profile on MEA lifetime. Cell conditions were as follows: 90 °C cell temperature, 70 °C gas inlet dew points, H_2/air, 7 psig anode overpressure

3 Statistical Lifetime Analysis and Modeling

A significant amount of data is generated during the lifetime tests and it is extremely important to analyze the data correctly. To illustrate this fact, the following example is provided (Fig. 8). Six identical samples were tested to failure and lasted 391, 525, 658, 919, and 994 h, respectively. The average lifetime of the six samples is 695 h and the standard deviation is 295 h as shown in Fig. 8a. But what does this mean? If a seventh sample is tested, will it last 695 h? It is impossible to say, because the wrong analysis was performed. Lifetime is not defined by a single data point; lifetime is a statistical distribution. As such, the data in this example must be analyzed in terms of probability – what is the probability that a sample will fail at any point in time? The probability analysis using a Weibull distribution is shown in Fig. 8b. Using the probability plot, it is clear that the average of 695 h is not the lifetime of the samples; instead, 695 h represents the time by which 50% of the samples will have failed. In other words, if a fuel cell stack consists of 100 MEAs, 50 of them will have failed by 695 h. Since it only takes one failed MEA to mark the end of the fuel cell stack lifetime, a more realistic definition of lifetime for the 100-MEA fuel cell stack occurs at a percentage failure of 1%, which corresponds to a lifetime of approximately 225 h.

Another benefit of using statistical analysis is the ability to incorporate censored data, or test data from a test that is still ongoing. When using statistical analysis, it is just as useful to the analysis to know that a sample failed at 2,000 h as it is to know that a sample did not fail at 2,000 h and is still under test.

A lifetime model is needed to predict lifetimes under "end-use" conditions from the "accelerated" lifetime test results. In the model, it is important to use the correct mathematical relationships to fit system variables listed in Sect. 2.3 between the

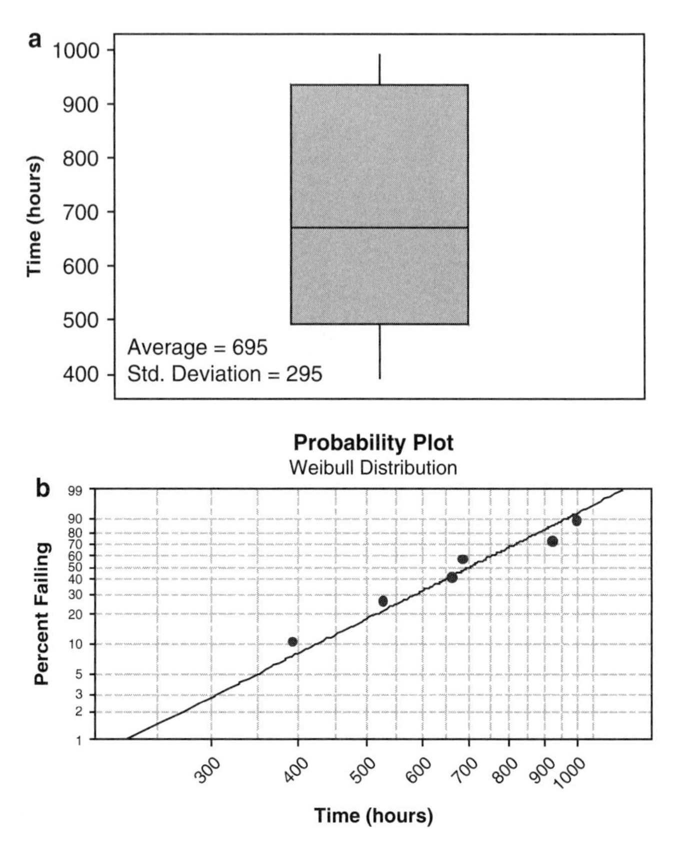

Fig. 8 Analysis of lifetime data using (**a**) average and standard deviation and (**b**) statistical probability distribution

different operating conditions. For example, the Arrhenius relationship can be used to account for the impact of temperature at the different test conditions. To predict the lifetime distribution at "end-use" conditions from "accelerated" lifetime test conditions, multivariate-nonlinear statistical methods need to be utilized. One such program is SPLIDA, which is built within an S-PLUS platform (Meeker and Escobar 1998) SPLIDA can account for nonlinear data, censored data, and also separate failure modes.

Previously, the data presented in Fig. 7 were analyzed in terms of averages and standard deviations to determine lifetime. However, averages and standard deviations do not capture the entire picture when it involves lifetime, so the data need to be reanalyzed using SPLIDA. The results of the SPLIDA analysis are shown in Fig. 9. Not only can SPLIDA provide a lifetime probability distribution for the data in Fig. 7, it can also predict the lifetime of the samples at "end-use" conditions (70 °C cell temperature and 100% inlet gas relative humidity). When SPLIDA was used for statistical analysis, the following relationships were selected: Weibull distribution for lifetime data, Arrhenius relationship for temperature, class relationship for load

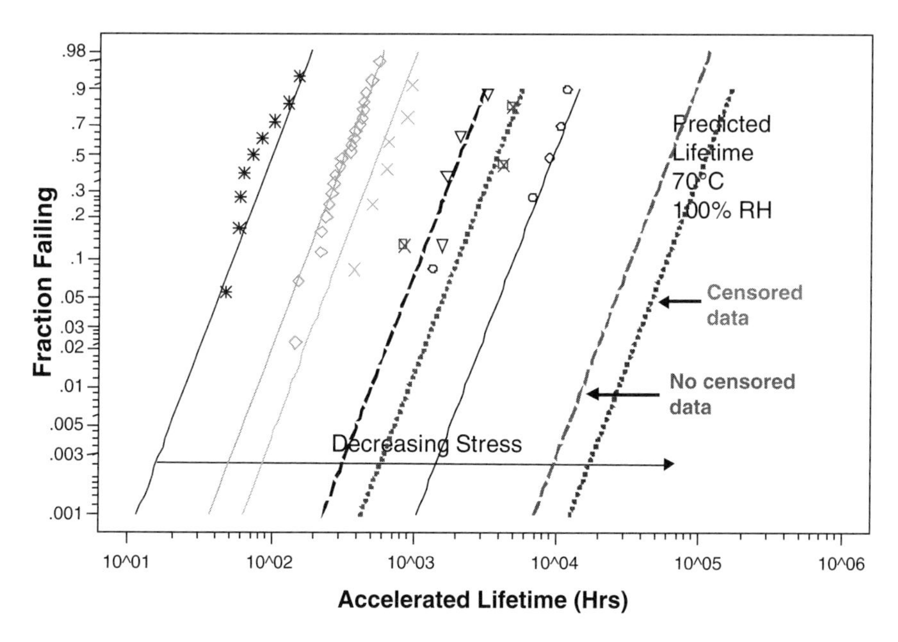

Fig. 9 Statistical MEA lifetime predictions from accelerated test data using SPLIDA. The *symbols* represent actual data points and the *lines* are the statistical model fit to the data: *solid lines* load profile A, *dashed lines* load profile B, *dotted lines* load profile C

profile, and humidity transformation for humidity effects. The Weibull distribution was chosen because it represents the "weakest link in a chain" failure model, which is what exists in an MEA – if any component fails, the MEA has failed. The Arrhenius relationship was selected as the means to fit the temperature dependence of the kinetics of the unknown degradation mechanism. The humidity transformation was utilized to normalize the relative humidity measurements. The class relationship was utilized for the different load profiles as a means to treat the load profiles as separate entities while still applying the same overall model to the different load profiles; i.e., we were able to find the "best fit" for the entire data set instead of generating a different model for each load profile. The data on the left in Fig. 9 were taken under highly accelerated conditions (hot and dry cell operation) and the data on right are model predictions at end-use conditions. At the end-use conditions, at a failure probability of 1%, the predicted lifetime of the samples under load profile C is approximately 20,000 h.

4 Summary

Utilizing tools such as accelerated tests and statistical lifetime modeling allow for early estimates of durability. There are three basic types of accelerated tests (1) screening tests are appropriate for measuring incremental improvements in component durability; (2) mechanistic tests are designed to provide a fundamental understanding of the degradation pathway; and (3) lifetime tests are used to determine the product's lifetime under a given set of operating conditions.

Lifetime is a probability distribution and therefore the data analysis needs to account for this fact. When the results of lifetime tests are combined with statistical lifetime analysis, it is possible to predict product lifetime at "end-use" conditions from "accelerated" degradation conditions. This can often lead to faster development of MEAs and components because critical lifetime information can be learned without waiting for lengthy lifetime tests to complete under "end use" conditions.

Acknowledgments This research was supported in part by the U.S. Department of Energy, Cooperative Agreement No. DE-FC36-03GO13098. Department of Energy support does not constitute an endorsement by the Department of Energy of the views expressed in this article. Thoughtful discussions with William Meeker of Iowa State University are acknowledged.

References

Curtin, D., Lousenberg, R., Henry, T., Tangeman, P. and Tisak, M. (2004) Advanced materials for improved PEMFC performance and life, J. Power Sources, 131(1–2), 41.

Escobedo, G. (2006) *Enabling Commercial PEM Fuel Cells with Breakthrough Lifetime Improvements*, 2006 DOE Hydrogen Program Review.

Frisk, J., Boand, W., Hicks, M., Kurkowski, M., Schmoeckel, A. and Atanasoski, R. (2004) *How 3M developed a new GDL construction for improved oxidative stability*, 2004 Fuel Cell Seminar, San Antonio, TX.

Gittleman, G., Lai, Y-H and Miller, D. (2005) *Durability of Perfluorosulfonic Acid Membranes for PEM Fuel Cells*, Fall AIChE Meeting.

Meeker, W.Q. and Escobar, L.A. (1998) *Statistical Methods for Reliability Data*, Wiley, New York, NY.

Schiraldi, D. (2006) Perfluorinated polymer electrolyte membrane durability, Polym. Rev., 46(3), 315.

Schiraldi, D., Zhou, C. and Zawodzinski, T. (2006) *Model Studies of Perfluorinated PEM Membrane Degradation*, 232nd ACS Meeting, San Francisco, CA.

Schwiebert, K., Raiford, K., Nagarajan, G., Principe, F. and Escobedo, G. (2005) *Strategies to Improve the Durability of Perfluorosulfonic Acid Membranes for PEM Fuel Cells*, Fuel Cell Durability.

Stevens, D., Hicks, M., Haugen, G. and Dahn, J. (2005) Ex situ and in situ stability studies of PEMFC catalysts: Effect of carbon type and humidification on degradation of the carbon, J. Electrochem. Soc., 152(12), A2309.

5
Stack Durability

Operating Requirements for Durable Polymer-Electrolyte Fuel Cell Stacks

Mike L. Perry, Robert M. Darling, Shampa Kandoi, Timothy W. Patterson, and Carl Reiser

Abstract Successful developers of fuel cells have learned that the keys to achieving excellent durability are controlling potential and temperature, as well as proper management of the electrolyte. While a polymer-electrolyte fuel cell (PEFC) has inherent advantages relative to other types of fuel cells, including low operating temperatures and an immobilized electrolyte, PEFC stacks also have unique durability challenges owing to the intended applications. These challenges include cyclic operation that can degrade materials owing to significant changes in potential, temperature, and relative humidity. The need for hydration of the membrane as well as the presence of water as both liquid and vapor within the cells also present complications. Therefore, the development of durable PEFC stacks requires careful attention to the operating conditions and effective water management.

1 Introduction

The durability of a polymer-electrolyte fuel cell (PEFC) strongly depends on operating conditions. Principal among these are potential, temperature, and relative humidity. For a given set of materials, fuel-cell stacks that are subjected to less aggressive operating conditions last longer and decay less. Therefore, consideration of both materials with improved stability and design strategies that minimize adverse operating conditions should provide the best path to developing PEFC systems that are optimized with respect to durability, cost, and performance.

The conditions that a fuel-cell stack will encounter depend upon the application. PEFCs are currently being considered for cars, buses, industrial vehicles, defense applications, portable electronics, back-up power, and other stationary applications. Describing all practical methods to maintain the desired operating conditions in these diverse applications is beyond the scope of this chapter. Instead, this chapter

M.L. Perry(✉)
United Technologies Research Centre, 411 Silver Lane, East Hartford, CT 06108, USA
e-mail: Perryml@utrc.utc.com

F.N. Büchi et al. (eds.), *Polymer Electrolyte Fuel Cell Durability*,

focuses on conditions that should be avoided in order to enhance stack durability. Illustrative examples of effective operating strategies and stack design are also described. This chapter concentrates on the durability of the membrane and catalyst layers, which together constitute the membrane-electrode assembly (MEA). Less attention is devoted here to the gas-diffusion layers (GDLs), bipolar plates, and seals.

This chapter is organized into five main sections. Sections 2, 3, and 4 are concerned with how potential, temperature, and relative humidity affect durability, and Sect. 5 gives conclusions. These sections discuss conditions commonly encountered during operation, what conditions are likely to cause degradation, and what may be done to prevent or minimize decay.

2 Potential

The electrode potential is the main driving force for many of the degradation modes present in PEFCs, including carbon corrosion and platinum dissolution; therefore, it is important to understand what electrode potentials are likely to be encountered in operation. Table 1 lists the gases present and approximate maximum electrode potentials at the two electrodes of a PEFC under different conditions. All potentials reported in this work are with respect to a reversible-hydrogen electrode. The terms "fuel electrode" and "air electrode" are used here because anodic and cathodic reactions can occur at both electrodes depending upon the local conditions. The electrode potentials can be controlled under some of these conditions; however, the values in Table 1 assume no attempt to control the potential. The cell potentials shown in Table 1 do not rigorously account for ohmic resistance and are intended as a guide only. Boxes containing two numbers separated by a slash are used when two regions with different electrode potentials are present.

The normal condition prevails when excess hydrogen and air are present and power is being produced. The idle state refers to a minimal power output, corresponding

Table 1 Gases present and approximate electrode potentials during various operating conditions of a polymer-electrolyte fuel cell

	Fuel electrode		Air electrode		
Condition	Gas	Potential (V)	Gas	Potential (V)	Cell potential (V)
Normal	H_2	0.05	Air	0.85	0.8
Idle	H_2	0.0	Air	0.90	0.9
Open circuit	H_2	0.0	Air	1.0	1.0
Off	Air	1.1	Air	1.1	0
Start	H_2/air	0.0/1.1	Air	1.0/1.5	1.0
Stop	Air/H_2	1.1/0.0	Air	1.5/1.0	1.0
Partial H_2 coverage	H_2/inert	0.05/1.1	Air	0.85/1.5	0.8
Fuel starvation	H_2/inert	>1.5	Air	0.85	<−0.65
Air starvation	H_2	0.05	Air/H_2	0.85/0.05	−0.1

to parasitic loads such as pumps and blowers, with ample hydrogen and air present. Open circuit occurs when there is no load on a cell with hydrogen and air on the fuel and air electrodes, respectively. The off state occurs when a cell is not used for a long time, unless an inert gas is intentionally added to the electrode compartments. During the off state, the electrodes achieve a mixed potential that is set by oxygen reduction and corrosion or oxidation reactions. The start condition refers to the replacement of air on the fuel electrode by hydrogen. The stop condition is the opposite of the start condition. The cause of the high electrode potentials when the cell is started or stopped has been described in the literature (Reiser et al. 2005; Meyers and Darling 2006). This mechanism can take place whenever there is partial hydrogen coverage on one electrode and air on the other electrode.

Two hydrogen-deficient conditions are possible; one is partial hydrogen coverage that damages the air electrode and the other is fuel starvation that damages the fuel electrode. These two conditions can be distinguished by the relationship between fuel flow and current. Specifically, if

$$N < \frac{I}{2F} \tag{1}$$

then fuel starvation, causing damage to the fuel electrode, will occur. In (1), N is the molar flow of hydrogen, I is the current, and F is Faraday's constant. Since there is not enough hydrogen present on the fuel electrode to support the current, the cell responds by oxidizing other materials such as water and carbon at the fuel electrode. This condition can occur during galvanostatic operation of a single cell or in a cell within a stack that is driven by other cells with sufficient hydrogen. The partial hydrogen coverage condition is possible when the limiting current for the hydrogen-oxidation reaction on some portion of the active area falls below the average current density. Stated mathematically, partial hydrogen coverage may cause damage to occur when

$$N > \frac{I}{2F} \text{ but } \frac{I}{A} > i_{\lim}(p_{H_2}) \text{ somewhere on the active area,} \tag{2}$$

where A is the active area. Thus, enough hydrogen is being fed to the fuel compartment to support the total current, but access of hydrogen to portions of the electrode is inadequate. This allows oxygen crossing through the membrane from the air electrode to set the potential on portions of the fuel electrode. Equations (1) and (2) apply at steady state, and do not account for hydrogen crossing through the membrane or leaking through the seals. Thus, these equations are intended as guides.

The air-starvation condition occurs when the current exceeds the limiting current for the oxygen-reduction reaction (ORR). In this case, hydrogen evolves from the air electrode and the potential of the air electrode approaches the reversible-hydrogen potential. The potential of the cell is slightly negative under this condition.

Summarizing Table 1, the potentials typically experienced on PEFC electrodes range from approximately 0 to 1.5 V or more. Since higher potentials promote degradation modes such as carbon corrosion and platinum dissolution, one should strive to minimize exposure to high potentials. Before discussing ways of accomplishing this, we briefly review how potential affects platinum and carbon.

2.1 *Effect of Potential and Potential Cycles on Platinum*

Commercial PEFC electrodes contain dispersed platinum or platinum-alloy catalysts supported on high surface area carbon. High platinum surface area is required to minimize the overpotential for the ORR. Examination of the Pourbaix diagram for platinum indicates that dissolution is expected to occur in a triangular region where pH < 0 and the electrode potential is between approximately 1 and 1.2 V at 25 °C (Pourbaix 1974).

The solubility of platinum in phosphoric acid was measured at 176 and 196 °C and was found to be consistent with the Nernst equation (Bindra et al. 1979). However, recent measurements of platinum solubility in 0.5 M sulfuric acid at 80 °C (Ferreira et al. 2005) and in 0.57 M perchloric acid at 23 °C do not follow the Nernst equation (Wang et al. 2006).

Cycling the potential of a platinum electrode in acid electrolyte causes higher dissolution rates than potentiostatic experiments at similar potentials (Kinoshita et al. 1973; Woods 1976; Patterson 2002; Mitsushima et al. 2007). This is an important consideration for transportation applications that require frequent and rapid load changes and is receiving increasing attention. For example, a study of the effects of frequency, wave form, and lower potential on the rate of surface area loss from a Pt/C catalyst in a PEFC was recently published (Paik et al. 2007). These researchers found that rates of surface-area loss could not be explained solely by cumulative time at high potential; rather, the rate of area loss also increased with increasing cycle frequency and decreasing lower potential. The rate of surface area loss decreased when the lower potential was set above the potential needed to reduce platinum oxide. The current density associated with charging the double layer increased with potential cycling, indicating changes to the carbon support, and the electrodes became more hydrophilic.

Temporary performance losses have been observed in PEFCs after long galvanostatic or potentiostatic experiments (Uribe and Zawodzinski 2002; Jarvi et al. 2003; Eickes et al. 2006; Donahue et al. 2002). All of these researchers found that this decay could be recovered by lowering the potential of the air electrode below approximately 0.6 V. Much of this decay has been attributed to oxidation of platinum, and reconstruction of these oxides, causing a reduction in catalytic activity for the ORR. Although recoverable decay can be problematic, especially for devices intended to operate at relatively high potentials for long times, it is a lesser concern than permanent loss of platinum surface area.

Historically, platinum alloys have been used in phosphoric acid fuel cells (PAFCs) to increase the performance and durability of the air electrode (Luzack and Landsman 1983). Interest in using platinum alloys to address similar problems in PEFCs has been increasing recently. Supported PtCo and PtIrCo alloy catalysts, for example, appear to lose less surface area than platinum when cycled to high potentials (Yu et al. 2005b; Mathias et al. 2005; Protsailo 2006). Compatibility of the alloying elements with the membrane should be considered when contemplating replacement of platinum with a platinum alloy, because many alloying elements are more soluble than platinum. Additionally, one should measure the propensity of any alternative catalyst to generate species that could contribute to chemical attack of the membrane.

2.2 *Effect of Potential on Carbon*

Carbon is commonly used in both PEFCs and PAFCs as a catalyst support, in GDLs, and as a bipolar-plate material. Carbon is not thermodynamically stable under all conditions encountered by an acid fuel cell. The electrochemical oxidation of carbon is accelerated by increasing temperature and potential; hence, carbon corrosion can be lowered to acceptable levels by minimizing the time at high temperature and potential. Because the operating temperature of a PEFC is lower than that of a PAFC, the corrosion rate of carbon is generally lower. However, during transportation applications, there are certain operating conditions under which carbon corrosion can be accelerated: (1) operation at, or near, open circuit, (2) fuel starvation, (3) partial hydrogen coverage, as shown in Table 1, and (4) cyclic operation.

Binder et al. proposed a mechanism that describes the formation of CO_2 and surface oxides (Binder et al. 1964). This mechanism involves the oxidation of carbon in the lattice structure followed by hydrolysis and disproportionation. Their experiments indicated that 80% of the charge is consumed by the evolution of CO_2, while 20% goes to the formation of surface oxides. Later studies demonstrated that the current efficiency for CO_2 production depends on the type of carbon, as well as temperature, potential, and time (Kinoshita and Bett 1974; Kinoshita 1988).

Kangasniemi et al. (2004) studied the surface electrochemical oxidation of Vulcan XC-72, a carbon black, using a combination of cyclic voltammetry, thermogravimetric analysis-mass spectrometry (TGA-MS), X-ray photoelectron spectroscopy, and contact-angle measurements. Vulcan XC-72 was exposed to potentials from 0.6 to 1.2 V for up to 120 h at both room temperature and at 65 °C, which better emulates the fuel-cell environment. The analytical techniques indicated that Vulcan XC-72 undergoes surface oxidation at potentials above 1.0 V at room temperature and above 0.8 V at 65 °C. At room temperature, they found that significant surface oxidation occurs during the first 16 h of the hold and several oxide groups (ether, carboxyl, carbonyl) were identified by X-ray photoelectron spectroscopy and were correlated to TGA-MS evolution peaks of CO_2. TGA-MS spectra of CO showed an increase in CO evolution above 800 °C, which is consistent with cyclic voltammogram peak currents corresponding to the hydroquinone/quinone species. However, these electroactive species were found to be only a small fraction of the electrochemically generated surface oxides. Contact-angle measurements showed a decrease in the hydrophobicity as a result of carbon oxidation, which may have negative implications on the long-term stability of a PEFC.

Researchers at General Motors estimate that a PEFC in an automotive application will undergo approximately 30,000 start/stop cycles (Mathias et al. 2005).

Assuming 10 s of hot time per stop, this necessitates catalyst support durability at 1.2 V of around 100 h with less than 30 mV drop in performance at 1.5 A cm^{-2}. Similarly, it was estimated that the idle time of the stack at cathode potentials of approximately 0.9 V could amount to several thousand hours over the vehicle life. The data at 80 °C predict that standard Vulcan carbon supports do not meet automotive requirements with respect to start/stop, as well as prolonged idle conditions.

Figure 1 shows electron-microprobe images of a PEFC after start/stop cycling. This stack had a Pt/C cathode and a PtRu/C anode. Thinning and brightening of the cathode catalyst layer is observed near the fuel exit, whereas the inlet appears to be undamaged. Furthermore, the band of platinum in the membrane is more developed near the exit because this region has been subjected to higher potentials. A band of ruthenium is visible outside the MEA at the exit. This occurs because the anode has been subjected to higher potentials at the exit. These patterns are generally consistent with the reverse-current mechanism that results in high cathode potentials in parts of the cell where hydrogen is absent (Reiser et al. 2005).

Rapid carbon corrosion can occur during what may appear to be normal operation if partial hydrogen coverage exists (Patterson and Darling 2006). During local fuel starvation, the cell potential may appear normal despite the presence of the reverse-current mechanism in hydrogen-depleted regions of the cell. Partial hydrogen coverage can result from poor fuel distribution across the active area, which could be caused by liquid water in the fuel-flow path blocking hydrogen access to the anode catalyst layer.

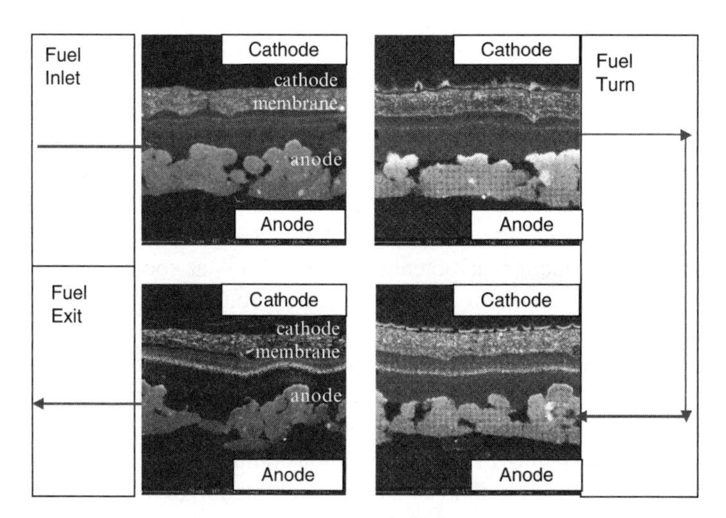

Fig. 1 Cross-sectional electron microprobe images of four locations of a membrane electrode assembly (MEA) from a polymer-electrolyte fuel cell (PEFC) stack that was subjected to 1,994 uncontrolled start/stop cycles. The stack utilized two fuel passes, as shown. As expected by the reverse-current mechanism, the amount of damage depends on the distance from the fuel inlet. Note the changes in the cathode catalyst layer and the presence of platinum in the membrane, especially in the second pass

Unacceptable corrosion of catalyst supports in a PAFC led to the development of more corrosion-resistant carbons (Kinoshita 1988; Landsman and Luzack 2003). However, these graphitized carbons have low surface area, which typically necessitates lower catalyst-to-carbon ratios than currently employed in PEFC. For example, approximately 20 wt% Pt/C is typical in a PAFC, whereas approximately 50 wt% Pt/C is common in a PEFC. A lower ratio of platinum to carbon results in thicker catalyst layers for a given amount of catalyst, increasing ohmic and transport losses. Additionally, PAFC units with graphitized supports cannot meet the durability requirements without the implementation of controlled procedures for starting and stopping that minimize the time spent at elevated temperature and potential. Analogously, system-mitigation strategies in PEFC products can enable acceptable results with materials that would otherwise be unacceptable. For example, the start/stop decay mechanism can be mitigated by discharging the stack through a resistor when the fuel is being introduced to minimize the potential of the air electrode during the fuel transition; over 12,000 start/stop cycles have been demonstrated with this approach and the performance degradation was reduced substantially (Perry et al. 2006). Similarly, local fuel starvation can be prevented by designing cells and systems that ensure proper fuel distribution to all of the cells in a stack and to all of the active area within each cell.

Extended periods at open circuit and idle conditions should be avoided to improve catalyst layer stability. In a fuel-cell hybrid vehicle, this can be accomplished by utilizing the energy-storage system (e.g., batteries or capacitors) in a manner analogous to that employed in hybrids that use internal combustion engines. Owing to parasitic power requirements, fuel-cell power plants are inefficient at extremely low power; therefore, the time spent at these conditions should be minimized to maximize efficiency and life.

As shown in Table 1, the range of potentials that the electrodes in a PEFC can be exposed to is quite large, if no attempts are made to control the potential. Therefore, the best approach is to balance the use of materials that tolerate high potentials, perhaps at the cost of lower performance, with strategies that control the potentials experienced by the electrodes.

3 Temperature

The operating temperatures of a PEFC are typically between 65 and 80 °C. The low operating temperature enables quick starting and enhances power density, because it reduces the need for thermal insulation. These attributes are key reasons why PEFCs are attractive for transportation applications. However, the low operating temperature makes rejection of heat to ambient surroundings more difficult than it is for internal combustion engines of comparable power. For this reason, the automotive application would prefer higher operating temperatures, and maintenance or improvement of the performance and durability characteristics of the fuel-cell system. Additionally, a higher operating temperature should improve tolerance to impurities such as CO, which would be beneficial for applications operating on reformed hydrocarbons. Phosphoric acid is used as a model to illustrate issues associated with operation above 120 °C.

In general, the rates of thermally activated processes such as chemical reactions and transport in condensed phases increase exponentially with temperature. Thus, one might expect that the rates of proton transport in the electrolyte and oxygen reduction at the cathode should increase dramatically as the operating temperature is increased. The expected benefits associated with increasing the operating temperature are, however, difficult to obtain in systems based on the current class of perfluorosulfonic acid (PFSA) membranes because system water balance deteriorates as temperature increases. Dehydration of the electrolyte membrane negatively affects proton transport (Gottesfeld 1997), oxygen-reduction kinetics (Neyerlin et al. 2005), and generally results in reduced cell performance (Gasteiger et al. 2003).

Presently, PEFC power plants that incorporate PFSA-based membranes are capable of starting from ambient temperatures. This is accomplished by using the heat generated by the inefficiency of the PEFC to raise the temperature of the stack. Although starting is more challenging when the temperature is below 0 °C, it has been demonstrated on complete PEFC systems (Meyers 2005; Blair 2005). Currently, many research groups are investigating what limits subfreezing starts of PEFC stacks, as well as if repeated cold starts or freeze/thaw cycles result in permanent damage. How water is managed throughout subfreezing starts and stops affects the success of these transitions (Meyers 2005; Cho et al. 2004; Perry et al. 2007).

Another reason that a low operating temperature is an inherent advantage is that the influence of thermal cycles is reduced for smaller temperature ranges. Thermal cycling is an inevitable aspect of discontinuous operation, and small differences between operating and ambient temperatures help minimize thermally induced stresses. The effects of thermal cycles on the membrane, as well as related changes in relative humidity, are discussed in Sect. 4. In addition, the durability of seals depends upon the number and magnitude of thermal cycles; this issue is not discussed further in this chapter.

3.1 Effect of Temperature on Platinum Sintering

In addition to platinum dissolution discussed in Sect. 2, platinum area may also be lost by sintering. Two sintering mechanisms have been postulated for hot, concentrated phosphoric acid (1) the Smoluchowski collision model in which crystallites migrate, collide, and coalesce on the support and (2) Ostwald ripening, in which crystallites dissociate into metal atoms that diffuse to and associate with larger particles (Bett et al. 1976).

Gruver et al. (1980) measured the surface area versus time for platinum supported on Vulcan in phosphoric acid at 191 °C for more than 20,000 h. They showed that a plot of ln (S) versus ln (t) was linear over the time range from 100 to 20,000 h. The intercept of the lines was 4.8 and the slope was –0.186, with platinum area in

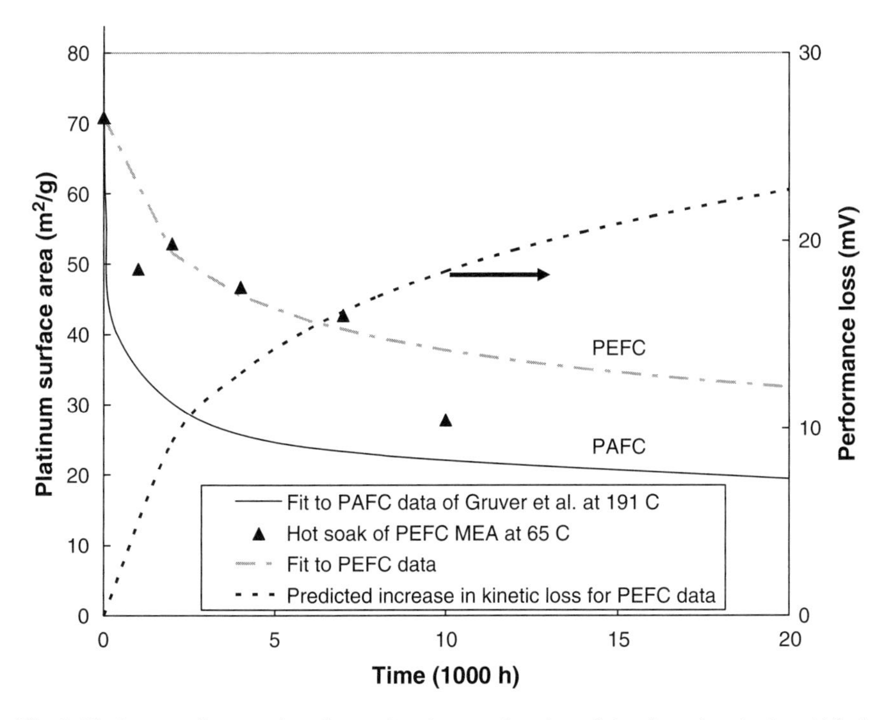

Fig. 2 Platinum surface area loss due to sintering as a function of time in a phosphoric acid fuel cell (PAFC) and a PEFC. Also included is the performance loss due to the loss in platinum area, as predicted by equation (4) (PAFC data from Gruver et al. 1980)

square meters per gram and time in hours. Figure 2 shows a fit to their data on a linear scale. The surface area drops by more than half within the first 3,000 h and then decreases at a much slower rate for the remainder of the experiment.

Bett et al. (1976) determined an activation energy of 88 kJ mol^{-1} for the rate constant for platinum particles sintering on carbon in hot, concentrated phosphoric acid. Applying this activation energy to estimate the rate of sintering in PEFC at 65 °C would indicate a very low rate of surface area loss. However, measurements of surface area loss for a commercial PEFC MEA soaked in 65 °C deionized water, shown in Fig. 2, indicate that the rate of surface area loss is significant. The PEFC data were fit to the equation

$$\frac{1}{S^n} = \frac{1}{S_0^{\ n}} + kt \tag{3}$$

with time in hours and area in square meters per gram. $S_0 = 70.9$, $n = 4.45$, and $k = 8.98 \times 10^{-12}$.

Loss of platinum surface area causes performance loss. The kinetic loss takes the form

$$\Delta V = \frac{RT}{F} \ln\left(\frac{S}{S_0}\right) \tag{4}$$

provided that the ORR is first order in oxygen concentration, follows Tafel kinetics with a cathodic transfer coefficient of 1, and has a constant exchange current density. S and S_0 are the present and initial values of the catalytic area, respectively. The estimated effect of platinum surface area loss on cell polarization using (4) for the PEFC data is also shown in Fig. 2. It has been shown that the rate constant for the ORR increases with increasing platinum particle size in phosphoric acid (Bregoli 1978), which would result in a lower kinetic loss than predicted by (4).

Analytical solutions have been developed for the polarization of porous electrodes limited by either mass-transport or ohmic losses through the depth of the electrode (Perry et al. 1998). When either of these processes is important, a porous electrode displays a double Tafel slope. It can be shown from (32)–(34) in Perry et al. (1998) that electrodes limited by either kinetic, transport, or ohmic losses have the same dependence on surface area loss. This argument holds only if the platinum particles remain uniformly distributed through the depth of the electrode.

3.2 Effect of Temperature on Corrosion of PEFC Components

The durability of the constituents of the catalyst layer is expected to be negatively affected by higher temperatures. The rates of carbon corrosion and platinum dissolution and sintering increase with increasing temperature. In many ways, phosphoric acid can be considered to be a model high-temperature membrane, and an examination of the similarities between the decay of PEFC and PAFC catalyst layers should provide guidance for the development of high-temperature membrane cells (Fuller et al. 2005). The approach taken in this section is to combine data for carbon corrosion and platinum dissolution with empirical observations regarding the behavior of PEFCs and PAFCs as functions of temperature and potential. The intent is that the data will allow us to make inferences about the behavior of acid cells at intermediate temperatures.

Figure 3 shows an approximate map of maximum potential versus temperature for an electrode in concentrated acid. The temperature range covers normal operation of PEFCs and PAFCs. Regions 1–3 represent operating regimes of PEFCs described in Table 1. Region 1 is the off condition, where oxygen reduction and corrosion or oxidation occurs. This region covers a wide temperature range because the cell may be stopped while hot and then allowed to cool to ambient temperature with both electrodes exposed to air. Region 2 includes idle and hydrogen/air open-circuit conditions. Region 3 is the cathode potential under normal conditions. Both electrodes in PEFCs are likely to be exposed to Region 1 condition, while only the air cathode should be exposed to the Region 2 and 3 conditions. The temperature range from 65 to 80 °C represents the region where water balance is possible, at pressures of interest for transportation applications. Region 4 represents the operating envelope for a PAFC cathode in a stationary application. The potential of PAFC electrodes is typically controlled to below approximately 0.8 V to prevent degradation (Landsman and Luzack 2003). This is represented by the line labeled "dummy load applied" in Fig. 3. PAFC stacks maintained in this range can operate for in excess of 50,000 h.

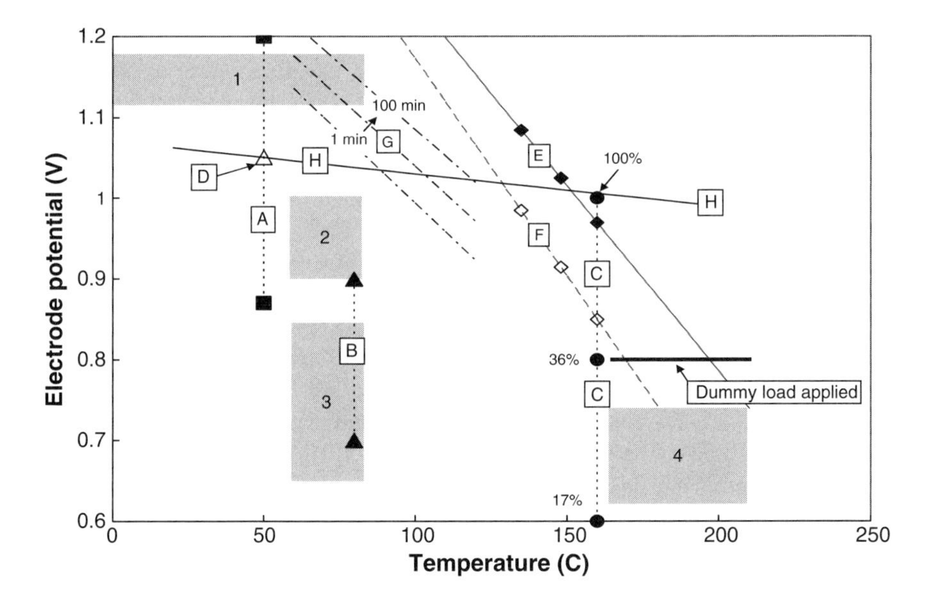

Fig. 3 Potential versus temperature highlighting various decay modes encountered in the operation of PEFC and PAFC stacks

In Fig. 3, line A, which terminates in squares, represents the potential-cycling experiments of Patterson (Patterson 2002). These were square-wave cycles with a period of 60 s between 0.87 and 1.2 V for 100 h at 50 °C resulting in a loss of 50% of the initial surface area. Line B, which terminates in filled triangles, represents the potential-cycling experiments of Mathias et al. (2005). They found a surface area loss of approximately 50% for Pt/C after 100 h of square-wave cycling the potential of the electrode between 0.7 and 0.9 V with a period of 60 s at 80 °C. Comparison of lines A and B suggests that the rate of surface area loss is a strong function of temperature.

The filled circles and line C correspond to experiments conducted in phosphoric acid (Passalacqua et al. 1992). These authors held 20% Pt/C electrodes at 0.6, 0.8, and 1.0 V in 98% phosphoric acid at 160 °C for 1,000 min. The amount of platinum surface area lost was 17, 36, and 100%, in order of increasing potential. Sintering may be responsible for some of the area loss at lower potentials (Kinoshita 1988). For comparison, Patterson reported an 18% loss in platinum surface area after 300 h at 1.05 V and 50 °C for a PEFC (Patterson 2002). This point is the open triangle labeled "D" in Fig. 3. Clearly, Pt/C is much more stable under the lower temperature conditions of the PEFC.

Line E represents a constant rate of carbon corrosion that intersects the upper-right corner of the PAFC operating window. The data were extrapolated using data reported in the literature for corrosion of Vulcan XC-72 heat-treated at 2,700 °C in concentrated phosphoric acid (Kinoshita 1988). The experiments were done near the temperatures and potentials indicated by the filled diamonds. Kinoshita reported the corrosion rates measured 100 min after the start of the experiments. When the data were extrapolated to the curve shown in Fig. 2, the dependence of corrosion rate on temperature and potential was assumed to be independent of time. The experimental data were fit to a Tafel expression with an anodic transfer coefficient, $\alpha = (1-\beta)n$, of 0.277, which is equivalent to a Tafel slope of 310 mV per decade at 160 °C. The best-fit activation energy was 79.7 kJ mol^{-1}.

Line F is an extrapolation of data from Kinoshita for Vulcan XC-72 at the same mass-specific corrosion rate. The open diamonds refer to the experimental data. Heat-treated Vulcan can be operated approximately 0.1 V higher than Vulcan at a given temperature and corrosion rate, according to the figure.

The three nearly parallel dot-dashed curves labeled "G" represent carbon corrosion data reported by researchers at General Motors for Pt/C in a polymer electrolyte (Mathias et al. 2005). The curves were constructed for the same instantaneous corrosion rate as for lines E and F. The three curves correspond to 1, 10, and 100 min after the start of the experiment. At a fixed temperature, the potential needed to drive carbon corrosion at a constant rate increases with time. Equivalently, the rate of carbon corrosion decreases with time at given temperature and potential. The magnitude of the slope of lines G is less than that of line E or line F. Therefore, using PAFC data may underestimate carbon degradation below 140 °C.

Line H is a line of constant platinum solubility based on the Nernst equation assuming that standard potential does not change with temperature:

$$U = 1.188 + \frac{RT}{2F}\ln\left(c_{\mathrm{pt}^{2+}}\right) \tag{5}$$

This line was drawn through point D and intersects line C near its apex. If the rate of platinum dissolution were simply proportional to the concentration in solution, and the solubility followed the Nernst equation, then approximately 100% of the platinum should be lost at point D in 1,000 min to maintain consistency between the PEFC and PAFC data. The observed rate of platinum dissolution in the PEFC is considerably lower than this simple approach indicates. This could be because the dissolution or transport of platinum is strongly activated by temperature.

Figure 3 demonstrates that as the operating temperature of a fuel cell increases, the maximum safe operating potential decreases. The safe operating domains depend on the longevity requirements and duty cycle of the intended application, as well as the materials employed in the cells. However, it is clear from Fig. 3 that if one wishes to operate at elevated temperatures, then one must develop more corrosion-resistant materials and use strategies to minimize the time spent at elevated potentials.

4 Humidity

Water is not only the product of the PEFC reaction; it is also critical for stable operation. PFSA membranes require water to transport protons since they reside in aqueous clusters within the polymer and their mobility depends on the characteristics of the aqueous network (Hsu and Gierke 1983; Halim et al. 1994). The more hydrated the membrane, the higher the ionic conductivity and the higher the performance (Doyle and Rajendran 2003). On the other hand, the presence of excessive liquid water can restrict access of reactant gases to the electrodes and result in significant performance losses. The best operating condition for a PFSA-based PEFC is fully saturated reactants without excessive liquid water.

Maintaining a proper level of humidification can be difficult in actual power plants. The reactants are typically humidified before entering the cell to keep the membrane well hydrated. Condensation often occurs in the cell owing to effects including water production, gas consumption, pressure drop, electro-osmotic drag, diffusion, and temperature changes. This liquid water can cause nonuniform gas distribution from cell to cell and limit gas access within the cells. Additionally, pushing liquid water along the gas channels increases pressure drops, which reduces system efficiency.

To operate continuously without adding water, a PEFC system must operate in water balance, such that the amount of water leaving the system is equal to the water being generated by the stack. Table 2 gives the dew points of the air exhaust of a PEFC under different conditions. These values can be calculated by applying a mass balance and a correlation of temperature versus vapor pressure (Masten and Bosco 2003). The design and operation of a stack affects the relationship between water balance and heat rejection (Perry and Darling 2004). Table 2 was constructed assuming 50% excess air flow and neglecting water in the fuel exit.

Water-recovery devices are usually employed to humidify the air entering the stack, which enables operation at higher coolant temperature while maintaining similar humidification levels. The third column in Table 2 shows the dew point of the air exiting the stack when 50% of the water leaving the stack is recovered and returned to the inlet. This figure may not be realistic in all cases. Table 2 indicates that operation with a well-humidified membrane is limited to temperatures below approximately 85 °C. Higher temperatures are possible if pressure or air utilization is increased. Higher temperatures will also occur in an evaporatively cooled stack, as shown in the final column of Table 2, which assumes the average cell potential is 0.6 V. This approach requires effective water distribution within the cell to avoid reactant starvation because relatively large amounts of water must be introduced into the active area (Meyers et al. 2006). Table 2 makes clear that operation above approximately 90 °C will require membranes that are less sensitive to water content than PFSA membranes, because the gas will necessarily have low water activity.

Table 2 Dew points of the air exhaust of a polymer-electrolyte fuel cell under different conditions

Air exit pressure (bar)	Water balance temperature (°C)	Temperature with recycle (°C)	Evaporative temperature (°C)
1	59	71	80
1.5	68	80	90
2	75	88	98

Development of an electrolyte membrane with good performance under hot, dry conditions is an area of active research (Kopasz 2007).

Maintaining the desired humidification level throughout the cell is especially challenging in applications that require frequent changes in power. These power changes cause near instantaneous changes in heat and water generation. If the design of the power plant incorporates external humidification of the reactant gas, then the capability of the control scheme and the mass of the humidification equipment will limit the time to attain a steady state at the new power level. Therefore, during transients dehydration or flooding can occur depending on how the power is changed. This can result in reduced membrane life owing to the combined effects of chemical and mechanical stress.

Water can condense in the gas manifolds because they are generally located at the periphery of the stack where heat loss to ambient drives condensation. The condensate can block individual cell passageways, resulting in reactant gas maldistribution or, in the extreme, insufficient flow to support the current produced by the stack. This starvation condition can cause irreversible damage if it occurs on the fuel side. Condensation can also occur within the active area when reactant gases flow from relatively high temperature zones into lower-temperature zones. The liquid water can block flow in the channels; however, unlike manifold condensation, this condition rarely results in starvation of an entire cell. However, the blockage can cause local reactant gas starvation, resulting in a decrease in cell potential.

Thermal cycling presents another water-management challenge, namely, to control water migration when the cell is stopped to allow the subsequent start to be successful. After the PEFC is turned off, water may be driven from the active area to the stack perimeter by temperature gradients during the cooling period. The worst condition occurs when the ambient temperature is below 0 °C because the temperature gradients are the largest and frost heave can take place (Meyers 2005; He and Mench 2006). Additionally, water residing in the reactant gas channels or large pores in the GDLs can fall to the bottom of the stack. A successful start requires that water migration occurring when the cell is off does not block gas channels or GDL pores that provide reactant access to the catalyst layers. During a start under subfreezing conditions, this water may be in the form of ice. If blockage does occur during the start, then the current distribution will be distorted by poor local reactant gas access to the catalyst layers. Accordingly, the start power level will be reduced, resulting in extended start time, at best, and possibly irreversible cell damage due to local fuel starvation.

Generally, the durability of polymer-electrolyte membranes improves with increasing hydration. Membrane life is defined as the operating time at which the membrane loses the ability to separate the reactants owing to pinholes. Continued operation beyond that point causes local reactions that lead to further increase in reactant gas mixing, which eventually results in a loss of cell efficiency as well as a potentially unsafe condition. There have been many experiments, ranging from rotating-disk laboratory-scale tests to full-size (more than 400 cm^2) stack tests, showing degradation of the membrane, as measured by fluoride emissions, thinning, and life, to be directly related to hydration level (LaConti et al. 2003; Yu et al. 2005a; Collier et al. 2006; Mittal et al. 2006). Dry membrane locations usually fail first.

Membranes subjected to repeated changes in relative humidity undergo mechanical degradation (Huang et al. 2006), and it has been shown that exposing membranes to hydration cycles will result in failure (Mathias et al. 2005). This degradation mechanism is not surprising since the PFSA membranes undergo relatively large dimensional changes when they are exposed to varying hydration levels. For example, an unconstrained, dry PFSA membrane will increase in volume by about 74% when equilibrated with liquid water (Gebel et al. 1993). These dimensional changes are much larger than those experienced by materials subjected to thermal cycles in the temperature range of a PEFC. In a cell, where the membrane is constrained, hydration changes result in significant mechanical stress (Weber and Newman 2004). The degree of mechanical stress depends upon the change in hydration level (i.e., the change in λ, which is the molar ratio of water molecules to ionic groups). For example, a PFSA membrane that is cycled from a supersaturated condition to 80% relative humidity undergoes more rapid degradation than a membrane cycled between 80 and 30% relative humidity, since $\Delta\lambda$ is greater in the former case and the swelling-induced dimensional change is roughly proportion to $\Delta\lambda$ (Huang et al. 2006).

Since humidification affects both chemical and mechanical degradation in membranes, controlling the humidification is the key to maximizing membrane life. For example, Fig. 4 shows membrane life for identical MEAs operated under accelerated conditions with different humidification strategies; the difference in longevity among these otherwise identical cells is dramatic. The porous-plate cell utilizes an internal-humidification technology that has been described elsewhere (Reiser 1997; Wheeler et al. 2001, Weber and Darling 2007). Figure 4 also demonstrates the need for durable seal materials, since in the case where the membrane was kept well hydrated the seal was the first component to fail.

All membranes currently being deployed operate best when fully hydrated; thus, systems which maintain these conditions should enable PEFCs to achieve the cost, life, and power-density requirements for commercial use. Unless a membrane is developed that does not require a relatively high level of hydration, some means of storing and providing water should be provided to ensure stable operation. All other fuel-cell technologies, with the single exception of solid oxide, employ internal electrolyte reservoirs to stabilize the three-phase interface between the electrolyte, reactant gas, and catalyst. Without these reservoirs, cell performance can suffer from changes in electrolyte concentration or hydration caused by changes in operating conditions. Lifetime can also be limited by loss of electrolyte with time. Multicollector fuel cells accommodate the loss of electrolyte, due to evaporation and corrosion, by employing relatively thick porous electrodes as electrolyte reservoirs. PAFCs use an electrolyte reservoir integral with the separator plate to provide additional acid for evaporative losses and for expansion by hydration during a safe shutdown. Alkaline fuel cells employ internal reservoirs and flooded-porous electrodes to accommodate the relatively large electrolyte-volume variations associated with changing power conditions. PEFCs also benefit from the use of reservoirs that minimize variations within the cells, and thereby stabilize performance and extend life. The reservoir strategy provides a robust solution for achieving the required hydration level over a wide range of power plant operating conditions (Reiser 1997; Wheeler et al. 2001; Weber and Darling 2007).

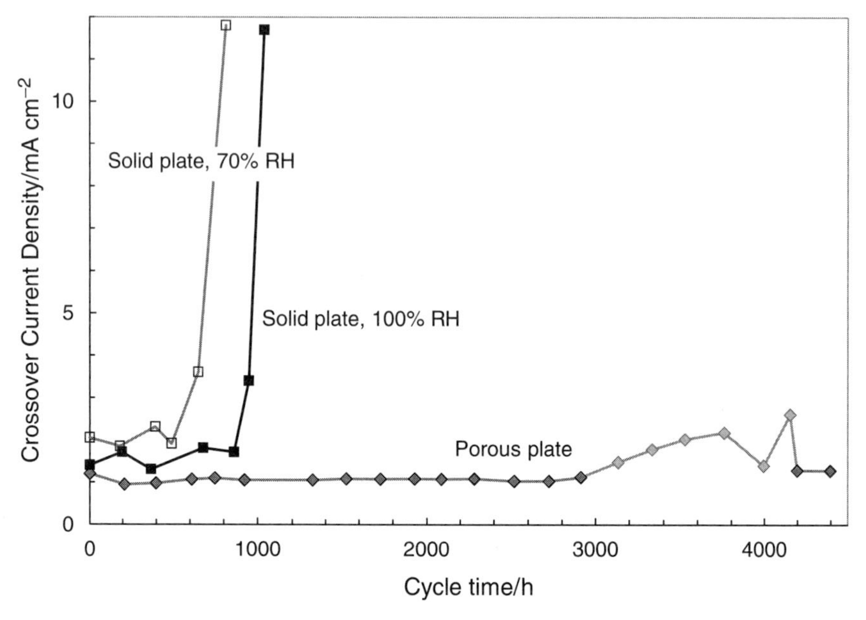

Fig. 4 Durability of identical MEAs operating under accelerated, cyclic conditions with different methods of humidification. Cycle conditions were as follows: 50 min at 0.8 A cm^{-2}, 10 min at 0.1 A cm^{-2}; 80 °C; anode reactants H_2 with 2.5% air bleed, 70% relative humidity (RH) (except where noted); cathode reactants 42% O_2 in N_2, 70% RH (except were noted). The membrane in the porous-plate cell did not fail; the test was discontinued after approximately 4,500 h. The seal started leaking at approximately 3,100 h and was subsequently replaced at approximately 4,200 h (the period with the leaky seal is indicated by *lighter diamonds*)

5 Conclusions

The range of conditions that a PEFC can be exposed to is quite large. Transients, both start/stop cycles and load cycles, can result in significant changes in potential, temperature, and relative humidity. Although the exact conditions will depend on the application, all applications require some transients and a PEFC system must be designed to either withstand or minimize the conditions that can be encountered during these transients. Existing membranes require hydration and the stability of these membranes in a dynamic fuel-cell environment, especially one that includes large humidity cycles, is a serious concern. In addition, the presence of liquid water within the cells can make uniform reactant delivery, which is necessary to achieve stable performance, challenging. Therefore, effective water management is imperative and can be achieved by incorporating liquid reservoirs within the cells. Operation at higher temperatures (e.g., approximately 120 °C) offers important advantages, including elimination of the liquid phase. However, to be practical, it also requires a membrane electrolyte that functions well in dry environments, as well as even more stringent control of the potentials experienced by the cells. In any case, consideration of both materials with improved stability and system mitigations provides the best path to designing PEFC systems that are optimized with respect to durability, cost, and performance.

References

Bett, J.A.S. Kinoshita, K., and Stonehart, P. (1976) Crystallite growth of platinum dispersed on graphitized carbon black, *J. Catal.*, **41**, 124–133.

Binder, H., Kohling, A., Richter, K., and Sandstede, G. (1964) Über die anodische oxydation von Aktivkohlen in wässrigen elektrolyten, *Electrochim. Acta*, **9**, 255–274

Bindra, P., Clouser, S.J., and Yeager, E. (1979) Platinum dissolution in concentrated phosphoric acid, *J. Electrochem. Soc.* **126**, 1631–1632.

Blair, L. (2005) PEMFC Freeze Start, U.S. DOE's Fuel Cell Operations at Sub-Freezing Temperatures Workshop, Phoenix, AZ, USA.

Bregoli, L.J. (1978) The influence of platinum crystallite size on the electrochemical reduction of oxygen in phosphoric acid, *Electrochim. Acta*, **23**, 489–492.

Cho, E.A., Ko, J.J., Ha, H.Y., Hong, S.A., Lee, K.Y., Lim, T.W., and Oh, I.H., (2004) Effects of water removal on the performance degradation of PEMFCs repetitively brought to $<0\,°C$, *J. Electrochem. Soc.*, **151**, A661–A665.

Collier, A., Wang, H., Yuan, X.Z., Zhang, J., and Wilkinson, D.P. (2006) Degradation of polymer electrolyte membranes, *Int. J. Hydrogen Energy*, **31**, 1838–1854.

Donahue, J., Fuller, T.F., Yang, D., and Yi, J.S. (2002) Method and apparatus for regenerating the performance of PEM fuel cell, *U.S. Patent 6,399,231.*

Doyle, M. and Rajendran, G. (2003) Perfluorinated membranes in: *Handbook of Fuel Cells Fundamentals, Technology, and Applications*, Edited by Vielstich, W., Gasteiger, H.A., and Lamm, A., vol. 3, Wiley, Hoboken, NJ, pp. 351–395.

Eickes, C., Piela, P., Davey, J., and Zelenay, P. (2006) Recoverbale cathode performance loss in direct methanol fuel cells, *J. Electrochem. Soc.*, **153**, A171–A178.

Ferreira, P.J., la O', G.J., Shao-Horn, Y., Morgan, D., Makharia, R., Kocha, S., and Gasteiger, H.A. (2005) Instability of Pt/C electrocatalysts in proton exchange membrane fuel cells, *J. Electrochem. Soc.*, **152**, A2256–A2271.

Fuller, T.F., Perry, M.L., and Reiser, C. (2005) Applying the lessons learned from PAFC to PEM fuel cells, *Electrochem. Soc. Proc. Vol.*, **8**, 337–345.

Gasteiger, H.A., Gu, W., Makharia, R., Mathias, M.F., and Sompalli, B. (2003) Beginning-of-life MEA performance – Efficiency loss contributions in: *Handbook of Fuel Cells Fundamentals, Technology, and Applications*, Edited by Vielstich, W., Gasteiger, H.A., and Lamm, A., vol. 3, Wiley, Hoboken, NJ, pp. 593–610.

Gebel, G., Aldebert, P., and Pineri, M. (1993) Swelling study of pefluorosulphonated ionomer membranes, *Polymer*, **34**, 333–339.

Gottesfeld, S. (1997) Polymer electrolyte fuel cells in *Advances in Electrochemical Science and Engineering*, Edited by Alkire, R., Gerischer, H., Kolb, D., and Tobias, C., vol. 5, Wiley, Germany, pp. 195–301.

Gruver, G.A., Pascoe, R.F., and Kunz, H. R. (1980) Surface area loss of platinum supported on carbon in phosphoric acid electrolyte, *J. Electrochem. Soc.*, **127**, 1219–1224.

Halim, J., Büchi, F.N., Haas, O., Stamm, M., and Scherer, G.G. (1994) Characterization of perfluorosulfonic acid membranes by conductivity measurements and small-angle X-ray scattering, *Electrochim. Acta*, **39**, 1303–1307.

He, S. and Mench, M. (2006) One dimensional transient model for frost heave in polymer electrolyte fuel cells, *J. Electrochem. Soc.*, **153**, A1724–A1731.

Hsu, W.Y. and Gierke, T.D. (1983) Ion transport and clustering in Nafion perfluorinated membranes, *J. Membr. Sci.*, **13**, 307–326

Huang, X.Y., Solasi, R., Zou, Y., Feshler, M., Reifsnider, K., Condit, D., Burlatshy, S., and Madden, T. (2006) Mechanical endurance of polymer electrolyte membrane and PEM fuel cell durability, *J. Polym. Sci. Part B – Polym. Phys.*, **44**, 2346–2357.

Jarvi, T., Patterson, T., Cipollini, N., Hertzberg, J., and Perry, M. (2003) Recoverable performance losses in PEM fuel cells, *Electrochemical Society Meeting Abstracts*, Paris.

Kangasniemi, K.H., Condit, D.A. and Jarvi, T.D. (2004) Characterization of vulcan electrochemically oxidized under simulated PEM fuel cell conditions, *J. Electrochem. Soc.*, **151**, E125–E132.

Kinoshita, K. (1988) *Carbon: Electrochemical and Physiochemical Properties*, Wiley, New York, NY.

Kinoshita, K. and Bett, J.A.S. (1974) Effects of graphitization on the corrosion of carbon blacks, *Proceedings of the Symposium on Corrosion Problems in Energy Conversion and Generation*, pp. 43–55.

Kinoshita, K., Lundquist, J.T., and Stonehart, P. (1973) Potential cycling effects on platinum electrocatalyst surfaces', *J. Electroanal. Chem. Interfacial Electrochem*, **48**, 57.

Kopasz, J. (2007) High temperature membrane working group, *U.S. DOE Annual Hydrogen Program Review*, Arlington VA, USA.

LaConti, A.B., Hamdan, M., and McDonald, R.C. (2003) Mechanisms of membrane degradation in: *Handbook of Fuel Cells Fundamentals, Technology, and Applications*, Edited by Vielstich, W. Gasteiger, H.A., and Lamm, A., vol. 3, Wiley, Hoboken, NJ, pp. 647–662.

Landsman, D.A. and Luzack, F.J. (2003) Catalyst studies and coating technologies in: *Handbook of Fuel Cells Fundamentals, Technology, and Applications*, Edited by Vielstich, W., Gasteiger, H.A., and Lamm, A., vol. 4, Wiley, Hoboken, NJ, pp. 811–831.

Luzack, F.J. and Landsman, D.A. (1983) Ternary fuel cell catalysts containing platinum, cobalt and chromium, *U.S. Patent 4,447,506*.

Masten, D.A. and Bosco, A.D. (2003) System design for vehicle applications: GM/Opel in: *Handbook of Fuel Cells Fundamentals, Technology, and Applications*, Edited by Vielstich, W., Gasteiger, H.A., and Lamm, A., vol. 3, Wiley, Hoboken, NJ, pp. 714–724.

Mathias, M.F., Makharia, M., Gasteiger, H.A., Conley, J.J., Fuller, T.J., Gittleman, C.J., Kocha, S.S., Miller, D.P., Mittelsteadt, C.K., Xie, T., Yan, S.G., and Yu, P.T. (2005) Two fuel cell cars in every garage?, *Electrochem. Soc. Interface*, **14**, 24–35.

Meyers, J.P (2005) Fundamental issues in subzero PEMFC startup and operation, *U.S. DOE's Fuel Cell Operations at Sub-Freezing Temperatures Workshop*, Phoenix AZ, USA.

Meyers, J.P. and Darling, R.D. (2006) Model of carbon corrosion in PEM fuel cells, *J. Electrochem. Soc.*, **153**, A1432–A1442.

Meyers, J.P., Darling, R.D., Evans, C., Balliet, R., and Perry, M.L. (2006) Evaporatively-cooled PEM fuel-cell stack and system, *ECS Trans.*, **3**, 1207–1214.

Mitsushima, S., Kawahara, S., Ota, K., and Kamiya, N. (2007) Consumption rate of Pt under potential cycling, *J. Electrochem. Soc.*, **154**, B153–B158.

Mittal, V.O., Kunz, H.R., and Fenton, J.M. (2006) Effects of catalyst properties on membrane degradation rate and the underlying degradation mechanism in PEMFCs, *J. Electrochem. Soc.*, **153**, A1755–A1759.

Neyerlin, K.C., Gasteiger, H.A., Mittelsteadt, C.K., Jorne, J., and Gu, W. (2005) Effect of relative humidity on oxygen reduction kinetics in a PEMFC, *J. Electrochem. Soc.*, **152**, A1073–A1080.

Paik, C.H., Saloka, G.S., and Graham, G.W. (2007) Influence of cyclic operation on PEM fuel cell catalyst stability, *Electrochem. Solid-State Lett.*, **10**, B39–B42.

Passalacqua, E., Antonucci, P. L., Vivaldi, A., Patti, A., Antonucci, V., Giordano, N., and Kinoshita, K. (1992) The influence of Pt on the electrooxidation behavior of carbon in phosphoric acid, *Electrochim. Acta*, **37**, 2725–2730.

Patterson, T. (2002) Effect of potential cycling on loss of electrochemical surface area of platinum catalyst in polymer electrolyte membrane fuel cell, *AIChE National Conference*, New Orleans, LA, USA.

Patterson, T.W. and Darling, R.D. (2006) Damage to cathode catalyst of a PEM fuel cell caused by localized fuel starvation, *Electrochem. Solid-State Lett.*, **9**, A183–A185.

Perry, M.L. and Darling R.M. (2004) Optimizing PEFC stack design and operation for energy and water balance in transportation systems, *Electrochem. Soc. Proc.*, ***PV2004-21***, 634.

Perry, M.L., Newman, J., and Cairns, E.J. (1998) Mass transport in gas-diffusion electrodes: A diagnostic tool for fuel-cell cathodes, *J. Electrochem. Soc.*, **145**, 5–15.
Perry, M.L., Patterson, T.W., and Reiser, C. (2006) System strategies to mitigate carbon corrosion in fuel cells, *ECS Trans.*, **3**, 783–795.
Perry, M., Patterson, T., and O'Neill, J. (2007) PEM fuel cell freeze durability and cold start project, *U.S. DOE Annual Hydrogen Program Review*, Arlington VA, USA.
Pourbaix, M. (1974) *Atlas of Electrochemical Equilibria in Aqueous Solutions*, National Association of Corrosion Engineers, Houston, TX, USA.
Protsailo, L. (2006) Development of high temperature membranes and improved cathode catalysts for PEM fuel cells, *U.S. DOE Annual Hydrogen Program Review*, Arlington VA, USA.
Reiser, C. (1997), Ion exchange membrane fuel cell with water management pressure differentials, *U.S. Patent 5,700,595*.
Reiser, C.A., Bregoli, L., Patterson, T.W., Yi, J.S., Yang, J.D., Perry, M.L., and Jarvi, T.D. (2005) A reverse-current decay mechanism for fuel cells, *Electrochem. Solid-State Lett.*, **8**, A273–A276.
Uribe, F.A. and Zawodzinski, T.A. (2002) A study of polymer electrolyte fuel cell performance at high voltages; dependence on cathode catalyst layer composition and on voltage conditioning, *Electrochim. Acta*, **47**, 3799–3806.
Wang, X., Kumar, R., and Myers, D. J. (2006) Effect of voltage on platinum dissolution, *Electrochem. Solid-State Lett.*, **9**, A225–A227.
Weber, A.Z. and Darling, R.M. (2007) Understanding porous water-transport plates in polymer-electrolyte fuel cells, *J. Power Sources*, **168**, 191–199.
Weber, A.W. and Newman, J. (2004) A theoretical study of membrane constraint in polymer-electrolyte fuel cells, *AIChE J.*, **50**, 3215–3226.
Wheeler, D.J., Yi, J.S., Fredley, R., Yang, D., Patterson, T., and VanDine, L. (2001) Advancements in fuel cell stack technology at international fuel cells, *J. New Mater. Electrochem. Syst.*, **4**, 233–238.
Woods, R. (1976) Chemisorption at electrodes in: *Electroanalytical Chemistry*, Edited by A. J. Bard, vol. 9, Marcel Dekker, New York, NY, p. 1.
Yu, J., Matsuura, T., Yoshikawa, Y., Islam, M.N., and Hori, M. (2005a) In situ analysis of performance degradation of a PEMFC under nonsaturated humidification, *Electrochem. Solid-State Lett.*, **8**, A156–A158.
Yu, P., Pemberton, M., and Plasse, P. (2005b) PtCo/C cathode catalyst for improved durability in PEMFCs, *J. Power Sources*, **144**, 11–20.

Design Requirements for Bipolar Plates and Stack Hardware for Durable Operation

Felix Blank

Abstract The main requirements for the bipolar plate are electrical contacting of electrodes, current conduction, supply of gases and cooling media, removal of products, and the separation of reactant gases and cooling media. Another important factor for efficient operation and high durability is a homogeneous gas distribution. The homogeneity of the local reactant partial pressures is influenced by the depletion of oxygen and the accumulation of water and is locally affected by covered contact areas (landings) of the bipolar plate. Degradation effects in connection with hydrogen or oxygen undersupply and starvation are strongly influenced by bipolar plate design. Bipolar plate design may be influenced by the material, the production process, the gas diffusion layer properties, the membrane properties, the temperature profile requirements, the temperature level requirements, the interaction of the anode and the cathode, and especially by the liquid water transport through the membrane electrode assembly. As a result, different cell types have been established, e.g., gases can be conducted by a porous structure or in channels. Nonoptimized bipolar plate structures reduce the performance or involve the danger of increased degradation of the membrane electrode assembly and the bipolar plate. Requirements for an efficient and durable stack operation are a homogeneous distribution of fluids and a homogeneous and defined cell compression. The manifold of a fuel cell stack has the function to carry reactants and cooling media to the stack or to different cell rows and to carry off products and cooling media; therefore, the design of the manifold has an impact on the distribution of gas and cooling media. The stack-compression hardware has the function to lock the stack components into position with a defined and homogeneous pressure. Too high, too low, or inhomogeneous compressions have negative effects on the performance and durability of the stack.

F. Blank
GR/VFS
Daimler AG
e-mail: felix.blank@daimler.com

F.N. Büchi et al. (eds.), *Polymer Electrolyte Fuel Cell Durability,*

1 Influence of the Bipolar Plate Design on Degradation

The design of the bipolar plate has a strong influence on the cell performance and on degradation effects. All effects in connection with hydrogen or oxygen undersupply are caused by the bipolar plate design. This applies also to degradation effects in connection with humidification or drying, and furthermore to effects of high or inhomogeneous voltage or current distribution and to high or inhomogeneous temperature and freezing effects. Degradation effects are not described here in detail. Influences of the bipolar plate design on degradation effects will be discussed.

In the following the requirements for the bipolar plate are described as this is the basis for the understanding of bipolar plate designs. Then different types of flow fields are described, illustrating the difficulty to fulfill all requirements at the same time.

1.1 Bipolar Plate Requirements

For best performance and durability, the material and the design of the bipolar plate have to fulfill a series of requirements:

- *Current conduction.* The bipolar plate has to conduct the electrons from one electrode through the plate to the other electrode; therefore, the design needs contact areas with a low contact resistance and a plate material with a low electrical resistance. A homogeneous current distribution is important as inhomogeneous current distributions will aggravate degradation effects of the membrane electrode assembly (MEA) and the plate. This results in the requirement of homogenous material and of a design accepting and emitting the current homogenously.
- *Separation of media.* To separate the different media, the plate material has to be tight against air, hydrogen, and cooling fluid (e.g., water and glycol) at different pressures of the fluids.
- *Gas distribution.* To enable a highly efficient electrochemical reaction, the complete active area has to be supplied with oxygen and hydrogen. Additionally the inert gas fractions and the liquid water have to be removed. To allow for a wide range of operation, the design of the bipolar plate has to enable the removal of liquid water for different gas flows, also under low mass flow conditions and at low temperatures. In contrast the pressure drop must not be too high in the case of high gas flows and high loads as this will negatively affect the efficiency of the system.
- *Mechanical stability.* To reduce the contact resistance, to have the necessary compression pressure for the sealing of the cells, and to resist different pressures of the media, the plate must be mechanically stable. Changing strain results from different pressures and different temperatures between idling and full-load operation. Degradation by changing mechanical strain has to be avoided by proper dimensioning of the materials.
- *Cooling.* To prevent overheating and to control the cell temperature, the distribution of coolant media has to be homogeneous. Under operation, there always exists

a temperature difference between coolant media inlet and outlet. However, ideally the temperature of the cell should be homogeneous as excessive temperatures may trigger or aggravate degradation effects of the MEA or the bipolar plate. Furthermore by inhomogeneous cooling, the cooling system becomes inefficient. The distribution of coolant media also influences the freeze start capability of the stack.

- *Weight*. Depending on the application a low weight of the plate is essential.
- *Cost*. Depending on the application the plate requires low costs.
- *Material*. As a result of the above-mentioned conditions, requirements for material groups were established. One material group is carbon-based materials, i.e., carbon compound materials with a high carbon fraction. The other material group is metals, for cost reasons, normally those based on stainless steel. Both material groups have specific characteristics and thus they have a great influence on the design.
- *Packaging*. Depending on its application, the bipolar plate has specific requirements for the size of the active cell area and possibly for the maximum width in one or two dimensions.
- *Applications*. Different applications require different operating conditions and focus on the above-mentioned requirements, which results in different bipolar plate designs.

For automotive powertrain applications the following factors are important: cost, weight, volume, freeze start ability, and high efficiency. The important factors for stationary applications, where the heat might be used besides electric power, are volume, weight, freeze start, and less electric efficiency. However, here the requirement for a long life is more relevant.

1.2 Standard Channel-Based Flow Fields

The cross section of a standard cell is shown in Fig. 1. A balanced dimension of channel cross section and contact area has to be found for the channel design. To permit a high power density, the channel depth is limited. A high pressure drop supports water removal and gas distribution between the channels. A low pressure drop ensures good system efficiency. One limiting factor of mass transport is the width of the contact area. With increased production of water in the contact area, oxygen diffusivity is hindered more and more, i.e., these areas become more and more inactive.

Figure 2 illustrates this aspect for varying water load along the channel over the active cell area. Very narrow contact areas are required; however, this has the disadvantage that the mechanical stress in this area rises, which can have a negative influence on the porosity of the gas-diffusion layer (GDL). A reduction of the contact area is accompanied by a rise of electric contact resistance as well as by increased thermal flow from the GDL.

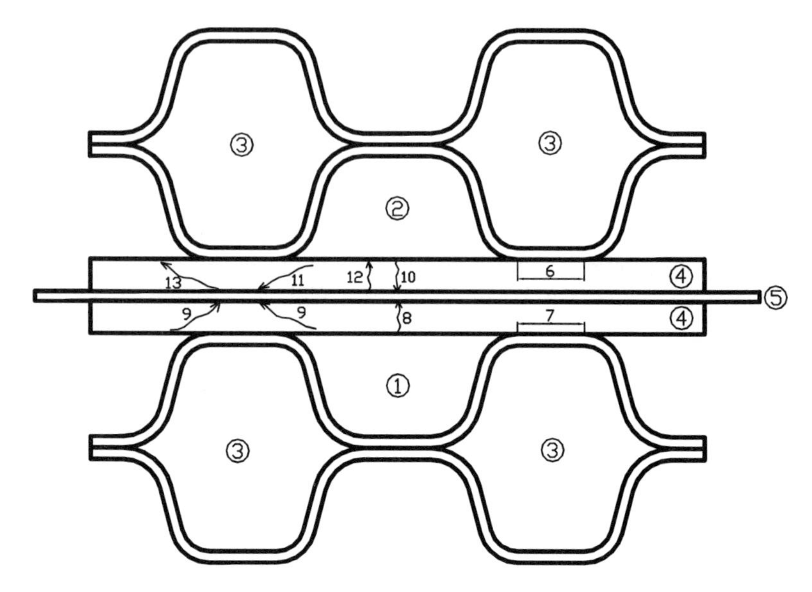

Fig. 1 Cross section of a standard cell: (1) anode channel; (2) cathode channel; (3) cooling channel; (4) gas-diffusion layer; (5) membrane; (6) cathode contact area; (7) anode contact area; (8) hydrogen diffusion length at the channel; (9) hydrogen diffusion length at the contact area; (10) oxygen diffusion length at the channel; (11) oxygen diffusion length at the contact areas; (12) water diffusion length at the channel; (13) water diffusion length at the contact areas

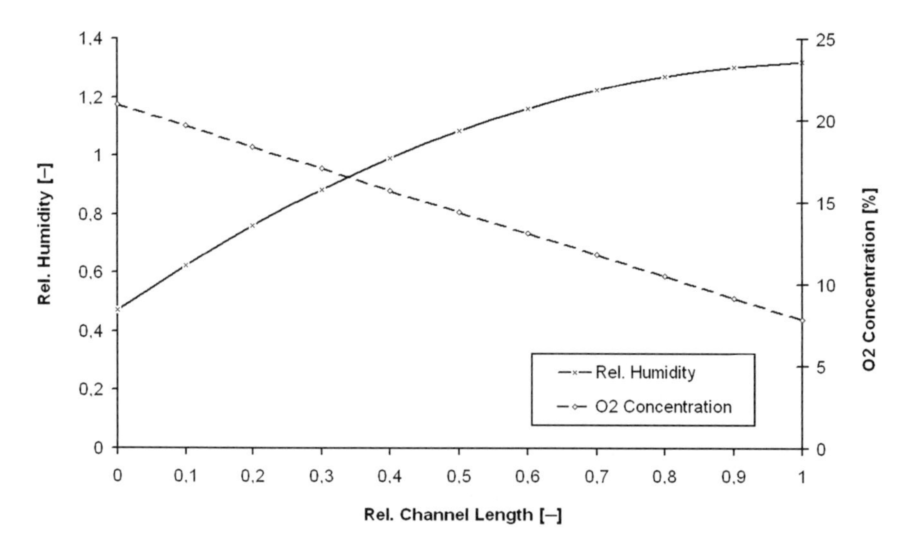

Fig. 2 Variation of oxygen concentration and relative humidity along the channel, calculated for a constant cell temperature of about 75 °C and an air stoichiometry of 1.8. The relative humidity of air at the inlet is 0.5. At a relative humidity in excess of 1, liquid water is formed

Figure 3 shows a schematic top view of a bipolar plate with a straight flow field. The reactant gases flow unidirectionally from one end of the plate to the other. Figure 2

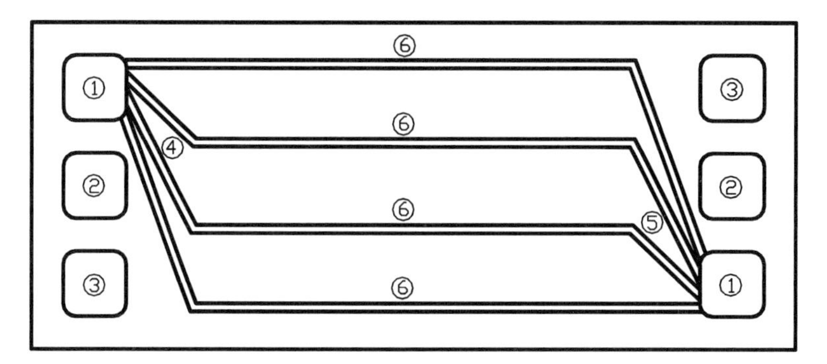

Fig. 3 Cathode straight flow field plate: (1) cathode ports; (2) anode ports; (3) coolant ports; (4) distribution area; (5) collecting area; (6) cathode channel (active cell area)

illustrates the concentration distribution in the cathode channel. The electrochemical reaction depletes oxygen and produces water. The reaction under the contact areas becomes increasingly difficult towards the outlet because of the lower oxygen partial pressure. Especially in the case of high current densities, the reaction may also be impeded by increasing accumulation of liquid water under the contact areas. This effect also becomes more pronounced towards the outlet as the water concentration in the cathode gas increases during its course along the channel and thus impedes the water removal. On the anode side there is less liquid water, but also here the same problem may occur for very humid conditions.

As hydrogen has a considerably higher diffusion coefficient than oxygen, the reaction is usually less impeded by depletion. However, the reactant concentration on the anode side decreases because of the usually smaller stochiometries. Pure (humidified) hydrogen operation results in great fractions of water vapor in the gas at the cell outlet. If the anode gas carries additional gas species (e.g., nitrogen or carbon dioxide), the problem of an impeded reaction also increases considerably at the anode.

In conclusion, this shows that degradation mechanisms depending on undersupply of reactant gases occur increasingly towards the gas outlet.

As ordinary membranes require a certain water content for proton conduction, humidification of the reactant gases prior to entry into the cell is necessary. The higher the temperature of the gases and the higher the relative humidity, the greater is the humidification effort. Especially in the case of mobile systems, gas humidification is limited. If the gas is not completely saturated when flowing into the active area, it will withdraw water from the membrane. In the case of great volume flows, the membrane at the inlet area of the cell may dry. If gases are fed to the cell while no current is drawn, depending on the temperature, relative humidity, and mass flow of the gases, desiccations may occur in the entire active area.

The straight flow field shown in Fig. 3 has the advantage that it allows for an adjusted temperature gradient. Cathode and cooling water may be arranged in an almost perfect co-flow. Thus, the relative humidity of the gas is least reduced at the cell inlet because of the lower inlet coolant temperature and additionally the relative

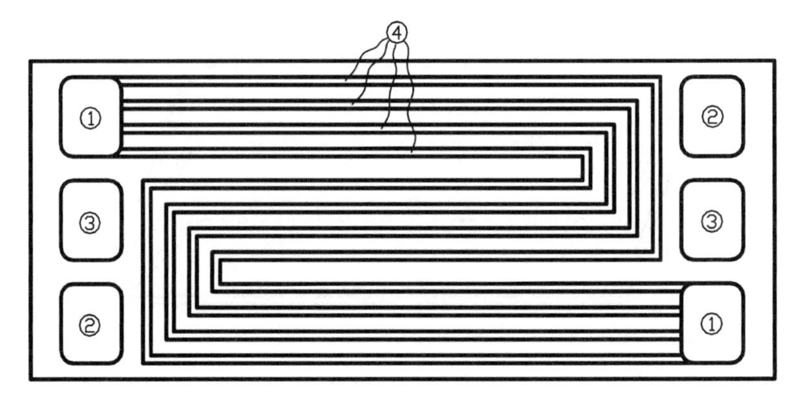

Fig. 4 Cathode serpentine flow field plate: (1) cathode ports; (2) anode ports; (3) coolant ports; (4) cathode channel (active cell area)

humidity is somewhat reduced at the outlet because of the higher cell temperature. The problem described above of liquid water accumulation is thus attenuated. A straight flow field has the disadvantage that it is generally designed to be quite long and narrow. This is because for good removal of liquid water, a certain pressure drop is required in the channels. To realize this, the channel needs to be of a certain length. However, long and narrow dimensions of bipolar plates can be problematic with respect to production and assembly. Lower aspect ratios of the active cell area result in a greater number of channels and consequently the mass flow in each channel is reduced and thus the pressure drops. Therefore, in general the water removal will be impeded when the number of channels is increased.

These problems can be alleviated by a serpentine flow field (Fig. 4). To a certain extent, also for a serpentine flow field gases and coolant media may be arranged in co-flow to take advantage of the temperature gradient.

Of course other channel-based flow-field designs are possible. In general attention has to be paid to the temperature gradient and good water removal. Flow fields with meandering channels involve the risk of liquid water collecting in the bends.

1.3 Interdigitated Flow Field

The bipolar plate shown in Fig. 5 has a so-called interdigitated flow field (Wood et al. 1998; Wang and Liu 2004; Hwang and Liu 2006). Reactant gases are conducted over the active cell area in channels which are closed at their ends. Between every two feeding channels there is a channel for gas dissipation. Thus, the gas is forced to flow through the domains under the contact areas. This will bring about a better supply under the contact areas, which is critical in the standard designs. Under ideal conditions, such a flow field operates successfully. However, the removal of accumulated liquid water through the porous layer is critical because of the density differences of liquid and gas.

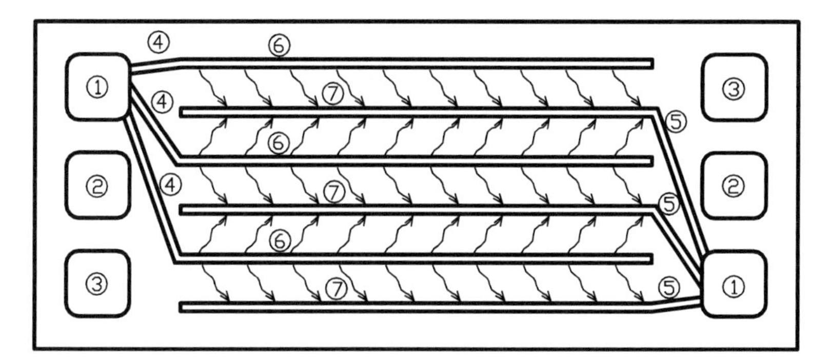

Fig. 5 Cathode interdigitated flow field plate: (1) cathode ports; (2) anode ports; (3) coolant ports; (4) distribution area; (5) collecting area; (6) cathode feeding channel (active cell area); (7) cathode outlet channel (active cell area)

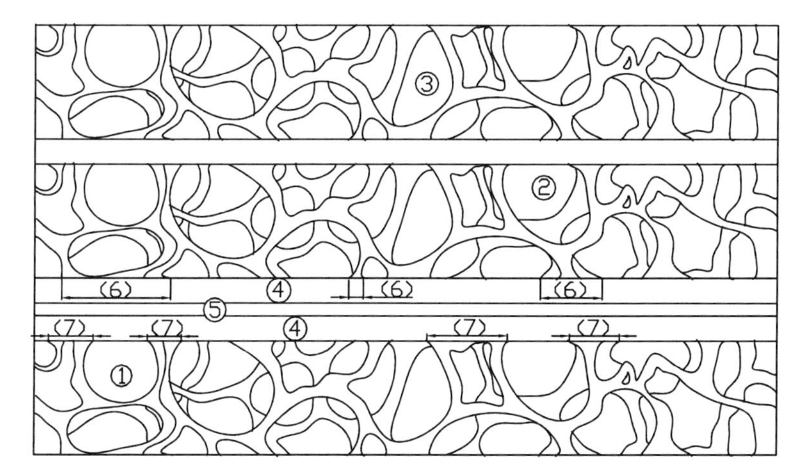

Fig. 6 Cross section of a cell with a porous gas distribution structure: (1) anode channel; (2) cathode channel; (3) cooling channel; (4) gas-diffusion layer; (5) membrane; (6) cathode contact area; (7) anode contact area

1.4 Porous Gas Distribution Structures

The structure represented in Fig. 6 uses a porous layer for carrying the media. These media are additionally separated from each other by a gastight layer.

The width of the contact area can thus be reduced, i.e., the risk of undersupply is locally reduced by shorter diffusion ways. As the gases are not forced to overflow the whole active cell area through channels between the inlet and the outlet of the active cell area, they take the way with the lowest resistance. Then, corners are streamed/passed through much less. This may result directly in undersupply or it may result in collecting liquid water ensued by undersupply. Therefore, cell structures with porous layers usually have several inlets and outlets to avoid dead zones, but then the stack often has greater dimensions. Another disadvantage is a greater pressure drop with identical stack height or a greater cell pitch with identical pressure drop.

1.5 Bipolar Plates with Porous Cathode/Cooling Plate

An elegant method for water management is to use porous plates, to allow for water exchange between the cooling circuit and the gas streams (Reiser 1997; Grasso et al. 2005; Weber and Darling 2007; Fig. 7). The cell in Fig. 8 has a porous plate between the cooling medium and the cathode. On the one hand water humidifies the gases by evaporation when the gas is not saturated and on the other hand liquid water collecting

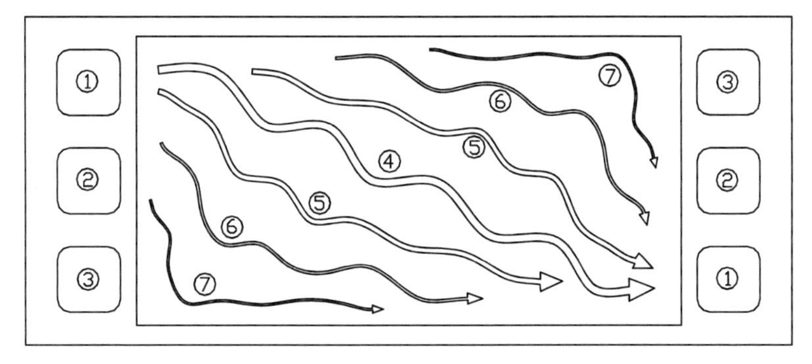

Fig. 7 Flow distribution in a cathode plate with a porous gas distribution structure: (1) cathode ports; (2) anode ports; (3) coolant ports; (4) main flow; (5) reduced flow; (6) strongly reduced flow; (7) minimum flow

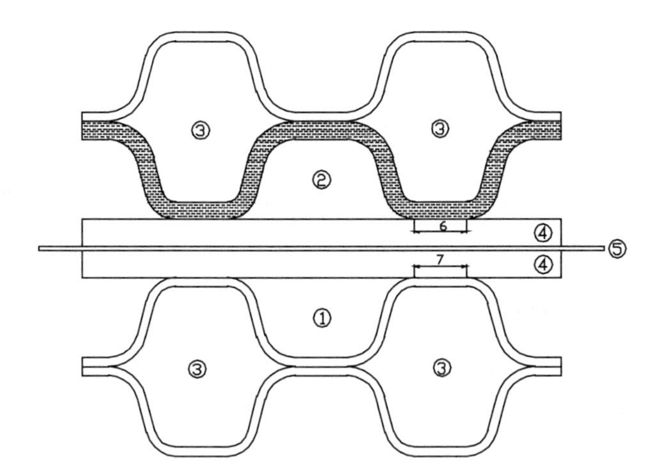

Fig. 8 Cross section of a cell with a porous bipolar plate on the cathode side: (1) anode channel; (2) cathode channel; (3) coolant channel; (4) gas-diffusion layer; (5) membrane; (6) anode contact area; (7) cathode contact area

at the cathode will be removed through the porous structure. This reduces the problem of liquid water in the channels. However, difficulties arise if the cell is to be suitable for freeze starts because the electrochemical components are sensitive to contact with common cooling media. Species other than water, contained in the coolant, can pass through the porous structure into the cathode compartment.

1.6 Bipolar Plate Structure with Integrated Air Injection

Figure 9 shows a bipolar plate whose cathode is split into two sections. In the inlet area of the cell only a small part of the gas is in direct contact with the MEA. Thus, the volume flow of the air is lower and hence high local relative humidity is obtained much faster than with the full-volume flow. In turn, oxygen is also depleted faster because of the smaller volume of gas in direct contact with the MEA. When the concentration of oxygen reaches a critical value, new gas is fed from the injection channels to the cathode channels through holes. The characteristics of humidity and the characteristics of the oxygen concentration are illustrated in Fig. 10 in comparison with those for a standard flow field. In Fig. 10 it is assumed that the cathode gas is not humidified and that the cell operates with a relatively high temperature (75 °C). The air injection cell shows a considerably higher relative humidity in the first third of the channel than for the standard flow field. The membrane

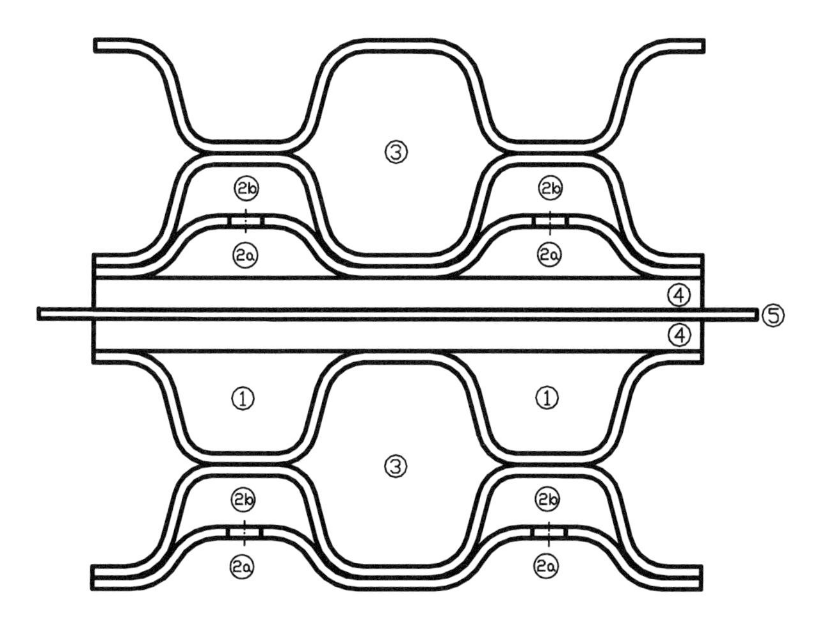

Fig. 9 Cross section of a cell with a bipolar plate with an integrated air injection: (1) anode channel; (2a) cathode channel; (2b) cathode feed channel; (3) coolant channel; (4) gas-diffusion layer; (5) membrane

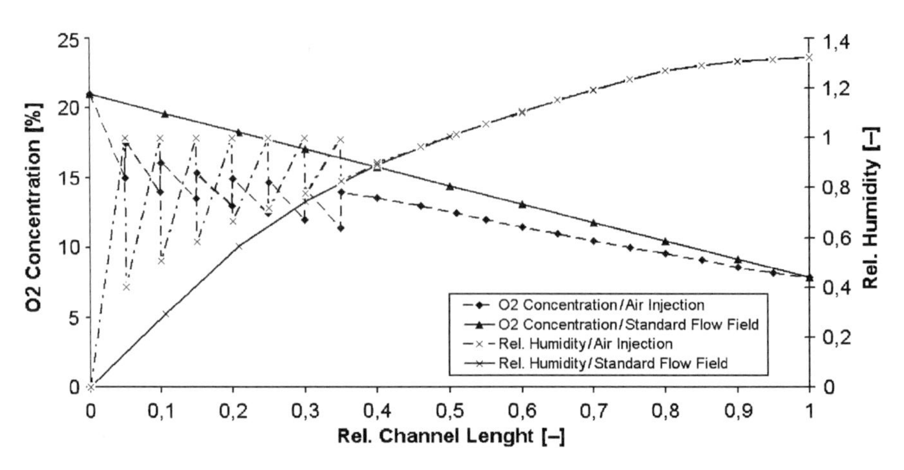

Fig. 10 Humidity distribution and oxygen concentration along the channel for a standard cell and a cell with reduced air flow in the beginning and sevenfold subsequent air injection during the first 35% of the length of the air channel. Calculated for a cell temperature of about 75 °C and an air stoichiometry of 1.8

of the standard cell would soon run dry, i.e., long-term operation would not be possible. As a consequence however, the oxygen partial pressure is slightly lower over the entire channel length.

2 Media Manifolding

The manifold has the function to distribute reactants and cooling media to one or more stacks and within the stack to a high number of cells; therefore, the design of the manifold influences the distribution of gas and cooling media. The design of the manifolds also has an influence on all degradation effects in connection with undersupply of a reactant gas or cooling media. In addition to the design requirement of not limiting the fluid distribution, the manifolds have to be designed to avoid accumulations of liquid water. The latter can also have an impact on reactant distribution and degradation effects in connection with freezing and corrosion.

3 Stack-Compression Hardware

The function of the stack-compression hardware is to fasten stack components with a defined and homogeneous pressure. If these requirements are not accurately fulfilled, the function and the durability of the stack will be influenced negatively. The MEAs and the bipolar plates need to be fixed in accurate positions and the compression of gaskets needs to be safe and homogeneous. If the pressure is too high, it will cause mechanical failure of the membrane or of the bipolar plate. If the pressure is too low,

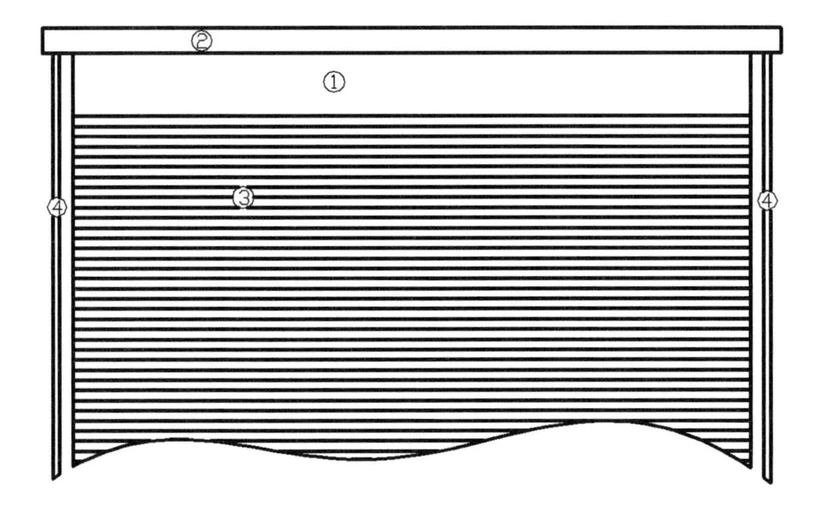

Fig. 11 End plate hardware design with beams: (1) end plate; (2) beam; (3) cells; (4) tie rods

it may cause gas or cooling fluid leakage. Another effect of low compression is the increased contact resistance between the GDL and the bipolar plate, which will result in an inhomogeneous current distribution and thus in a reduced lifetime of the MEA.

The challenge for the compression hardware design is to fulfill the functions in a large temperature window from far below freezing to maximum temperatures near 100 °C. Additionally there are strong limitations for volume, weight, and costs, especially in the case of automotive applications.

The main mechanical challenge for end plates of the compression hardware is shown in Fig. 11. Even very high beam strengths will result in bending of the end plate and consequently uneven pressure distribution to the cells. Even if the deflection of the end plate is very small, it will cause a change to the electrical conductivity and contact resistances in the active area. Additionally it can result in mechanical failure of areas under high pressure.

Figure 12 shows an example of compact compression hardware with improved force distribution. The force is transferred to the end plate hardware by steel or plastic bands. This permits a very short distance from the stack. The band lies on a leaf spring which is nearly flat when the force is applied. As a result the force is not transferred to the end plate at a single point but over a wide area. As this area lies in the center of the plate, the deflection of the plate is much reduced.

4 Conclusions

Assuming that the fuel cell stack is homogeneously fixed by the compression hardware, the main influence on degradation effects on the MEA is an inhomogeneous fluid distribution. If there are no influences by inhomogeneous cells, the cell with the

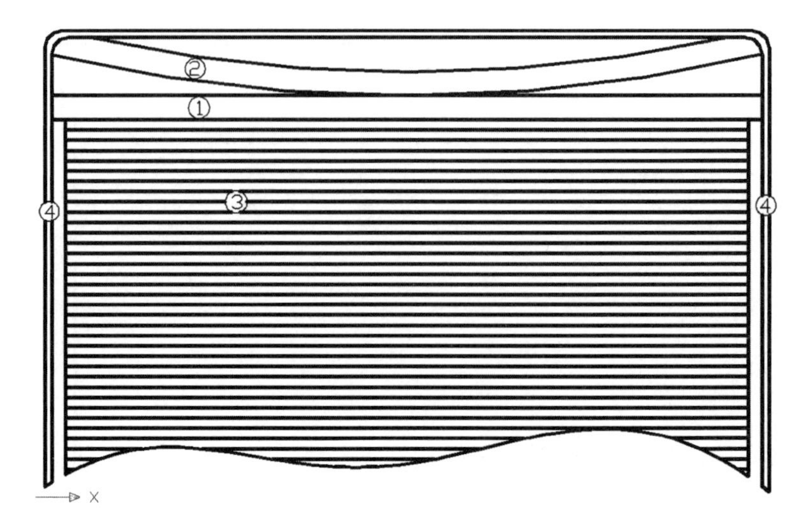

Fig. 12 End plate hardware design with leaf springs: (1) end plate; (2) leaf spring; (3) cells; (4) tie band

lowest gas supply will be most vulnerable to starvation effects. Assuming that there is no gas shorting, the region with the lowest reactant concentration is the contact area near the outlet of the channel with the lowest gas supply. These effects of local reactant distribution are superposed by the cooling media distribution and by the local temperature. In other words, the more any effect leading to inhomogeneity is reduced and the more the total active area is supplied in a homogeneous manner, the more durable the cells will be.

References

Grasso, A.P., Scheffler, G.W., Van Dine, L.L., Dufner, B.F., and Breault, R. (2005) PEM fuel cell passive water management. *United States Patent* 6,916,571.

Hwang, J.J., and Liu, S.J. (2006) Comparison of temperature distributions inside a PEM fuel cell with parallel and interdigitated gas distributors, J. *Power Sources*, 162, 1203–1212.

Reiser, C., (1997) Ion exchange membrane fuel cell power plant with water management pressure differentials. *United States Patent* 5,700,595.

Wang, L., and Liu, H. (2004) Performance studies of PEM fuel cells with interdigitated flow fields, J. *Power Sources*, 134, 185–196.

Weber, A.Z., and Darling, R.M. (2007) Understanding porous water-transport plates in polymer-electrolyte fuel cells, J. *Power Sources*, 168, 191–199.

Wood, D.L., Yi, J.S., and Nguyen, T.V. (1998) Effect of direct liquid water injection and interdigitated flow field on the performance of proton exchange membrane fuel cells, *Electrochim. Acta*, 43, 3795–3809.

Heterogeneous Cell Ageing in Polymer Electrolyte Fuel Cell Stacks

Felix N. Büchi

Abstract Polymer electrolyte fuel cell stacks, in the commonly used bipolar arrangement, consist of multiple stacked single cells in a filter-press-type arrangement. The bipolar arrangement connects the cells in series electrically and in parallel for the reactant and coolant flows; therefore, all cells have to carry the same current but they can receive different reactant mass flows. The reactant and the coolant supply may be different owing to statistically varying percolation resistances of the fluids and owing to the position of the cells in the stack. Therefore, the commonly made assumption that individual cells perform equally is valid neither for normal operation nor for the degradation of individual cells. Differences between cells can be of systematic or stochastic nature and translate into differences in the degradation rate under operation or start/stop conditions. The four main cases are discussed.

1 Introduction

For technical applications of polymer electrolyte fuel cells (PEFC), individual cells are commonly combined in a bipolar arrangement in filter-press-type stacks. While for low-power applications (less than 10 W) other arrangements of the electrical series connection of single cells are investigated (Angstrompower 2001, Heinzel et al. 1998, Jiang and Chu 2001), in applications which require high specific power densities the concept of the low-resistance bipolar arrangement is generally exploited. In this arrangement individual cells are connected electrically in series and for the media supply in parallel (Fig. 1). This arrangement results in all cells carrying the same current, but being subject to variations with respect to reactant and coolant mass flow. These disparities stem from systematic and stochastic differences of plates, seals, and electrochemical components, are due to the location in the stack (i.e., end or center), and most importantly are due to (small) variations in operating conditions resulting in temperature differences, stochastic condensation of water, and differences in reactant mass flows.

F.N. Büchi
Laboratory for Electrochemistry, Paul Scherrer Institute, 5232 Villigen PSI, Switzerland
e-mail: felix buechi@psi.ch

F.N. Büchi et al. (eds.), *Polymer Electrolyte Fuel Cell Durability*,

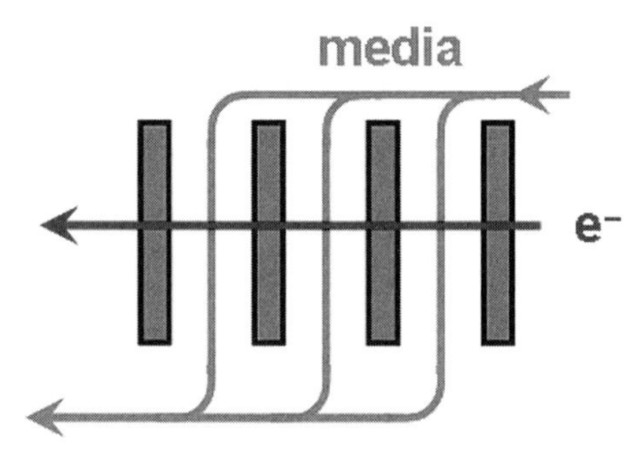

Fig. 1 Classic bipolar arrangement in fuel cell stacks. Cells are arranged electrically in series and in parallel for the reactant gas and coolant streams

These differences lead to the fact that in a stack the individual cells carrying the same current never have completely identical cell voltages. Such differences do not only show up during normal operation, they can also affect the rate of degradation of individual cells. This topic of heterogeneous cell ageing in stacks is of importance because it is not the average decay in cell performance that is lifetime-limiting, but the performance of the weakest cell in a stack. At the latest when the weakest cell suffers from a (catastrophic) failure, even if all other cells are performing well, the stack has reached the end of life (EoL), or at least needs to be repaired.

The intrinsic degradation mechanisms of all components have been described in detail in previous chapters. This chapter investigates the possibilities and effects of unequal cell operation and degradation in stacks. The chapter is organized as follows: the heterogeneities between cells are discriminated into systematic and stochastic reasons and the effects apply to the time during operation and during the shutdown/ startup periods. A section is devoted to each of these cases and conclusions are drawn in the final section.

2 Heterogeneous Cell Operation in Stacks

When current is drawn from a stack composed from a number of cells (two to more than 200) the individual cells do not produce an exactly equal voltage. This is due to systematic and stochastic differences of individual components and operating conditions.

The problem of differences between individual cell operation in the passive series connection of stacks is as old as the concept of the bipolar stack (Blair and Dircks 1992), or in general of the series connection of galvanic cells itself. Therefore developers and manufacturers measure the voltage of individual cells (or groups of cells) during operation (Lacy 2001, Webb and Moller-Holst 2001). Although this procedure is a burden with respect to complexity and cost of the fuel cell system, it seems to be necessary to prevent unwanted events with respect to degradation of individual cells. However, in the bipolar arrangement of fuel cell stacks only measures for all the cells can be taken when an unwanted state is observed in a single cell because in general there is no external control handle on individual cell operation.

The deviations of single cell voltages from the average value can have systematic or stochastic causes. Four general cases of the combinations of stochastic and systematic deviations and the periods of operation and start/stop are discussed in detail in the following sections.

2.1 Systematic Differences Under Operation

Systematic deviations during operation stem from the location of the cell in the stack (i.e., near inlets of air, fuel, or coolant or far from these). Because the media are fed in parallel to the cells and the cross sections of the manifolds are of limited size mainly differences in pressure drop across the cells lead to differences in reactant and coolant flow through cells in different locations.

A thorough analysis of these effects has been given by Chang et al. (2006). It shows that not only the fluid dynamics of the reactants and coolant in the manifold can have a distinct influence on the cell homogeneity, but also the resistivity of the bus plates can lead to systematic current density distribution and cell voltage differences of individual cells.

Depending on the location of the cells in the stack, and on the geometry and arrangement of the manifolds, cells are fed with different mass flows of reactant and coolant. The fluid dynamics in the manifolds not only creates differences in the reactant flow through individual cells under steady-state conditions, but is also a critical issue when high dynamics of the applied load are applied to the stack, as is required, for example, for automotive application. Cells at the far end of the gas inlets can suffer from gas starvation, leading to systematic accelerated catalyst degradation (see Part II Chapter 5.1) in these cells.

The individual performance of the cells, however, cannot only be influenced by their position in the stack, but also by differences of the electrochemical components. Owing to manufacturing tolerances, such as slight thickness or density variations of gas-diffusion layers, catalyst layers, or the membrane, also cells in the center of the stack can have slightly different electrochemical behavior, even when being operated under perfectly identical conditions with respect to mass flow of reactants and temperature.

All these factors can lead to small systematic differences in the degradation rate of individual cells, being prescribed by either their position in the stack or the differences given by tolerances in the electrochemical components.

Figure 2 shows the cell voltage distribution of a 240-cell stack at 0.65 A/cm^2 at the beginning of life (BoL) and at the EoL. At the BoL, the cell voltages show a small scatter, but are constant along the length of the stack (Fig. 2, top). At the EoL, the voltages have degraded from an average cell voltage at the BoL of 0.769 ± 0.005 V to 0.641 ± 0.052 V. The scatter between the individual cells has increased

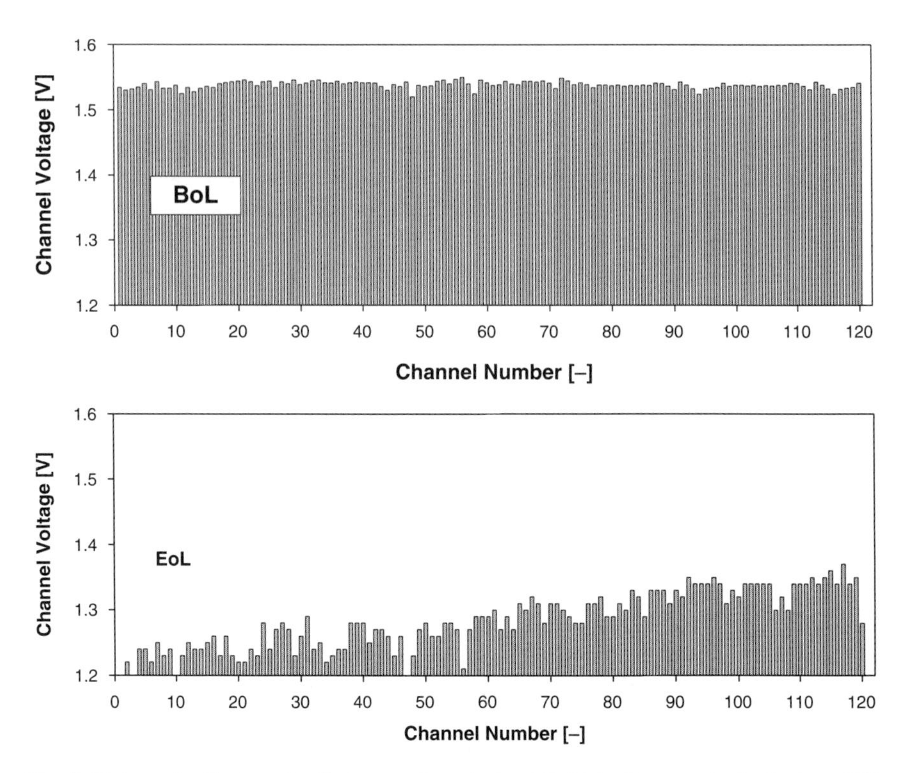

Fig. 2 Voltage distribution of cells in a stack under load (0.5 A/cm^2) top at the beginning of life (*BoL; top*) and at the end of life (*EoL; bottom*) of stack operation. The cell voltage is shown as the sum of the voltages of two adjacent cells. Gas inlets are on the left side at cell/channel number 0

and a distinct slope (0.6 mV/cell) of the cell voltages towards the end of the stack (gas inlets at cell 1) is observed (Fig. 2, bottom).

When investigating the reason for this type of systematic degradation also the metal contamination of the membrane electrode assemblies (MEAs) at different stack locations (entrance, middle, end) was analyzed. It was found that not only a gradient along the flow channels within each cell had evolved for the contaminants lead and zinc (Fig. 3, top), but that a gradient had also evolved along the stack, when the same locations in the individual cells were compared (Fig. 3, bottom). A welding spot in the gas supply system was identified as the source of the metal contamination. The higher loading of contaminants at the front end of the stack corroborates with the voltage losses observed in this section.

The gas streams can carry water droplets from the humidification process. These droplets, which are able to carry contaminants much more easily than the gas phase, have a different distribution pattern in the manifold owing to the influence of gravity and varying gas speed along the length of the manifold. Commonly it is expected that the bigger the droplets the shorter the travel path within the manifold. Therefore, if the droplets carry contaminants, cells close to the inlet will suffer from a higher contaminant load, leading to a higher degradation rate as observed in the plot at the bottom of Fig. 2 for the cells closer to the gas inlet.

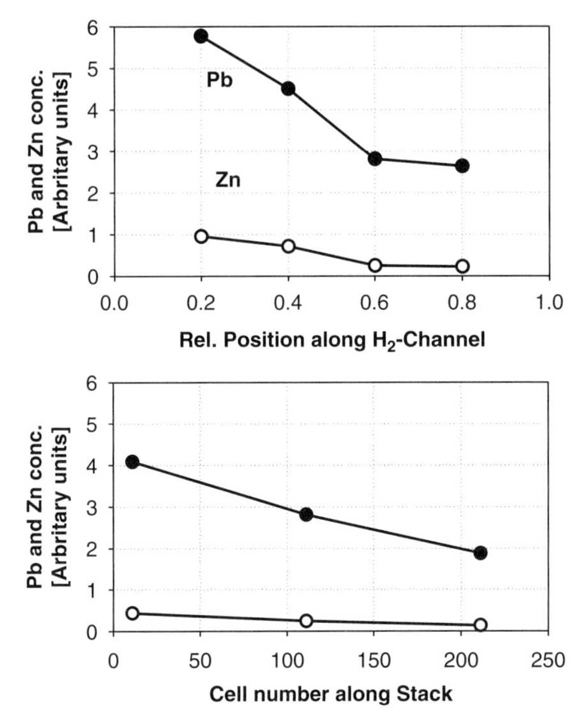

Fig. 3 *Top*: Concentration of zinc and lead in aqueous extractions from membrane electrode assembly (MEA) samples measured by inductively coupled plasma mass spectrometry, obtained from different locations along the hydrogen channel from an MEA of a cell at the center of the stack. *Bottom*: Concentration of zinc and lead in MEAs (at the same position along the gas channel) of different cells along the stack. Gas inlets are at cell number 0

End cells in a stack can have a distinctly different thermal behavior owing to higher heat losses than the center cells. Lower temperature can lead to enhanced condensation of water and thus change the percolation resistance of the gas streams in an unfavorable manner. Inspection of the plot at the bottom of Fig. 2 shows that indeed the cells at the far end of the stack have a considerably lower voltage (the voltage of the last channel is 74 mV lower than the average voltage of the preceding five channels), also indicating accelerated degradation.

The results in this section show that a systematic degradation of cells in a stack related to their position can have various causes influenced by the fluid dynamics in the manifolds or the thermal behavior.

2.2 *Systematic Differences at Start/Stop*

At start and stop systematic differences along the stack with respect to fluid dynamics and temperature can occur. While the fluid dynamics issues seem to be controllable during the stop and start phases, the development of a thermal profile along the stack after shutdown is more difficult to prevent if no or limited auxiliary power is available for continued operation of the thermal management. Generally, after shutdown end cells will be subjected to higher heat loss than cells in the center of the stack and thus temperature gradients along the stack will evolve.

A recent study of Bradean et al. (2006) has shown that temperature gradients along the stack at shutdown can be deliberately generated and exploited. The concept takes advantage of the migration of water through the MEA in a temperature gradient. When a gradient is enforced on a cell with dry plate channels and a wet MEA, the water migrates towards the colder plate and the water content of the MEA decreases to a minimum. This effect can be exploited to dry the MEA after shutdown and thus enhance the subfreezing durability.

When a stack is insulated on all sides except for one end, even with natural cooling a systematic temperature gradient along the stack evolves, driving the water of the MEA towards the colder plate. However, with a pristine stack and natural cooling the temperature gradients along the stack are comparatively small. A heat reservoir at the insulated end of the stack can therefore be used to generate well-defined temperature gradients. In this case the water migrates reliably in all cells out of the MEA into the channels of the plate. With this water distribution freeze startup is considerably improved, without the need for auxiliary, external power.

This concept is an example for generating and exploiting systematic temperature differences along the stack to reduce or control degradation.

2.3 Stochastic Differences Under Operation

During operation, cells in different locations in a stack do not only experience systematic difference in operation conditions, but are also affected by stochastic deviations form average conditions.

Stochastic changes of the fuel or air mass flow through individual cells mainly stem from water condensation phenomena in the gas flow path in the cells. Liquid water, despite homogeneous production, is released into the gas channel in an irregular manner with respect to droplet size and location on the active area. And the liquid water is removed from the gas channels in irregular slugs or rivulets (Pekula et al. 2005) leading to fluctuations in the current density distribution and individual cell voltage (Owejan et al. 2007a) caused by a fluctuation of the reactant mass flows. Individual cells thus experience stochastically different operating conditions with respect to mass flow during operation. Especially under dynamic operation protocols this can lead to unexpected and unobserved starvation conditions in stochastic locations in individual cells with temporarily high liquid water content. This in turn will lead to stochastic degradation of cells.

The plot at the bottom of Fig. 2 gives an example also for the stochastic ageing of cells in a stack, as observed by the steady-state cell voltages. The scatter observed between cells at the EoL has considerably increased compared with that at the BoL. In fact the standard deviation increased from ±5 mV per measured channel of two cells to ±52 mV.

When a cell degrades in manner different from its neighbors, its current density distribution will change. As the cells are connected electrically by bipolar plates, not having infinite conductivity, the change of current density distribution in one cell will also affect the current density distribution in neighboring cells (Freunberger et al. 2007, Kim et al. 2005, Santis et al. 2006). The effect of coupling is strongly governed by the geometry of the cell and the conductivity of the bipolar plate (Freunberger et al. 2007). This means that adjacent cells are not operating independently of their neighbors and the coupling opens up the possibility that cells suffer from degradation, i.e., by locally enhanced current density just by having a neighboring cell having an abnormal current density distribution owing to, for example, excess condensation. This subject has been investigated experimentally and modeled on the stack level by Wang et al. (2006). A similar mechanism, with stochastically distributed resistive spots in different cells in a stack and subsequent promotion of accelerated ageing, was proposed by Kulikovsky (2006). However, not only electrical irregularities will affect neighboring cells; this phenomena exists also on the thermal level as shown by Promislow and Wetton (2005).

In conclusion, stochastic operating conditions and electrical coupling between cells in the stack will contribute to different degradation rates of the cells. This again means that if the weakest cell limits the life of the stack, the EoL of the stack will be reached before the average cell performance reaches EoL conditions.

2.4 Stochastic Differences at Start/Stop

The occurrence of stochastic cell voltage deviations under operation is the most frequent case. Under operation, this is predominantly due to changes in percolation resistances for the reactant streams owing to nonhomogeneous condensation of water. Similar reasoning applies to operating differences at start and stop. Especially the automotive application will need frequent stop and start cycles. Even though shutdown will generally occur from well-defined operating conditions and temperatures, depending on the shutdown procedure cells are likely not to experience identical voltage conditions, again mainly owing to stochastic distribution of condensed water. It has been shown that if shutdown or startup procedures are not chosen carefully, severe damage may be induced by reverse currents (Reiser et al. 2005). In a multicell stack, this mechanism will also likely be of random nature.

Severe differences between cells can occur during startup of a stack from frozen conditions owing to inhomogeneous formation of ice in the cells and an inhomogeneous increase in temperature (Reiser 2004). When individual cells experience inverted voltage conditions they will suffer more rapidly from ageing than those cells not exposed to negative voltages.

In summary, stop and start procedures of the fuel cell system need to be designed carefully so as not to induce random differences in cells which can contribute to accelerated degradation.

3 Conclusions

Individual cells in stacks, even though carrying the same current, have different cell voltages. This is due to systematic and stochastic materials, tolerances, mass transport, and temperature differences which consequently also lead to differences in current density distribution in different cells.

The differences in local current density affect not only operation, but can also influence local ageing processes. Because the EoL of a fuel cell stack has to be defined by the weakest cell in the stack, keeping the rates of degradation as homogeneous as possible will increase the total life of the stack. Therefore, design and engineering need to carefully take into account the factors for heterogeneous cell operation to minimize deviations between cells. With given materials and degradation processes, this is a necessary means for obtaining the maximum life of a fuel cell stack.

While operating differences along the stack, be they stochastically or systematically, are generally regarded as unfavorable or damaging, it has been shown that under shutdown conditions situations exist where a gradient along the stack can be beneficial and can be exploited to control or reduce degradation.

Acknowledgements The results in Sect. 2.1 were obtained in a common project of the Paul Scherrer Institute and Michelin Recherche et Technique SA, Givisiez, Switzerland. The author would like to thank in particular A. Delfino and G. Paganelli from Michelin.

References

Blair, J.D., and Dircks, K. (1992) Method and apparatus for monitoring fuel cell performance, *US Patent* 5,170,124

Bradean, R., Haas, H., Eggen, K., Richards, C., and Vrba, T. (2006) Stack Models and Designs for Improving Fuel Cell Startup from Freezing Temperatures, *ECS Transactions*, 3, 1159–1168

Chang, P.A.C., St-Pierre, J., Stumper, J., and Wetton, B. (2006) Flow distribution in proton exchange membrane fuel cell stacks, *J. Power Sources*, 162, 340–355

Freunberger, S.A., Schneider, I.A., Sui, P.-C., Wokaun, A., Djilali, N., and Büchi, F.N. (2007) Cell Interaction Phenomena in Polymer Electrolyte Fuel Cell Stacks, *J. Electrochem. Soc.*, submitted

Heinzel, A., Nolte, R., Ledjeff-Hey, K., and Zedda, M. (1998) Membrane fuel cells – concepts and system design, *Electrochim. Acta*, 43, 3817–3820

http://www.angstrompower.com

Jiang, R., and Chu, D. (2001) Stack design and performance of polymer electrolyte membrane fuel cells, *J. Power Sources*, 93, 25–31

Kim, G.S., St-Pierre, J., Promislow, K., and Wetton, B. (2005) Electrical coupling in proton exchange membrane fuel cell stacks, *J. Power Sources*, 152, 210–217

Kulikovsky, A.A. (2006) Electrostatic broadening of current-free spots in a fuel cell stack: The mechanism of stack aging?, *Electrochem. Comm.*, 8, 1225–1228

Lacy, R.A. (2001) Measuring cell voltages of a fuel cell stack, US Patent 6,313,750

Owejan, J.P., Trabold, T.A., Gagliardo, J.J., Jacobson, D.L., Carter, R.N., Hussey, D.S., and Arif, M. (2007a) Voltage instability in a simulated fuel cell stack correlated to cathode water accumulation, *J. Power Sources*, 171, 626–633

Owejan, J.P., Trabold, T.A., Jacobson, D.L., Arif, M., and Kandlikar, S.G. (2007b) Effects of flow field and diffusion layer properties on water accumulation in a PEM fuel cell, *Int. J. Hydrogen Energy*, 32, 4489–4502

Pekula, N., Heller, K., Chuang, P.A., Turhan, A., Mench, M.M., Brenizer, J.S., and Unlu, K. (2005) Study of water distribution and transport in a polymer electrolyte fuel cell using neutron imaging, *Nucl. Instrum. Methods Phys. Res. Sect. A: Accelerators Spectrometers Detectors Associated Equipment*, 542, 134–141

Promislow, K., and Wetton, B. (2005) A simple, mathematical model of thermal coupling in fuel cell stacks, *J. Power Sources*, 150, 129–135

Reiser, C. (2004) Battery-boosted, rapid startup of frozen fuel cell *United States Patent* 6,777,115

Reiser, C.A., Bregoli, L., Patterson, T.W., Yi, J.S., Yang, J.D., Perry, M., L., and Jarvi, T.D., A Reverse-Current Decay Mechanism for Fuel Cells, *J. Electrochem. Solid-State Lett.*, 8, A273–A276, (2005)

Santis, M., Freunberger, S.A., Papra, M., Wokaun, A., and Büchi, F.N. (2006) Experimental investigation of coupling phenomena in polymer electrolyte fuel cell stacks, *J. Power Sources*, 161, 1076–1083

Wang, G., Ramani, M., and Eldrid, S. (2006) Plate In-Plane Electrical Resistance Impact to Stack Performance, *ECS Transactions*, 3, 1049–1056

Webb, D., and Moller-Holst, S. (2001) Measuring individual cell voltages in fuel cell stacks, *J. Power Sources*, 103, 54–60

Part III
System Perspectives

1. Introduction

Editors

The British say, THE PROOF OF THE PUDDING IS THE EATING. In the fuel cell world this means that it is finally in the application and not in the laboratory where the systems need to demonstrate their durability. The 2010/1015 U.S. DOE lifetime target for automotive applications is 5,000 h, which is equivalent to 150,000 driven miles, and the Japanese NEDO's lifetime targets for stationary applications are 40,000 and 90,000 h at 2010 and 2015, respectively.

Therefore testing under real the world conditions of the most important applications of the automotive and stationary CHP areas is of high importance. Real world operating conditions include start/stop cycles, dynamics operation and load cycling as well as the use of ambient air for the cathode supply including all of it's impurities, or hydrogen obtained from reformation of hydrocarbons where again a number of side products of the reforming reactions pose challenges for the durability of the fuel cell system. In addition to harsh operating conditions, in many applications, the system cannot be tailored to provide favourable operating conditions for durability, such as i.e. fully humidified feeds, slow dynamics or inert-gas purging upon start/stop cycles, due to economic, space or weight boundary conditions. One further important and sometimes under-estimated challenge is the interplay of the different sub-systemes (e.g., reformer, humidifier, stack, tailgas burner, a.s.f.) forming an functioning fuel cell system. On the system level therefore, the control algorithms and feedback loops of the system need also need detailed elaboration to allow for durable system operation.

In Part III, degradation phenomena and mitigation strategies for the systems used in stationary CHP and automotive applications are discussed by some of the most prominent industrial developers in the respective field.

F.N. Büchi et al. (eds.), *Polymer Electrolyte Fuel Cell Durability*,

2
Stationary

Degradation Factors of Polymer Electrolyte Fuel Cells in Residential Cogeneration Systems

Takeshi Tabata, Osamu Yamazaki, Hideki Shintaku, and Yasuharu Oomori

Abstract Characteristics of operation conditions of polymer electrolyte fuel cells (PEFCs) for stationary use compared with automobile use are reviewed in terms of durability, and the degradation factors found in PEFCs for residential cogeneration systems are described on the basis of long-term operation data. It was observed that degradation of the membrane, loss of electrochemical surface area mainly due to sintering of noble metals, decrease in carbon monoxide tolerance, and decrease in gas diffusivity due to loss of hydrophobicity of the catalyst layer are the main degradation factors. It was demonstrated that the degradation of the membrane is greatly suppressed by saturated humidification conditions and that the sintering is limited under the operation conditions of cogeneration systems. Although the decreases in CO tolerance and gas diffusivity are the most important factors for long-term durability, the potential of the durability of the existing PEFCs has also been demonstrated by long-term operation of single cells for more than 50,000 h as well as that of actual cogeneration systems for 18,000 h.

1 Introduction

Polymer electrolyte fuel cells (PEFCs) are expected to be an environmentally friendly new power source not only for automobile but also for stationary applications. In Japan, a cogeneration system which generates power and heat from city gas or oil at a customer site is wide spread mainly for industrial and commercial use because of its high energy saving effect. Recently, a residential cogeneration system based on a gas engine was introduced by Osaka Gas and the number of installations have already exceeded 30,000 units in Japan, but the applicable market is limited because of its low power generation efficiency (22.5%) owing to the stoichiometric

T. Tabata (✉)
Fuel Cell Development Department, Osaka Gas Co. Ltd, 3-4,
Hokko Shiratsu 1-Chome, Konohana-ku, Osaka 554-0041, Japan
e-mail: ttabata @ osakagas.co.jp

F.N. Büchi et al. (eds.), *Polymer Electrolyte Fuel Cell Durability*,

Table 1 Required specification of residential polymer electrolyte fuel cell cogeneration systems

	0.75-kW class	1.0-kW class
Turn down (kW)	0.25–0.75	0.3–1.0
Power generation efficiency (%)[a]	30–35	30–35
Total thermal efficiency (%)[a]	60–80	60–82
Temperature of the storage tank	60°C	
Startup and shutdown	Daily (summer) or continuous (other seasons)	
Load change	Continuous load-following	
Life (h)	40,000–90,000	
System price	US $4,500–5,000 including storage tank and back-up boiler	

[a]Efficiencies are represented in terms of the lower heating value

gas engine used as a power source. Since higher power generation efficiency can be expected for a PEFC even on a small scale (0.7–1 kWe) and its heat-to-power ratio is quite suitable for residential use, a residential cogeneration system has been regarded as a main target of PEFC applications and many Japanese manufacturers have been developing it from the beginning of their development of PEFCs. A typical required specification of a residential PEFC cogeneration system is shown in Table 1 and leading manufacturers have achieved such performance targets (Nishizaki et al. 2005; Kusama 2005) except for durability and manufacturing cost.

The basic construction and materials of a PEFC stack for residential use are not so different from those for automobile use and so the durability issues are basically the same as those for automobiles. However, quite long life (40,000–90,000 h), which cannot be easily verified by actual operation, is required for residential use for economic reasons, which means the durability issue is difficult solve. Although the operation conditions of PEFCs in residential cogeneration systems are generally milder than those in automobile applications, some of them are specifically severe and important for residential use. A comparison of the operation conditions and their influence on the durability of cell stacks for residential and automobile use is summarized in Table 2.

CO tolerance is still an issue for residential PEFC cogeneration systems in which hydrogen-rich gas reformed from city gas or oil is utilized as a fuel for the PEFC. Although the allowable CO concentration for a Pt–Ru anode catalyst, as low as 10 ppm, is stably achieved by a recent fuel processing system (Echigo et al. 2006) and so CO tolerance is not an apparent problem for achievement of the required performance, there are quite a few reports on the long-term durability of CO tolerance. Furthermore, it has not been clarified how the addition of breed air to the fuel which is widely applied for a reformed-gas-fueled PEFC influences the long-term durability of the anode and the membrane.

On the other hand, since a PEFC in residential systems is operated under ambient pressure to maximize power generation efficiency by reducing power loss at pumps and blowers, its performance is more easily influenced by gas diffusivity, and the effect of the suppression from flooding by squeezing water is also weaker than for

Table 2 Comparison of operation conditions and their influence on the durability of the cell stack for residential and automobile use

Item	Residential	Automobile	Issues affecting durability of cell stacks for residential use
Fuel	Reformed gas	Pure H_2	Decrease in CO tolerance of anode due to dealloying of Pt–Ru
			Influence of the addition of the breed air
Temperature	<75°C	>80°C	Lower risks of thermal degradation and insufficient humidification compared with automobile use
Pressure	Ambient	Pressurized	Lower risks of mechanical degradation and cross-leakage than for automobile use
			Higher risk of flooding owing to low pressure drop than for automobile use
			Large voltage drop derived from a decrease in gas diffusivity in the cathode
Startup and shutdown	Less than daily	More than several times per day	Same influence of OCV, high cathode potential, (local) potential cycle, but lower risk owing to lower cycle number than for automobile use
			Freeze–melt cycle is not now considered
Minimum load	30% of nominal	Idling	Lower risks of cathode corrosion and Pt dissolution owing to lower cathode potential than for automobile use
			Lower risk of membrane degradation owing to scarce near-OCV operation
Load change	Small and slow	Large and rapid	Lower risks of mechanical degradation and cross-leakage derived from the changes in temperature and water content than for automobile use
Required life	90,000 h	5,000 h	Accelerated degradation method is required to predict life

OCV Open circuit voltage

a pressurized PEFC. Although such disadvantages have been overcome in terms of initial performance, the durability is far from being verified.

In this chapter, centered on such specific factors for residential application, degradation phenomena observed in long-term operation experiments assuming residential cogeneration systems at Osaka Gas are reviewed and analyzed. In addition, the present status and the potential of the durability of PEFCs for residential use are briefly reported.

2 Degradation Phenomena Observed in Cells for Residential Use and Their Contributing Factors

2.1 Degradation of Membrane Under Unsaturated Humidification

In residential cogeneration systems, waste heat is stored in a storage tank and utilized as hot tap water and for floor heating, etc. To minimize the volume of the storage tank and to use hot water in various appliances, the temperature of the hot water is desired to be as high as possible. Since the waste heat is mainly generated in the cell stack, higher operation temperature of the cell stack is required for this purpose. However, there is an upper limit to the operation temperature assuming a saturated humidification condition since the heat required for humidification also increases with operation temperature. If the cell stack can be operated under unsaturated humidification, this limit is removed. Operation under unsaturated humidification was explored in 2000–2002.

To confirm the durability of PEFCs under unsaturated humidification conditions, membrane electrode assemblies (MEAs) are operated in a single cell module supplying simulated reformed gas (SRG) and air with various relative humidities. Although some MEAs survived after 12,000 h of operation at a relative humidity of 100% (i.e., saturated humidification), eight of eight MEAs at a relative humidity below 85% could not generate power within 3,000–12,000 h (Ibe et al. 2003). The example of the longest operation is shown in Fig. 1. Although the dew points of the supplied air and fuel are lower than the operation temperature of the cell only by several degrees, the cell voltage dropped greatly and fluctuated after 10,000 h. Since the voltage drop was temporarily recovered by decreasing the dew point of the supplied air, it is considered that flooding at the cathode causes the fluctuation and drop of the cell voltage. However, the cell voltage dropped more rapidly even at a lower dew point, and finally the power generation could not be continued. Since the open circuit voltage also became lower during the cell voltage drop under operation and obvious holes were found in the membrane after the experiment, it was clarified that the degradation of the membrane is accelerated by unsaturated humidification. The detailed mechanism of this phenomenon is explained in the chapter "Chemical Degradation of Perfluorinated Sulfonic Acid Membranes." The authors note that

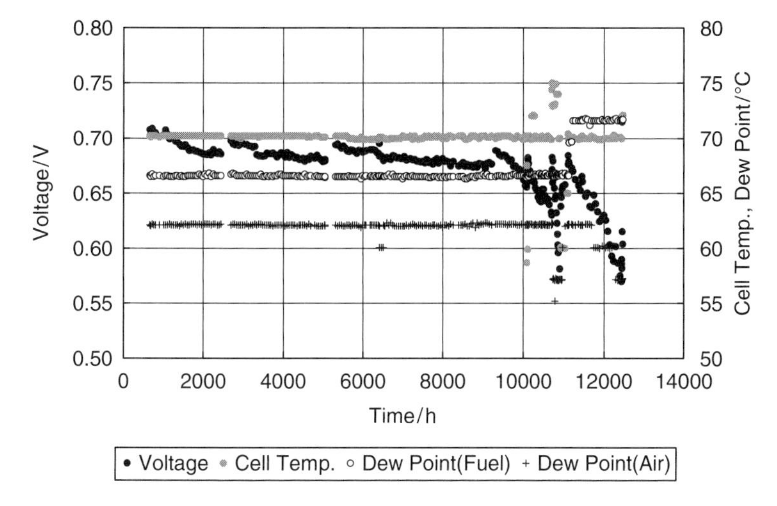

Fig. 1 Durability of a membrane electrode assembly (MEA) under unsaturated humidification conditions. $T = 70°C$, current density 300 mA cm^{-2}, Uf = 60%, Uo = 40%. Anode/cathode, simulated reformed gas (SRG) (20% CO_2, 10 ppm CO, H_2 balance)/air

this result was obtained in 2000–2001 using MEAs available at the time and that stable operation was demonstrated for more than 12,000 h under unsaturated humidification using an MEA with an improved membrane. The current results for the improvement of the membrane which is tolerant against unsaturated humidification are described in the chapter "Improvement of Membrane and Membrane Electrode Assembly Durability." Nonetheless, unsaturated humidification operation should be noted as a potential factor for degradation of long-term durability of a polymer electrolyte.

2.2 Decrease in Effective Surface Area of Electrode Catalysts

The overvoltage of the activation at the cathode, generally considered to contribute the most to the overall overvoltage, increases with operation time owing to a decrease in the effective surface area of the cathode catalyst. Figure 2 shows the change in cell voltage of a single cell under saturated humidification conditions using SRG and pure hydrogen. Although the magnitude of the voltage drop seems to be smaller for pure hydrogen than for SRG, which will be discussed later, significant degradation is observed even for pure hydrogen.

Current–voltage curves of this MEA before and after long-term operation are shown in Fig. 3. The slope in the Tafel region seems to be unchanged and to slide downward by 50 mV. Since the Tafel slope is approximately −87 mV per decade, the voltage drop of 50 mV corresponds to the decrease in effective surface area by approximately 1/3.8. Various factors are considered as a reason for the decrease in effective surface area: sintering of platinum particles, dissolution of platinum, electrical isolation, and isolation from reactants. In Fig. 3, the current density at which the

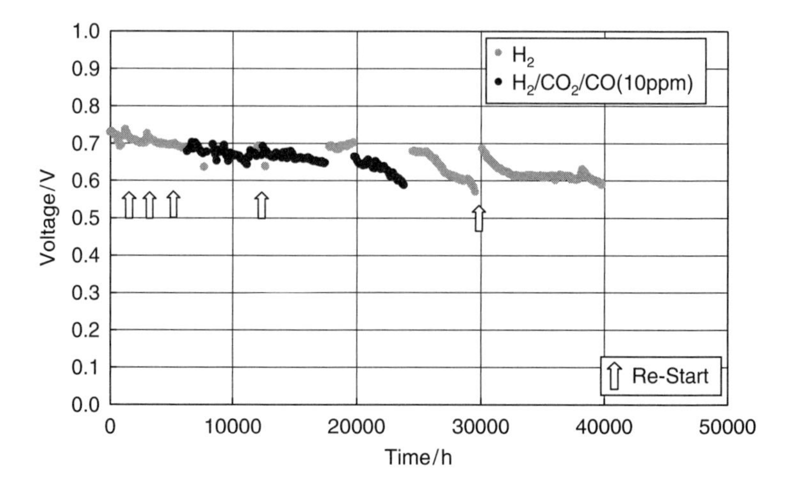

Fig. 2 Durability of a MEA under saturated humidification conditions. $T = 70°C$, current density 300 mA cm^{-2}, relative humidity 100%, Uf = 60%, Uo = 40%. Anode/cathode, SRG (20% CO_2, 10 ppm CO, H_2 balance) or pure H_2/air

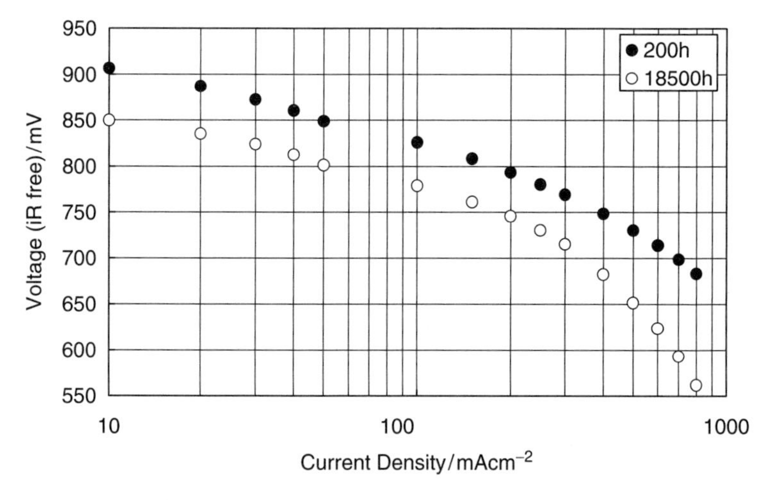

Fig. 3 *I–V* curves of the MEA represented in Fig. 2 before and after long-term operation. The *IR* drop was subtracted. Anode/cathode H_2/air. The other conditions were the same as for Fig. 2, except Uf and Uo were below 50 mA cm^{-2} where the flow rate is constant at 50 mA cm^{-2}

cell voltage drops off the Tafel slope also decreased with operation time. This means that the increase in the overvoltage of the mass transfer also contributes to the cell voltage drop in long-term operation. Although the microscopic explanation for this overvoltage of the mass transfer is uncertain, such mass-transfer limitation will occur if the accessible cathode catalyst is spatially limited even at low current density. For instance, if half of the geometrical cathode area is immersed in water where the cathode catalyst is never utilized, the essential current density is double the apparent current density. In such a case, the current–voltage curve will slide left by half the current density, and as a result, the Tafel slope will slide downward, corresponding to half of the effective surface area of the cathode catalyst.

To investigate the factors contributing to the decrease in effective surface area of the cathode catalyst, AC impedance analysis was also carried out during the operation represented in Fig. 2. The change in the capacity of the cathode activation, which corresponds to electrochemical surface area (ECSA) of the cathode catalyst at the operation current density, is shown in Fig. 4. The ECSA from the AC impedance analysis decreased by approximately a half throughout the operation from 200 to 18,500 h. This change is smaller than the value obtained from the current–voltage curve. This is because the ECSA from the AC impedance analysis is considered to contain the surface area of the cathode catalyst which is not effectively utilized for power generation owing to a mass-transfer problem such as flooding but which is electrochemically active.

Furthermore, to clarify the reason for the decrease in ECSA, transmission electron microscopy analyses were carried out for the MEA samples (the same specification and operated under conditions similar to those for Fig. 2), and the results are summarized in Fig. 5. The tendency for the particle size of the cathode catalyst to increase agrees well with the tendency for the ECSA to decrease as shown in Fig. 4. The extent of the decrease in the surface area calculated from Fig. 5 is approximately 1/1.8 after 17,000 h of operation. Considering that the operation conditions of the samples represented in Fig. 5 were not exactly the same as those for the samples represented in Fig. 4 in addition to experimental errors, it may be concluded that the decrease in the ECSA of the cathode is mainly due to the increase in the catalyst particle size of the cathode catalyst, i.e., sintering.

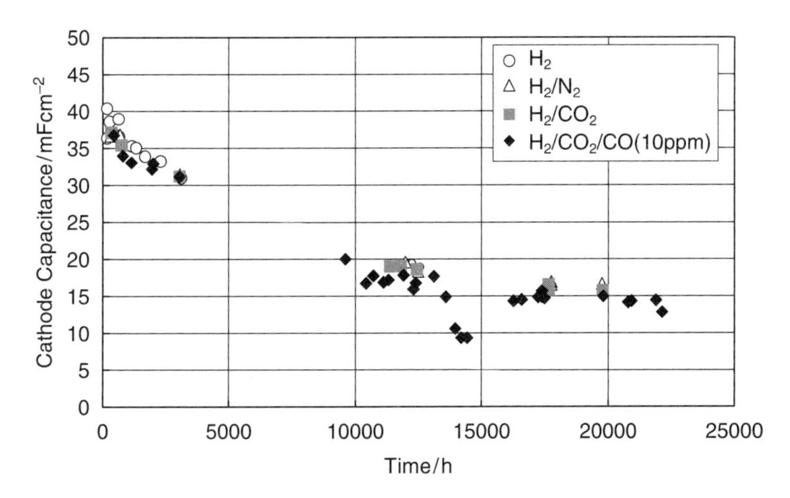

Fig. 4 The change in the capacitance of the cathode derived from AC impedance analyses during the operation represented in Fig. 2. Anode/cathode, pure H_2, H_2–N_2(20%), H_2–CO_2(20%), or SRG (20% CO_2, 10 ppm CO, H_2 balance)/air. The other experimental conditions are the same as for Fig. 2

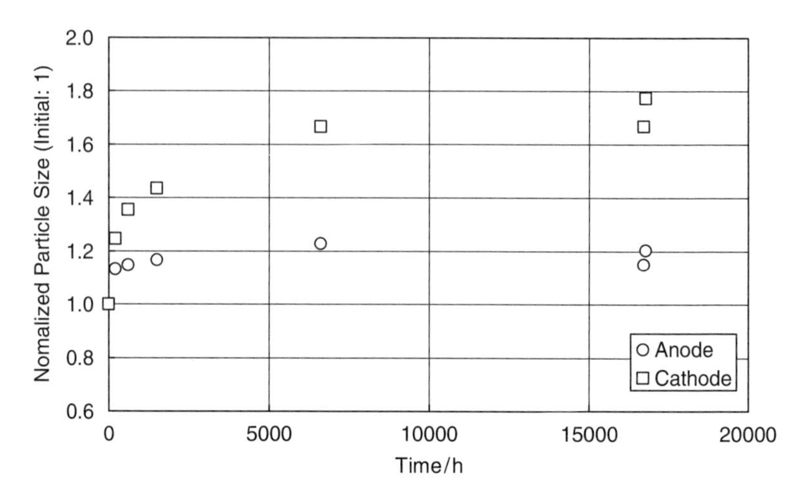

Fig. 5 The changes after long-term operation in the catalyst particle size measured by transmission electron microscopy in MEAs of the same specification as in Fig. 2. The operation conditions are similar to those for Fig. 2

Although migrated platinum particles were sometimes observed in the membrane as reported for automobile operation (Ferreira et al. 2005), the quantity is limited for continuous operation under the conditions of residential application. Therefore, dissolution and migration of platinum is not considered to be a major issue for residential use.

Figures 4 and 5 also show that the surface area of the cathode catalyst largely decreased before 10,000 h of operation and was saturated afterwards. Such tendency is similar to that for general nonelectrochemical noble metal catalysts, and therefore an empirical method to predict the activity after degradation for a general platinum catalyst might be applicable to the cathode catalyst assuming continuous operation.

On the other hand, the sintering of anode catalyst is quite limited from Fig. 5. In fact, the impact of the effective surface area of the anode catalyst on the cell performance has only been observed in combination with the CO tolerance discussed later.

2.3 Decrease in Gas Diffusivity

Since liquid water evolved in the gas channel as a product of the reaction, in a PEFC there is always a risk of inhibition of gas diffusion by the water. Especially, the cell stack for residential use must be designed to reduce the pressure drop; its flow field is carefully designed to avoid flooding. Nonetheless, flooding has sometimes been observed after several thousand hours of operation, which is considered to be caused by the decrease in hydrophobicity of the surface of the path of gas (from the catalyst layer to the separator) due to degradation.

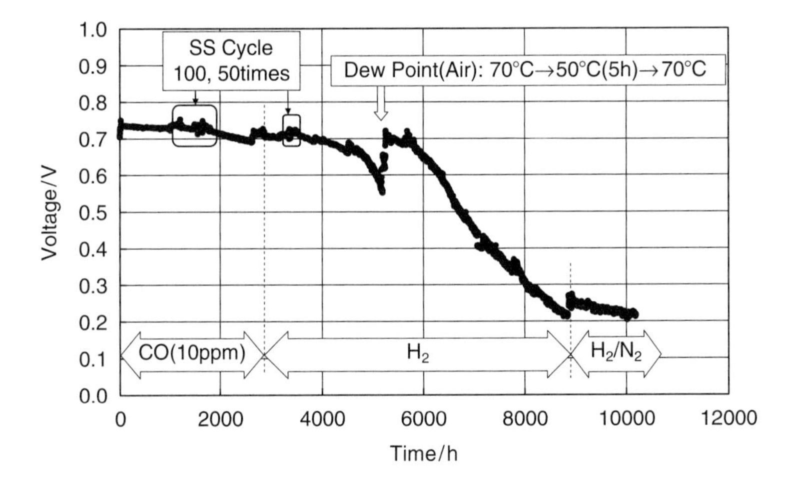

Fig. 6 Durability of a different type of MEA under saturated humidification conditions. The experimental conditions are the same as for Fig. 2

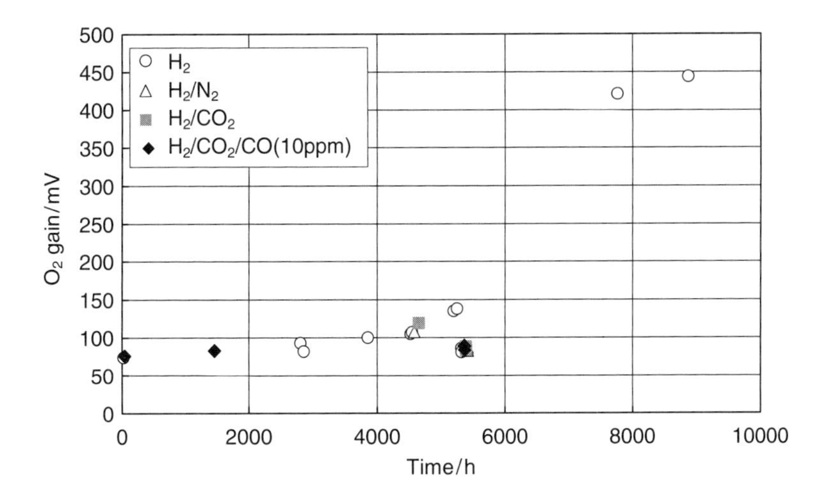

Fig. 7 Change in O_2 gain during the test represented in Fig. 6. O_2 gain is defined as the increase in the voltage by switching the cathode gas from air to pure O_2

Figure 6 shows a typical example in which the cell voltage decreased after many hours of operation owing to the flooding in the cathode. Although the cell voltage began dropping at 4,000 h, it was completely recovered by dry treatment: lowering the dew point of the inlet air. However, the voltage decrease accelerated after the dew point was reset to 70°C. Such behavior clearly indicates that the voltage drop is caused by the flooding. Since the anode gas is pure H_2, the flooding is not affected by the fuel gas composition; this is discussed in Sect. 2.4.

It is more clearly shown in Fig. 7 that the flooding occurred at the cathode side. The change in O_2 gain, defined as the increase in the cell voltage by switching the cathode gas from air to pure O_2, during the operation shown in Fig. 6 is plotted in Fig. 7. The increase in O_2 gain in Fig. 7 corresponds to the voltage drop in Fig. 6, and so most of the voltage loss can be accounted for by the diffusion overvoltage at the cathode.

The overvoltage by flooding originates in the concentration of the current into the area where the active area is not macroscopically immersed in water. The intraplanar distribution of current density is considered to be gradually changed with the accumulation of water even if there is no change in hydrophobicity of the materials. For example, in Fig. 6, after the dry treatment, it takes more than 1,000h to reach the voltage just before the treatment. Thus, there is ground to discuss whether the material degradation occurred during the decay of the voltage around 6,000–9,000h in Fig. 6, but there is a possibility that some trace change affected the current distribution.

On the basis of these considerations, the apparent "accelerated" characteristics of the flooding should not be considered as accelerated degradation even when it takes several thousand hours but as the nonlinear properties of the overvoltage by the decrease in the limiting current density. However, since such a voltage drop due to flooding as in Fig. 6 is not necessarily observed within a couple of tens of thousands of hours, it is likely that some degradation factors accelerate flooding. Recently, it was reported that the carbon support is electrochemically oxidized during the potential cycle under startup and shutdown and it leads to the increase in the overvoltage of gas diffusion at the cathode (Chizawa et al. 2006). In the experiment represented in Fig. 6, startup and shutdown were repeated at an early stage, and so this may cause the degradation of hydrophobicity.

Anyhow, to guarantee the durability of the MEA, it is necessary to declare that such a voltage decay as in Fig. 6 does not appear within 40,000 or 90,000h. However, it is not easy to define the steady-state voltage when the flooding occurs, because it may take more than several thousand hours to reach the steady state; therefore, it is considered that evaluation methods as well as a theoretical approach based on phenomenological modeling must be developed to meet the purpose.

2.4 Decrease in CO Tolerance

As mentioned already, the voltage difference between pure H_2 and SRG increased after 10,000h in Fig. 2, where the breed air was not added. The difference is essentially due to the presence of CO from the results of the experiments varying the gas composition of the anode gas (Yamazaki et al. 2007a), i.e., degradation of CO tolerance occurred. As an index of CO tolerance, H_2 gain, defined as the voltage increase by switching the anode gas from SRG (containing 10ppm CO) to pure H_2, was measured during the operation represented in Fig. 2, and is shown in Fig. 8. It is clear that the H_2 gain accelerated after 10,000h, and that most of the voltage drop in the case of SRG can be explained by the increase in H_2 gain due to the degradation of CO tolerance. As a result, the MEA of Fig. 2 could not be operated using SRG after 25,000h without the addition of breed air. Therefore, the CO tolerance is one of the most important factors to achieve long-term durability, especially because it also shows "accelerated" characteristics like the gas diffusion problem explained earlier.

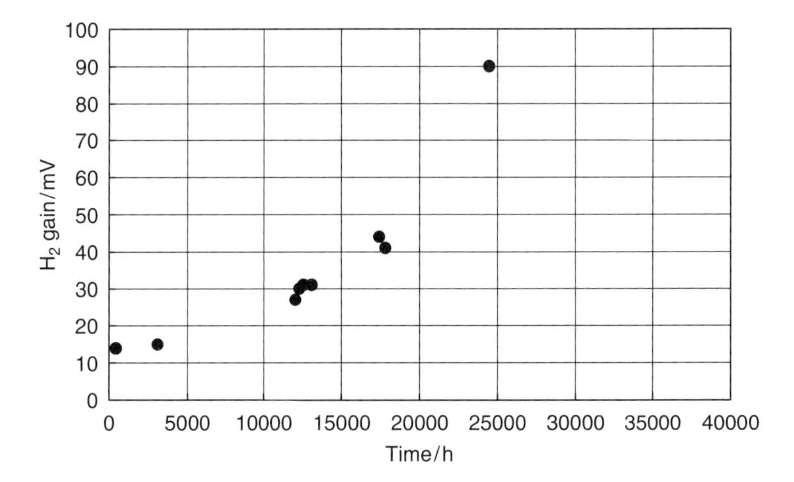

Fig. 8 Change in H_2 gain during the test of Fig. 2. H_2 gain is defined as the increase in the voltage by switching the anode gas from SRG (containing 10 ppm CO and 20% CO_2) to pure H_2

The physical nature of the degradation of CO tolerance is reasonably considered to be related to the state of ruthenium which is added and alloyed with platinum to endow the anode with the CO tolerance. In fact, the amount of ruthenium decreased in the MEA which had the same specification as that for Fig. 2 and more or less lost its CO tolerance after several thousand hours of operation. However, it was also observed that the ruthenium content was almost unchanged though the CO tolerance was clearly degraded. It has been reported that the physicochemical properties of the surface of the anode catalyst vary much and depend on the initial ruthenium to platinum ratio, the degree of alloying, and so on. Therefore, it may not be appropriate to relate the decrease in CO tolerance directly to the decrease in the amount of ruthenium. Nonetheless, the change in the state of ruthenium in Pt–Ru particles is considered to cause the degradation of CO tolerance since the degraded MEA shows even better CO tolerance than the MEA of the pure platinum anode. Detailed studies are required to clarify the nature of the state of ruthenium which shows CO tolerance and its degradation mechanism.

On the other hand, the "accelerated" characteristics of CO tolerance have not been explained reasonably well. The authors recently proposed a model in which limiting current exists near operation current density in the presence of CO and it is determined by the rate of H_2 adsorption onto the trace amount of empty site of surface platinum uncovered by CO (Tabata et al. 2007). According to the model, the limiting current density i_{max} can be expressed by

$$i_{max} = \alpha(T)\frac{p_{H_2}}{p_{CO}}\exp\left(-\frac{E_{CO}}{kT}\right)L_{EC}, \qquad (1)$$

where p_{H2}, p_{CO}, E_{CO}, and L_{EC} are partial pressures of H_2 and CO, the adsorption heat of CO, and the ECSA of the anode, respectively. $\alpha(T)$, the pre-exponential factor and only dependent on temperature, can be predicted by the theory of absolute reaction rate, and it is reported that the predicted i_{max} has a value comparable to that obtained by experiment. By the model, the relation between the current density i and the overvoltage due to CO poisoning η_a can be expressed as

$$\frac{i}{i_{max}} = \frac{1-\exp\left(-\frac{2e\eta_a}{kT}\right)}{\left(1+\sqrt{\beta(T)\exp\left(\frac{E_{H_2}}{kT}\right)p_{H_2}\exp\left(-\frac{2e\eta_a}{kT}\right)}\right)^2}, \quad (2)$$

where E_{H2} and $\beta(T)$ are the adsorption heat of H_2 and another pre-exponential factor, respectively. This formula is essentially the same as for the overvoltage of gas diffusion determined by limiting current. Therefore, a small decrease in the limiting current density may cause a large increase in the overvoltage at a current density near i_{max}, and its nonlinear dependence leads the "accelerated" characteristics as shown in Figs. 2 and 8. According to this model (1), the physical nature of the degradation of CO tolerance is the increase in the adsorption heat of CO, but the decrease in the ECSA of the anode also proportionally decreases i_{max}. In the case of adding breed air, i_{max} is greatly increased by reducing p_{CO}, but the decrease in the activity of CO oxidation becomes also one of the degradation factors. Anyway, further examinations are required to verify the adequacy of this model.

On the other hand, although H_2 gain is a convenient index of CO tolerance, it is reported that H_2 gain sometimes increases while the essential CO tolerance is unchanged (Yamazaki et al. 2007b). Figure 9 shows the voltage trend during the

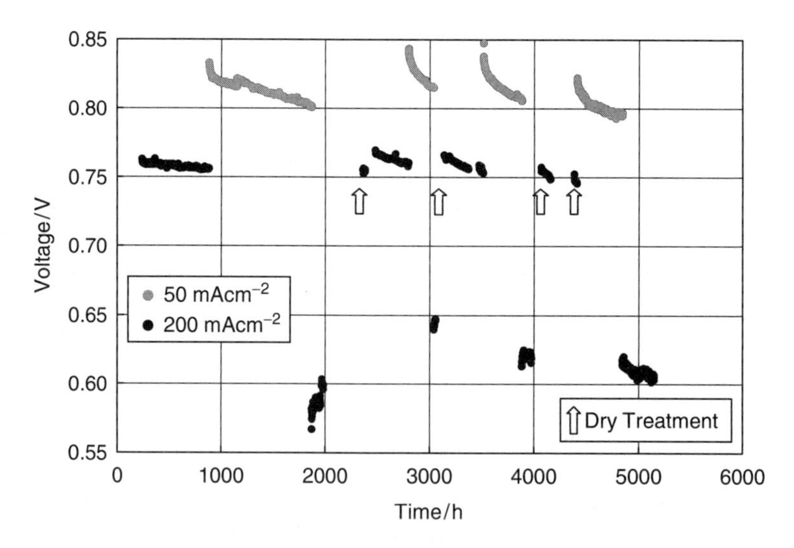

Fig. 9 Durability test of a MEA with periodic operation at low current density. The experimental conditions are the same as for Fig. 2 except for the current density

operation including low current density. By a change in the current density from 200 to 50 mA cm^{-2} (around 900 h), the voltage was lifted but decreased more rapidly than at 200 mA cm^{-2}. After the current density had been reset to 200 mA cm^{-2} (around 1,900 h), the voltage decreased considerably from its value just before the operation at 50 mA cm^{-2} as shown in Fig. 9. At this point, H_2 gain increased by 150 mV, which corresponds to the decrease in the voltage in Fig. 9 (Yamazaki et al. 2007b). However, the voltage loss was almost completely recovered by dry treatment similar to that for Fig. 6 with recovery of H_2 gain. These phenomena were reproduced repeatedly as shown in Fig. 9. On the other hand, O_2 gain was unchanged during the series of the operations, and so the diffusion at the cathode is not considered to affect these phenomena.

If the CO tolerance of the MEA was essentially degraded, it is strange that the voltage was almost fully recovered by the dry treatment. To investigate the reason for the increase in H_2 gain in this case, AC impedance analyses were carried out. The results at the point where H_2 gain increased are shown in Fig. 10. In the experiment, the impedance of the diffusion at the cathode was excluded by using pure O_2. Extra impedance was observed around 1 Hz in the presence of CO, but the impedance was also found in the case of H_2(40%)–N_2(60%) as shown in Fig. 10. Furthermore, the electrochemical impedance spectrum was almost identical to that of fresh MEA after dry treatment in both cases.

On the other hand, AC impedance of the MEA which showed increased H_2 gain after 22,000 h of operation was also measured for comparison, and is shown in Fig. 11. Extra impedance was also found in the presence of CO (more clearly in the case of 100 ppm CO), but the frequency was around 100 Hz, which is clearly different from the extra impedance found in the case of H_2(40%)–N_2(60%) of around 1 Hz.

From these results, the reason for the increase in H_2 gain in Fig. 9 is not considered to be essential degradation of CO tolerance owing to the change in the state of ruthenium, but should be considered to be related to the diffusion matter at the anode. Detailed analyses indicated that the frequency around 1 Hz did not represent an essential physical property but was determined by the experimental configuration, and that the impedance around 1 Hz increases under the condition where the

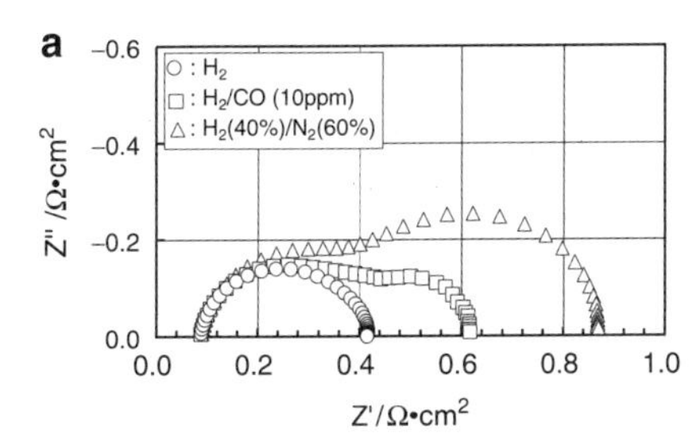

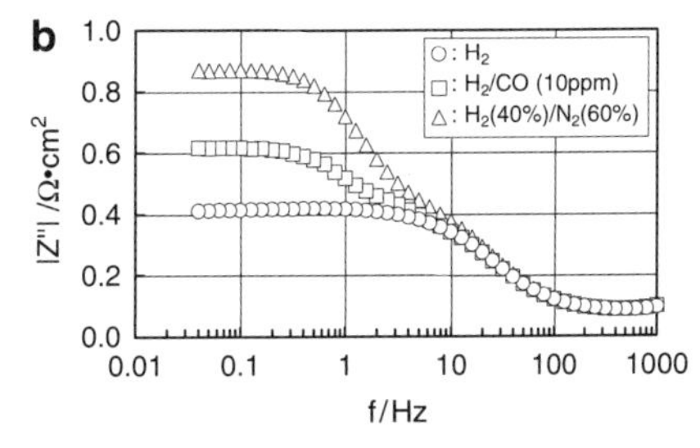

Fig. 10 Cole–Cole plot (**a**) and Bode plot (**b**) of electrochemical impedance spectroscopy (EIS) analyses of the MEA represented in Fig. 9 after low current density operation before dry treatment. Anode/cathode, pure H_2, H_2(40%)–N_2(60%), or H_2–CO (10 ppm)/O_2. Current density, 200 mA cm^{-2}. The other experimental conditions are the same as for Fig. 2

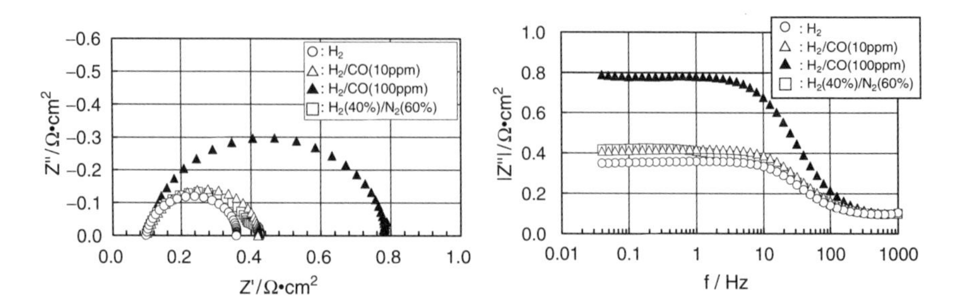

Fig. 11 Cole–Cole plot (**a**) and Bode plot (**b**) of EIS analyses of the MEA after continuous operation for 22,000 h. Anode/cathode, pure H_2, H_2(40%)–N_2(60%), H_2–CO(10 ppm), H_2–CO(100 ppm)/O_2. The other experimental conditions are the same as for Fig. 10

dependence of Uf of the voltage become large (Tabata et al. 2007). Combined with the fact that the voltage drop was almost fully recovered by the dry treatment, it is considered that the flooding at the anode might occur at the stage when H_2 gain increased. From the amplitude of the extra impedance in Fig. 10, the concentration at the surface of the anode catalyst around the outlet of the cell was calculated to be 5% H_2 for the experiment using H_2(40%)–N_2(60%), or 300 ppm CO for the 10-ppm CO experiment (Tabata et al. 2007). In both cases, the calculated concentration shows very different from the concentration at the flow field on the separator estimated to be 21% H_2 or 25 ppm CO, which indicates that there is strong inhibition of gas diffusion at the anode in this case. It has also been suggested that the increased H_2 gain in Fig. 9 was due to the flooding at the anode from the result that such an increase in H_2 gain was not observed using the separator designed to achieve high linear velocity of anode gas to avoid flooding (Yamazaki et al. 2007b).

On the other hand, it is also observed that the increase in H_2 gain is recovered to some extent even for the MEA in which CO tolerance is essentially degraded. This phenomenon may originate in the disproportional degradation of CO tolerance within the plane of the active area. If the area where CO tolerance is not degraded is immersed in water and remains unused during normal operation while the area actually used loses CO tolerance, the dry treatment apparently recovers the degradation of CO tolerance by utilizing the immersed area as the active area. It is very difficult to distinguish whether CO tolerance has been degraded or not after such dry treatment. However, the important fact is the CO tolerance under the condition of normal operation, and it can be distinguished by impedance analyses whether the increase in H_2 gain is due to essential degradation or flooding at the anode as mentioned above.

3 Present Status of the Durability of PEFCs for Residential Use

The major degradation factors of the MEA for the residential PEFC system discussed in the previous sections are based on the results of long-term durability tests. However, during such long-term tests, the improvement of materials used in the

MEAs was performed ceaselessly to prolong their lives. The present status of the durability of PEFCs for residential use is briefly reported here.

Figure 12 shows the result of a long-term durability test of a MEA (single cell). It is clearly demonstrated that the degradation was kept within the allowance (10% of initial voltage) even after 50,000 h using CO_2(20%)–H_2 balance gas without the addition of breed air. Although this MEA was fabricated in 2001, it was shown that the durability over 40,000 h had been already endowed except for CO tolerance at that time. The MEA is still in operation.

An example of the result of the durability test of a PEFC cell stack in an actual cogeneration system for residential use is shown in Fig. 13. No degradation tendency

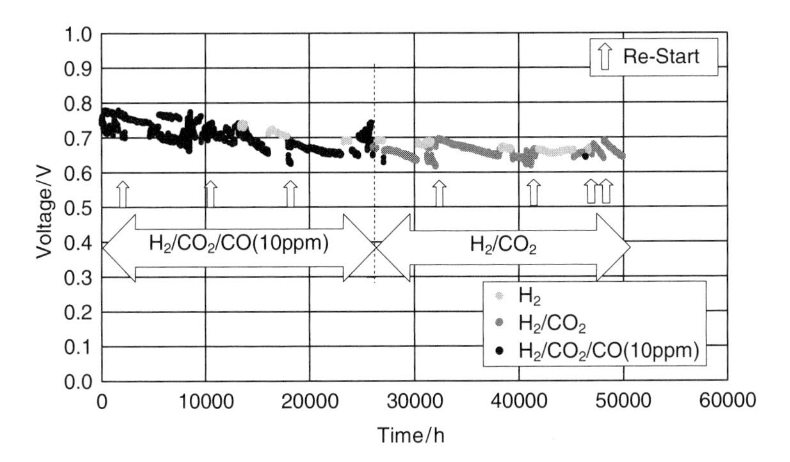

Fig. 12 The change in the voltage of MEA during the long-term durability test. The experimental conditions are the same as for Fig. 2 except for the gas composition of the fuel

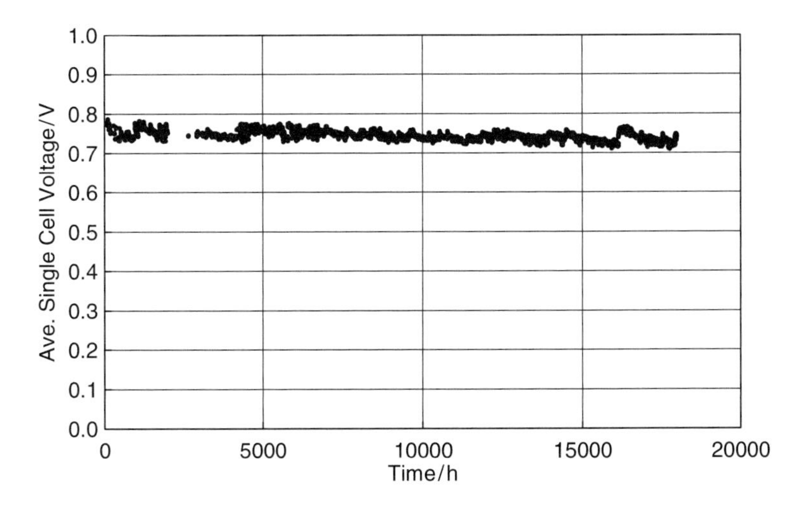

Fig. 13 Change in the voltage of the cell stack operated in an actual 0.75-kW class polymer electrolyte fuel cell cogeneration system

was observed in the voltage trend for approximately 18,000 h, which is considered to be caused by the improvement of the accuracy of system control to keep the required operation conditions of the cell stack to ensure its durability in the system as well as by the improvement of the materials constituting the MEAs.

Standing at the stage just before the commercialization of residential cogeneration systems planned in 2009, there is an optimistic forecast for the durability of the cell stack among the engineers working on this issue. However, to guarantee the life of the cell stack upon commercialization, it is required to predict theoretically that its life is longer than the required number of hours. Especially, the influence of breed air on the durability of the membrane and other materials is still unclear although many manufactures of PEFC systems have employed breed air addition to avoid CO poisoning (Inaba et al. 2008).

To predict the life of cell stacks by establishing methods of accelerated degradation especially concerning gas diffusivity and CO tolerance, the project "Fundamental Research of the Degradation of PEFC Stacks" started in 2004 in Japan, subsidized by the New Energy and Industrial Technology Development Organization (NEDO). In the project, leading cell stack manufacturers, academic authorities on PEFCs, and users experienced in the evaluation of the durability of PEFCs provide their own knowledge on the durability of PEFCs and discuss the mechanism of degradation as well as the method for its evaluation. The test methods of accelerated degradation were proposed in financial year 2007, and it is expected that the results will be applied to the commercial cell stacks to guarantee their life for commercialization.

4 Conclusions

Major degradation factors of the MEA for residential PEFC cogeneration systems were reviewed on the basis of the results of long-term operation data, and are summarized as follows.

Fatal degradation of the membrane occurred within 10,000 h under unsaturated humidification conditions, but it was greatly suppressed by saturated humidification and the MEA can be operated for over 40,000 h.

Although considerable voltage drop is derived from the loss of the ECSA owing to the sintering of the cathode catalyst, the decrease in surface area seems to be saturated before 10,000 h at about half of the initial surface area.

The voltage drop caused by the decrease in gas diffusivity at the cathode owing to flooding is sometimes observed after several thousand hours of operation. Although the voltage drop takes several thousand hours and apparently has "accelerated" character, it is not clear if the voltage drop is caused by material degradation or by reversible change in the current distribution. Nonetheless, there is a mode to accelerate the decrease in hydrophobicity to lead flooding, and so the issue of gas diffusivity is considered to be one of the most important factors for the durability.

The decrease in CO tolerance is usually observed after 10,000 h and also shows “accelerated” characteristics. However, such characteristics are derived from the nonlinear dependence of the overvoltage of the anode on the current density during the decrease in the limiting current caused by CO poisoning. Although the overvoltage caused by the existence of CO is also observed when the flooding occurs at the anode, it can be distinguished by AC impedance analyses whether the overvoltage originates from the essential degradation of CO tolerance or the flooding at the anode.

Although the degradation of gas diffusivity and that of CO tolerance are not completely understood and their countermeasures have not been established, the potential durability of the existing PEFC has been also demonstrated by long-term operation of single cells for more than 50,000 h as well as of actual cogeneration systems for 18,000 h.

Acknowledgment Part of this chapter contains the results of the project “Fundamental Research of Degradation of PEFC Stacks” subsidized by NEDO.

References

Chizawa, H., Ogami, Y., Naka, H., Matsunaga, A., Aoki, N., and Aoki, T. (2006) ECS Trans. 3(1), 645–655.

Echigo, M., Shinke, N., Yasuda, M., and Tabata, T. (2006) in Abstract 2006 Fuel Cell Seminar, Poster Session 3 No. 49.

Ferreira, P.J., la O’, G.J., Shao-Horn, Y., Morgan, D., Makharia, R., Kocha, S., and Gasteiger (2005) J. Electrochem. Soc. 152, 2256.

Ibe, S., Hirai, K., Shinke, N., Yamazaki, O., Higashiguchi, S., Yasuhara, K., Hamabashiri, M., and Tabata, T. (2003) in Abstract 2003 Fuel Cell Seminar, p. 941.

Inaba, M., Sugishita, M., Wada, K., Yamada, H., and Tasaka, A. (2008) J. Power Sources. 178(2), 699.

Kusama, N. (2005) in Abstract 2005 Fuel Cell Seminar, p. 303.

Nishizaki, K., Kawamura, M., Osaka, N., Ito, K., Fujiwara, N., Nishizaka, Y., and Kitazawa, H. (2005) in Abstract 2005 Fuel Cell Seminar, p. 299.

Tabata, T., Yamazaki, O., Oomori, Y., and Shintaku, H. (2007) ECS Trans. 11(1), 279–285.

Yamazaki, O., Oomori, Y., Shintaku, H., and Tabata, T. (2007a) ECS Trans. 11(1), 287–295.

Yamazaki, O., Oomori, Y., Shintaku, H., and Tabata, T. (2007b) ECS Trans. 11(1), 297–308.

3
Automotive

Fuel Cell Stack Durability for Vehicle Application

Shinji Yamamoto, Seiho Sugawara, and Kazuhiko Shinohara

Abstract In recent years, the importance of fuel cell vehicles has been increasing in the North American, European, and Japanese markets amid desires to reduce CO_2 emissions and resolve energy problems. Therefore, improving stack durability has become an increasingly important issue. However, many membrane electrode assembly degradation phenomena occur in the stack under various vehicle operating conditions. This chapter presents an analysis of membrane electrode assembly degradation phenomena and the results obtained with several durability improvement measures.

1 Introduction

Along with a growing awareness of global environmental issues in recent years, the reduction of atmospheric emissions of CO_2 has emerged as a societal demand. In this regard, there are strong desires for greater diffusion of technologies that can reduce environment-impacting substances. Fuel cells have attracted interest as a next-generation technology because they provide higher energy conversion efficiency and have less impact on the environment than existing power supply systems. Vigorous research and development efforts are under way to implement fuel cells in a variety of applications, ranging from large-scale power-generating systems to compact power supplies for portable devices and also as the power source for motor vehicles.

A fuel cell stack that produces electrical energy for driving a traction motor consists of multiple layers comprising a membrane electrode assembly (MEA) and a bipolar separator plate (Fig. 1). The MEA consists of such materials as a

S. Yamamoto(✉)
Fuel Cell Laboratory, Nissan Research Center, Nissan Motor Co. Ltd.,
1 Natsushima-cho, Yokosuka-shi, Kanagawa 237-8523, Japan,
e-mail:yama-shin@mail.nissan.co.jp

F.N. Büchi et al. (eds.), *Polymer Electrolyte Fuel Cell Durability*,

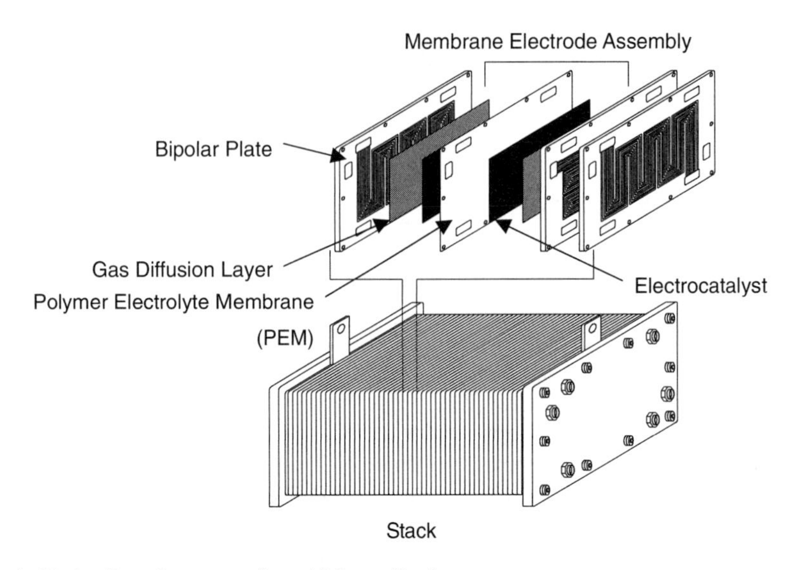

Fig. 1 Fuel cell stack structure for vehicle application

polymer electrolyte membrane (hydrogen ion exchange membrane), electrocatalysts, (platinum supported on carbon), and a gas-diffusion layer made of carbon paper or other materials.

Many different types of phenomena that can lead to MEA degradation occur under the diverse operating conditions of vehicles. For this reason, there are numerous durability issues that differ from those of a similar MEA used in a stationary generating system. In addition, with the present level of fuel cell technology, the volume (weight) of the stack and auxiliary units is too large relative to that of an internal combustion engine and a large quantity of platinum must also be used. Accordingly, there are issues related to cost and supply of the required resources that must also be addressed to facilitate practical use of fuel cells in vehicles. One pressing issue in particular is to improve stack durability, in addition to achieving acceptable MEA cost and performance levels. As one example of some target values, Table 1 shows the technical targets proposed by the US Department of Energy for transportation applications in 2010 (U.S. DOE 2005).

At present, sufficient stack durability has yet to be attained with respect to the specific operating conditions of vehicles. Furthermore, methods for accurately and reliably evaluating durability phenomena have yet to be clearly established. Taking MEA degradation as a typical example, this chapter presents an analysis of degradation phenomena and some proposed measures for mitigating degradation, based on what is generally known at present.

Table 1 US Department of Energy technical targets for transportation applications in 2010

Cost	Membrane electrode assembly	$\$10\,kW^{-1}$
	Electrocatalyst	$\$5\,kW^{-1}$
	Membrane	$\$20\,m^{-2}$
Performance	Operating temperature	≤120°C
	Inlet water vapor partial pressure	≤1.5 kPa(abs)
	Membrane electrode assembly power density (at rated power)	$1{,}000\,mW\ cm^{-2}$
	Catalyst mass activity	$0.44\,A\ mg_{Pt}^{-1}$ at 900 mV iRfree
	Catalyst specific activity	$720\,\mu A\ cm^{-2}$ at 900 mV iRfree
	Membrane area specific resistance	$0.02\,\Omega\ cm^{2}$
Durability	At operating temp of ≤80°C	5,000 h
	At operating temp of >80°C	2,000 h
	Extent of performance (power density) degradation over lifetime	10%
	Electrochemical area loss	<40%

Table 2 Fuel cell operating modes and major types of degradation

Operation mode	Degradation	Cause
Start–stop	Cathode catalyst surface area loss	Catalyst particle agglomeration due to carbon support corrosion
	Catalyst layer water accumulation	Catalyst layer morphology change due to carbon support corrosion
	Membrane pinhole formation	Mechanical stress by hydration/dehydration
Load cycling	Cathode catalyst surface area loss	Catalyst dissolution by potential cycle
	Membrane pinhole formation	Mechanical stress by hydration, pressure and thermal cycle
Idling	Membrane pinhole formation	Chemical decomposition by peroxide attack
	Membrane proton conductivity loss	Chemical decomposition by peroxide attack
	Catalyst activity loss	Poisoning by membrane fragments
High load	Catalyst surface area loss	Catalyst particle ripening at higher temperature
Severe environmental condition	Catalyst activity loss	Poisoning by air/fuel impurities
	Membrane proton conductivity loss	Cation contaminants exchanged with protons
	Gas-diffusion layer gas permeability loss	Gas flow blocked by dust accumulation

2 Degradation Phenomena Under Actual Operating Conditions

Table 2 lists typical fuel cell operating modes under actual vehicle operating conditions along with various types and causes of degradation observed (Iiyama 2007;

Yamamoto and Shinohara 2006). Characteristic fuel cell operating modes corresponding to vehicle operating conditions can be broadly classified as start–stop (stack startup/shutdown), load cycling (acceleration/deceleration while driving), high-load operation (steady high-speed driving), and low-load operation (continued idling). Another important degradation factor is the atmospheric environment of the area where a vehicle is driven. Compared with the operation of existing stationary power-generating equipment, the operation of a fuel cell vehicle is characterized by a much larger number of starts/stops and by frequent and large changes in operating conditions (load fluctuations) while on the road. Such operating modes are thought to be factors that accelerate performance degradation.

Typical examples of degradation phenomena include (1) carbon support oxidation/corrosion, (2) platinum dissolution/agglomeration, (3) chemical degradation of the electrolyte membrane, and (4) three-dimensional structural changes in the electrocatalyst layers, among others. These phenomena do not occur individually, but rather simultaneously and in a compound manner.

Naturally, the durability level demanded of the MEA will vary depending on the stack operating conditions and operating modes. Accordingly, analyses of the factors causing degradation phenomena and estimates of degradation rates are especially important parameters in designing stack durability. That makes it essential to analyze degradation phenomena thoroughly and to devise measures for controlling degradation rates. At present, there are virtually no standard test procedures that have been clearly defined for accurately simulating degradation modes, so such methods must also be developed.

The following sections present the results of an analysis of various degradation phenomena along with some examples of measures for mitigating degradation.

3 Typical Degradation Phenomena and Measures for Mitigating Degradation

3.1 Start–Stop Degradation (Carbon Corrosion)

This section first describes degradation phenomena associated with start–stop operation, which is one of the most typical operating modes of fuel cell vehicles. During stack startup, when hydrogen is introduced into the anode where air is already present, the electrode potential of the portion of the cathode opposite the part of the anode where hydrogen is absent can exceed 1.4 V, causing carbon corrosion. A similar phenomenon is also observed during shutdown (Reiser et al. 2005; Meyers and Darling 2006).

The equations for the oxidation reaction of carbon are described as

$$\mathrm{C} + 2\mathrm{H_2O} \rightarrow \mathrm{CO_2} + 4\mathrm{H^+} + 4e^- \ (0.207\,\mathrm{V})$$

and

$$C + H_2O \rightarrow CO + 2H^+ + 2e^- \ (0.518\,V)$$

These equations indicate that the carbon support undergoes an oxidation reaction, albeit at a very slow rate, in parallel with the oxidation reduction reaction at the cathode even during ordinary power generation. This means that a higher electrode potential promotes greater carbon corrosion. Typical examples include anode carbon corrosion due to fuel (H_2) starvation and cathode carbon corrosion due to start–stop operation, both of which are caused by a sharp increase in electrode potential. In the case of vehicle applications in particular, the frequency of start–stop operation is especially high, which has a huge impact on durability. That makes it extremely important to analyze degradation phenomena carefully and to find suitable countermeasures.

Figure 2 shows scanning electron microscope images of the changes in the cathode catalyst layer before and after a start–stop test. The cathode catalyst layer shows pronounced signs of carbon corrosion following the test. The thickness of the catalyst layer was reduced, and the profiles of the carbon particles are no longer distinct. The results of an electron probe microanalysis of the elemental composition revealed that platinum dissolved and migrated from the catalyst layer into the electrolyte membrane, with the diffusion being driven by the concentration gradient; platinum then recombined with H_2 that crossed over from the anode and was redeposited.

One effective measure for suppressing such carbon corrosion is to use a carbon support with high crystallinity and a low specific surface area. However, that causes a marked drop in power-generation performance owing to coarsening of the platinum particles and their agglomeration. Accordingly, at this point improvement of the carbon material is still not effective in mitigating degradation permanently.

As noted earlier, carbon oxidation is strongly dependent on the electrode potential, which suggests that manipulation of cell operation so as to control the electrode potential at startup would be a very effective way of suppressing carbon corrosion. The envisioned operating method supplies H_2 to the air-filled anode while simultaneously controlling the voltage by forcibly extracting current so as to suppress the rise in the cathode potential; after that air is supplied to the cathode to initiate power generation (Scheffler et al. 2003).

Figure 3 shows the configuration of the testing apparatus for evaluating durability by simulating the start–stop operation of an actual vehicle. It is structured so that air can also be introduced to the anode to simulate the inflow of air to the anode following stack shutdown. A voltage-limiting circuit is also provided. This configuration makes it possible to control the cathode potential at startup and shutdown.

The protocol used with this testing apparatus is shown in Fig. 4. When hydrogen is supplied to the anode at startup, the voltage is controlled so that oxygen is consumed at the cathode, thereby limiting the voltage to a slight rise. After hydrogen has been allowed to flow for 10 s, the voltage limitation is cancelled and

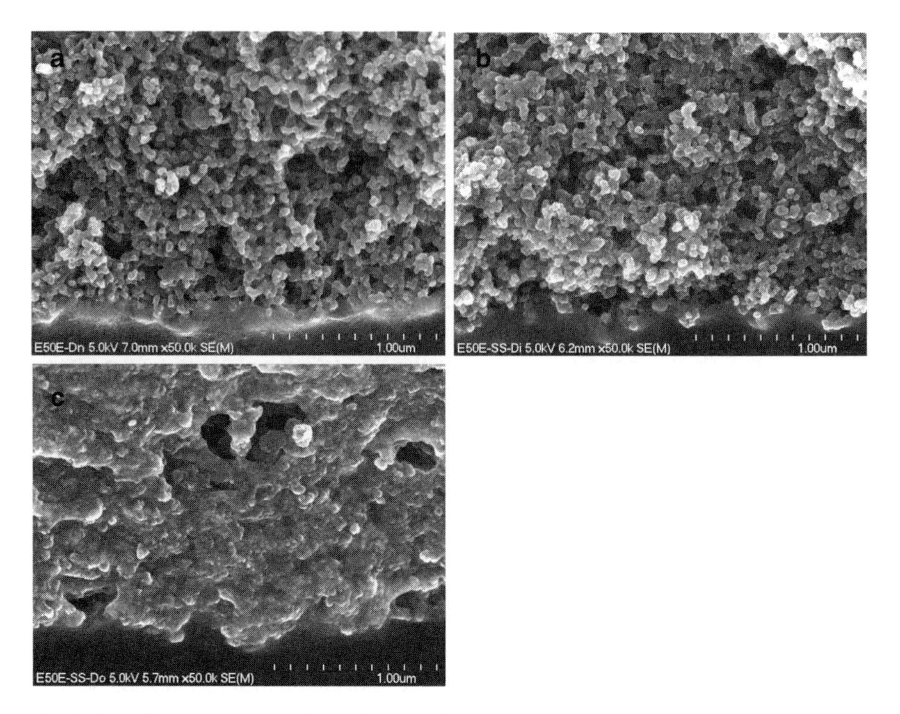

Fig. 2 Scanning electron microscope images of cathode catalyst layer. (**a**) Before start–stop test, (**b**) after start–stop cycle test (inlet), and (**c**) after start–stop cycle test (outlet)

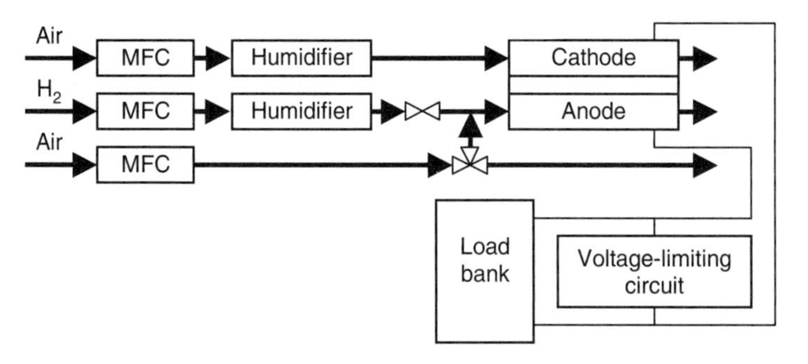

Fig. 3 Testing apparatus for start–stop durability evaluation. Mass Flow Controller (MFC)

simultaneously the load device applies current for 30 s to ramp up the current to the specified setting, after which steady-state power generation is performed for 30 s. The load is then removed and the hydrogen stream to the anode and the oxygen stream to the cathode are stopped. The remaining hydrogen at the anode is simultaneously purged by the air flow. When the voltage is limited, the voltage-limiting circuit is closed at the same timing.

Figure 5 shows the degradation mitigation effect of this voltage manipulation in a start–stop durability test conducted with the testing apparatus and protocol to simulate the above-mentioned degradation phenomenon involving carbon corrosion. When the voltage was not manipulated at startup and shutdown, the voltage drop rate was very large, with the voltage dropping sharply as the number of cycles increased. Even when a carbon support with relatively high crystallinity was used,

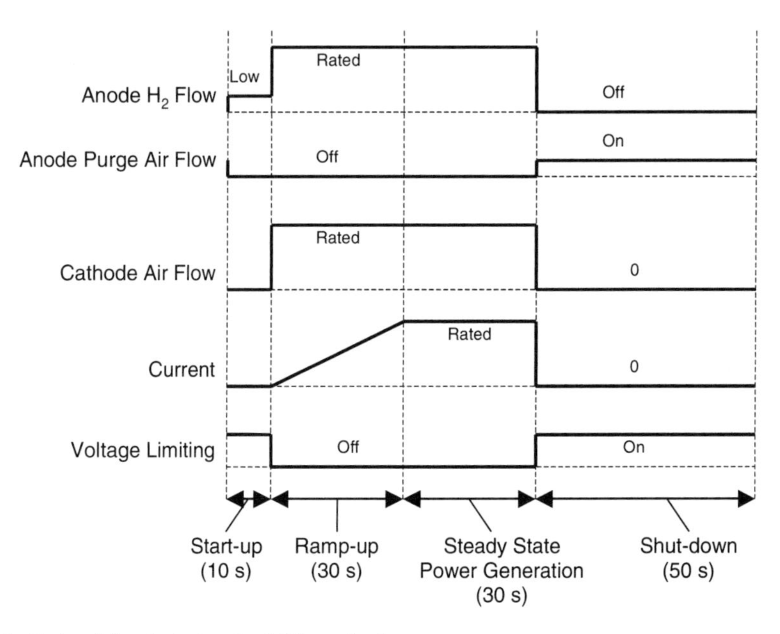

Fig. 4 Protocol for start–stop durability evaluation

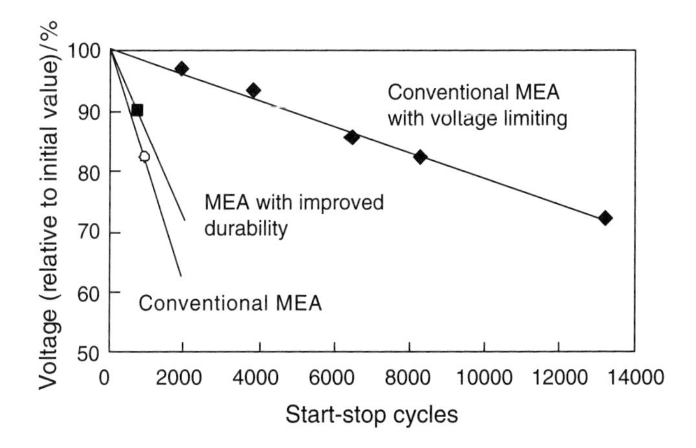

Fig. 5 Degradation mitigation effect of voltage limiting in start–stop cycle test. *MEA* membrane electrode assembly

the effect on improving durability was small. In contrast, when the voltage was manipulated at startup and shutdown, the voltage drop rate was noticeably reduced, thereby confirming that carbon corrosion was amply mitigated.

Figure 6 shows examples of potential cycling test protocols that are used to simplify this test of carbon cathode durability when it is conducted on MEAs. These test protocols do not require the air supply system to the anode or the voltage-limiting circuit shown in Fig. 3. As one example of their application, Fig. 7shows the results that were obtained when a potential-cycling test was conducted under the conditions in Fig. 6a. An analysis of the results indicates that CO_2 was generated by potential cycling and that it was accompanied by a decline in cell performance. This suggests that the test protocol is one effective method of evaluating start–stop degradation.

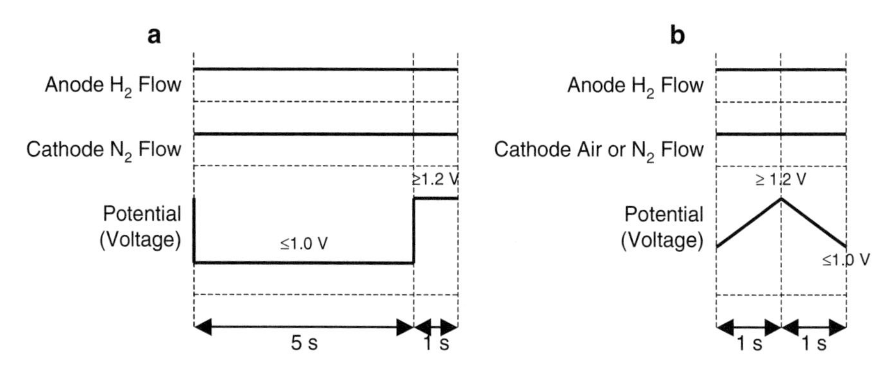

Fig. 6 Examples of potential cycling test protocols: (**a**) rectangular potential wave; (**b**) triangular potential wave

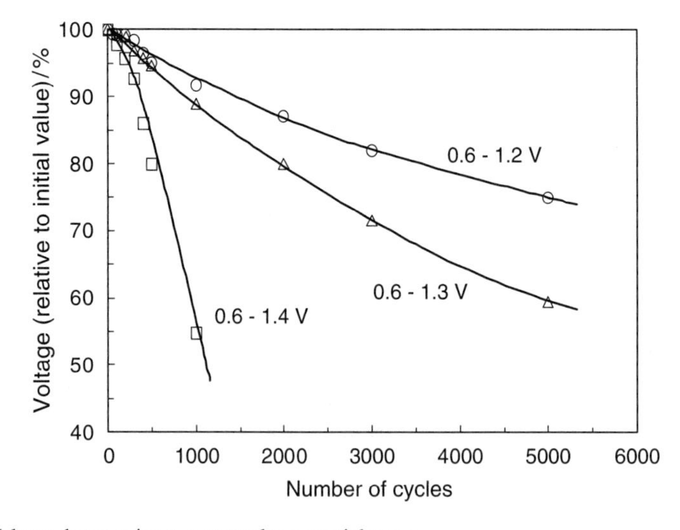

Fig. 7 Voltage decay using a rectangular potential wave

3.2 Load-Cycling Degradation (Platinum Dissolution)

Loss of the electrochemically active surface area (ECSA) of the cathode catalyst has been observed depending on the operating conditions (temperature and humidity) and potential-cycling conditions during vehicle operation. Degradation that occurs during high potential holds (0.9–1.2 V) typical of no-load operation and idling and degradation that takes place during load cycling (0.6–0.9 V) under ordinary power generation are two factors that cause a marked loss of ECSA (Los Alamos National Laboratory 2005).

Although platinum dissolution occurs while holding a high potential during no-load operation and idling, load cycling accompanied by sudden potential changes has a larger influence on platinum dissolution. For that reason, it is essential to suppress platinum dissolution during load cycling.

Figure 8 shows transmission electron microscope (TEM) images of an MEA after a load-cycling test. Platinum dissolution caused a pronounced loss of the cathode catalyst layer, to the extent that platinum particles in the layer cannot be clearly discerned. This TEM observation especially revealed that dissolved platinum ions migrated from the catalyst layer into the electrolyte membrane, with the diffusion being driven by the platinum ion concentration gradient; the platinum ions combined with crossed-over H_2 and platinum was redeposited near the interface between the electrocatalyst layer and the electrolyte membrane. In addition, the platinum particles redeposited in the electrolyte membrane were in the form of agglomerates (ranging in size from several tens to several hundreds of nanometers) that resulted from the coalescing of platinum particles several nanometers in size. Although no distinct carbon corrosion was observed, it can be inferred that the main cause of load-cycling degradation is the loss of ECSA stemming from platinum dissolution and subsequent particle agglomeration. Furthermore, it was also found that the loss of ECSA is strongly dependent on the potential-cycling conditions and

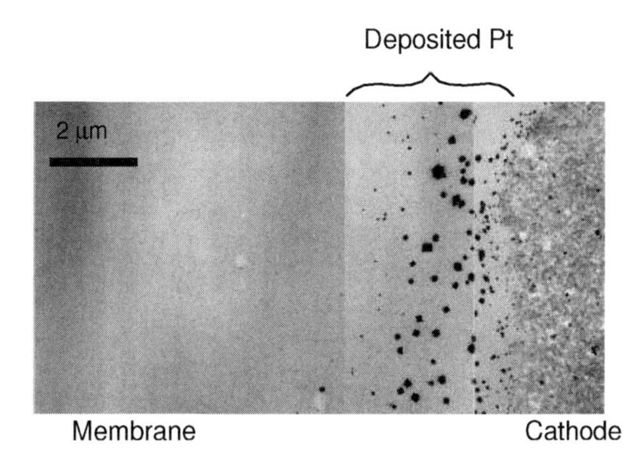

Fig. 8 Transmission electron microscope image of the MEA after the load-cycling test

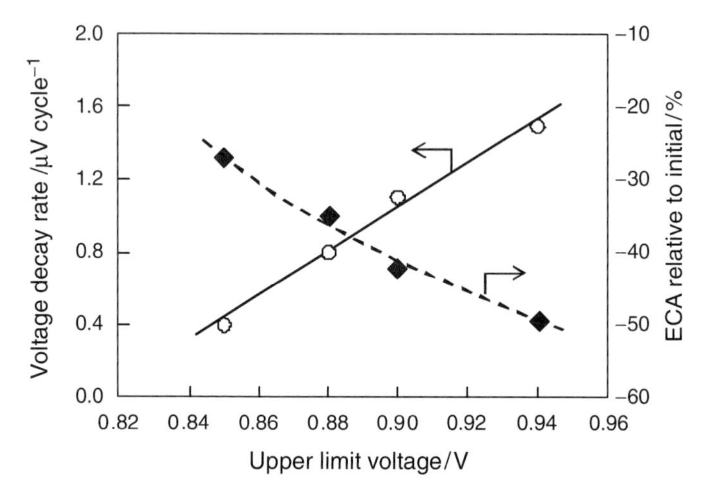

Fig. 9 Load-cycling degradation as a function of upper-limit voltage

that it can be significantly mitigated by controlling the upper-limit voltage and shortening the high-voltage exposure time, among other measures.

Figure 9 shows the effect of controlling the upper-limit voltage on mitigating potential-cycling degradation in a load-cycling test. The figure compares the voltage decay rate for operation without upper limit voltage control (equivalent to 0.94 V) and operation with manipulation of the upper-limit voltage. The lower-limit voltage was set at 0.6 V (equivalent to a rating of 1 A cm^{-2}). Manipulation of the upper-limit voltage eased the degradation rate, and the effect became more pronounced as the upper-limit voltage was further restricted. The results indicate that the degradation rate was suppressed to one-quarter to one-third of that seen for operation without any voltage manipulation.

3.3 *Idling Degradation (Electrolyte Membrane Degradation)*

Because the electrolyte membrane cannot block the reactant gases, small quantities of H_2 diffuse from the anode to the cathode as well as O_2 and N_2 from the cathode to the anode. Additionally, under an idling condition, a high voltage (0.9–1.0 V) resembling that of no-load operation is held for a long time because the extractable load current is virtually 0 A cm^{-2}. The oxygen reduction reactions for this operating mode are shown below:

$$O_2 + 4H^+ + 4e^- \rightarrow 2H_2O \; (1.229\,V)$$

$$O_2 + 2H^+ + 2e^- \rightarrow H_2O_2 \; (0.695\,V)$$

While H_2O is produced by the oxygen reduction reaction at the cathode, H_2O_2 is readily produced at the low-potential anode and diffuses through the electrolyte membrane from the anode to the cathode. Diffusion of H_2O_2 presumably causes oxidation degradation of the electrolyte membrane, or the incursion of cation impurities generates radicals that accelerate chemical degradation (LaConti et al. 2003).

Figure 10 shows TEM images of an MEA following an open-circuit endurance test in which H_2 was supplied to the anode and O_2 to the cathode. The test conditions were a cell temperature of 90 °C, 30% relative humidity, anode atmosphere of H_2, and cathode atmosphere of O_2. Similar to the results of the load-cycling test, it was found that platinum from the cathode catalyst layer dissolved and was redeposited in the electrolyte membrane. Under these test conditions, redeposited platinum particles were observed near the center of the electrolyte membrane. The position of redeposited platinum particles is determined by a balance between the mixed potential of the electrolyte membrane and the partial pressures of the anode H_2 and cathode O_2. It was estimated that platinum particles would be redeposited near the center of the electrolyte membrane under the conditions used in this test (Fig. 11). Chemical degradation of the electrolyte membrane was observed centered on the band of redeposited platinum particles. An analysis was made of the drain water discharged from the MEA during the test and fluoride ions were detected, which suggests that the electrolyte membrane was partially decomposed (Ohma et al. 2007).

Voltage manipulation also has a large effect on suppressing idling degradation, similar to the effect described earlier for load-cycling degradation. It was found that

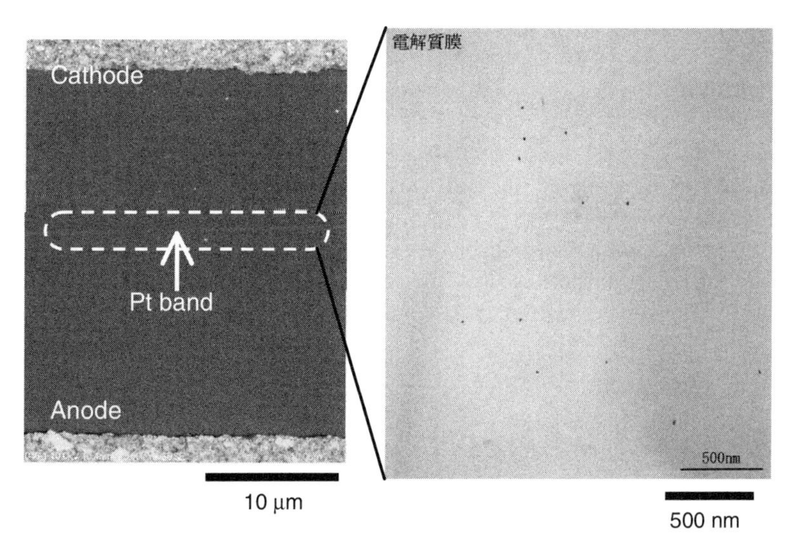

Fig. 10 Load-cycling degradation as a function of upper-limit voltage. (Reproduced with permission from Ohma et al. (2007), copyright 2007, The Electrochemical Society, Inc.)

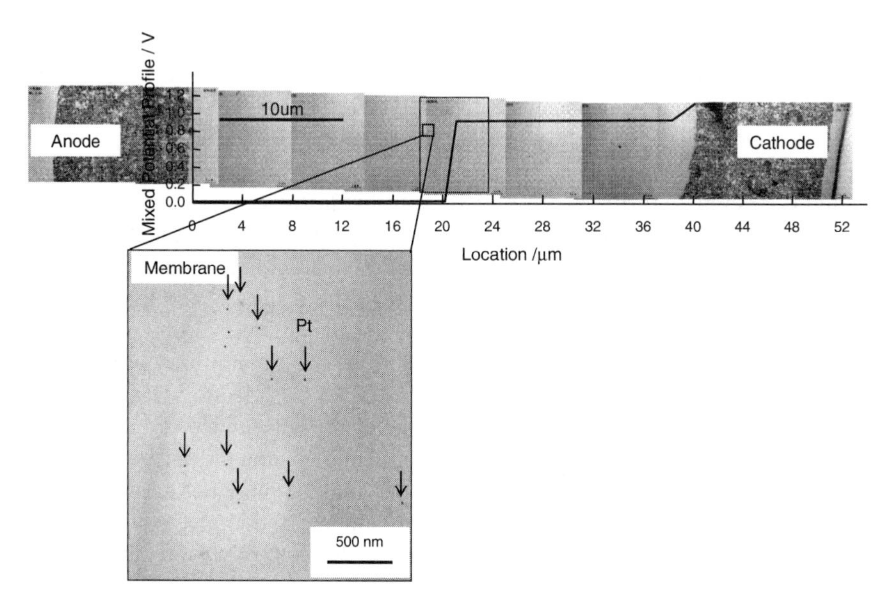

Fig. 11 Hypothesis of the relation between deposited platinum and mixed potential profile. (Reproduced with permission from Ohma et al. (2007), copyright 2007, The Electrochemical Society, Inc.)

controlling the upper-limit voltage of the test cell suppressed the degradation rate to one-fifth to one-quarter of that for operation without any voltage manipulation.

It was also observed that electrolyte membrane degradation was not limited to simply the formation of pinholes in the membrane. The decomposition products poisoned the electrocatalyst layers, causing cell performance to decline. Figure 12 shows the changes in the cyclic voltammograms before and after an open-circuit endurance test. The results clearly indicate a loss of ECSA following the test. The ECSA was then recovered through power generation under constant-current operation. Sulfur was detected in the drain water discharged from the MEA during power generation. Sulfur was probably one product of the decomposition of the solid polymer membrane during the open-circuit endurance test. It is likely that sulfur remained in the membrane after the test and was one factor contributing to catalyst poisoning.

3.4 Cathode Degradation Due To Impurity Contamination

As shown in Table 3, various types of impurities are known to be present in the ambient environment in which vehicles are operated (Ministry of the Environment, Japan 2004). Naturally, filters and other devices are used to prevent such impurities

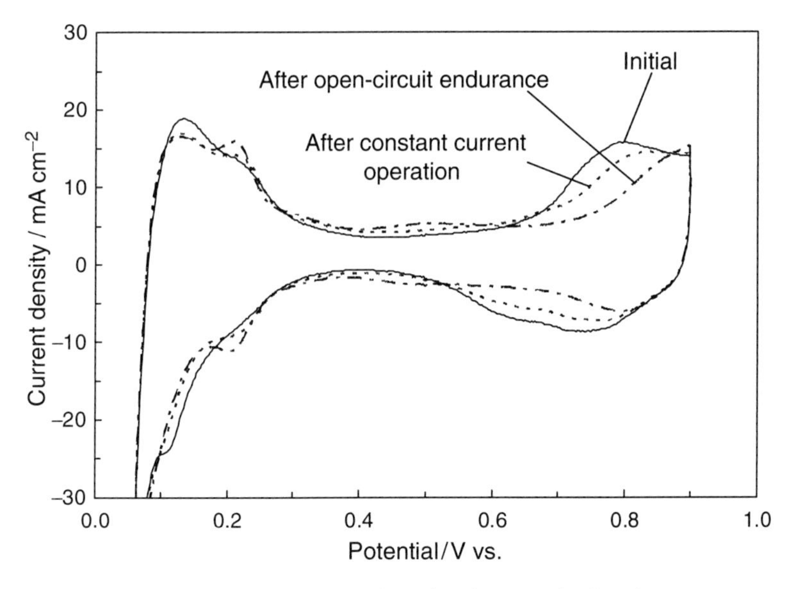

Fig. 12 Changes in cyclic voltammograms before/after the open-circuit endurance test

Table 3 Financial year 2004 status of air pollution (annual averages in Japan)

Air pollutant	Units	Ambient air pollution monitoring stations	Roadside air pollution monitoring stations
Sulfur dioxide	ppm	0.004	0.004
Nitrogen dioxide	ppm	0.015	0.028
Nitrogen monoxide	ppm	0.008	0.035
Carbon monoxide	ppm	0.4	0.6
Non-methane hydrocarbons	ppmC	0.21	0.29
Suspended particulate matter	mg m^{-3}	0.025	0.031

in the air from entering fuel cells through the air supply system, but it is virtually impossible to completely remove all impurities permanently. SO_2 is one type of impurity that has an especially large impact on MEA performance. Figure 13 shows one example of an evaluation of the performance degradation that can be caused by SO_2 contamination of MEAs. The results are for the decline in cell voltage in relation to the cumulative SO_2 dose input into the test cell. It is seen that the voltage declined with an increasing dose. From the cyclic voltammograms in Fig. 14 showing the changes before and after the dose was input, an extreme loss of ECSA can be discerned. These changes very closely resemble the results of the open-circuit endurance test, suggesting that they both indicate the same type of degradation.

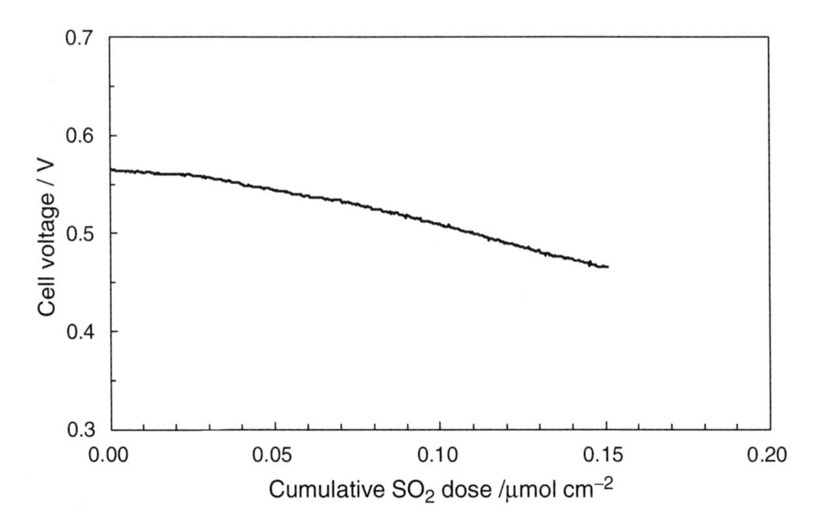

Fig. 13 Cell voltage as a function of cumulative SO_2 dose. Current density, 1 A cm^{-2}; anode gas, H_2; cathode gas, air +50 ppb SO_2

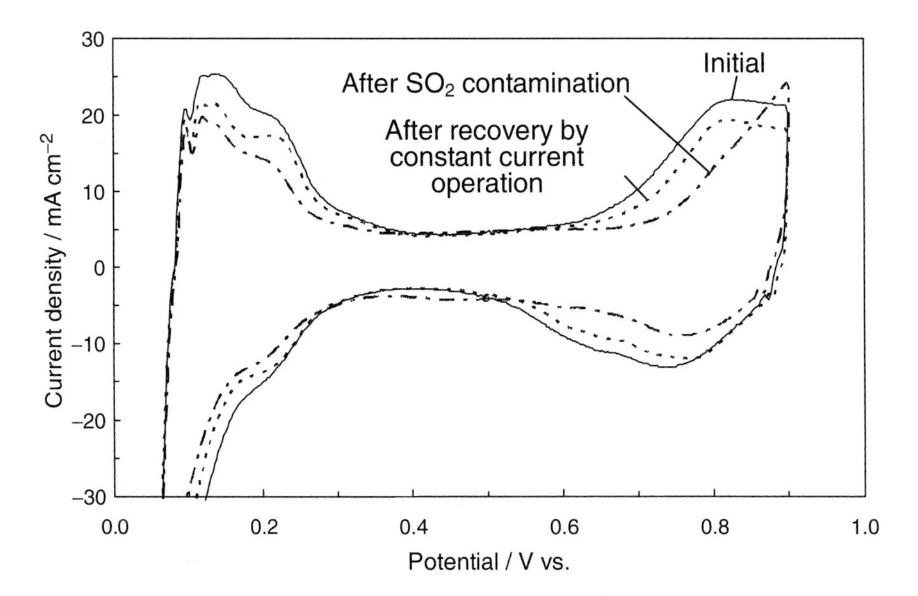

Fig. 14 Changes in cyclic voltammograms due to SO_2 contamination

4 Future Issues

The preceding section presented an analysis of certain degradation factors (start–stop operation, load cycling, idling, and impurity contamination) specific to automotive fuel cells and discussed possible solutions. The following degradation factors are especially important for improving MEA durability:

1. Subzero startup cycles (including freezing/thawing cycles)
2. Anode contamination
3. Degradation over time during long-term storage
4. Time-related decline in power-generation performance during load cycling

These unresolved degradation factors must be analyzed and measures need to be devised for improving durability to attain stack durability on a par with the durability of internal combustion engines. The impact on performance of the principal types of degradation discussed in this chapter must also be reduced further. Having an accurate understanding of degradation phenomena is an effective basis for designing more suitable operating conditions as one part of measures for mitigating degradation. Enhancing the performance of the constituent materials of MEAs is also extremely important.

5 Summary

This chapter has presented the results of an analysis of major MEA degradation phenomena related to the operating conditions of fuel cells for vehicle application. On the basis of the results, the following measures were identified for mitigating degradation:

1. The principal cause of start–stop degradation induced by carbon corrosion is a sharp rise in the cathode potential; therefore, controlling the potential is an effective way of improving stack durability.
2. Load-cycling degradation induced by platinum dissolution and idling degradation related to electrolyte membrane decomposition are both mainly caused by exposure of the cathode to high potentials. An effective measure for mitigating such degradation is to manipulate the operating mode so as to control the upper-limit voltage.

The above-mentioned degradation factors that have a substantial impact on durability were analyzed in this study and effective measures were found for mitigating their influence. To improve stack durability further, the phenomena involved in various degradation factors must be thoroughly analyzed in future work and suitable improvement measures need to be developed.

6 Concluding Remarks

Because automotive fuel cells are operated under such a wide range of conditions, the requirements for stack durability are especially rigorous. The constituent materials of MEAs have been widely used for fuel cell components for many years, so it will not be easy to enhance durability through material substitutions while still maintaining the same level of performance. However, it is also estimated that CO_2 emissions from vehicles must be reduced in the future by 50–70% from the current levels, if the automotive industry is to achieve sustainable development while still protecting the environment (The Institute of Applied Energy, Japan 2005). The practical use of fuel cell vehicles will be indispensable to the accomplishment of that goal, which means that improvement of stack durability is a critical objective that must be attained.

In future work, the authors plan to conduct further analyses of MEA degradation phenomena and devise measures for suppressing degradation, in order to promote the application of fuel cells to production vehicles. Additionally, efforts will also be made to develop new or substitute materials and to develop operating methods that are effective in inhibiting degradation.

References

Iiyama, A. (2007) Proceedings of the FC ExpoTechnical Conference, Tokyo.

LaConti, A., Hamdan, M., and McDonald, R. (2003) In *Handbook of Fuel Cells-Fundamentals, Technology and Application*, Vol. 3, W. Vielstich, A. Lamm, H. A. Gasteiger (Eds.), Chap. 49, Wiley, London, pp. 647–662.

Los Alamos National Laboratory (2005) DOE Hydrogen Program Review.

Meyers, J. and Darling, R. (2006) J. Electrochem. Soc., 153, A1432–A1442.

Ministry of the Environment, Government of Japan (2004) FY 2004 Report on the State of Air Pollution.

Ohma, A., Suga, S., Yamamoto, S., and Shinohara, K. (2007) J. Electrochem. Soc., 154, B757–B760.

Reiser, C., Bregoli, L., Patterson, T., Yi, J., Yang, J., Perry, M., and Jarvi, T. (2005) Electrochem. Solid-State Lett., 8, A273–A276.

Scheffler, G.W., Reiser, C.A., Van Dine, L.L., and Steinbugler, M.M. (2003) US Patent, 6,514,635.

The Institute of Applied Energy, Japan (2005) Strategic Technology Roadmap in the Energy Field–Energy Technology Vision 2100.

U.S. DOE (2005) Hydrogen, Fuel Cells and Infrastructure Technologies Program Multi-Year Research, Development and Demonstration Plan, EERE.

Yamamoto, S. and Shinohara, K. (2006) Nissan Technical Review, 59, 70–74 (in Japanese with English summary).

Part IV
R&D Status

1. Introduction

Editors

Fuel cell development and commercialization advances every year to new frontiers and is brought to the next level by R&D teams in academia, institutes, and industries throughout the entire world. However, R&D programs offered by the funding agencies are setting the scene for the scientists and engineers working on the technology on a daily basis. This includes not only funds to perform the developmental work but also formulating technical targets and, most importantly, teaming up R&D groups from different institutions to create joint developmental efforts. Besides national and local governments and institutions whose smaller-scale funding programs are sometimes not as visible, the main developmental programs are offered by the three main funding agencies, viz., the *European Union*, the *US Department of Energy* (US DOE), and the Japanese *New Energy and Industrial Technology Development Organization* (NEDO). All three organizations are running multimillion euro programs to advance fuel cell development. In the USA, the Hydrogen, Fuel Cells & Infrastructure Technologies Program (US DOE; http://www1.eere.energy.gov/hydrogenandfuelcells/about.html) is funding the development of fuel cell components, stacks, and systems and hydrogen safety and infrastructure. Within the European Commission's Framework Programmes 6 and 7, projects such as AutoBrane (automotive high-temperature membrane) or NextGenCell are funded, just to name a few. Especially the NextGenCell program needs some consideration since it is the first fuel cell program jointly funded and organized by the European Union and the US DOE in order to develop MEAs, stacks, and stationary combined heat and power fuel cell systems in a collaborative effort from development institutions on both sides of the Atlantic (https://www.hfpeurope.org/uploads/2233/3515/NextGenCell_ReviewDays07_10–11OCT2007.pdf). Finally, the Japanese NEDO (http://www.nedo.go.jp) is focusing its activities on the funding of fuel cell activities for both stationary and automotive power systems, which is outlined in more detail in the chapter "Durability Targets for Stationary and Automotive Applications in Japan."

F.N. Büchi et al. (eds.), *Polymer Electrolyte Fuel Cell Durability*,

2
R&D Status

Durability Targets for Stationary and Automotive Applications in Japan

Kazuaki Yasuda and Seizo Miyata

Abstract The New Energy and Industrial Technology Development Organization (NEDO) has been promoting the national development of polymer electrolyte fuel cells under the direction of the Ministry of Economy, Trade, and Industry. NEDO proposed an R&D target road map for the technical development of stationary and vehicle systems in 2005 and made some revisions in 2008. This road map shows the technical development themes and target values to be achieved in each stage of development. This chapter describes the polymer electrolyte fuel cell R&D targets of NEDO.

1 Expectations for Polymer Electrolyte Fuels Cells as a New Energy Source

Fuel cells and hydrogen-related technologies are core technologies for the realization of hydrogen-energy utilization to achieve energy conservation, reduction of environmental loads, petroleum substitutes, diversification of energy supplies, and creation of new businesses. The importance of such technologies has increased since the Kyoto Protocol was enacted, and it has been accelerated by steep increases in crude oil prices, instability of supply, and other factors. In the midst of these circumstances, to realize the practical application of fuel cells, both the public and the private sectors have been tackling the development of the infrastructure for fuel cells, hydrogen-related business development, verification studies, and the establishment of rules and standards intended primarily for polymer electrolyte fuel cells (PEFCs) in Japan.

Auto manufacturers and stack makers have been developing PEFCs and have recently achieved significant advances in PEFC technology. The Ministry of Economy, Trade, and Industry (METI) of Japan has been strongly funding and

K. Yasuda (✉)
Advanced Fuel Cell Research Group, Research Institute for Ubiquitous Energy Devices, National Institute of Advanced Industrial Science and Technology (AIST), 1-8-31 Midorigaoka, Ikeda, Osaka 563-8577, Japan
e-mail: k-yasuda@aist.go.jp

F.N. Büchi et al. (eds.), *Polymer Electrolyte Fuel Cell Durability,*

Fig. 1 Residential polymer electrolyte fuel cell cogeneration system operated in a customer's house

supporting the development of fuel cells. The world's first commercialized residential PEFC cogeneration system was delivered to the prime minister's residential area in April 2005. From that day, the commercialization of residential PEFC cogeneration systems was started on a limited scale. Figure 1 shows an example of an installed residential PEFC cogeneration system for commercialization. This system consists of a fuel cell unit containing a natural gas fuel processor and a hot water storage tank, and generates electricity with hot water stored in the tank for use as needed. Some fuel cell cars have already been delivered through leasing to limited consumers such as government-related entities. Fuel cell buses have also been serving on an experimental basis. At Aichi Expo 2005, eight fuel cell hybrid buses were used for on-site transportation. Fuel cells have also been actively studied for use in portable appliances such as laptop computers because of their high energy density.

2 National R&D Projects in Japan

The New Energy and Industrial Technology Development Organization (NEDO) has been promoting the national development of PEFCs under the direction of the Agency of Natural Resources and Energy of METI. NEDO is managing about 60% of METI's entire budget for fuel cells and hydrogen. NEDO has been making progress in R&D of fuel cells and hydrogen technologies, establishing codes, regulations, and standards and demonstrating stationary PEFC systems in field tests.

In financial year 2005 some projects were renewed or started (NEDO 2007). "Strategic Development of PEFC Technologies for Practical Application" is a particularly important project for R&D on PEFCs. This project promotes the development of technology for the practical application at the initial-introduction stage, the development of elemental technology at the full-introduction stage, and the development of next-generation technology at the full-dissemination stage to comprehensively develop highly efficient, highly reliable, and low-cost PEFCs. Fundamental information and tools for a better understanding of PEFCs are needed to make dramatic advances in durability. For this reason, the "Development of Technology on Basic and Common Issues" was started by different consortiums of businesses, universities, and research institutes in this project. The objective is to establish a technology to estimate the durability of PEFC stacks beyond 40,000 h for the commercialization of PEFC systems beginning in 2009 based on the analysis of deterioration behavior, origin, and mechanism. Under the direction of Seizo Miyata, the senior program manager of NEDO, five consortiums have been making intense efforts to clarify the degradation mechanism and develop accelerated test methods. Loss of resistance to CO poisoning of the anode catalyst, mass-transport decay in the cathode, and carbon corrosion have been studied by a consortium consisting of stack makers for home cogeneration use. Electrolyte degradation, platinum and platinum alloy dissolution, and the effects of impurities have been the subjects of fundamental studies by universities and research institutes. The development of analytical technologies has been promoted to elucidate the mechanism of PEFC degradation. A technique for the visualization of fuel cell stacks has also been pursued. Evaluation techniques using neutron radiography, nuclear magnetic resonance spectroscopy, electron spin resonance spectroscopy, X-ray absorption fine structure spectroscopy, and transmission electron microscopy tomography have been examined. New technologies for the creation of highly durable PEFCs have been actively developed in the "Development of Elemental Technology" project. Membrane makers have been developing highly durable membranes that operate above 100°C and under low-humidity conditions. A membrane with a radical trap layer has also been studied by a stack maker.

NEDO launched a project to demonstrate a 1-kW-class stationary PEFC system. To date, 3,307 1-kW-class PEFC systems have been installed in residences with subsidization. This project is considered a demonstration study. The aims of the project are to verify the performance of stationary fuel cell systems, to obtain operational data on fuel cells under a variety of actual usage conditions, to identify problems to be solved for early market introduction, and to help reduce the cost of production. The "Development for Safe Utilization and Infrastructure of Hydrogen" project encompasses R&D related to the production, transportation, and storage of hydrogen. The "Establishment of Codes & Standards for Hydrogen Economy Society" project is for the review of regulations and codes and standardization. In the area of international standards, NEDO continues to work with the International Organization for Standardization and the International Electrotechnical Commission. "Development of Standards for Advanced Application of Fuel Cells" was launched in financial year 2006 to promote the rapid dissemination of portable equipment. The "Development of Solid Oxide Fuel Cell (SOFC) System Technology" project has also been pursued.

3 NEDO Road Map for PEFC Development

To participate in joint operations by government and the private sector, NEDO is acting as an agent to promote the development of fuel cells and hydrogen-related technologies with collaboration among businesses and universities. For the appropriate promotion of a technology-development program, a "technology development scenario" should always be shared among stakeholders and the program should be implemented effectively and efficiently in accordance with the scenario. In May 2005, a "technology development road map," which defines technical themes to be addressed and the expected time of realization, was established with a view to achieve completion by around 2020. The "Fuel Cell and Hydrogen Technology Development Road Map Committee" (chairman, M. Watanabe, Yamanashi University), which includes representatives from industry, government, and academia, contributed to the design and establishment of the road map. The first committee meeting was held on 27 January 2005. A "Technical Trend Research Committee" was also established in Osaka Science & Technology Center and made a detailed road map and developed a final plan by holding a second meeting in early April. Meanwhile, the circumstances surrounding this technology change quickly, and updating is necessary. Therefore, the road map was reviewed in 2006 and 2008.

The "Fuel Cell and Hydrogen Technology Development Road Map" is separate from similar plans prepared by businesses and specific industries associated with future business development and technology strategies of particular products and summarizes technical themes and expected times of realization for practical fuel cells in a gradual manner. This road map is designed to deal with technical development options to be examined for the substantial practical application of fuel cells, technical targets, and the expected time of realization of each technical theme, and to coordinate them systematically to the greatest degree possible. The contents of the technical development to be promoted by NEDO, their respective times of realization, and their rankings are identified in this system.

Figures 2 and 3 summarize NEDO's road map for the technical development of stationary and vehicle systems proposed in 2005 and the revisions made in 2008. As described before, this road map is a target map that shows the technical development themes and target values to be achieved in each stage of development. Figure 2 is for a stationary 1-kW-class PEFC system for home-use cogeneration. PEFC systems for commercial and light-industrial use are technically similar to those for residential use, but it is somewhat easier to achieve the cost target of the former. This road map shows the targets for electrical generation efficiency, durability, operation temperature, and cost at each R&D stage. The efficiency is shown in terms of the higher heating value (HHV) with the lower heating value (LHV). The system cost means the shipment price from the system developer.

At present, the electrical generation efficiency is about 33%, the operating temperature is approximately 70°C, the durability is about 20,000 h, and the system cost is over several million yen per kilowatt. A first-generation system for the commercialization of stationary fuel cells will be introduced in 2009 at a target cost of ¥2-2.5 million per kilowatt. We think that a lower cost and improved durability are necessary for successful commercialization. A durability of more than 10 years (about 90,000 h) will be needed for home-use cogeneration PEMFC systems. A cost of below ¥0.4 million per kilowatt is also necessary for widespread commercialization.

To increase the operating temperature of a PEFC to approximately 90°C, the development of advanced technologies, such as materials for high-temperature operation in low-humidity conditions, has to be promoted. Long-term advanced technologies, such as a non-precious-metal catalyst and a novel electrolyte for operation without humidification, have also been studied. We expect that electrical generation efficiency should exceed 36% in terms of the HHV in the future.

Figure 3 shows the targets for fuel cells in vehicles. A family car is assumed here. Other vehicles such as a delivery car or bus are technically similar and the cost target is easier to reach. For a vehicle system, the efficiency value is shown in terms of the LHV. The HHV is given in parentheses for reference. Since NEDO has not been conducting system development for vehicles, the stack production cost is given on this map. "Operation temperature" includes a lower temperature value as the starting temperature.

At present, vehicle efficiency is considered to be about 50%, the operating temperature is approximately 80°C, the durability is about 1,000 h, and the system cost is several hundred thousand yen per kilowatt. For the commercial success of fuel cell systems in vehicles, their durability and performance have to be enhanced. A durability of more than 5,000 h has been established as a target for 2015. Simultaneously, the road map shows that vehicle efficiency should be 60% (LHV),

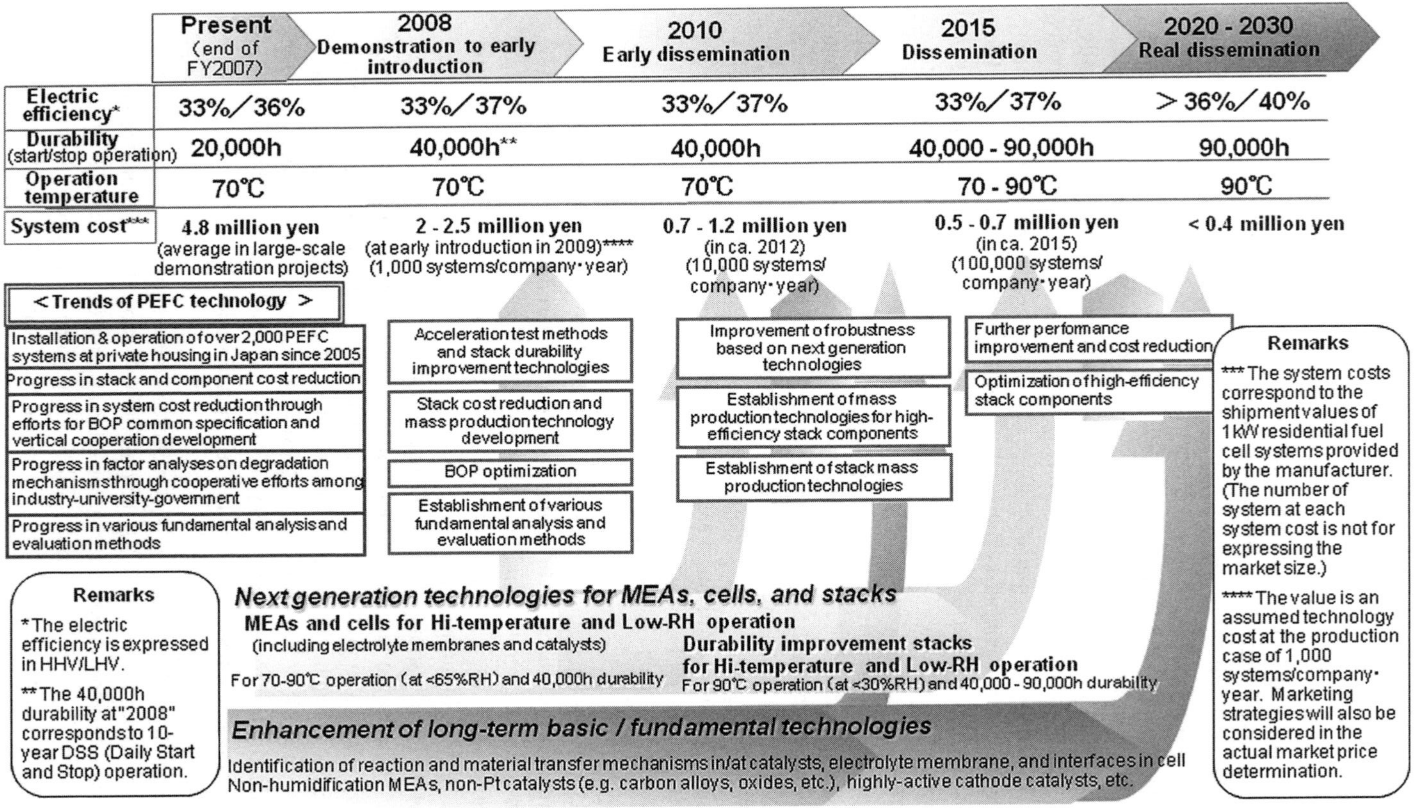

	Present (end of FY2007)	2008 Demonstration to early introduction	2010 Early dissemination	2015 Dissemination	2020 - 2030 Real dissemination
Electric efficiency*	33%/36%	33%/37%	33%/37%	33%/37%	> 36%/40%
Durability (start/stop operation)	20,000h	40,000h**	40,000h	40,000 - 90,000h	90,000h
Operation temperature	70℃	70℃	70℃	70 - 90℃	90℃
System cost***	**4.8 million yen** (average in large-scale demonstration projects)	**2 - 2.5 million yen** (at early introduction in 2009)**** (1,000 systems/company·year)	**0.7 - 1.2 million yen** (in ca. 2012) (10,000 systems/ company·year)	**0.5 - 0.7 million yen** (in ca. 2015) (100,000 systems/ company·year)	**< 0.4 million yen**

< Trends of PEFC technology >

- Installation & operation of over 2,000 PEFC systems at private housing in Japan since 2005
- Progress in stack and component cost reduction
- Progress in system cost reduction through efforts for BOP common specification and vertical cooperation development
- Progress in factor analyses on degradation mechanisms through cooperative efforts among industry-university-government
- Progress in various fundamental analysis and evaluation methods

2008:
- Acceleration test methods and stack durability improvement technologies
- Stack cost reduction and mass production technology development
- BOP optimization
- Establishment of various fundamental analysis and evaluation methods

2010:
- Improvement of robustness based on next generation technologies
- Establishment of mass production technologies for high-efficiency stack components
- Establishment of stack mass production technologies

2015:
- Further performance improvement and cost reduction
- Optimization of high-efficiency stack components

Remarks

*The electric efficiency is expressed in HHV/LHV.

**The 40,000h durability at "2008" corresponds to 10-year DSS (Daily Start and Stop) operation.

Next generation technologies for MEAs, cells, and stacks

MEAs and cells for Hi-temperature and Low-RH operation
(including electrolyte membranes and catalysts)
For 70-90℃ operation (at <65%RH) and 40,000h durability

Durability improvement stacks for Hi-temperature and Low-RH operation
For 90℃ operation (at <30%RH) and 40,000 - 90,000h durability

Enhancement of long-term basic / fundamental technologies

Identification of reaction and material transfer mechanisms in/at catalysts, electrolyte membrane, and interfaces in cell
Non-humidification MEAs, non-Pt catalysts (e.g. carbon alloys, oxides, etc.), highly-active cathode catalysts, etc.

Remarks

***The system costs correspond to the shipment values of 1 kW residential fuel cell systems provided by the manufacturer. (The number of system at each system cost is not for expressing the market size.)

****The value is an assumed technology cost at the production case of 1,000 systems/company·year. Marketing strategies will also be considered in the actual market price determination.

Fig. 2 Summary of the road map of the New Energy and Industrial Technology Development Organization (NEDO) for the technical development of a stationary 1-kW-class polymer electrolyte fuel cell system for home-use cogeneration

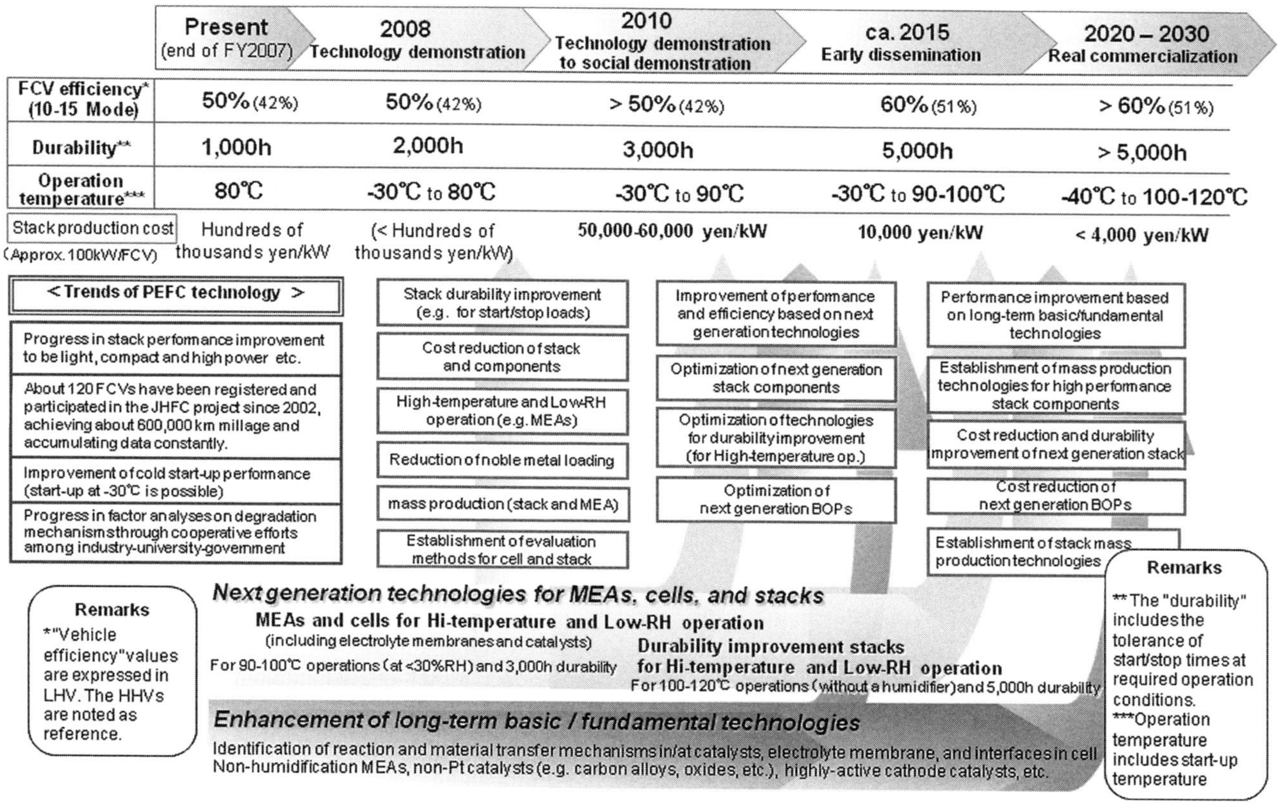

	Present (end of FY2007)	2008 Technology demonstration	2010 Technology demonstration to social demonstration	ca. 2015 Early dissemination	2020 – 2030 Real commercialization
FCV efficiency* (10-15 Mode)	50% (42%)	50% (42%)	> 50% (42%)	60% (51%)	> 60% (51%)
Durability**	1,000h	2,000h	3,000h	5,000h	> 5,000h
Operation temperature***	80℃	-30℃ to 80℃	-30℃ to 90℃	-30℃ to 90-100℃	-40℃ to 100-120℃
Stack production cost (Approx. 100kW/FCV)	Hundreds of thousands yen/kW	(< Hundreds of thousands yen/kW)	50,000-60,000 yen/kW	10,000 yen/kW	< 4,000 yen/kW

<Trends of PEFC technology>

- Progress in stack performance improvement to be light, compact and high power etc.
- About 120 FCVs have been registered and participated in the JHFC project since 2002, achieving about 600,000 km millage and accumulating data constantly.
- Improvement of cold start-up performance (start-up at -30℃ is possible)
- Progress in factor analyses on degradation mechanisms through cooperative efforts among industry-university-government

2008–2010:

- Stack durability improvement (e.g. for start/stop loads)
- Cost reduction of stack and components
- High-temperature and Low-RH operation (e.g. MEAs)
- Reduction of noble metal loading
- mass production (stack and MEA)
- Establishment of evaluation methods for cell and stack

ca. 2015:

- Improvement of performance and efficiency based on next generation technologies
- Optimization of next generation stack components
- Optimization of technologies for durability improvement (for High-temperature op.)
- Optimization of next generation BOPs

2020–2030:

- Performance improvement based on long-term basic/fundamental technologies
- Establishment of mass production technologies for high performance stack components
- Cost reduction and durability improvement of next generation stack
- Cost reduction of next generation BOPs
- Establishment of stack mass production technologies

Next generation technologies for MEAs, cells, and stacks

MEAs and cells for Hi-temperature and Low-RH operation (including electrolyte membranes and catalysts)
For 90-100℃ operations (at <30%RH) and 3,000h durability

Durability improvement stacks for Hi-temperature and Low-RH operation
For 100-120℃ operations (without a humidifier) and 5,000h durability

Enhancement of long-term basic / fundamental technologies

Identification of reaction and material transfer mechanisms in/at catalysts, electrolyte membrane, and interfaces in cell
Non-humidification MEAs, non-Pt catalysts (e.g. carbon alloys, oxides, etc.), highly-active cathode catalysts, etc.

Remarks
*"Vehicle efficiency" values are expressed in LHV. The HHVs are noted as reference.

Remarks
** The "durability" includes the tolerance of start/stop times at required operation conditions.
***Operation temperature includes start-up temperature

Fig. 3 Summary of NEDO's road map for the technical development of fuel cells for use in vehicles

the operating temperature should be more than 90–100°C, and the cost of stack production should be approximately ¥10,000 per kilowatt. NEDO is not directly promoting the development of vehicle fuel cell systems but rather is supporting the basic technologies needed to build fuel cells as well as the elucidation of the degradation mechanism. After 2020, NEDO hopes to see an increase in the operating temperature to approximately 100–120°C without a loading humidifier. Fuel cells have to be started at a low temperature and −40 °C is a target for widespread commercial use after 2020.

Reference

NEDO (2007) Outline of NEDO New Energy and Industrial Technology Development Organization 2007–2008, pp. 80–86. (http://www.nedo.go.jp/kankobutsu/pamphlets/kouhou/2007gaiyo_e/79_86.pdf)

Index

E

Printed in the United States of America